4th International Conference on Mechanical Engineering and Automation Science (ICMEAS 2018)

IOP Conference Series: Materials Science and Engineering
Volume 470

Beijing, China
12 – 14 October 2018

ISBN: 978-1-5108-8097-9
ISSN: 1757-8981

Printed from e-media with permission by:

Curran Associates, Inc.
57 Morehouse Lane
Red Hook, NY 12571

Some format issues inherent in the e-media version may also appear in this print version.

Printed by Curran Associates, Inc. (2019)

For permission requests, please contact the Institute of Physics at the address below.

Institute of Physics
Dirac House, Temple Back
Bristol BS1 6BE UK

Phone: 44 1 17 929 7481
Fax: 44 1 17 920 0979

techtracking@iop.org

Additional copies of this publication are available from:

Curran Associates, Inc.
57 Morehouse Lane
Red Hook, NY 12571 USA
Phone: 845-758-0400
Fax: 845-758-2633
Email: curran@proceedings.com
Web: www.proceedings.com

Table of contents

Volume 470

2018 4th International Conference on Mechanical Engineering and Automation Science (ICMEAS 2018)

12–14 October 2018, Beijing, China

Accepted papers received: 23 November 2018
Published online: 24 January 2019

Preface

2018 4th International Conference on Mechanical Engineering and Automation Science (ICMEAS 2018)

Committees

Peer review statement

Papers

Investigation of the Propulsion Properties of an M-shaped Bending-surface Feather Fan under Different Wind Velocities

Luo Qing, Nie Xiaofang and Zhou Ximing.....1

Improved Heat Transfer Correction Method for Turbocharger Compressor Performance Measurements

Liyong Huang, Chaochen Ma, Yanzhao Li, Jianbing Gao and Mingxu Qi.....11

Clutch Control Strategy of Driving Mode Transition for P2 Hybrid Electric Vehicle

Shuhan Wang, Yanjing Liu and Tianyuan Cai.....19

Finite Difference Method for Simulating Transverse Forced Vibration of Vertical Rope with Multi-Rope Winding Hoister

Guoying Wang, Xingming Xiao, Yansong Ma, Xiao Di and Kaixuan Wang.....25

Multi-body Dynamic Analysis of the Braking Process for Megawatt Wind Turbine Disc Brake

S Zhang, C Xue, J Yin, Z Sha, F Ma, D Yang and Y Liu.....31

Analysis on Vibration Characteristics of the Stringed Bead Wire of an Underwater Diamond Wire Saw

L Zhang, M Liu, L Liu, G L Zhang, S H Gao and L Q Wang.....37

Fault Diagnosis for Brake System in High-Speed Trains Using the Phased Features and Multi-layer Perceptron

Changzheng Fang, Xiaoyong Zhang, Yijun Cheng, Shengnan Wang, Li Zhang and Yingze Yang.....44

Unmanned Powered Paraglider Flight Path Control Based on PID Neutral Network

S T Qu.....50

Effect of Pendulum Parameters on Performance of Electromagnetic Energy Harvester

Hongyan Wang, Fengjuan Miao, Liying Zou and Tao Zhou.....60

Effect of Surface Waviness on Aerostatic Bearing with Speed

Chou Fan, Shanghan Gao and Zhuang Ni.....66

Failure Strain Measurement of the Piping System of a Nuclear Power Plant Using Image Signals

S W Kim, B G Jeon, D U Park and D W Yun.....75

Finite Element Analysis for the Lifting Platform of a Self-Propelled Multi-Functional Machine Used in Orchard

Bai Zongchun, Lv Xiaolan and Xia Liru.....82

Performance Differences of a Solid Attitude Control Thruster under Different Working Media

Zhong Yang, Hongwei Yan, Quanzhong Zhang and Wubin Li.....90

Influence of Yield Stress Mismatch on the Creep Stress Field around Crack Tip of Stress Corrosion Cracking

Zhang Jianlong and Xue He.....96

The Study of Imperfect Maintenance Strategy based on Improved Age Reduction Factor

Duan Junhang, Zhu Ling and Cao Li.....102

Multi-objective Optimization of the Rapidly Assembled Helipad Deck

Dan Wei, Xudong Deng, Lixun Zhang and Jiru Gong.....112

On-Policy Learning for the Swing Process Control of a Cutter Suction Dredger

Changyun Wei, Xiujing Chen and Fusheng Ni.....122

A Path Planning Method for Robotic Wire and Arc Additive Manufacturing of Thin-Walled Structures with Varying Thickness

Guocai Ma, Gang Zhao, Zhihao Li and Wenlei Xiao.....128

Optimization of Key Parameters of Two-way Inner and Outer Tube Differential Pressure Flowmeter

Bao-tong Fan and Jiang Meng.....136

Applications of Spray Cooling Technology in Aerospace Field

Ran Liu, Lei Zhang and Xian Zhang.....142

Simulation Study on the Influence Law of Two-body Wear in Sliding Guide Ways

Shengfang Zhang, Qiang Hao, Zhihua Sha, Yu Liu, Jian Yin, Fujian Ma and Dapeng Yang.....149

Backlash Analysis of Compound Planetary Gear Set in Rotary Actuators

Y Wang and F Yang.....155

Simulation Calculation of Flow Characteristics of Air Separation System Based on Flowmaster

Yu Zeng, Bing Sun and Haoyang Sun.....161

Modal Analysis of Double-Bolted Connection under Different Operating Conditions

Chao Cao, Xueyan Zhao, Weikang Du, Shourong Liu and Dehao Wu.....170

Flexible Assembly Tooling Structure and Positioning Accuracy Analysis of Aircraft Fuselage Framework

Q Bai, W Tian, Z H Shi, Y J Sun, J X Shen, W H Liao and H B Shi.....177

Influence of Micro-texture Parameters on Bone Drilling Force and Torque: A Finite Element Model

Yahui Hu, Xiangjun Li, Chuncheng Chen, Lili Su and Qingchun Zheng.....186

Estimation of Grinding Force with Consideration of Rupture Factor

Jianliang Guo, Jun Chi and Lianqing Chen.....195

Effect of the Aging Treatment on the Tensile Strength and Impact Toughness of Friction Welded Super304H Joints

Lu Yuan, Zhang Jianlong, Li Debiao, Bi Cheng and Ning Tao.....204

Effect of the Aging Treatment on the Precipitates and Vickers Hardness of Friction Welded Super304H Joints

Lu Yuan, Li Debiao, Zhang Jianlong, Shang Kang, Yang Yang and Cao Bo.....209

Optimized Control Based on Nonlinear Modelling and Parameter Estimation of the Electro-Hydraulic Servo System

Meng Li and Yong Zhang.....215

Adaptive Neuro-Fuzzy Inference System Based Orientation Control of an Intra-operative Ultrasound Robot

Shuangyi Wang, James Housden, Davinder Singh and Kawal Rhode.....225

Research of High Precision Humidity Fuzzy Control Algorithm of One Kind Humidity Verification Device

Guangda Lu, Qiuyue Zhang, Tongtong Qie and Qihui Feng.....231

A Robot Odor Source Localization Strategy Based on Bionic Behavior

Guangda Lu, Qiuyue Zhang, Tongtong Qie and Qihui Feng.....239

A Modeling Method for Abrupt Change Shafts in Gearboxes and Experiment Verification

Z C Liu, W Song and Q Lu.....247

Fault Diagnosis Method of Autonomous Underwater Vehicle Based on Deep Learning

Yushan Sun, Zikai Wang and Guocheng Zhang.....256

Piezoelectric Energy Harvesting Based on Vertical Vibration of Car Body Induced by Road Surface Roughness

Hongyan Wang, Fengjuan Miao, Zhilong Sun and Liqiu Jing.....262

Friction-thermal-vibration Coupling Analysis of Large-megawatt Wind Turbine Brake

S Zhang, P Xin, J Yin, Y Liu, Z Sha, F Ma and D Yang.....268

The Longitudinal Dynamic Characteristics of a V/STOL Aircraft Based on Deflected Slipstream during the Hovering Phase

Shi Chen and Weijun Wang.....274

Adaptive Control Algorithm for Steady Flight of Mars UAV with Four Rotors

Yangyang Liu, Lei Zhang and Jingyi Shao.....283

Magnetic Circuit Design and Finite Element Analysis of Ferrofluid Seal of Engineering Machinery Hydraulic Cylinder

Fan Chen, Xiaolong Yang and Shanghan Gao.....290

Effect of Crack on Dynamic Characteristics of Simple Supported Beam

Xueyan Zhao, Weikang Du, Chao Cao, Yifan Chen and Lijuan Yin.....297

The Critical Axial Force Model in Drilling of Carbon Fiber Reinforced Composites under Different Stacking Modes

M Li, W Tian, H L Sun, Z Liu, L Zhang, W H Liao and F F Hua.....304

Design and Simulation Technology of Piezoelectric Ultrasonic Transducer with Sandwich Composite Horn

Yi Xuan, Aimin Wang and Niansong Zhang.....312

Design of Data Acquisition System for Silicon Micro Machined Resonant Accelerometer

Yun Xu, Yaohui Qu and Zhihang Luo.....319

Increase of Refueling Systems Efficiency of Rocket Launching Sites by Utilization of Cold Energy of Cryogen

Daria Aksenova, Artem Novikov, Aleksandra Shimanova, Artem Shimanov and Dmitry Uglanov.....325

Automatic Text System Based on PLC for Walk-Through Metal Detector

Zheng Haidong, Jiang Lingling, Wu Jianyi, Wang Wei, Wang Xiaokan and Li Zhiguo.....333

PREFACE

ICMEAS 2018 was held in Beijing, China during 12-14 October, 2018, which is organized by Engineering Science and Hong Kong Society of Mechanical Engineers (HKSME), supported by Capitol Technology University. The conference provides a useful and wide platform both for displaying the latest research and for exchanging of research results and thoughts in broad Mechanical Engineering, Manufacturing Science and Engineering, Automation Science and other topics. The participants of the conference were from almost every part of the world, with background of either academia or industry, even well-known enterprise. The success and prosperity of the conference is reflected by high level of the papers received.

The proceeding is a compilation of the accepted papers which represents an interesting outcome of the conference. This book covers 3 chapters: Mechanical; Manufacturing; Engineering Science and Materials Science.

We would like to acknowledge all of those who supported ICMEAS 2018. Each individual and institutional help was very important for the success of this conference. Especially we would like to thank the organizing committee for their valuable advices in the organization and helpful peer review of the papers.

We sincerely hope that ICMEAS 2018 will be a forum for excellent discussions that will put forward new ideas and promote collaborative researches. We are sure that the proceedings will serve as an important research source of references and the knowledge, which will lead to not only scientific and engineering progress but also other new products and processes.

Conference Committee Co-chairs
Du Ruxu,The Chinese University of Hong Kong, Hong Kong
Ian McAndrew, Embry Riddle Aeronautical University, UK
Kai Cheng, Brunel University, UK
Mohd Hamdi Bin Abd Shukor, University of Malaya, Malaysia

IOP Conf. Series: Materials Science and Engineering **470** (2019) 011002 doi:10.1088/1757-899X/470/1/011002

COMMITTEES

Conference Committee Co-chairs

Prof. Du Ruxu,The Chinese University of Hong Kong, Hong Kong

Prof. Ian McAndrew, Embry Riddle Aeronautical University, UK

Prof. Kai Cheng, Brunel University, UK

Prof. Mohd Hamdi Bin Abd Shukor, University of Malaya, Malaysia

Program Committee Chairs

Prof. Stefan Dimo, University of Birmingham, UK

Prof. Yong Zhao, Nazarbayev University, Kazakhstan

Advisory Commitees

Prof. Hui Ding, Harbin Institute of Technology, China

Prof. Enrico Vezzetti, Politecnico di Torino, Italy

Prof. Egon Müller, Technische Universität Chemnitz, Germany

International Technical Committee

Prof. Josua Meyer, University of Pretoria, South Africa

Prof. Sandy To,Hong Kong Polytechnic University, Hong Kong

Prof. Antonis I.G. Vardulakis, Aristotle University of Thessaloniki, Greece

Prof. David Barton, University of Leedsm, UK

Prof. Shinn-Liang Chang, National Formosa University, Taiwan

Prof. Tang Sai Hong, Universiti Putra Malaysia, Malaysia

IOP Conf. Series: Materials Science and Engineering **470** (2019) 011002 doi:10.1088/1757-899X/470/1/011002

Prof. M. Chandrasekaran, Vels University, India

Prof. C. P. Jiang, National Formosa University, Taiwan

Prof. Harijono Djojodihardjo, UPM, Malaysia

Dr. Darwin Lau, the Chinese University of Hong Kong, Hongkong

Dr.T.Muthuramalingam, SRM University, chennai, India

Dr. Hassan Ghadbeigi, The University of Sheffield, UK

Dr. Mohammadreza Behi, University of Sydney, Australia

Dr. Pierre Kerfriden, Cardiff University, UK

Peer review statement

All papers published in this volume of *IOP Conference Series: Materials Science and Engineering* have been peer reviewed through processes administered by the proceedings Editors. Reviews were conducted by expert referees to the professional and scientific standards expected of a proceedings journal published by IOP Publishing.

Investigation of the Propulsion Properties of an M-shaped Bending-surface Feather Fan under Different Wind Velocities

Luo Qing, Nie Xiaofang※ and Zhou Ximing

Lilienthal laboratory, Beijing Implant Aircraft Co., Ltd.

※2377811041@qq.com

Abstract. Flight speed is an important factor that affects flapping-wing propulsion. This study conducted tests on a self-designed rotatory force-measuring platform and simulated different inflow conditions by changing the rotatory angular velocity. An M-shaped bending-surface feather fan was selected as the test model, and the rotatory platform was rotated at different speeds under different propulsion so that the flapping-wing propulsion at different inflow velocities could be measured. Moreover, the wind speed corresponding to the maximum propulsion was determined. The results show that as wind velocity increases within a certain range, the propulsion first increases, then decreases, and finally increases until reaching a maximum wind velocity value of 7.3 m/s.

1. Introduction

Flapping wing propulsion is a novel propulsion mode that differs from traditional rotary machine propulsion and jet propulsion. In recent years, more attention has been paid to ornithopters and underwater bio-robots, and it was found that flapping-wing propulsion has more advantages. At lower Reynolds numbers, flapping non-stationary motion can provide larger lifting forces and also exhibits higher propulsion efficiency, therefore, flapping-wing air vehicles can achieve long distance flights by consuming less energy [1].

Numerous research groups have investigated the generation mechanism of flapping-wing propulsion. Knoller and Bertz found that flapping wings can produce an equivalent angle of attack, and the projection of the generated normal force along the inflow direction can thus be regarded as the propulsion force [2][3]. Jones et al. experimentally investigated the effects of the flapping wing aspect ratio on propulsion properties [4]. Moreover, Ho et al. conducted related tests on the effects of flapping wing stiffness [5]. Furthermore, Jones et al. developed a flapping-wing system with a wingspan of 1270 mm, in which two pairs of flapping wings at the top and bottom can realize both flapping and pitching motion, and wind tunnel tests were performed [6]. Wang et al., from Dalian University of Technology, adopted a non-stationary dual-time Navier-Stokes (NS) equation for pre-processing and examining the propulsion properties of a two-wing ground-effect propeller [7]. According to their results, the propulsion force and efficiency exhibit a similar variation with velocity, i.e., both propulsion and efficiency can achieve a maximum value by varying the velocity at a fixed frequency. Yu` et al. performed numerical simulations to study the aerodynamic performance of a flexible micro-ornithopter and explored the influence of wing stiffness on lifting and propulsion forces [8]. In addition, Qian et al. performed numerical simulations to investigate the effect of flapping thin-plate wings with elastic trailing edges on propulsion [9].

The main factors affecting flapping wing propulsion force and efficiency are flight speed, frequency, flapping amplitude, shape, rules of motion, and bending and torsional deformation of the flapping wings [10]. However, the effect of velocity on flapping-wing propulsion has been poorly investigated. Therefore, to explore the relationship between the flapping-wing propulsion properties and velocity at a fixed flapping frequency, cantilever tests were used in this study and aerodynamic tests were performed on a flapping wing model to investigate propulsion forces at different wind velocities. This study provides a theoretical foundation for the design and practical application of large-propulsion flapping wing feather fans for different cruising velocities.

2. Experimental set up and method

2.1. Experimental set up

Figure 1. Rotary platform for force measurement.

We use the self-designed rotary platform for force measurement to finish this experiment, as shown in Fig. 1, and consists of a supporting structure, spindle, rotary supply circuit, and model mounting plate. The platform is not powered but can freely rotate under the action of external forces, thus achieving the advantages of low friction, stable operation, small volume, and a simple structure. It can also be powered when the platform rotates.

A transmission mechanism was installed on the rotary platform to achieve wing flapping, driven by a 1-kW motor and independently powered by a 5S lithium battery. A crank connecting rod structure was adopted, as shown in Fig. 2, and achieved flapping through the reciprocating up-and-down motions of the flapping rods via a sliding rock. The two flapping rods had an identical length of 1.5 m and were mounted on the platform in a symmetrical pattern.

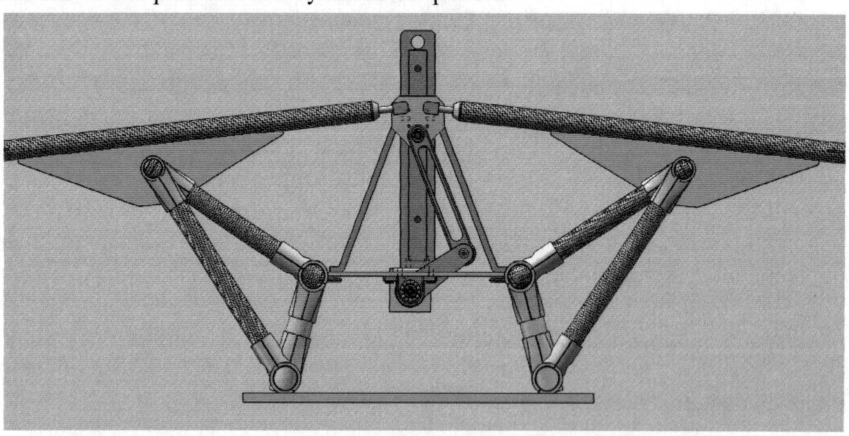

Figure 2. Illustration of the transmission mechanism structure.

A cantilever was fixed on the platform, perpendicular to the flapping rods, and a propeller was installed at a distance of 1.3 m away from the spindle, driven by a 3-kW motor. The motor was powered by a DC supply and provided the rotational power via an electric brush. Once the propeller was switched on, the platform began to rotate and the rotational speed was adjusted by varying the rotational speed of the propeller using a remote controller. The flapping model was mounted on the cantilever and then rotated with the platform while flapping. Therefore, relative motion between the flapping model and air was created and the model was used to simulate the flight process of an ornithopter. The relative velocity between the model and air was calculated based on the angular rotational speed of the platform, and the angular rotating speed was measured using a non-contact tachometer (UT372, UNI-T) with a measurement range of 10~99999 ± 0.04 r/min. The tachometer included a communication module for real-time monitoring via a desktop computer.

2.2. Experimental model

The experimental model was based on the measured propulsion of flapping-wing feather fans with different corners. Since the cantilevers of the rotating force measurement platform flapped up and down, the mounted feather fan model underwent a synchronous movement, and the air on the surface of the feather fan simultaneously exhibited spanwise and chordwise movement. However, the flapping propulsion is a function of the vorticity, mass, velocity, and direction of the shedding vortex, and the spanwise movement of the airflow is perpendicular to the forward direction of the model, and therefore makes no contribution to the flapping propulsion. Assuming the airflow on the surface of the feather fan only exhibis chordwise motion, the airflow disturbances produced during the upward and downward flapping during each flapping cycle develops backwards. In addition, the chordwise vorticity of the shedding vortex resulted in large propulsion magnitudes. On the other hand, spanwise motion of the airflow decreased both the vorticity and mass of the backward-moving shedding vortex, and the propulsion was reduced. Spanwise and chordwise motion of the airflow is illustrated in Fig. 3.

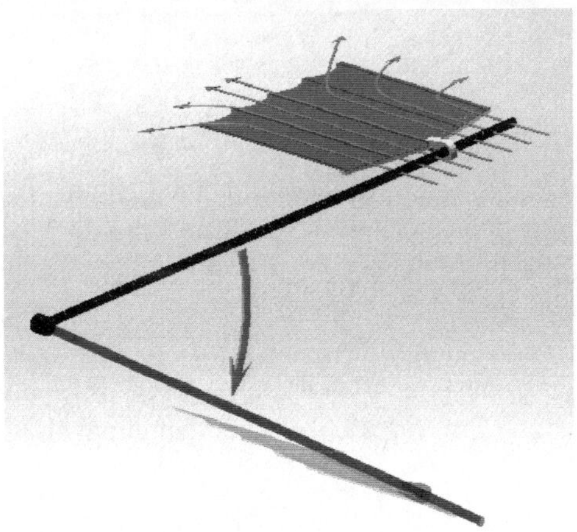

Figure 3. Illustration of spanwise and chordwise motion of airflow.

To avoid spanwise motion, the airflow should move backwards along the chordwise direction. Nine types of flapping weather fans with nine different bending angles (Designed by Lilienthal Laboratory in Beijing Implant Science and Technology Development Co., Ltd.) were used and propulsion was measured at a wind velocity of 15 m/s. The shape and bending angle of the flapping wing feather fan are shown in Fig. 4.

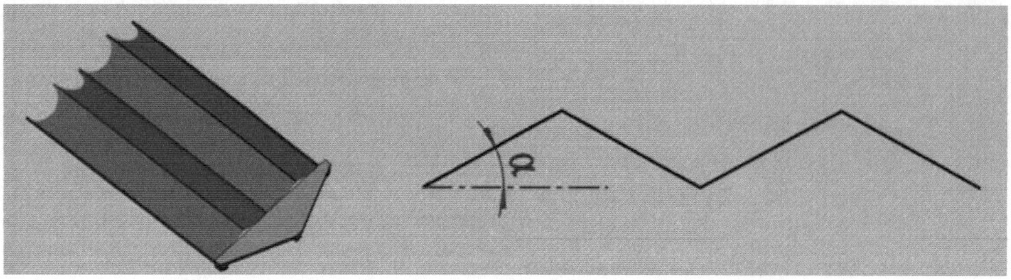

Figure 4. Shape and bending angle of flapping feather fan.

According to the experimental results, an increase of α causes an initial increase in the measured propulsion, followed by a decrease, and the maximum value was reached at $\alpha = 12°$. This is due to the fact that during the up-and-down flapping wing of the feather fan, the M-shaped groove hinders spanwise motion of the surface airflow and plays the role of chordwise diversion (Fig. 5), therefore, the propulsion increases. When the bending angle exceeded a certain value, the area of the fan increased creating drag, caused by an increase in the chordwise flexible deformation. When $\alpha > 12°$, propulsion increased. Thus, an M-shaped wing with a bending surface was used as the experimental model in this study.

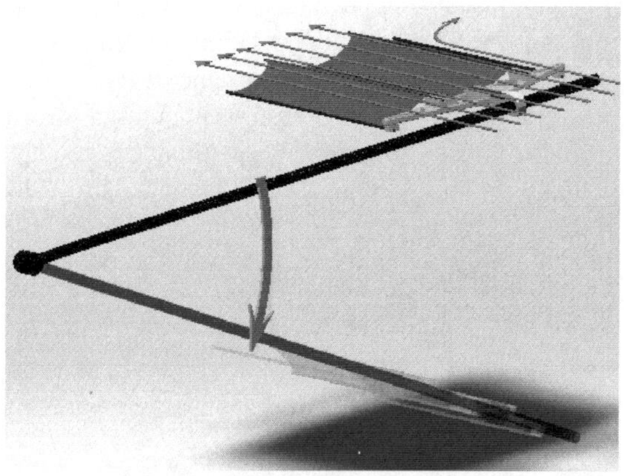

Figure 5. Diversion function of M-shaped feather fan.

As shown in Figure 6, the model consisted of a rectangle with a chord length of 850 mm and a width of 450 mm. The skeleton was composed of five carbon fibre tubes with particular tapering and the tube thickness was 0.4 mm. Ultralight nylon fabric was used as the skin material. The five carbon fibre tubes were alternately arranged along the spanwise direction in a uniformly-spaced pattern, and the angle of intersection between the line connecting adjacent tubes and the plane where the lower pipes were located was 15°, thereby forming an M-shaped bending surface.

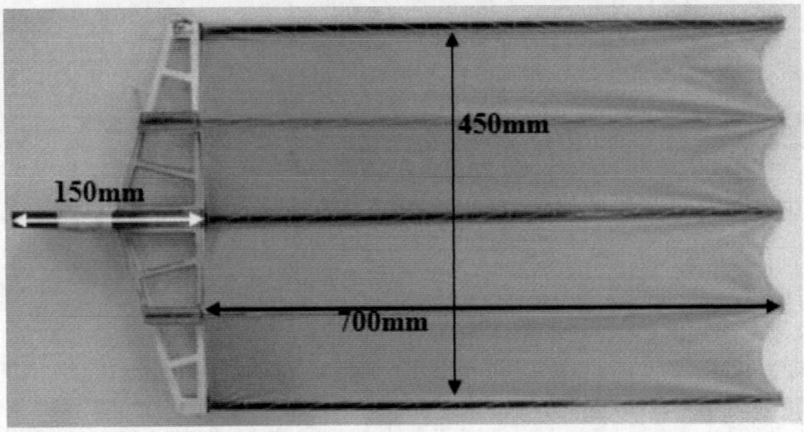

Figure 6. Image of flapping wing feather fan.

2.3. Experimental method

Two flapping-wing feature fans with identical parameters were installed on the flapping rods on both two sides of the platform, at a distance of 1.35 m from the center. The angle of intersection between the feather fan and horizontal plane was zero, and the installation direction was identical to the rotating direction of the platform. First, the flapping frequency was set to 2.5 Hz and the feather fan flapped up and down at a flapping angle of 15°. Then, the platform began to rotate, and the rotational speed of the propeller was adjusted so that the platform could rotate at different angular velocities, allowing for the relative velocity between the feather fan and air to be adjusted. Once the rotational speed of the platform was stable, the flapping stopped and the angular velocity of the rotary platform decreased to a fixed value. The propulsion at a flapping frequency of 2.5 Hz was calculated based on the variation of angular velocity. The preset wind velocity and corresponding angular velocity values are listed in Table 1.

Table 1. Corresponding wind speed and angular velocity values.

Wind velocity (m/s)	Angular velocity at flapping velocity of 2.5 Hz (r/min)
6	42.46
7	49.54
8	56.62
9	63.69
10	70.77
12	84.93
15	106.16

To reduce error, the experiment was repeated 6 times at each wind velocity.

2.4. Experimental principle

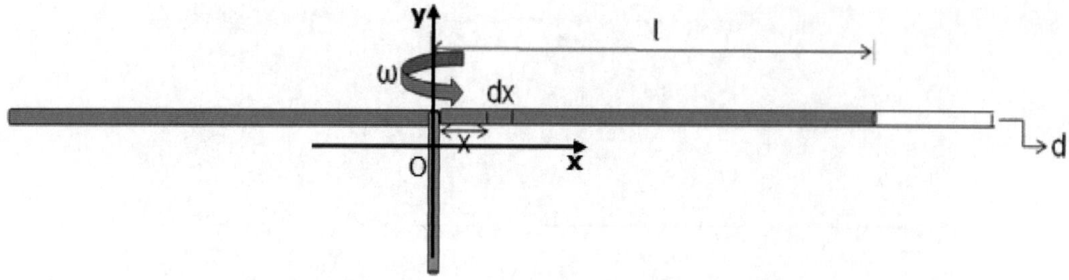

Figure 7. Forces acting on the simplified model.

First, the model was simplified and the applied forces were analyzed without considering flapping.

When the platform rotated at a fixed velocity, the propulsion was equal to the air resistance. Since the platform rotates around its axis, the linear velocities at different points on the rod are different, and the applied resistance exhibits a non-uniform distribution along the length of the rod. The resistance can then be calculated using integration, and a rectangular plane coordinate system can be established, as shown in Fig. 7. An infinitesimal area on the x-axis near the origin was assumed, with the horizontal ordinate denoted as x and the rotating angular velocity denoted as ω. Therefore, the resistance acting on an infinitesimal area can be written as

$$dF = \frac{1}{2}\rho(xw)^2 C_D d * dx \tag{1}$$

$$\frac{1}{2}F = \int_0^l \frac{1}{2}\rho(xw)^2 C_D d * dx \tag{2}$$

$$F = \frac{1}{3} d\rho C_D l \upsilon^2 \tag{3}$$

where l is the length of the cantilever in meters, C_D denotes the average resistance coefficient, ρ is the density of air in kg/m^3, d is the diameter of the cantilever in meters, w denotes the rotatory angular velocity of the platform in rad/s, and υ is the linear velocity of the model installation point in m/s.

In Eq. (3), F is related to C_D. Based on test results, IJ Ritchiek [11] concluded that the resistance coefficient of a smooth cylinder is a function of Reynolds number, i.e., $C_D = f(\text{Re})$. The relationship between the resistance coefficient and Reynolds number is illustrated in Figure 8.

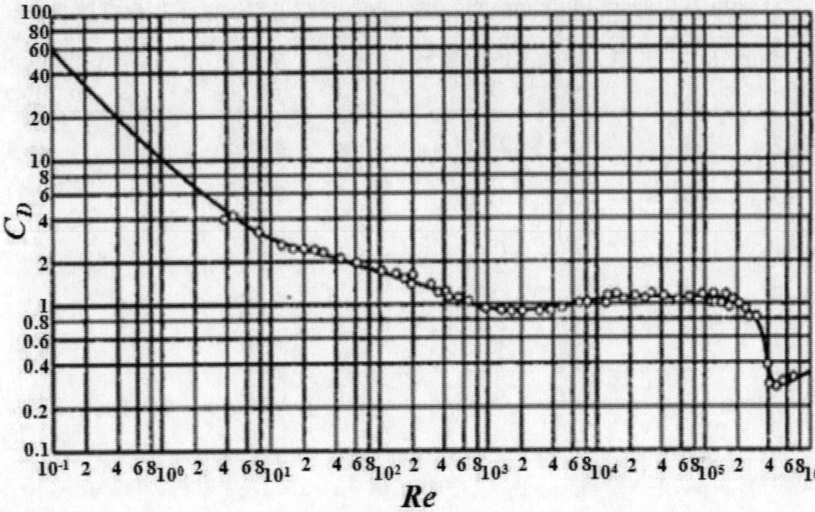

Figure 8. Relationship between resistance coefficient and Reynolds number.

Since the linear velocity of the cantilever at different points from the inside to the outside were different (Fig. 7), in theory, the resistance coefficients on the cylindrical cantilever will also be completely different. Reynolds number can be expressed as

$$\mathrm{Re} = \frac{\rho(xw)d}{\mu} \tag{4}$$

The mean Reynolds number (i.e., the equivalent Reynolds number) on the cantilever can be written as

$$\mathrm{Re} = \frac{1}{l}\int_0^l \frac{\rho w d}{\mu} dx \tag{5}$$

$$\mathrm{Re} = \frac{\rho(\frac{1}{2}lw)d}{\mu} \tag{6}$$

As described in Eq. (6), the Reynolds number at the center of the cantilever is equal to the average Reynolds number. After the value was determined, the corresponding resistance coefficient was acquired from Fig. 8, and resistance was calculated.

Considering that flapping also exists during forward movement of the cantilever, the flapping resistance may vary. Thus, validation tests were performed. After removing the flapping-wing feather fan, tests were conducted according to the method described in Section 1.3. It can be observed that after the flapping stops at a wind velocity of 6 m/s, the linear velocity at the installation point increases by 0.061 m/s. Moreover, as the wind velocity increased to 9 m/s and 10.3 m/s, the linear velocity at this point increased by 0.013 m/s and 0.021 m/s, respectively. Therefore, the effect of flapping on resistance can be neglected.

3. Experimental data and results

Using the rotary platform, linear velocities and average Reynolds numbers of the M-shaped bending-surface feather fan, while flapping or not flapping, were measured at different flow velocities. The results are listed in Table 2.

Table 2. Measured linear velocities and average Reynolds numbers.

Flapping + Propeller (m/s)	Average Reynolds number (Re)	Propeller (m/s)	Average Reynolds number (Re)
6.002	5458.27	4.942	4494.30
7.013	6377.69	6.607	6008.47
8.001	7276.19	7.209	6555.93
8.943	8132.85	8.36	7602.66
9.947	9045.90	9.516	8653.94
11.95	10867.45	11.79	10721.94
15.06	13695.71	14.85	13504.74

Compared to Fig. 8, the resistance coefficient C_D fluctuated only slightly around 1 throughout the test, with a minimum value of 0.9 and a maximum value of 1.1. Therefore, resistance was only related to linear velocity at the installation point of the feather fan. At each wind velocity, the flapping propulsion was calculated by

$$\Delta F = \frac{1}{3} d\rho C_D l(v_1^2 - v_2^2) \tag{7}$$

The form of Eq. (7) is similar to the kinetic energy variation of the system. In other words, flapping led to an increase in the kinetic energy of the system that is proportional to propulsion. According to the test, Δv^2 caused by flapping wing at each wind velocity is shown in Table 3.

Table 3. Linear velocity and Δv^2.

Flapping + Propeller (m/s)	Propeller (m/s)	ΔV^2 (m^2/s^2)
6.002	4.942	11.60064
7.013	6.067	12.37368
8.001	7.209	12.04632
8.943	8.36	10.087649
9.947	9.516	8.388553
11.95	11.79	3.7984
15.06	14.85	6.2811

The relationship between the wind velocity and Δv^2 was plotted, as shown in Fig. 9.

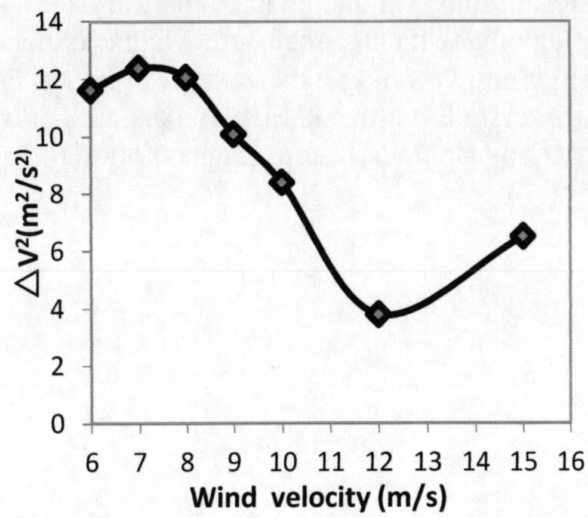

Figure 9. Relationship between wind velocity and $\triangle V^2$.

From Fig. 9, the propulsion of the flapping-wing feather fan first increased and then decreased, finally reaching the maximum at a wind velocity of 7.3 m/s and a minimum at a wind velocity of 12 m/s.

4. Conclusions

(1) As the wind velocity increases, the propulsion of the five-rod M-shaped feather fan first increases and then decreases, followed by a final increase. The optimal wind speed of the flapping-wing feather fan is 7.3 m/s.

(2) Under a particular flapping frequency and stiffness, the flapping-wing feather fan exhibits different degrees of deformation at different inflow velocities, and propulsion of the flapping-wing feather fan reaches a maximum value at an appropriate deformation angle.

(3) The measurement principle of the rotary force-measuring platform was theoretically analyzed, and reliability of the platform was validated. It can also be concluded that, for flapping-wing feather fans of different sizes, shapes, and areas, an optimal inflow velocity exists at which the feather fan can produce maximum propulsion at an optimal deformation angle during the flapping process.

In conclusion, a maximum propulsion velocity exists for each flexible flapping-wing feather fan according to different parameters.

References

[1] Zhang Weiwei, Xu Yang, Jiang Yuewen, et al.Analysis of the characteristics of the flapping-wing propulsion based on the CFD method [J] .Air Dynamics, 2014, 32 (4): 446-452.

[2] Knoller R. Die gesetze des luftwiderstandes[J].Flugund Motortechnik (Wien),1909,3(21):1-7.

[3] Betz, A. Ein beitrag zur erklarung des segeluges[J].zeitschrift fuer Flugtechnik und Motorluftschiffahrt,1912,3:269 - 272.

[4] K. D. Jonesy and M. F. Platzerz [J]. Flapping-wing propulsion for a micro air vehicle. AIAA, 2000.

[5] Steven Ho Unsteady aerodynamics and adaptive flow control of micro air vehicles[D].Los Angeles:University of California 2003:40-52

[6] K. D. Jonesy and M. F. Platzerz [C]. Numerical computation of flapping-wing propulsion and power extraction, Aiaa Aerospace Sciences Meeting and Exhibit. 1997.

[7] Tao W. Performance investigation of symmetrical plunging dual-foils propulsor[J]. Journal of Hydrodynamics, 2009.

[8] Chun-Jin Y U, Hai-Song A. Numerical study of aerodynamics for flexible membrane flapping-wing MAV[J]. Journal of University of Science & Technology of China, 2009, 39(12):1305-1310.

[9] Qian Jing, Zhang Zhengke, Luo Shijun, et al.Numerical study on flow characteristics of two-dimensional elastic flapping wing [J] .Acta Aerodynamica Sinica, 2012, 30 (1): 113-119.
[10] Shao Limin, Song Bifeng, Song Wenping.Experimental Study on Propulsion Characteristics of Micro Flapping Wing [J] .Act on Experimental Hydrodynamics, 2009, 23 (1): 1-6.
[11] IJ Ritchiek, Hydraulic friction manual[M]. Aero Engine Editorial Department, 1985

IOP Conf. Series: Materials Science and Engineering **470** (2019) 012002 doi:10.1088/1757-899X/470/1/012002

Improved Heat Transfer Correction Method for Turbocharger Compressor Performance Measurements

Liyong Huang[a], Chaochen Ma[a, *], Yanzhao Li[b], Jianbing Gao[a] and Mingxu Qi[a]

[a] School of Mechanical Engineering, Beijing Institute of Technology, Beijing 100081, China

[b] Shouguang Kangyue Turbocharger Ltd, Weifang 262711, China

* Corresponding author: E-mail : machaochen1900@163.com (Chaochen Ma)

Abstract. The compressor adiabatic performance map is important for turbocharger matching with engine. Its accuracy was affected by the heat transfer phenomenon in the turbocharger. The corrected method for compressor performance measurement which affected by heat transfer was widely investigated by some authors. The aim of this work was to measure compressor adiabatic efficiency by simplified method, and estimated the effects of internal heat transfer on compressor efficiency. This approach only added five temperature sensors on the turbocharger based on the standard instruments and doesn't need other special instruments, and the heat transfer properties of turbocharger and compressor adiabatic efficiency could be calculated by lumped model. This method was validated by other experiments. Finally, the heat absorbed by compressor under different operating conditions was analyzed by this method. The results showed water-cooling was main influenced factor, and water-cooling was a barrier for heat transfer from turbine to compressor..

1. Introduction

Turbocharger plays an dominant part in emission and energy reduction of road vehicles. The precise adiabatic efficiency of compressor is very important for engine simulating at the stage of engine design. The efficiency of compressor which was measured at standard test bench is not exacted due to heat transfer in turbocharger [1]. The traditional way to calculate the compressor efficiency, based on a ratio of adiabatic and actual of compressor enthalpy from measurements of a test-bench showed in Eq.1, the detailed about Eq.1 was presented in the reference [1], provides inaccurate results. This inaccuracy comes from the no consideration of heat fluxes effect in the turbocharger [1]. In order to corrected results of measured, a lot of experts and scholars did a great deal of researches for behavior of turbocharger heat transfer in the last few decades [2].

$$\eta = \frac{\Delta h_{is,C}}{\Delta h_{ac,C}} = \frac{C_{P,air} T_1 (\Pi^{(k_{air}-1)/k_{air}} - 1)}{C_{P,air}(T_{2,adi} - T_1)} \quad (1)$$

One dimensional lumped model of turbocharger heat transfer was studied by some authors. Turbochargers were simplified and divided into different one-dimensional nodes in the lumped model, convective or thermal conductance were used to connected different nodes of turbocharger [3]. The

IOP Conf. Series: Materials Science and Engineering **470** (2019) 012002 doi:10.1088/1757-899X/470/1/012002

model of calculation was validated by some experiments. The modeling methodology of one dimensional lumped model was given by Reyes-Belmonte et.al [4] and Shaaban et.al [5] .The heat transfer property of turbocharger was needed to measure by some particular instruments. The special oil as work medium was used in turbine and compressor in order to measure heat transfer properties of turbocharger. Turbocharger was often independent tested on test bench. In Ref. [6], a turbocharger was installed on a diesel engine and experiments performed at different engine speeds and loads. The test results showed that the temperature of the engine exhaust gases has a large impact for turbocharger internal heat transfer.

The lumped model was analyzed and validated by some researchers, and some other authors applied this method to correct the turbocharger performance of measurement[7-14]. The special and expensive instruments were required in the lumped model solving process.

The measured procedure of compressor adiabatic efficiency that applied in the previous studies is complex. Some simple and exact method was needed to investigate. The exacted and simplified methodology to measure adiabatic efficiency for compressor was presented in this paper. Experimental tasks was described in the first part, focusing on the description of sensors and equipment used, the testing methodology and the analysis of testing results were also reported in this part. The second part of this paper is focusing on validation for the testing methodology. The quantities of heat absorbed by compressor was analyzed for different operating conditions. The main conclusions of this work are summarized in the final part of this paper.

2. Experiment investigation

2.1. Test rig

The turbocharger used in this study was for gasoline engine, with turbine and compressor wheel diameters of 37 mm and 44 mm respectively. The experimental activity was developed at Shouguang Kangyue Turbocharger company. A schematic of the test rig is presented in Figure 1. This test bench is able to simulate the conditions of engine intake and exhaust system. The mass flow rate of 0.02-1.0 kg/s fresh air supplied by compression engine to turbine. The compressor sucks fresh air from environments and the operating states were set by valve (up and down stream of compressor) controlling. The hot air of turbine was heated by neutral gas in the combustion chamber. An combustor allows performing experimental investigations with turbine temperature up to 1100°C. An independent lubrication and water supply system is installed. System inlet pressure and temperature could be independent modified. The efficiency of compressor is calculated from the temperature of inlet and outlet, and the turbine efficiency is calculated by compressor power consumption. Data acquisition and actor controlling of this test rig was performed by NI (National instruments) hardware. The software of this test rig is developed by the company of KRATZER. Efficiency and other main operating parameters of turbocharger was recorded by automatic measured system of test rig.

The static air pressure was estimated by the piezo-resistive transducers with an accuracy of ±0.3% of full scale. The air temperature was measured by K type thermocouples (accuracy ± 0.4% of full scale). The thermal mass flow meter with an accuracy of ±1% at full scale was used to measure fresh air mass flow rate of compressor and turbine. The turbocharger rational speed was measured by an sensor (ACAM PICOTURN-BM) installed on the compressor volute and was able to detect the compressor wheel revolution with an accuracy of ±0.2% of full scale.

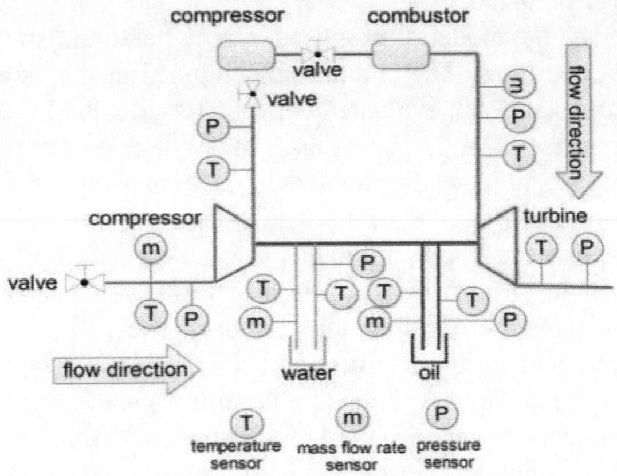

Figure. 1 Layout of turbocharger test bench

Additional temperature sensors were installed on components of turbocharger due to the process of turbocharger heat transfer is complex. The static temperature of nodes were measured by these sensors. Values of those sensors measured were used to reflect temperature distribution of turbocharger under different conditions, the heat transfer of model was validated by those values to a certain extent. In order to measure temperature of nodes C, three K type thermocouples (accuracy ± 0.4 % of full scale) were installed at the volute compressor. Two metal temperatures were also measured on the bearing casing (nodes of H1 and H3), one nearing the turbine casing and another nearing the compressor casing. Details of the thermocouple installation are given in Figure 2 and Figure 3. The temperature of turbine surface was measured by infrared thermometer (accuracy ± 1.5% of full scale) due to K type thermocouple was difficult to fix on the high temperature surface.

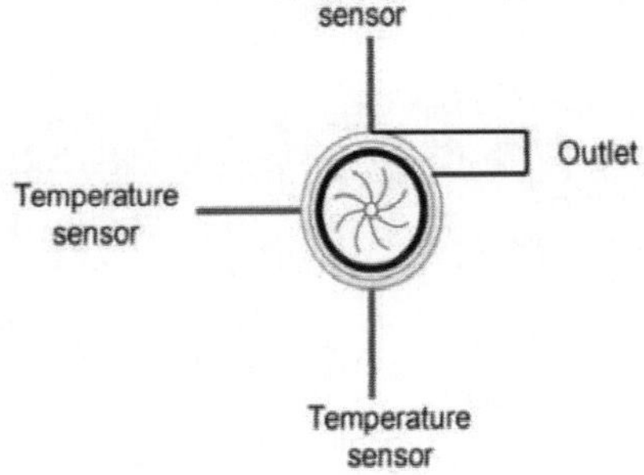

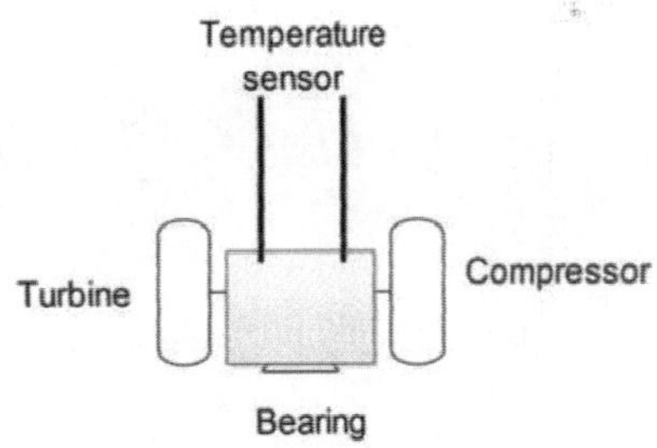

Figure 2. Compressor volute **Figure 3.** Bearing casing

2.2. Experimental study

All performed measurements were processed in order to obtain compressor adiabatic-efficiency (AE). The definition of AE as seen in Ref [12]. To ensure the accuracy of AE, the heat transfer was considered in the $T_{2,adi}$ of measured, $T_{2,adi}$ is an important parameter for AE, these shown in Eq.(2)-(3):

$$T'_{2,adi} = T_{2,adi} - \triangle T \tag{2}$$

$$m * C_{P,air} * \triangle T = Q_L \tag{3}$$

where: $T'_{2,adi}$: corrected temperature of compressor outlet;

$\triangle$ T: temperature difference between measured and corrected;

m: mass flow rates of compressor;

Q_L: heat of compressor obtained;

In order to set and validate the model of proposed, main operating parameters of turbocharger were measured under different conditions. The compressor obtained heat was affected by compressor and lubricating system operating states. To evaluate heat transfer affected by compressor operating states, the compressor work at different rotational speed and compression ratio was performed in the experiments. The lubricating states is also related to heat transfer from turbine to compressor, so measurements were performed for oil at different inlet temperature and pressure. The turbocharger was run with and without water in the experiments due to water is important factor for compressor obtained heat from turbine. In order to woken the heat transfer in the turbocharger, the water inlet temperature maintain the same as oil inlet temperature. The turbine inlet temperature plays an important role in the turbocharger internal heat transfer, three different inlet temperature (500°C, 600°C, 700°C) of turbine were given under the other conditions maintained constant.

Additional temperature sensors were installed on components of turbocharger due to the process of turbocharger heat transfer is complex. The static temperature of nodes were measured by these sensors. Values of those sensors measured were used to reflect temperature distribution of turbocharger under different conditions, the heat transfer of model was validated by those values to a certain extent. In order to measure temperature of nodes C, three K type thermocouples (accuracy ± 0.4 % of full scale) were installed at the volute compressor. Two metal temperatures were also measured on the bearing casing (nodes of H1 and H3), one nearing the turbine casing and another nearing the compressor casing. Details of the thermocouple installation are given in Figure 2-3. The temperature of turbine surface was measured by infrared thermometer (accuracy ± 1.5% of full scale) due to K type thermocouple was difficult to fix on the high temperature surface.

2.3. Experimental results

In order to evaluated the effects of heat transfer in the turbocharger, the section of experimental results will be make comparing for results with or without water cooling.

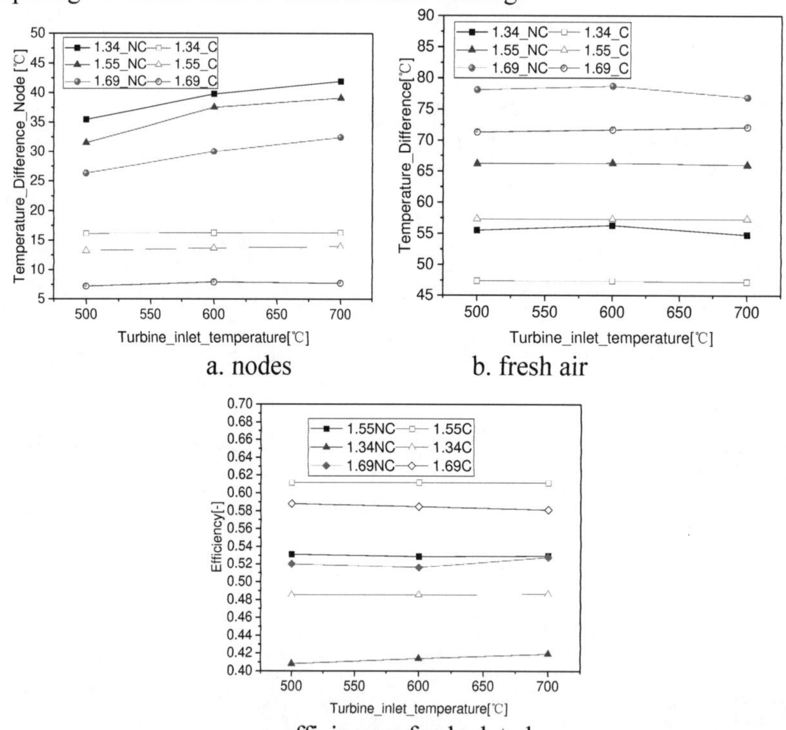

a. nodes b. fresh air

c. efficiency of calculated

Figure 4. Temperature difference cooling or no cooling

Difference value of temperature between nodes of C and H3 was presented in Figure 4 a (NC is no cooling, C is cooling). It's shown that temperature difference between nodes of C and H3 was affected by the compression ratio, air temperature of turbine inlet and water cooling states. It was direct proportion to air temperature of turbine inlet, inverse ratio to compression ratio. This phenomenon was caused by internal heat transfer of turbocharger. The quantity of heat obtained by compressor can be directed reflected by the temperature difference between nodes of C and H3, so the heat transfer from turbine to compressor is higher when compression ratio decrease and turbine inlet temperature increase. That heat flux will be partly removed by the coolant liquid in case it exists, so the difference is higher at the no water cooling. b and c of Figure 4 show difference value of fresh air temperature between compressor inlet and outlet and efficiency of calculated at different conditions. Compressor adiabatic efficiency was determined by rational speed and compression ratio of compressor. The traditional way to calculate the compressor efficiency, based on compression ratio and outlet temperature of compressor, therefore, the same trend was observed in these two pictures. More heat was absorbed by compressor under no water cooling due to water is the barrier of heat from turbine to compressor. The value of no-cooling is larger than cooling. The compressor outlet temperature and efficiency increased when compressor compression ratio increases and it will cause the temperature difference between C and H3 decreasing, and the heat transfer from turbine to compressor was woken by this phenomenon, therefore, the heat of compressor absorbed is lower when compressor work at higher compression ratio operating points. From above analysis, the heat transfer can be separate reflected by two different parameters of difference temperature of nodes C and H3 and difference value of fresh air temperature between compressor inlet and outlet. The error of calculated efficiency was lead by heat transfer in the turbocharger. The heat absorbed by compressor can be calculated by solved equation about k and temperature difference of nodes C and H3, difference value of fresh air temperature between compressor inlet and outlet. The more detailed solution process was described in the section 3.

3. Turbocharger thermal model

In order to simplified the complex heat transfer phenomena in the turbocharger, heat has been assumed extracted from the exhaust gas of turbine to the fresh air in the compressor before it is expanded in the turbine stator (see Figure 5). Water and oil is the barrier to prevent heat transfer from turbine to compressor. Turbocharger was simplified into different nodes. The bearing casing was divided into three part in order to make sure accuracy of model. The heat transfer in the turbocharger is very complex and its affected by the water and oil operating states, the energy obtained from surroundings was omitted due to the surface of compressor is small in area and coefficient of heat transfer by free convection is little [13], therefore, the heat obtained by fresh air in compressor equalled to volute of compressor transfer from bearing casing due to the energy balance. The heat of compressor obtained is sampled to nodes H3 transfer to C and heat absorbed by compressor was calculated according to Eq.(4.) The temperature of nodes were measured by thermocouples. Details of the thermocouple installation are given in Figure 1. In order to solve the value of heat transfer coefficients (K) , the different heat transfer conditions were given under compressor operating at the same rotational speed and compression ratio (see Figure 5). The form of equation set was presented as Eq. (5)-(6). Because compressor adiabatic efficiency only determined by rotational speed and compression ratio, heat transfer coefficients (K) of bear casing to compressor volute can solved by Eq. (5)-(6) and not need some specific test bench, the value of K is 0.03467818 (unit was omitted and the unit of other variable is ISO). Q is the heat of turbine released or compressor obtained.

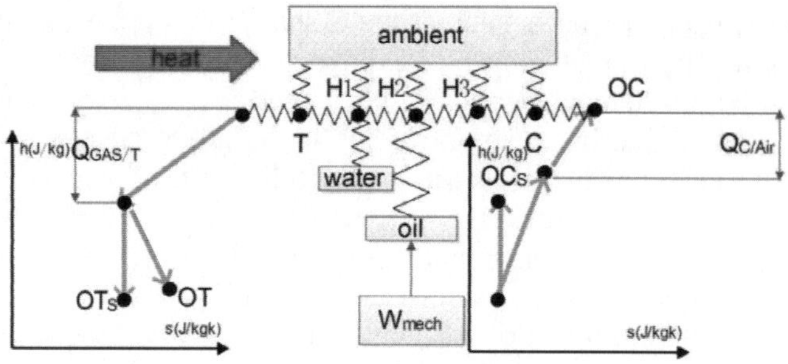

Figure 5. Schematic of turbocharger heat transfer

$$Q_{C/Air}=(T_{H3}-T_C)*K_{H1/C} \quad (4)$$

$$Q=(T_{H3}-T_C)*K_{H1/C} \quad (5)$$

$$Q'=(T'_{H3}-T'_C)*K_{H1/C} \quad (6)$$

4. Model validation

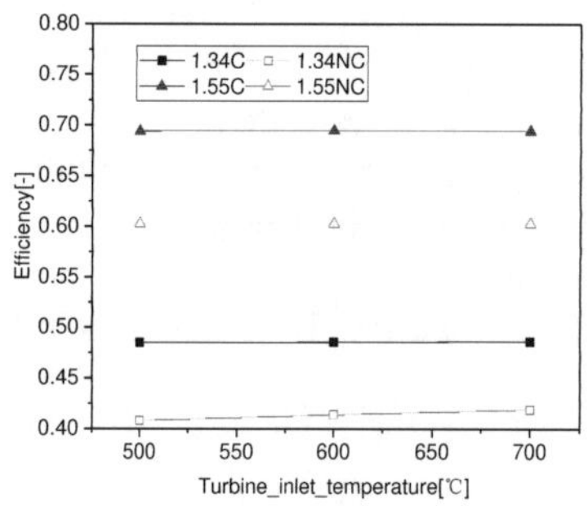

Figure 6. Efficiencies of no correction

Table1. Efficiency of correction

Turbine_inlet_T[°C]	1.34NC [-]	1.34C [-]	1.55NC [-]	1.55C [-]
500	0.556487974	0.556488	0.699807	0.699807443
600	0.556487974	0.556488	0.699807	0.699807443
700	0.556487974	0.556488	0.699807	0.699807443

The validation for corrected method was reported in the below. The comparison of efficiency at with or without cooling, different compression ratio and turbine inlet temperature is given in Figure 6. The compressor adiabatic efficiency was only determined by the rotational speed and compression ratio of compressor under ideal condition. Compressor adiabatic efficiency measured in hot condition is less than the cooling condition due to heat transfer in the turbocharger. The value of measured compressor adiabatic efficiency also was affected by other operating (e.g. oil operating states, turbine inlet temperature, rotational speeds of turbocharger) and geometry parameters. The heat transfer of turbocharger also was affected by these parameters. The discrepancy of corrected results between with and without cooling is less than 1e-4 as shown in Table 1. The discrepancy was shown in the Table 1 was caused by four main factors. First of all, some heat dissipated by bearing casing was omitted in corrected process. Second factor contributing to the overall discrepancy is that, some compressor absorbed heat of radiated by turbine was also neglected in corrected results. Third discrepancy influenced factor is that, compressor adiabatic efficiency has tiny difference when fresh air of compressor has different temperature state. Another overall discrepancy influenced factor is that, some measured error due to the temperature sensor was connected with turbocharger by high temperature glue and some other uncertainty factors in the measured process. This method can be improved in four sides of expressed above.

5. Conclusion

Traditionally, heat losses have been neglected in the compressor map measured process and the behaviour of the compressor has been predicted by direct use of manufacturer maps to interpolate without any corrected, therefore, the accuracy of map can't reach up to the requirement of engine design.

This paper reported the exacted methodology to measure efficiency and heat absorbed for compressor. This method isn't complex and can be performed at standard test bench of turbocharger .The following main conclusions can be drawn:

(1) Water cooling significantly impacts heat transfer between the compressor and turbine for gasoline turbocharger. Temperature of water is more close to temperature of compressor than turbine operating temperature. Qc of without water-cooling is three to four times for with water-cooling.

(2) The improved method gives exact measured results of adiabatic efficiency and Qc of compressor that only adding five temperature sensors based on standard test bench of turbocharger. Those sensors were used to measure temperature of nodes C and H3.

(3) Less quantities heat transfer from turbine to compressor at higher rational speeds due to the work range of compressor move to high compression ratio direction. The average temperature of fresh air in compressor was increasing when compression ratio increase. The temperature difference between nodes C and H3 was woken. Less quantities of heat was absorbed by compressor at higher compression ratio corresponding the same reason.

References

[1] Chesse P, Chalet D and Tauzia X. Impact of the Heat Transfer on the Performance Calculations of Automotive Turbocharger Compressor. 2011 *Oil & Gas Science & Technology.* **66** 514-5

[2] Fu J, Zhu G, Zhou F, Liu J, Xia Y and Wang S. Experimental investigation on the influences of exhaust gas recirculation coupling with intake tumble on gasoline engine economy and emission performance. 2016 *Energy Conversion & Management.* **127** 424-36.

[3] Olmeda P, Dolz V, Arnau F and Reyes-Belmonte M. Determination of heat flows inside turbochargers by means of a one dimensional lumped model. 2013 *Mathematical and Computer Modelling* **57** 1847-52.

[4] Reyes-Belmonte M. Contribution to the experimental characterization and 1-D modelling of turbochargers for IC engines [Ph.D. thesis]. 2013 *Valencia: Universitat Politecnica de Valencia.*

[5] Shaaban S. Experimental Investigation and Extended Simulation of Turbocharger Non-Adiabatic Performance [Ph.D. thesis]. 2013 *Hannover: Universitat Hannover*

[6] Romagnoli A, Martinez-Botas R. Heat transfer analysis in a turbocharger turbine: An experimental and computational evaluation. 2012 *Applied Thermal Engineering* **38** 58–77.

[7] Payri F, Olmeda P, Arnau FJ, Dombrovsky A and Smith L. External heat losses in small turbochargers: Model and experiments. 2014 *Energy.* **71** 534-46.

[8] Verstraete D, Bowkett C. Impact of heat transfer on the performance of micro gas turbines. 2015 *Applied Energy* **138** 445-9.

[9] Marelli S, Marmorato G and Capobianco M. Evaluation of heat transfer effects in small turbochargers by theoretical model and its experimental validation. 2016 *Energy* **112** 264-72.

[10] Schinnerl M, Seume JR, Ehrhard J and Bogner M. Heat transfer correction methods for turbocharger performance measurements. *ASME Turbo Expo 2016: Turbomachinery Technical Conference and Exposition.*

[11] Casey MV, Fesich TM. The Efficiency of Compressors With Diabatic Flows. 2010 *Journal of Engineering for Gas Turbines & Power* **132** 072302-1-9.

[12] Tanda G, Marelli S, Marmorato G and Capobianco M. An experimental investigation of internal heat transfer in an automotive turbocharger compressor. 2017 *Applied Energy* **44** 531–9.
[13] Aghaali H, Angstrom HE, Serrano JR. Evaluation of different heat transfer conditions on an automotive turbocharger. 2014 *International Journal of Engine Research* **16** 137-51.
[14] Baines N, Wygant KD and Dris A. The Analysis of Heat Transfer in Automotive Turbochargers. 2009 *Journal of Engineering for Gas Turbines & Power*. **132** 115-26.

Clutch Control Strategy of Driving Mode Transition for P2 Hybrid Electric Vehicle

Shuhan Wang [1], Yanjing Liu [1] and Tianyuan Cai [1]

[1] School of Transportation Science and Engineering, Beihang University, 100191, Beijing, China

liuyanjing@buaa.edu.cn

Abstract. Clutches are widely utilized for not only implementing gear shifting but also transmitting power in hybrid electric vehicles (HEVs). During the driving mode transition in a P2-structure HEV, the coordination of between power sources, such as engine and electric motor, is realized by a hybrid clutch. Therefore, it is of great significance to establish a control strategy for the hybrid clutch ensure drivability and driving comfort. In this study, a physical model of the hybrid clutch and other relevant components in the hybrid system is established in MATLAB/Simulink firstly. Then, a model-based coordinated control strategy is developed for the hybrid clutch during the driving mode transition. The hydraulic oil filling and torque transferring characteristics of the hybrid clutch are mainly considered in the control strategy. Finally, the proposed control strategy is validated by simulation and the results show that the quality of driving mode transition is obviously improved.

1. Introduction

As the energy shortage and environmental pollution problems are becoming increasingly serious, there is a worldwide trend for developing hybrid electric vehicles (HEVs) in the automotive industry. HEVs have multiple power sources and can be operated under variable conditions at different driving modes, such as electric motor driving mode, engine driving mode and hybrid driving mode. In a P2-structure HEV, shift clutches are utilized for gear shifting, while a hybrid clutch transmits power from engine or electric motor to the transmission. To ensure drivability and driving comfort during driving mode transition, the control of the hybrid clutch is necessary. Researches have been made to analyse the fluctuation in vehicle speed and engine torque during the connecting and disconnecting of hybrid clutch. There still exists some challenges in the control process of hybrid clutch, such as establishing specific control strategy for different phases in clutch fill process and reducing vehicle jerk during mode transition.

There are several studies that focus on this theme. Zhao Zhiguo proposed a robust control algorithm that resists external disturbance and model parameter fluctuation, which takes full advantage of fast and accurate torque (or speed) response of three electrical power sources and gets the clutch of DCT fully involved in the mode transition process [1]. In the study of Kumar Sushil, the control strategies for power flow management of a HEV were developed based on a supervisor controller which uses various torque and SOC checks to determine driving mode [2]. Su Yanzhao developed a dynamic coordinated control strategy (DCCS), including a staged engine torque feedforward and feedback estimation (ETFBC) and an active damping feedback compensation (ADBC) based on drive shaft torque estimation (DSTE) [3]. In the research of Schori Markusthe, theory of hybrid optimal control was used to automatically calculate lookup tables for optimal gear shifts, optimal torque-split between motor/generator was designed to perform the accurate position tracking control, especially when the parameters uncertainties and external disturbance occurred in the clutch actuating mechanism [4]. In the study of Yang Chao, a novel robust hierarchical mode transition control method for a plug-in hybrid electric bus (PHEB) with pre-transmission parallel hybrid powertrain was presented, in which a hierarchical control structure including two robust controllers in both upper layer and lower layer was proposed based on the divided five stages of clutch control [5]. Lei Zhenzhen proposed the coordinated control strategy (including fuzzy logic control and PID control) of clutches to implement the mode transition and shifting simultaneously, which can solve the conflict between mode transition and DCT shifting in a HEV [6].

This study established a physical model of the hybrid clutch and other components in the hybrid system of a HEV in MATLAB/Simulink firstly. Then, a model-based coordinated control strategy is developed for the hybrid clutch during the driving mode transition. In the proposed control strategy, the hydraulic oil filling and torque transferring characteristics of the hybrid clutch are mainly considered. Finally, the control strategy is validated through simulation. The results show that the quality of driving mode transition is successfully improved.

2. Configuration of P2-structure HEV

As shown in Figure 1, the P2 structure of HEV in this study mainly consists of an engine, a hybrid clutch, an 8-speed automatic transmission (8AT), a dual mass flywheel (DMF) and an electric motor. The engine and electric motor are controlled by engine control unit (ECU) and motor control unit (MCU) respectively. The control of 8AT and the hybrid clutch is implemented in transmission control unit (TCU). The driving mode transition is realized by the coordination of ECU, MCU, TCU and hybrid control unit (HCU).

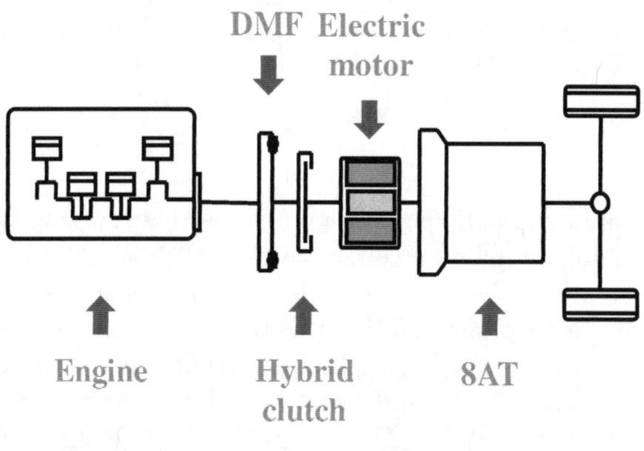

Figure 1. Schematic of P2 structure.

As presented in the schematic, the power generated from engine and electric motor can be transmitted to the drive shaft of HEV through the 8AT. The hybrid clutch is disconnected with the engine when electric motor driving mode is operated, which indicates the electric motor works as the power source. The hybrid clutch engages when there is a command signal that the transition to engine driving mode or hybrid driving mode need to be executed. During this process, the engagement of the hybrid clutch can be influenced by several factors, such as the characteristics of DMF, the ignition of the engine and the rotational speed of the electric motor.

3. Models and analysis of hybrid system

To achieve clutch control in the hybrid system, the models of the engine, the hybrid clutch, 8AT, DMF and the electric motor are developed and analysed.

3.1. Engine model

Considering the rigid-body motion of piston in the engine, the output torque of the engine (T_E) can be expressed as

$$T_E = T_{Ep} - T_{Ef} - T_{Ei} \tag{1}$$

where T_{Ep} is the cylinder pressure torque, T_{Ef} is the viscous friction torque, and T_{Ei} is the equivalent inertia torque.

3.2. Hybrid clutch model

A hyperbolic tangent function (*tanh*) model is utilized in the modeling of the hybrid clutch. The maximum torque that can be transferred by the clutch, which is also named as the torque capacity of the clutch (T_{C_cap}), is determined by the control pressure of the clutch. The torque capacity is calculated by

$$T_{C\_cap} = \mu_s \cdot (P_{C\_ctrl} - P_{C\_kp}) \cdot A_p \cdot r_C \cdot z \tag{2}$$

where μ_s is the static friction coefficient, P_{C_ctrl} is control pressure of the clutch, P_{C_kp} is the kiss-point pressure of the clutch, A_p is the cross-section area of the piston, r_c is the equivalent radius of the friction plates, z is the number of friction pairs in the clutch.

When the slip between the two axes that connected by the clutch equals to zero, the clutch is in the state of rigid connection. In this state, the actual torque that transferred by the clutch (T_C) is less than or equal to the T_{C_cap}. However, if the slip is not eliminated, the actual torque T_C can be calculated by the *tanh* model, which can be expressed as

$$T_C = \mu_d \cdot (P_{C\_ctrl} - P_{C\_kp}) \cdot A_p \cdot r_C \cdot z \tag{3}$$

$$\mu_d = \mu_s \cdot \tanh\left(\frac{\Delta\omega}{\Delta\alpha_C}\right) \tag{4}$$

where μ_d is the dynamic friction coefficient, $\Delta\omega$ is the rotation speed difference between the two axes, α_c is an adjustive parameter, whose function is shown in Figure 2.

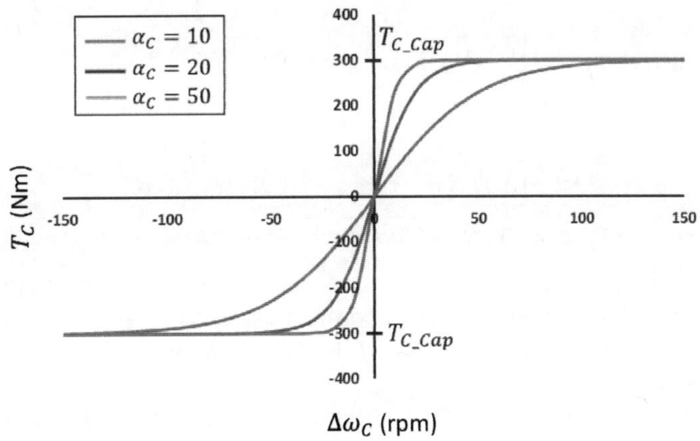

Figure 2. Different *tanh* models.

In the *tanh* models, the limitation for $\Delta\omega$ is $\Delta\omega_l = 5rpm$. The value of μ_d is determined to be 0.106 when $\Delta\omega \leq \Delta\omega_l$.

3.3. 8AT model

In this study, the input and output torque of 8AT can be calculated based on the kinematic and dynamic equations developed by the parameters and limitations in the transmission. The relationship between the input and output angular acceleration of the transmission ($\ddot{\theta}_{in}$ and $\ddot{\theta}_{out}$) can be expressed as

$$\ddot{\theta}_{in} = i_T \cdot \ddot{\theta}_{out} \tag{5}$$

where i_T is the total ratio of the transmission.

IOP Conf. Series: Materials Science and Engineering **470** (2019) 012003 doi:10.1088/1757-899X/470/1/012003

3.4. DMF model

The major function of DMF is reduce the fluctuation of the engine torque and maintain the response speed of the engine output. A simplified model of DMF is developed as

$$J_{D,pri}\ddot{\theta}_E = T_O - k_D\varphi_D - c_D(\dot{\theta}_E - \dot{\theta}_D) \quad (6)$$

$$J_{D,sec}\ddot{\theta}_D = -T_{C\_Fr} + k_D\varphi_D + c_D(\dot{\theta}_E - \dot{\theta}_D) \quad (7)$$

$$\varphi_D = \theta_E - \theta_D \quad (8)$$

where $J_{D,pri}$ and $J_{D,sec}$ are the inertia of primary and second mass, k_D is the torsional rigidity of the arc spring, c_D is the rotation damping coefficient, θ_E and θ_D are the rotation angle of the primary and second mass, φ_D is the displacement angle of DMF, T_{C_Fr} is the torque transferred to the front of hybrid clutch.

3.5. Electric motor model

In this study, the electric motor is directly controlled by MCU. The actual torque of electric motor can be expressed as

$$T_M = \begin{cases} T_{M,cmd} & if\left|T_{M,cmd}\right| < T_{M\_max}(n_M) \\ sign(T_{M,cmd})\, T_{M\_max}(n_M) & if\left|T_{M,cmd}\right| < T_{M\_max}(n_M) \end{cases} \quad (9)$$

where $T_{M,cmd}$ is the command torque of electric motor, $T_{M\_max}(n_M)$ is the maximum torque when motor speed is n_M.

4. Clutch control strategy

The dynamic control flowchart of clutch in this study can be illustrated as shown in Figure 3. The characteristics of torque-pressure transferring (T2P) and pressure-current transferring (P2C) are derived from the look-up tables stored in TCU. The clutch pressure is controlled by a direct proportional control valve.

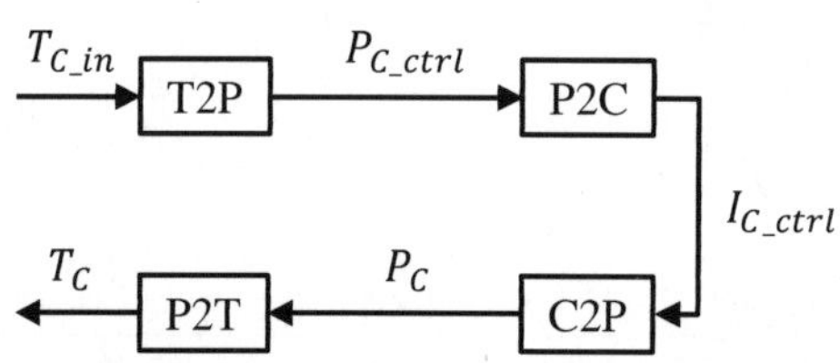

Figure 3. Control flowchart of clutch.

During the mode transition, the engine is started and engaged into the powertrain through the engagement of the hybrid clutch. And the driving torque for the vehicle operation and the engine-start is provided by the electric motor. To describe oil filling and torque transferring characteristics of the clutch during mode transition, this study divides this process into six stages, as shown in Figure 4.

Stage 1: The mode transition starts and the command pressure of the hybrid clutch (P_{C_ctrl}) is fixed at a high level to realize a fast-filling. Free displacement of clutch piston is diminished. A pulse of clutch torque (T_C) is produced due to the high-level pressure. The torque is used to start the engine by acting on the output shaft of the engine via DMF.

Stage 2: P_{C_ctrl} is fixed at a low level to reach the kiss-point. The engine speed (n_E) is regulated actively due to the ignition. The torque produced by the engine is negative for the input shaft of 8AT and the vibration produced by ignition can be transmitted to the transmission via the clutch. In order to ensure the comfort, the clutch can be disconnected temporarily. This stage ends when n_E is larger than the electric motor speed (n_M). The torque that the engine transmits to the input shaft is mainly determined by the clutch pressure in this stage.

Stage 3: P_{C_ctrl} increases to achieve the balance between the engine torque (T_E) and T_C. The value of n_E keeps rising when the torque that transferred by clutch is smaller than T_E. The peak of the engine speed is crucial for achieving high-quality control of the whole process.

Stage 4: The slip is eliminated via the control of clutch pressure. P_{C_ctrl} is adjusted to adapt to the continuous changes in engine torque.

Stage 5: A rapid increase in P_{C_ctrl} is executed to achieve the quick engagement of clutch. This stage ends until the pressure reaches the maximum and the clutch is locked.

Stage 6: The exchange of torque is realized between the engine and the electric motor. And the torque control strategy in the following phases can be implemented after this stage.

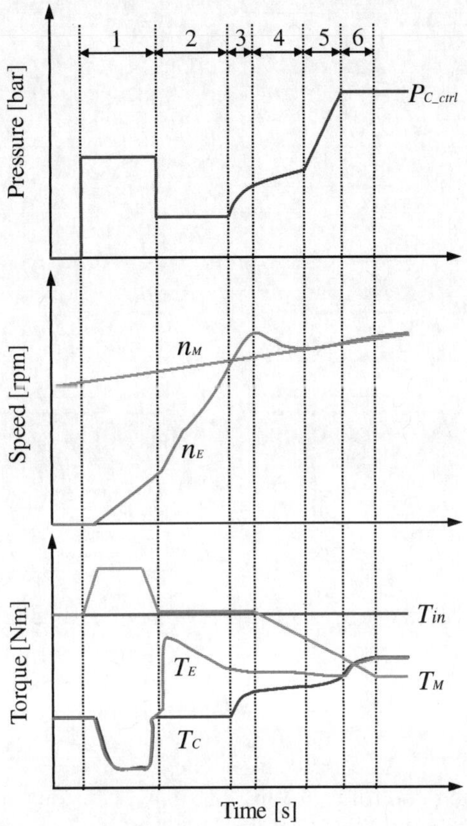

Figure 4. Schematic diagram of mode transition process.

5. Simulation results

The simulation results of the clutch control in this study are shown in Figure 5. In the simulation, the mode transition from the electric motor driving mode to the hybrid driving mode starts when the vehicle speed reaches the pre-set threshold value. According to the model of the hybrid clutch, the final compressed pressure of the clutch is 20bar.

Before the execution of the control strategy, the torque produced by the hybrid clutch does not properly match the output torque of the electric motor (T_E), as shown in Figure 5 (a). Meanwhile, a response delay in the actual pressure of the clutch (P_C) occurred in low-level filling process to reach the kiss-point pressure. Thus, this mismatching leads to the fluctuation in the input shaft torque of 8AT (T_{in}) and the flare of the engine speed n_E, which has a bad influence on the driving comfort. In the simulation, the speed flare of n_E is over 300 rpm.

Due to the occurred jerk during the mode transition, the adjustment in the clutch control is executed in the next mode-transiting case based on the control strategy. The control pressure of the hybrid clutch P_{C_ctrl} is adapted. As shown in in Figure 5 (b), the command value during the low-level filling process of the clutch increased. The torque exchange between the engine and the electric motor is smooth and fast, which shortens the period of mode transition. Moreover, it is obvious that the fluctuation in n_E is diminished to an acceptable level.

The jerk is relieved after the adjustment and the performance of the control strategy is proved to be effective. The simulation results indicate that the quality of driving mode transition is obviously improved.

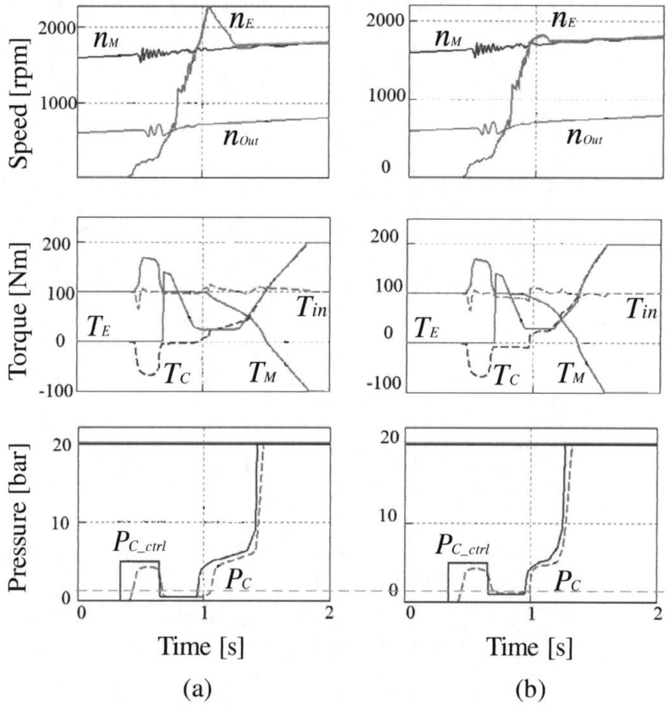

Figure 5. Simulation results.

6. Conclusions

This study focused on the clutch control problems that may occur during the driving mode transition for P2-HEV. The models of the hybrid clutch and other components in the hybrid system are established in MATLAB/Simulink. A model-based clutch control strategy is developed to achieve high-quality mode transitions in the hybrid system. In the proposed control scheme, the hydraulic oil filling and torque transferring characteristics of the hybrid clutch are mainly considered based on the previous jerks. The simulation results show that the quality of driving mode transition is successfully improved.

In the future work, the temperature of the hydraulic oil and the wear of the friction plates in the hybrid clutch will be considered in the design of clutch control strategy. Detailed adaptive methods will be used to achieve high-quality mode transitions when the driving mode changes frequently.

References

[1] Zhao Z, Lei D, Chen J and Li H 2018 *Mech. Sys. & Sig. Proc.* **105** 68
[2] Kumar S, West M, Conter J, Cawley M, Maady R and Mayyas A 2015 *IFAC-PapersOnLine* **48** 139
[3] Su Y, Hu M, Su L, Qin D, Zhang T and Fu C 2018 *Mech. Sys. & Sig. Proc.* **107** 221
[4] Schori M, Boehme T J, Frank B and Schultalbers M 2013 *IFAC Symp. on Nonlinear Control Systems* vol 46 (Toulouse: The International Federation of Automatic Control) p 475
[5] Yang C, Jiao X, Li L, Zhang Y and Chen Z 2018 *Mech. Sys. & Sig. Proc.* **99** 326
[6] Lei Z, Sun D, Liu Y, Qin D, Zhang Y, Yang Y and Chen L 2017 *Mech. & Mach. Theor.* **114** 125

Finite Difference Method for Simulating Transverse Forced Vibration of Vertical Rope with Multi-Rope Winding Hoister

Guoying Wang, Xingming Xiao*, Yansong Ma, Xiao Di and Kaixuan Wang
School of Mechatronic Engineering, China University of Mining and Technology, Xuzhou 221116, China

*Email address: xl6265@cumt.edu.cn

Abstract. During the high speed of the mine hoisting system, external excitation will produce large transverse vibrations on the vertical rope, which may eventually lead to dynamic instability. In this paper, the Hamiltonian principle is used to establish the partial differential equations transverse vibration of the vertical rope. By using the Finite Difference Method, the differential vibration control equation under lateral excitation is discretized and solved, and the transverse vibration characteristics of the vertical rope are improved. Taking the multi-cable winding hoisting system as an example, the effect of external excitation on the transverse vibration of the vertical rope was simulated and analysed. The simulation results show that the solution of vertical rope vibration equations based on the Finite Difference Method can improve the vibration results of the system.

1. Introduction

With the increasing focus on exploitation of deep mineral resources, stable design and operational reliability of hoisters for long distance transport have received increasing attention. The hoisters are used to transport mineral resources in deep mines to the surface platform. During the transport process, the vertical rope of the mine hoisting system will produce transverse vibrations, and these vibrations have a very significant impact on the performance of the hoisting system. With the gradual development of the coal mine in deeper areas, it will have more increased transverse vibration problems. Therefore, it is necessary to analyse the transverse vibration of the vertical rope with the mine hoisting system.

Many scholars have extensively studied the characteristics of mine hoisting systems. It involves the modelling, the kinematics and dynamics analysis of the hoist system. [1, 2] based on Galerkin method, study the transverse vibrations characteristics and the longitudinal-transverse coupling vibration analysis of rope for friction hoisting system. By considering the head sheave eccentricity and the defect of the rigid guide to the vibration influence of the rope, and analysing the vibration characteristic of the rope in different operational phases. Take the multi-cable winding hoisting system as the example, [3] analyses the longitudinal vibration characteristics of steel wire ropes in ultra-deep mine hoisting test bench. [4] analyse the modelling of rope transverse vibration for flexible hoisting systems with time varying length.

However, most scholars use the Galerkin method to solve discretion partial differential equations into ordinary differential equations. The Finite Difference Method has the following advantages compared with the Galerkin method: The Finite Difference Method includes all vibration modes of the rope vibration, and is more suitable for the complex constraint conditions. [5] uses the Finite

Difference Method to simulate transverse forced-vibration of elevator suspended system. Discrete differential equations are discretized into ordinary differential equations by using the Finite Difference Method, and then the step differential integration is used to solve the ordinary differential equations. The comparison with the Galerkin method shows that the effectiveness of the Finite Difference Method is verified. [6] uses the Finite Difference Method to simulate transverse forced-vibration of elevator suspended system. [7] uses the direct numerical method of the Finite Difference Method to discretize the partial differential equations into algebraic equations and solve the flexible hosting system.

In this paper, Finite Difference Method is used to study the influence of the transverse vibrations in the main rope during the transport of the mine hoisting system in external excitation. The dynamic model of a multi-cable winding hoister is derived via Hamilton principle. The infinite dimensional distributed-parameter system is discretized by the Finite Difference Method. The effect of external excitation on the transverse vibration of the vertical rope was simulated and analysed. It provides ways and methods for research on the safety of deep mine hoisting systems.

2. Mathematical model

The separation between the vertical rope and the head sheave is used as the fixed point of the vertical rope. The dynamic model of multi-cable winding hoisting system is shown in Figure 1.

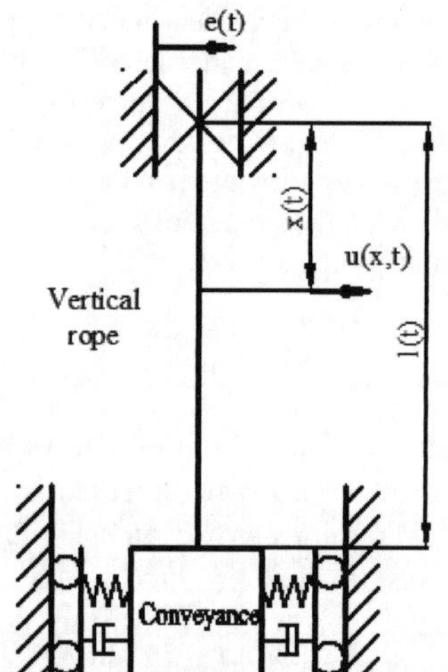

Figure1. Model of multi-cable winding hoisting system

Where $l(t)$ is the length of the vertical rope, $e(t)$ is the transverse interference excitation of the vertical rope. ρ is the rope linear density, A is the rope effective steel area, E is the rope effective Young's Modulus, m is the mass of conveyance. The vertical rope is seen as an axially moves string. The conveyance is seen as a rigid body suspended from the lower end of the vertical rope, and the transverse direction is constrained by the springs and dampers of the rigid guides. Due to the axial movement of the head sheave, a transverse interference excitation is generated on the upper end of the vertical rope. During the transport of vertical rope, the transverse vibration at $x(t)$ on the vertical rope is recorded as $u(x,t)$, and the vertical downward direction is the positive direction.

In this paper, the assumption of the establishment and solution model as follows:

- The slight elastic deformation of the vertical rope due to transverse vibration is far less than the length of the entire string.
- The material of the vertical rope obeys Hooke's law.

- Not consider the effect of longitudinal and torsional deformation of the vertical rope.
- Multi-rope is considered to have the same physical parameters as a single rope and the physical parameters remain unchanged.
- Ignore the effects of bending stiffness, air flow in shaft, damping and friction of rope.
- The conveyance as a rigid body.

The total kinetic energy of the vertical rope in transverse vibration can be represented as

$$E_k = \frac{1}{2}\rho\int_0^{l(t)}\left(v\partial u(x,t)/\partial x+\partial u(x,t)/\partial t\right)^2 dx+\frac{1}{2}m\left(v\partial u(x,t)/\partial x+\partial u(x,t)/\partial t\right)^2\Big|_{x=l(t)} \tag{1}$$

The elastic potential energy of the vertical rope in transverse vibration can be represented as

$$E_p = \frac{1}{2}\int_0^{l(t)}\left[\left(\rho(l-x)+m\right)g+EA\left(\partial u(x,t)/\partial x\right)^2/4\right]\left(\partial u(x,t)/\partial x\right)^2 dx \tag{2}$$

The gravity potential energy of the vertical rope in transverse vibration can be represented as

$$E_g = -mgu(l,t)-\int_0^{l(t)}\rho g u(x,t)dx \tag{3}$$

According to the generalized Hamilton's principle, the true motion of the system between $t = t1$ and $t = t2$ satisfy:

$$\int_{t1}^{t2}\left(\delta E_k-\delta E_p-\delta E_g\right)dt = 0 \tag{4}$$

Using the variation principle and the sub-integration method, the transverse vibration governs equation of the vertical rope can be obtained. For clarity, subscripts x,t represent partial derivatives are used throughout this paper.

$$\rho u_{tt}+\rho 2vu_{xt}+\rho(1+a)u_x+\left(\rho v^2-T\right)u_{xx}-3EAu_x^2u_{xx}/2=0 \tag{5}$$

The vertical rope transverse vibration boundary condition in $x = l(t)$ is obtained as Eq.(6)

$$mu_{tt}+2mvu_{xt}+\left(ma+T\right)u_x+mv^2u_{xx}+EAu_x^3/2=0 \tag{6}$$

The vertical rope transverse vibration boundary condition in $x = 0$ is obtained as Eq. (7)

$$u(x,t) = e(t) \tag{7}$$

Hypothesize the system is motionless state under initial conditions.

$$u(x,0) = 0, u_t(x,0) = 0 \tag{8}$$

3. Discrete mathematical model

The mathematical model of the multi-cable winding hoister is an infinite distributed-parameter system. There are no accurate analytic solutions. Using the Finite Difference Method, govern equations obtain the approximate solutions.

The lengths of vertical rope $l(t)$ are time-varying. By introducing a new independent variable $\xi = x/l(t)$, relative to x of the time-varying spatial domain $[0,l]$ is translate into relative to ξ of the fixed domain $[0,1]$. The relational expression can be obtained as

$$\hat{u}(\xi,t) = u(x,t) \tag{9}$$

The govern equation and boundary condition of the vertical rope can changed as

$$\begin{aligned}&\hat{u}_{tt}+\left(2v(1-\xi)/l\right)\hat{u}_{\xi t}+\left(\left(v^2(\xi-1)^2-l(1-\xi)g-mg/\rho\right)/l^2\right)\hat{u}_{\xi\xi}\\&+\left(\left(al-2v^2\right)(1-\xi)+l\right)\hat{u}_\xi-3EA\left(\hat{u}_\xi\right)^2\left(\hat{u}_{\xi\xi}\right)/2\rho l^4=0\end{aligned} \tag{10}$$

$$\begin{aligned}&\hat{u}_{tt}+\left(2v(1-\xi)/l\right)\hat{u}_{\xi t}+\left(v^2(\xi-1)^2/l^2\right)\hat{u}_{\xi\xi}\\&+\left(\left(m\left(al-2v^2\right)(1-\xi)+\rho gl^2(1-\xi)+mgl\right)/ml^2\right)\hat{u}_{\xi}+\left(EA/2ml^3\right)\hat{u}_{\xi}^{\ 3}=0\end{aligned}\tag{11}$$

$$\hat{u}(0,t)=e(t),\hat{u}(\xi,0)=0,\hat{u}_t(\xi,0)=0\tag{12}$$

Application of difference approximations to the time, space, and mixed partial derivatives leads to

$$\begin{aligned}&\hat{u}_{tt}(\xi,t)=\left(\hat{u}(\xi,t+1)-2\hat{u}(\xi,t)+\hat{u}(\xi,t-1)\right)/\tau^2\\&\hat{u}_{\xi t}(\xi,t)=\left(\hat{u}(\xi,t)-\hat{u}(\xi-1,t)-\hat{u}(\xi,t-1)+\hat{u}(\xi-1,t-1)\right)/h\tau\\&\hat{u}_{\xi}(\xi,t)=\left(\hat{u}(\xi+1,t)-\hat{u}(\xi,t)\right)/h\\&\hat{u}_{\xi\xi}(\xi,t)=\left(\hat{u}(\xi+1,t)-2\hat{u}(\xi,t)+\hat{u}(\xi-1,t)\right)/h^2\end{aligned}\tag{13}$$

Where h,τ are spatial and time intervals.

Substituting Eqs. (12) into the Eqs. (10)- (11),

$$\begin{aligned}\hat{u}(\xi,t+1)=&-2\tau v(1-\xi)\left(\hat{u}(\xi,t)-\hat{u}(\xi-1,t)-\hat{u}(\xi,t-1)+\hat{u}(\xi-1,t-1)\right)/lh\\&-\tau^2\left(v^2(\xi-1)^2-l(1-\xi)g-mg/\rho\right)\left(\hat{u}(\xi+1,t)-2\hat{u}(\xi,t)+\hat{u}(\xi-1,t)\right)/l^2h^2\\&-\tau^2\left(\left(al-2v^2\right)(1-\xi)+l\right)\left(\hat{u}(\xi+1,t)-\hat{u}(\xi,t)\right)/l^2h+2\hat{u}(\xi,t)-\hat{u}(\xi,t-1)\\&+3\tau^2EA\left(\hat{u}(\xi+1,t)-\hat{u}(\xi,t)\right)^2\left(\hat{u}(\xi+1,t)-2\hat{u}(\xi,t)+\hat{u}(\xi-1,t)\right)/2\rho l^4h^4\end{aligned}\tag{14}$$

$$\begin{aligned}\hat{u}(\xi,t+1)=&-2\tau v(1-\xi)\left(\hat{u}(\xi,t)-\hat{u}(x-1,t)-\hat{u}(\xi,t-1)+\hat{u}(\xi-1,t-1)\right)/lh\\&-\tau^2v^2(\xi-1)^2\left(\hat{u}(\xi+1,t)-2\hat{u}(\xi,t)+\hat{u}(\xi-1,t)\right)/l^2h^2\\&-\tau^2\left(m\left(al-2v^2\right)(1-\xi)+\rho gl^2(1-\xi)+mgl\right)\left(\hat{u}(\xi+1,t)-\hat{u}(\xi,t)\right)/l^2mh\\&-\tau^2EA\left(\hat{u}(\xi+1,t)-\hat{u}(\xi,t)\right)^3/2ml^3h^3+2\hat{u}(\xi,t)-\hat{u}(\xi,t-1)\end{aligned}\tag{15}$$

Combine Eqs. (12)- (15), the transverse vibrations numerical solution of the vertical rope is obtained.

4. Numerical simulation

In mine hoisting system, transport distance is larger and continuously being increased. Now in China, most of new mines with shaft depths over 1000 m have recently been considered. [8] Typical parameters of the multi-cable winding hoister used in simulation are showed in Table 1.

Table 1. Simulation parameters of the vertical rope values.

Simulation	Values
Vertical rope length $l(t)$ (m)	1100
Hosting time t (s)	90
Rope effective steel area A (m^2)	$3\text{x}10^{-3}$
Conveyance mass M (kg)	$5\text{x}10^{4}$
Rope effective Young's Modulus E (N/m^2)	$1.53\text{x}10^{11}$
Rope linear density ρ (kg/m)	8.02
Maximum hosting velocity v (m/s)	15
Maximum hosting acceleration a (m/s^2)	0.75

The head sheave eccentricity is used as the upper external disturbance excitation, and it can be represented as a trigonometric function $e(t) = 0.001 \times \sin(\pi t)$. The upward movement profiles of the multi-rope winding hoister are shown in Figure 2- 4.The transverse vibration displacement at the midpoint of the vertical rope is shown in Figure 5.

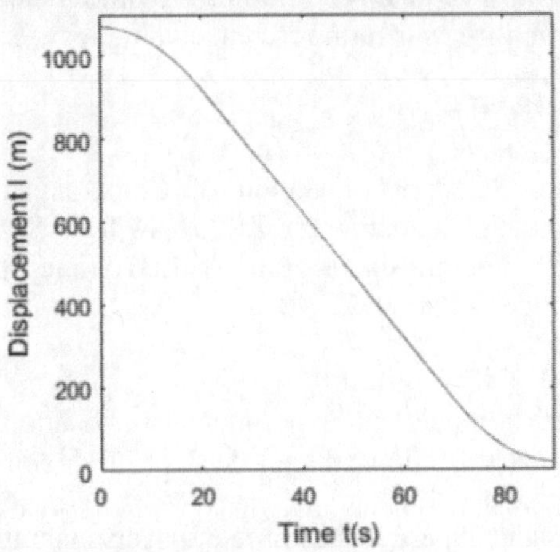

Figure 2. The movement displacement of multi-rope winding hoister.

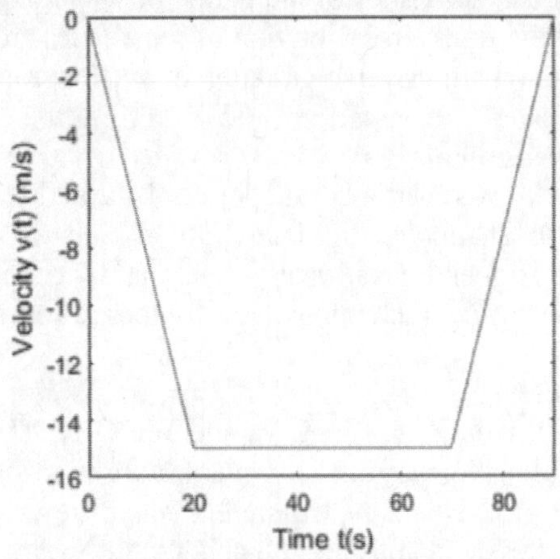

Figure 3. The movement velocity of multi-rope winding hoister.

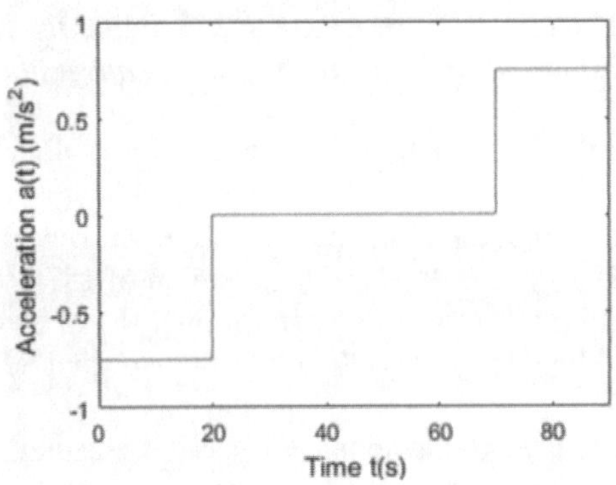

Figure 4. The movement acceleration of multi-rope winding hoister.

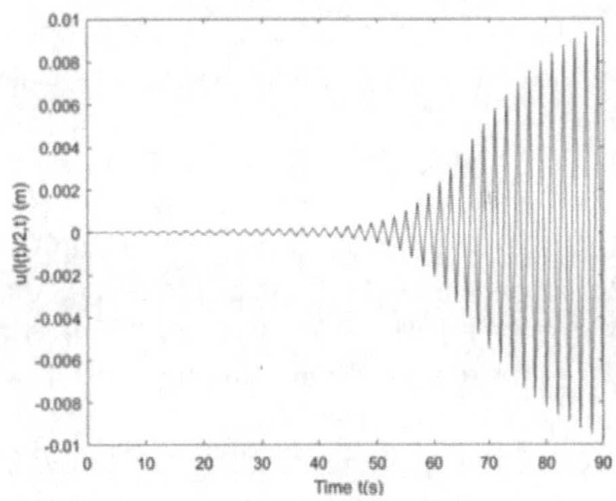

Figure 5. The transverse vibration displacement at the midpoint of the vertical rope

According to Figure 5, the length of the vertical rope gradually becomes shorter, and the transverse vibration displacement gradually increases in the upward movement stage of hoister. In the acceleration phase, the transverse vibration displacement of the vertical rope is small; in the uniform velocity phase, the transverse vibration displacement of the vertical rope sharp increase; in the deceleration phase, the transverse vibration displacement of the vertical rope increases slowly. The sharp increase in transverse vibration displacement of the vertical rope will lead to the dramatic increase in transverse vibration energy, which may ultimately lead to dynamic instability. The simulation results are compared with [4] and the conclusions are consistent, which verifies the validity of the method.

5. Conclusions

In this paper, based on the generalized Hamilton's principle, the partial differential equation for the transverse vibration of the vertical rope was established in the mine hoisting system. By using the Finite Difference Method, the continuous spatial and temporal variables of the partial differential equation are discretized. Through numerical simulation, the transverse vibrations at the upward movement of the multi-rope winding hoister vertical rope are obtained.The analysis of the transverse vibration characteristics of the multi-rope winding hoisting vertical rope studies provides a theoretical basis for the transverse vibration of the vertical rope, and it has certain reference value for the further study about the transverse vibration of vertical rope.

Acknowledgments

This work was partially supported National Natural Science Foundation of China under grant 51675520, Shandong Province Natural Science Foundation under grant ZR2017QF011, Shandong Province Higher Educational Science and Technology Program under grant J16LB10, and Weifang City Science and Technology Development Program under grant 2017GX017.

References

[1] Wu J, Kou Z M, Liang M, and Wu G X. Theoretical model and experimental verification of the coupling longitudinal-transverse vibration of rope for friction hoisting system 2015 *Journal of China University of Mining & Technology.* **44**(5) 1006-1013.

[2] Wu J, Kou Z M, Liang M, and Wu G X. Analysis and experiment of rope transverse vibration for multi-rope friction hoisting system 2015 *J. Huazhong Univ. of Sci. & Tech. (Natural Science Edition)* **43**(6) 12-21.

[3] Wang C M, Zhang S L, Wang J S, and Du B. Longitudinal vibration analysis of steel wire rope in ultra-deep lifting test bench 2017 *Machinery Design & Manufacture.* **6** 30-33.

[4] Bao J H, Zhang P, and Zhu C M. Modeling and analysis of rope transverse vibration for flexible hoisting systems with time-varying length 2012 *Journal of Shanghai Jiaotong University.* **46**(3) 341-345.

[5] Wang W and Qian J. Finite difference method for simulating transverse forced-vibration of elevator suspended system 2014 *Journal of Vibration Engineering.* **27**(3) 180-185.

[6] Gao W X, Wu J, and Zhang Q S. Longitudinal vibration characteristics of multi-rope friction hoisting system by the finite difference method 2017 *China Mining Magazine.***26**(6) 157-160.

[7] Wang J, Pi Y J, Hu Y M, and Gong X S. Modeling and dynamic behavior analysis of a coupled multi-cable double drum winding hoister with flexible guides 2017 *Mechanism and Machine Theory.* **108** (2017) 191-208.

[8] Cao G H, Cai X, Wang N, Peng W H, and Li J S. Dynamic response of parallel hoisting system under drive deviation between ropes with time varying length 2017 *Shock and Vibration.* 1-10.

Multi-body Dynamic Analysis of the Braking Process for Megawatt Wind Turbine Disc Brake

S Zhang, C Xue, J Yin, Z Sha, F Ma, D Yang and Y Liu*

School of Mechanical Engineering, Dalian Jiaotong University, Dalian, China 116028

*liuyu_ly12@126.com

Abstract. The dynamic characteristics of brake during the braking process are studied by the method of multi-body dynamics, in order to reduce the vibration of disc brake in the emergency braking process, improving the braking efficiency of the brake of the megawatt wind turbine. The flexible multi-body dynamics model of the mechanical assembly for the whole megawatt wind turbine disc brake is built, including the brake spring, active brake caliper, passive brake caliper and compensatory mechanism. The simulation results under the common working conditions are compared with the test results under the same working condition to verify the model. The influence trends of braking parameters, such as initial braking speed, spring braking force and friction coefficient between the brake pairs, on the dynamic characteristics of the brake are analyzed, which provides the theoretical basis for reducing the severe vibration during the emergency braking process in this paper.

1. Introduction

With the development of wind power industry, megawatt wind turbine has gradually become the key of wind turbine research in the world[1]. Its braking system is of vital importance in the efficient operation of wind power generation[2].The problem of vibration and other unstable braking is easy to be found in the urgent braking process of brake which is the main device in the brake system, because speed of main shaft is high and rotary inertia is large in the megawatt wind turbine, this brings the hidden danger to the normal operation of the wind turbine[3].As an important part of the wind turbine braking system, wind turbine brake's performance directly affects the safe operation of wind turbine.

It is necessary to do the further study on the dynamic characteristics of the wind turbine disc brake, in order to solve the problem of vibration and unstable braking during the braking process of the megawatt wind turbine disc brake. In the study of dynamic characteristics of disc brake, scholars in the world have made comparatively deep analysis using multi-body dynamics method. A multi-body dynamic model for vehicle brake was established by Kim et al, and the change of braking torque was measured by changing the thickness of the brake disc. It was found that the change amplitude of braking torque can be reduced by reducing the contact stiffness of the brake pair and increasing the torsional rigidity[4]. The parameters of the brake members were redesigned and the shape of the brake pad was optimized to make the force of brake pad uniform by Swift et al[5], which can be used to study the influence of contact area between the brake pairs on vibration of brake during the braking process . The change amplitude of brake torque was studied by using the model of disc brake with assembly error, and then the whole model of the brake was established to obtain the transmission way on vibration of brake and get the specific method of reducing the vibration by Gao et al·[6]. The vibration of the brake can be reduced by reducing the damping of the friction plate for the fixed caliper disc brake, in the study of the design of car brake by Ning Xiaobin and Zhang Wenming[7].

In this paper, the influence of the interaction among the brake components is taken into account in the braking process, the multi-body dynamic model of mechanical assembly for the whole megawatt wind turbine disc brake, including brake spring, active brake caliper, passive brake caliper and compensatory mechanism, is established, and the rationality of the model is verified. The influence trends of braking parameters on the dynamic characteristics of the brake are analyzed, which provides the theoretical basis for efficient braking of megawatt wind turbine disc brake.

2. Establishment of the multi-body dynamic model of the whole wind turbine disc brake

2.1. Establishment of the three-dimensional model for brake

Some components are included in the three-dimensional model of mechanical assembly for megawatt wind turbine disc brake, such as brake disc, up-pad and down-pad, compensation mechanism(includes compensation nut, compensation putter, compensation nut sleeve and piston), active brake caliper, passive brake caliper, end cover and brake spring. The outer diameter of the brake disc is 800mm, the inner diameter is 200mm, and the thickness is 40mm. The structural parameters of up-pad and down-pad are the same. The length of brake pad is 270mm and the width is 130mm. The size of guide grooves on both sides of the brake pad is 60mm×10mm. The material properties of the brake disc and brake pad are shown in Table 1[8]. The material of other components is Q345-b. Three-dimensional assembly model of mechanical assembly for wind turbine brake is shown in Figure. 1.

Table 1 Material attributes of brake disc and pads at room temperature[8]

brake element	material	density /(kg/m^3)	modulus of elasticity /MPa	Poisson ratio
brake pad	Copper based powder metallurgy	5250	1.80×10^5	0.30
brake disc	high thermal conductivity cast iron	7250	1.38×10^5	0.28

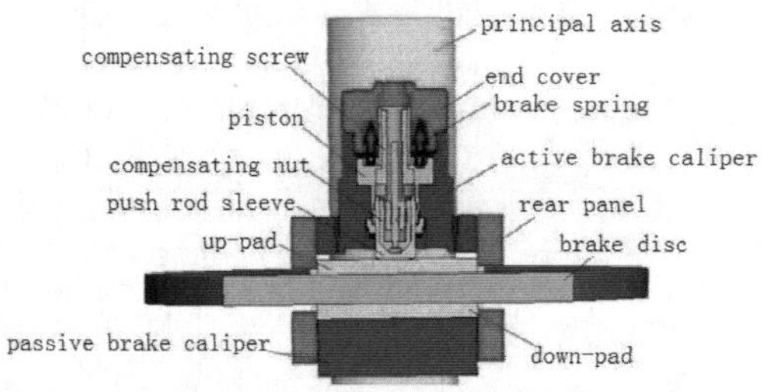

Figure 1. 3D solid model of mechanical assembly for wind turbine disc brake

2.2. Establishment of flexible body for the main components of brake

Vibration of brake disc is most obvious during braking process. Therefore, the brake disc is the most important represented component of braking vibration. Besides, the braking performance of the brake can be affected seriously by the contact situation between brake pairs of brake disc and brake pads. Brake disc and brake pads are the important components of brake which need to be flexible. The most critical step is to establish a rigid area for the components in the process. The flexible bodies of brake discs and pads, key points of connection with rigid bodies and rigid areas of components are shown in Figure. 2.

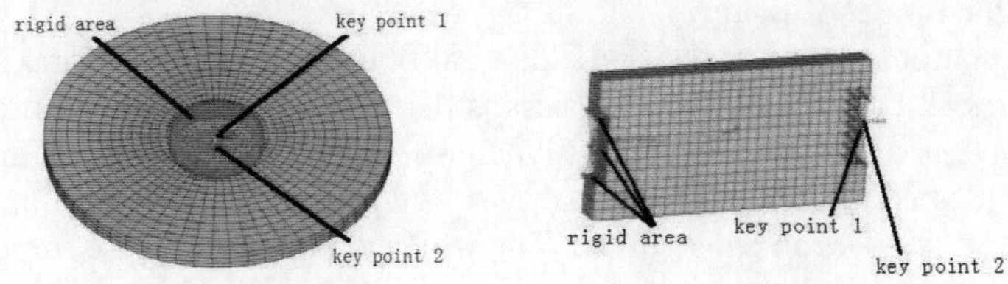

(a) flexible body of brake disc (b) flexible body of brake pads

Figure 2 Flexible bodies of main parts for brake

2.3. Establishment of the dynamic model for brake

Three-dimensional model of mechanical assembly for brake is imported into the multi-body dynamics software and the flexible bodies of main components are replaced. According to the braking principle of the brake, constraint relationships are set up. And the constraint relationship for dynamics model of wind turbine disc brake are shown in Table 2.

Table 2 Joints of dynamic model for wind turbine disc brake

constraint link	constraint type	constraint link	constraint type
up-pad /down-pad and ground	translation	up-pad/down-pad and brake disc	contact
end cover and piston	spring	Brake disc and ground	revolute
Push rod sleeve and up-pad	contact	Passive brake caliper and down-pad	contact

The setting of "contact" in the constraint relationship is particularly important, and some settings are included in the contact attribute, such as stiffness coefficient, force index, damping coefficient and penetration depth, which are given by the materials of contact components and contact conditions. Friction between brake pairs is calculated by coulomb force. The dynamic or static friction coefficient, and the dynamic or static slip speed must be selected based on the materials of brake pairs and the actual operating conditions. The multi-body dynamics model of mechanical assembly for megawatt wind turbine disc brake is shown in Figure 3. The up-pad, down-pad and the brake disc are flexible bodies, but the others are rigid bodies in this model.

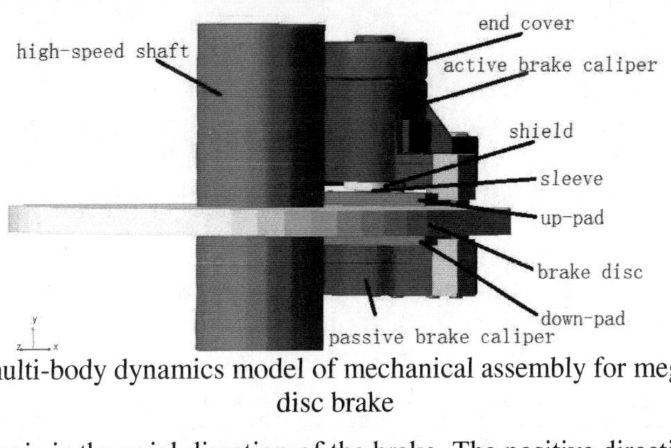

Figure 3 Flexible multi-body dynamics model of mechanical assembly for megawatt wind turbine disc brake

In Figure 3, the Y axis is the axial direction of the brake. The positive direction of Y is positive. It is found that the axial vibration of the brake is the most obvious during the braking process by simulation tests. Therefore, it is necessary to analyze the axial dynamic characteristics of the brake. Unless special instructions are given, the dynamic characteristics of the brake studied is about axial direction during the braking process in this paper.

3. Verification for the model of brake

The initial braking rotational speed of the brake disc is 1000r/min, the spring braking force is 17000N and the friction coefficient between the brake pairs is 0.3, under the common working condition. The simulation results obtained by the multi-body dynamics model of brake are compared with the test results obtained by test-bed to verify the rationality of the multi-body dynamics model of mechanical assembly for megawatt wind turbine disc brake. The working condition of the inertial test rig for wind turbine disc brake is that the environment temperature is 20°C, humidity is 30% and the moment of inertia is 600 kg/m^2. Through the acquisition device of test rig, such as sensor and amplifier device, the test curve of rotational speed-time and braking torque-time are reported in braking process, and the test results and simulation results are compared with each other. It is necessary to convert the unit of the rotational speed, since the unit of rotational speed in ADAMS is deg*s-1. The comparison curves between the simulation and test of braking results are shown in Figure 4 under the common working conditions.

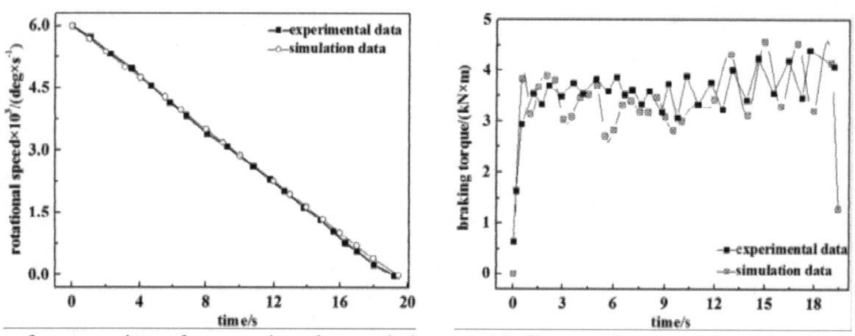

a) Curve of comparison for rotational speed-time b) Curve of comparison for braking-time

Figure 4 Curve of comparison between simulation and test under common condition

It can be seen from Figure 4(a) that the simulation value of braking time is 19.44s, the test value is 19.16s and the error rate is 1.5% in this working condition. Besides, the slope of the simulation and test curve is basically consistent and the braking is completed by constant deceleration. The average value of braking torque is about 3468 N·m, which is found in the curve of comparison between simulation and test for the braking torque-time in the braking process under the common working condition, and the curve are fluctuated in the form of sine curve in the Figure 4(b). Simulation and test curve get risen at the end of braking, because the friction between brake pairs is turned into the static friction from dynamic friction, and the static friction is always higher than dynamic friction at that time. According to the comparison results, the simulation results corresponded well with test results under the same working condition.

It can be shown by comprehensive analysis that the simulation results of the multi-body dynamics model are in conformity with the actual working condition, so the model is reasonable and can be used for the subsequent simulation research for dynamic characteristic in the braking process.

4. Analysis for dynamic characteristics of brake during the braking process

4.1. The influences of braking parameters on dynamic characteristics of brake caliper

The multi-body dynamics model is operated to study the influences on dynamic characteristics of braking parameters which are changed by equal gradient based on common braking condition. The simulation condition of different braking parameters are shown in Table 3.

Table 3 Simulation condition of different braking parameters

Simulation condition	initial braking rotational speed/(r/min)	friction coefficient between the brake pairs	spring braking force/N
1	800, 1000, 1200	0.3	17000
2	1000	0.3	13000, 17000, 21000
3	1000	0.25, 0.3, 0.35	17000

The influence trends of braking parameters on the dynamic characteristics of passive brake caliper are studied mainly, because the vibration of passive brake caliper is more severe than that of active brake caliper, according to the braking principle. Three kinds of working condition, as shown in Table 3, are respectively set in the simulations, such as initial braking rotational speed, spring braking force, friction coefficient between the brake pairs. The dynamic characteristic of passive brake caliper are shown in the Figure 5, under the conditions of different braking parameters.

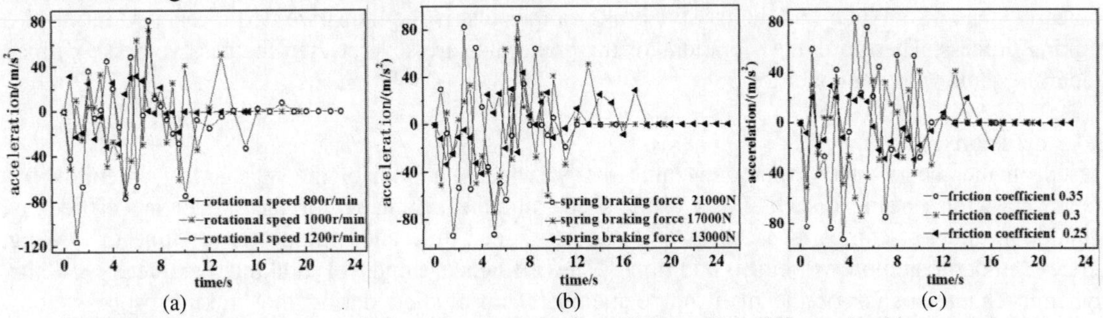

Figure 5 Dynamic characteristics of passive brake caliper under different braking parameters

It can be seen from Figure 5(a) that the higher the value of the initial braking rotational speed is, the longer the braking time is, and the larger the fluctuation for the axial acceleration of passive brake caliper is. The reason is that the higher initial braking rotational speed is, the larger the energy of the brake disc is, under the same external load. So it takes longer time to achieve braking. Besides, the larger the energy is, the severer the vibration of the brake is during the braking process. It can also be seen from Figure 5(b) that the larger the spring braking force is, the shorter the braking time of the brake is, and the larger the fluctuation for the axial acceleration of passive brake caliper is. The reason is that the larger the spring braking force is, the larger the external load is, and the braking time is obviously shorten. But the larger the spring braking force, the larger the deformation of the spring is, and the larger the instability of the spring braking force, and even the larger the impact to the brake disc is, which causes the severer vibration of the brake. Besides, the larger friction coefficient between the brake pairs is, the shorter the braking time of the brake is, and the larger the fluctuation for the axial acceleration of passive brake caliper is. The reason is that the larger the friction coefficient between the brake pairs is, the larger the friction resistance is, this shortens the braking time. But the larger friction resistance between the brake pairs is, the larger friction resistance will be resisted by the brake disc with the severer vibration, causing severer vibration of the brake, as shown in Figure 5(c).

4.2. Influence of brake parameters on the dynamic characteristics of brake disc

The dynamic characteristic of brake disc are shown in Figure 6, under the conditions of different braking parameters.

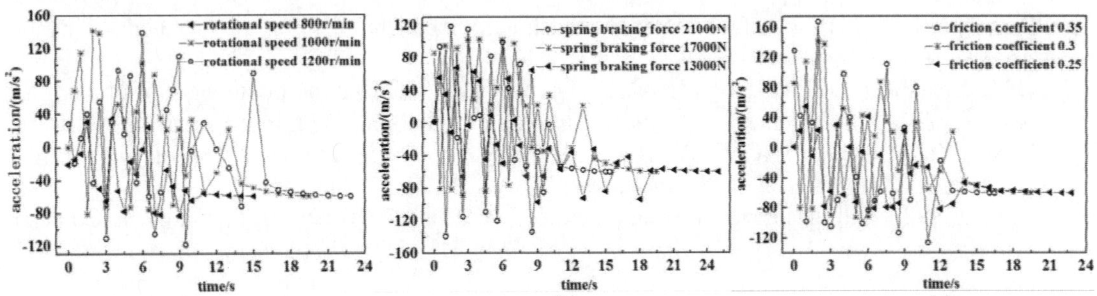

Figure 6 Dynamic characteristics of brake disc under the different braking parameters

It can be seen from Figure 6 that the larger the braking parameters are, such as initial braking rotational speed, the spring braking force and friction coefficient between the braking pairs, the larger the fluctuation for the axial acceleration of the brake disc is. Besides, the fluctuation degree for the axial acceleration of the brake disc in Figure 6 is larger than that of the passive brake caliper in Figure 5 under the same condition. Brake disc is the terminal component of the brake to achieve braking and the brake disc is equipped with a certain amount of energy at the early stage of braking owing to the higher initial braking rotational speed of the brake disc. But the passive brake caliper is stationary at the early stage of braking, and the passive brake caliper just is the transitive part of energy during the braking process. Therefore, the vibrations of the brake disc are severer than the passive brake caliper under the same working conditions.

5. Conclusions

Based on the theory of dynamics, the multi-body dynamic model of the mechanical assembly for megawatt wind turbine disc brake is established, and the rationality of the model is verified by simulation and test comparison curves of rotational speed-time and braking torque-time in braking process under common working conditions. The influence trends of braking parameters on the dynamic characteristics of the main brake members are studied during the braking process. It is concluded that with the increase of the initial braking rotational speed, the spring braking force and the friction coefficient between the brake pairs, the fluctuation tends for the axial acceleration of passive brake caliper and brake disc is increased roughly, indicating the vibration degree of the brake is also more severe. Besides, the vibration of the brake disc is severer than that of the passive brake caliper, indicating that the vibration of the brake can be reflected well by that of the brake disc during the braking process.

Acknowledgments

The research work is supported by National Natural Science Foundation of China under Grant No. 51475066 and No. 51675075 and Natural Science Foundation of Liaoning Province under Grant No. 2015020114.

References

[1] LV Da-Peng. 2017,Wind turbine control and operation maintenance technology *J* . *Information recording material*, **18**(5):51-53. In Chinese.
[2] YAO Xing-Jia, SONG Jun. 2016,Wind Turbine Principles and Applications. 3rd Edition *M. Machinery Industry Press* . In Chinese.
[3] Heilig J, Wauer J. 2003, Stability of a Nonlinear Brake System at High Operating Speeds *J. Nonlinear Dynamics*, **34**(3-4):235-247.
[4] Kim S H, Han E J, Kang S W, et al. 2008, Investigation of influential factors of a brake system to reduce brake torque variation *J* . *International Journal of Automotive Technology*, **9**(2):233-247.
[5] Penninger C L, Swift R A. 2004, Disc Brake Lining Shape Optimization by Multibody Dynamic Analysis *J* . *Sae Transactions* .
[6] GAO Xiao-Jie, YU Zhuo-Ping, ZHANG Li-jun, et al. 2006, Simulation analysis of brake jitter based on ADAMS *J* . *Journal of system simulation*, **18**(6):1668-1670. In Chinese.
[7] NING Xiao-Bin, ZHANG Wen-Ming. 2004, Multibody dynamic analysis of Disc Brake Vibration *J* . *Nonferrous Metal Engineering*, **56**(4):119-121. In Chinese.
[8] LIU Li. 1992, High heat conduction cast iron for brake disc *J* . *World car*, (5):62-67. In Chinese.

IOP Conf. Series: Materials Science and Engineering **470** (2019) 012006 doi:10.1088/1757-899X/470/1/012006

Analysis on Vibration Characteristics of the Stringed Bead Wire of an Underwater Diamond Wire Saw

L Zhang[1], M Liu[1,a], L Liu[1], G L Zhang[1], S H Gao[1] and L Q Wang[1]

[1]College of Mechanical and Electrical Engineering, Harbin Engineering University, Harbin 150001, China

[a] 424123587@qq.com

Abstract. Vibration is one of the most important factors when the diamond stringed bead wire cuts pipeline, which not only affects the quality and efficiency of cutting operations but also affects the life of the stringed bead wire. The finite element model of the stringed bead wire is established by ANSYS Workbench in the paper. The all kinds of vibration of the stringed bead wire is analyzed by the dynamic simulation of diamond bead vibration, and The natural frequency and vibration amplitude of each order are obtained. The results show that the third - order model of stringed bead wire has large amplitude. The influence of the central distance of the drive wheel and the guide wheel and the diameter of the stringed bead wire on the vibration of the bead rope is studied, which obtain the third-order vibration amplitude of the stringed bead wire with the change of the two parameters and determine the best value of the amplitude of the bead rope in the saw cutting. The experimental results show that the effective cutting time of the stringed bead wire is the longer when the vibration amplitude is the optimum value,which means that service life of stringed bead wire is the longer.

1. Introduction

The diamond wire saw has a wide range of applications not only in mining, stone plate processing, special-shaped processing, road or building demolition, but also can be applied to ocean engineering operations such as cutting of submarine pipelines and demolition of pipeline support on offshore oil platforms [1]. The stringed bead wire, the main cutting part of diamond wire saw, is a kind of flexible processing tool with unique processing characteristics [2]. However, the vibration of the stringed bead wire will always accompany the entire cutting process of diamond wire saw. Slight vibrations are good for machining [3]. Severe vibration will cause the bead rope to break during operation and the excessive exfoliation of the diamond grains, which will seriously affect the cutting efficiency and service life of diamond wire saw [4]. Therefore, the issue of vibration of this type of flexible element attracted the attention of researchers. Wickert [5] studied the elastic moving string by using the modal analysis method of structural dynamic characteristics, and analyzed the steady state of any simple harmonic excitation on the string. Foda [6] used the Green function to study the suppression of the vibration of the fixed-axis moving string at both ends of the hole, and obtained a method to suppress the vibration of the axially moving string. Prof. Chen Liqun conducted in-depth theoretical research on the vibration of the axially moving string, and simplified this type of flexible element into an axially moving string [7]. However, there are few theoretical studies related to the stringed bead wire vibration. What are the relationships between the vibration amplitude of the stringed bead wire and various parameters, and how to adjust the parameters to achieve the purpose of vibration reduction. These issues are worthy of in-depth research.

2. Experimental conditions and finite element model

2.1. Experimental platform and conditions

The vibration frequency measurement of the stringed diamond bead wire is carried out on a prototype machine of the diamond wire saw. The experiment used the stringed bead wire produced by the Italian Maritime company to cut reinforced concrete and steel materials. The experimental platform is shown in figure 1. This experimental platform is mainly used for underwater pipe cutting experiments. The outer diameter of the bead used in the experiment is 10.0mm. The diameter of the flywheels on both sides is 0.32m. The center distance of the two flywheels is 1.4m. The stringed bead wire tension is 2000N. The cutting speed of the diamond wire saw is 23m/s.

Figure 1. The stringed bead wire experiment bench

The natural frequency of free vibration allows the design of the mechanical structure to avoid resonance or to vibrate at a specific frequency, which is an important parameter in dynamic load structure design. The stringed bead wire vibration system mechanical model is shown in figure 2.

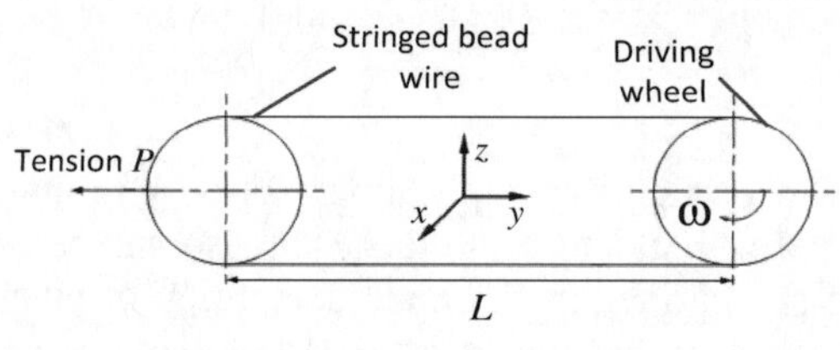

Figure 2. The stringed bead wire vibration system mechanical model

2.2. The establishment of the stringed bead wire finite element model

The study of longitudinal vibration of axially moving strings originated from the work of Skutch in 1897. When the axial displacement of the string is small, the linear vibration model can be used. In the previous study, the natural frequency of the stringed bead wire free vibration was obtained by analytical method, and the natural frequency is 771.5Hz. However, in practice, the structure of the stringed bead wire is complex and there are many influencing factors. Therefore, it is very important to establish the finite element vibration model of the stringed bead wire and to realize the free vibration simulation of the stringed bead wire [8].

The natural frequency finite element model of the bead string was established, which is shown in figure 3. In the model, the performance parameters of 45# steel are selected for the driving wheel, the driven wheel and the bead. Its elastic modulus is E=2.09e11Pa and its Poisson's ratio is μ=0.3. The wire rope belongs to the flexible body and its parameters are as follows: its elastic modulus is

E=0.981e11Pa and its Poisson's ratio is μ=0.26. In the contact setting, the annular section between the beading rope and the flywheels on both sides is in contact with the binding, so that the wire rope and the wheel are always in contact with each other within the angle of the wrapping angle. Then, meshing is performed. The meshing takes into account that the driving wheel, the driven wheel, and the beading rope are all in the rotating state. Therefore, the sweeping mesh is used for division. The set constraint is that the left drive wheel xyz is fixed in all three directions, the right driven wheel is fixed in the x and z directions, and there is a forced displacement along the y direction. The forced displacement of the driven wheel at the right end is obtained from the following formula (1):

$$\Delta l = \frac{TL}{EA} = \frac{2000\times 3.81}{2.06\times 10^{11}\times \pi \times \left(5\times 10^{-3}\right)^2} = 0.471\times 10^{-3} m \tag{1}$$

In this formula, T is the tension force of the stringed bead wire, L is the length of the stringed bead wire, E is the elastic modulus of the stringed bead wire, A is the cross-sectional area of the stringed bead wire.

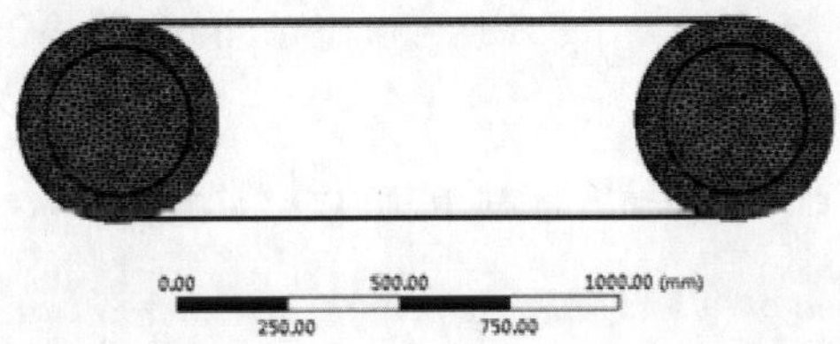

Figure 3. The stringed bead wire free vibration finite element model

3. Finite element result analysis

When Δl=0.471mm, the results are shown in figure 4, figure 5 and figure 6 respectively.

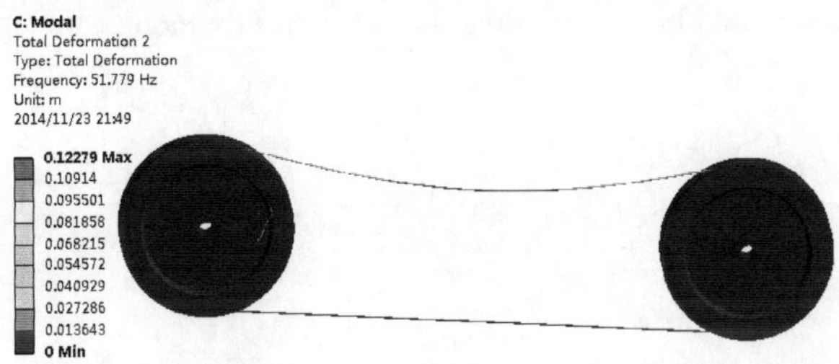

Figure 4. First-order model of the stringed bead wire modal analysis

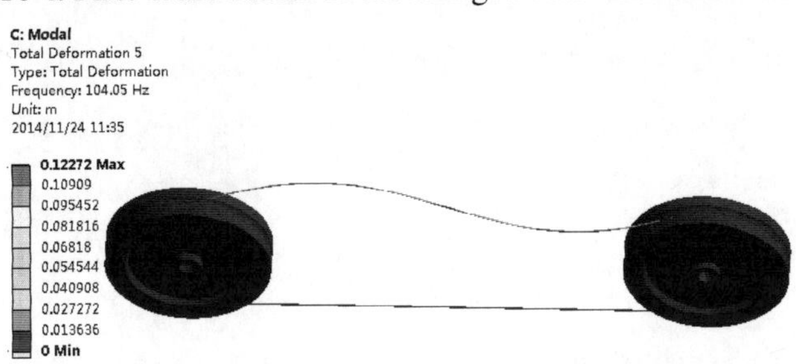

Figure 5. Second-order model of the stringed bead wire modal analysis

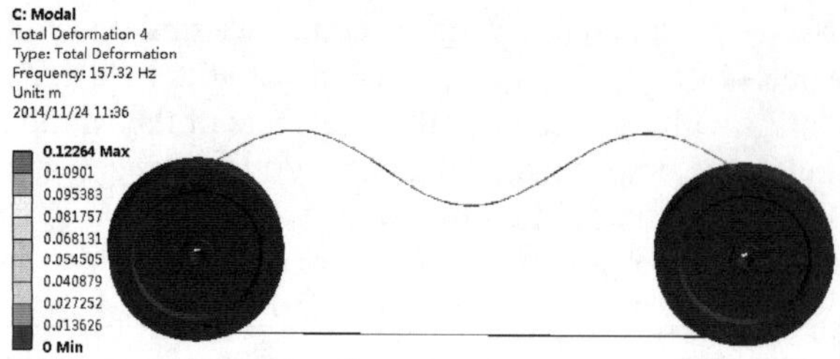

Figure 6. Third-order model of the stringed bead wire modal analysis

From figures 4, figures 5 and figures 6, it can be seen that the frequencies of the first, second and third order models of the stringed bead wire are 51.779 Hz, 104.05 Hz and 157.32 Hz respectively. It can be seen from figure 6 that the vibration amplitude of the third-order model during the stringed bead wire movement is relatively large. In view of this situation, this paper changes the two parameters of center distance L between the two rounds and bead diameter D, and focuses on the analysis of the variation law of the third order vibration amplitude of the beaded string with the corresponding parameters.

4. Influence of the parameters on vibratory amplitude of stringed bead wire

4.1. Influence of center distance on vibrational amplitude of stringed bead wire

In actual production, the distance L between the driving wheel and the driven wheel of the diamond wire saw varies with the cutting work size. In the simulation analysis, the nominal diameter of the wire rope is d=5.8mm, the outside diameter of the beads is 10mm, the number of beads per meter is 38, and the tension of the stringed bead wire is 2000N. figure 7 shows the amplitude values obtained when the parameters of L are set to 1.2 m, 1.3 m, 1.4 m, 1.5 m, and 1.6 m respectively. It is the third-order vibration amplitude of the stringed bead wire that changes with the center distance L.

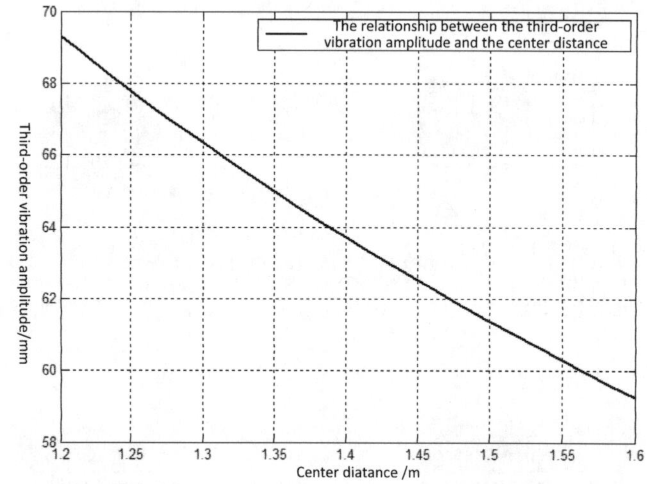

Figure 7. Change law of third-order vibration amplitude with central distance

From figure 7, we can see that when the center distance L is in the range of 1.2~1.6m, the vibration amplitude of the stringed bead wire third-order model decreases as the center distance increases, and the overall change trend is approximately a linear decrease. The main reason is that as the center distance increases, the length of the stringed bead wire becomes longer, so the vibration amplitude is reduced due to the energy consumption during the vibration process.

4.2. *Influence of bead diameter on vibration amplitude of stringed bead wire*

IOP Conf. Series: Materials Science and Engineering **470** (2019) 012006 doi:10.1088/1757-899X/470/1/012006

The bead diameter will decrease as the bead wears during beading. The nominal diameter of wire rope, the length of the stringed bead wire and the bead number per meter are set as in the previous paragraph. Bead diameter is taken as 9.0mm, 9.5mm, 10.0mm, 10.5mm and 11.0mm respectively. In this condition, simulation analysis is performed. The vibration amplitude of the stringed bead wire modal analysis is shown in figure 8.

From figure 8, the diameter of the bead is in the range of 9~11mm, and the other parameters remain unchanged. With the increase of the bead diameter, the vibration amplitude of the stringed bead wire third-order model also decreases, and the overall change trend is approximately linear.

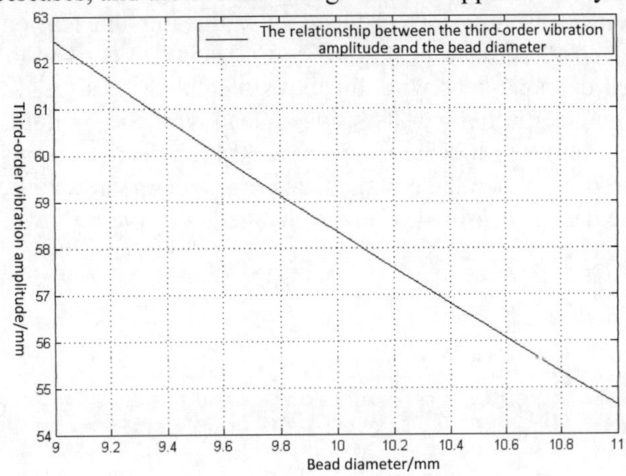

Figure 8. Change law of third-order vibration amplitude with bead diameter

4.3. *Influence of two-parameter coupling on vibrational amplitude of stringed bead wire*

From figure 9, the corresponding relationship between the third-order vibration amplitude of the stringed bead wire and the two parameters (the bead diameter and the center distance) is approximated as a curved surface, which is not a single linear proportional relationship.

As shown in figure 9, the sensitivity curves of the two parameters corresponding to the curved surface are displayed, where DS_1 is the bead diameter and DS_2 is the center distance.

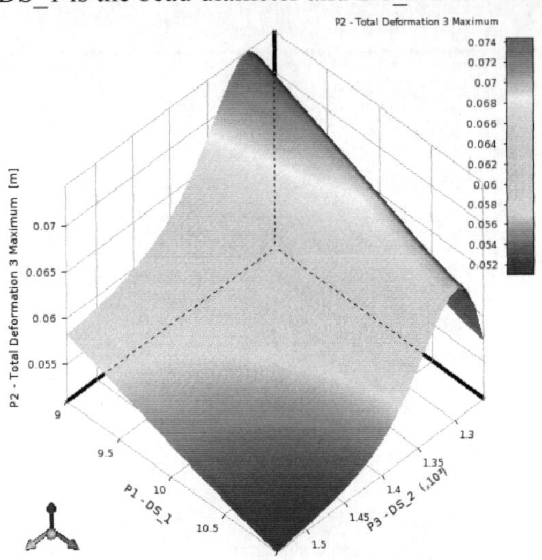

Figure 9. Correspondence between output and input parameters

From figure 10, when the sensitivity is less than 0.5, the curve DS_2 is above the curve DS_1, which indicates that the variation of the center distance has a greater influence on the vibration amplitude of the stringed bead wire. On the contrary, when the sensitivity exceeds 0.5, the curve DS_1 is above the curve DS_2, which indicates that the diameter of bead has a greater influence on the vibration amplitude of the stringed bead wire. When the sensitivity is 0.5, the two effects are equivalent, and the stringed bead wire vibration amplitude is about 0.0585m.

5. Influence of the parameters on vibratory amplitude of stringed bead wire

In this paper, based on the existing experimental platform, the experimental study, the influence of the center distance and the bead diameter on the stringed bead wire cutting life, was carried out. During the experiment, the center distance was changed by adjusting the tensioning device. The center distance is adjusted to 1.2m, 1.25m, 1.3m, 1.35m, 1.4m, 1.45m. According to the diameter size of the commonly used stringed diamond bead wire, the experimental stringed bead wire diameter is 9.0mm, 10.0mm, 10.5mm, 11.0mm respectively. The stringed bead wire cutting life is evaluated based on the effective cutting time per unit length, which is the ratio of the effective cutting time of the wire saw to the length of the beaded rope. When the cutting speed of the diamond wire saw is 23m/s and the feed speed is 2.5mm/min, the data is measured. And the change law curve is drawn.

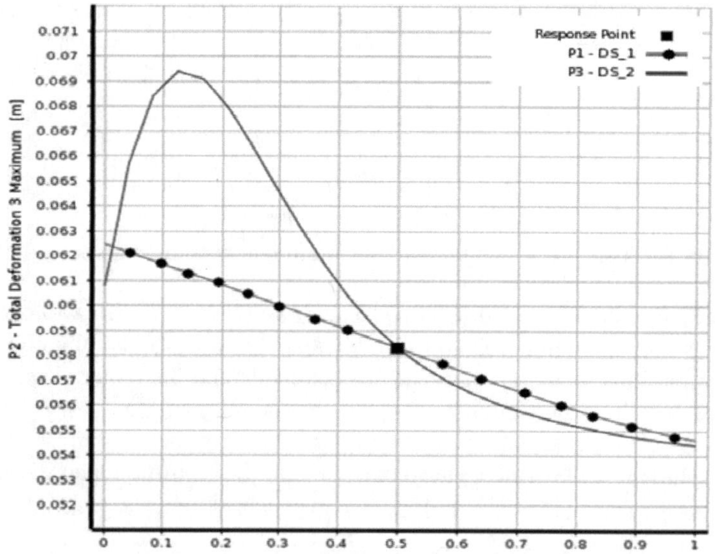

Figure 10. Parameters curve sensitivity correspondence

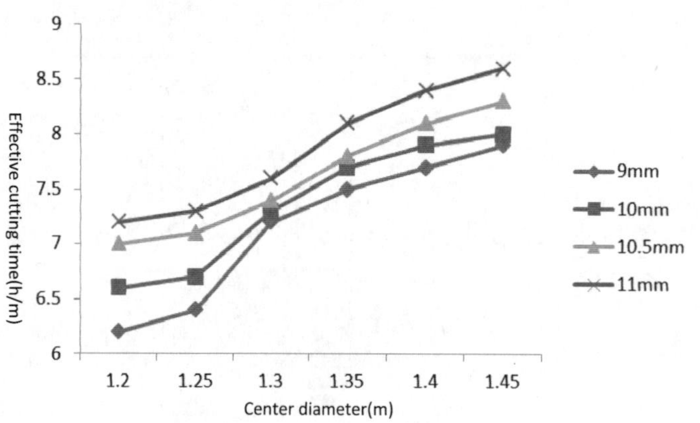

Figure 11. Cutting life test curve

The curve of the cutting life test change law is shown in figure 11. As can be seen from figure 11, when the center distance is constant, the effective cutting time of the stringed bead wire becomes longer as the bead diameter increases. And when the bead diameter is constant, the effective cutting time of the stringed bead wire increases with the center distance increasing. From the simulation results, it can be seen that as the center distance and bead diameter increase, the vibration amplitude of the stringed bead wire decreases. In summary, the stringed bead wire vibration amplitude of the diamond wire saw has an impact on the cutting life of the stringed bead wire. When the vibration amplitude affected by the center distance and bead diameter becomes small, the stringed bead wire has longer cutting times, which means the cutting life of the stringed bead wire becomes longer. The stringed bead wire cutting life test was completed and the experimental results were compared and analyzed. Draw conclusions as follows:

6. Conclusion

In this paper, the free vibration finite element model of the stringed diamond bead wire is established. When other parameters are fixed, the influence of center distance and bead diameter on vibration amplitude was analysed.

(1) Within a certain range of the center distance, if the center distance between the driving wheel and the driven wheel of the diamond wire saw machine increases, and the vibration amplitude of the stringed bead wire will decrease accordingly.

(2) If the bead diameter of the stringed diamond bead wire increases, the vibration amplitude will be reduced.

(3) If the bead diameter of the stringed diamond bead wire increases, the vibration amplitude will be reduced, the center distance has a great influence on its vibration amplitude during the initial vibration stage. Subsequently, the bead diameter gradually becomes the dominant influence factor of the vibration amplitude change.

(4) According to the sawing experiment of the diamond wire saw, the center distance and the bead diameter of the diamond wire saw will affect the cutting life of the diamond wire saw. Two parameters jointly affect the third-order vibration amplitude of the stringed bead wire. And the smaller the vibration amplitude, the longer the effective cutting time of the stringed bead wire.

References

[1] Cheng L, Experimental prototype development of underwater horizontal diamond wire saw 2008 , Harbin Engineering University.

[2] Tönshoff H K, Hillmann-Apmann H. Diamond tools for wire sawing metal components 2003 . *Key Engineering Materials*, 250(3–6) , 33-40.

[3] Brehl, D. E., & Dow, T. A.. Review of vibration-assisted machining 2008 . *Precision Engineering*, 32(3), 153-172.

[4] Guoqin H, Xipeng X. Technology of with diamond wire 2005 . *Tool Engineering*,39(8),7-11.

[5] Wickert, J. A., Mote, C. D., & Jr. Classical vibration analysis of axially moving continua 1990. *Trans Asme Journal of Applied Mechanics*, 57(3), 738-744.

[6] Foda, M. A. Vibration control and suppression of an axially moving string 2012 . *Journal of Vibration & Control*, 18(1), 58-75.

[7] Liqun C. Longitudinal vibration of axially moving strings and its control 2001 . *Advances In Mechanics*. (25) , 535-537.

[8] Shuai C, Hui H, Xipeng X. Influence of saswing parameters on the vibration characteristics of diamond wire saw 2011 . *Diamond & Abrasives Engineering*. (3) , 1-4.

Fault Diagnosis for Brake System in High-Speed Trains Using the Phased Features and Multi-layer Perceptron

Changzheng Fang[1], Xiaoyong Zhang[2,3,*], Yijun Cheng[2,3], Shengnan Wang[2,3], Li Zhang[2,3] and Yingze Yang[2,3]

[1] CRRC ZhuZhou Locomotive Co., Ltd., Zhuzhou, China
[2] School of Information Science and Engineering, Central South University, Changsha, China
[3] Hunan Engineering Laboratory of Rail Vehicles Braking Technology, Zhuzhou, China

*Email:zhangxy@csu.edu.cn

Abstract. Recently, High-Speed Train (HST) has been developing quickly. The brake system is the most important part in HST. Therefore its safety and reliability have attracted much attention from both academic and industrial community. Many data-driven methods are introduced to handle the fault diagnosis problem for brake system. However, the existing data-driven methods barely analyse the brake system. The used features lack of interpretability. In this paper, a fault diagnosis model using the phased features and multi-layer perceptron (MLP) is proposed for brake system. The proposed model is divided into two stages: offline and online stages. In the offline stage, after the analysis of brake system, the phased features are extracted. Then the MLP model is trained based on the phased features. In the online stage, the model learned in the offline stage is used to diagnosis the fault. The experimental platform for brake fault is constructed to validate the proposed model. The results shows the superiorities compared with other diagnosis methods in terms of precision, recall and F1-score.

1. Introduction

In recent years, High-Speed Train (HST) has been developing quickly in China [1]. In 2017, HSR extended to 29 of the country's 33 provincial-level administrative divisions and exceeded 25,000 km in total length, accounting for about two-thirds of the world's high-speed rail tracks in commercial service. According to China's Mid-to-Long Term Railway Network Plan, by 2030, the HST would link all provincial capitals (excluding Lhasa) and cities with more than half million people [2]. With the fast development of HST, its safety and reliability has attracted more and more attention.

Brake system is the most critical control components for the safety and reliability of the train operating at high speed. The efficient brake system is a guarantee that the train would decelerate within a reasonable distance even under the most adverse conditions [3]. If fault is diagnosed timely, maintenance decision can prevent serious damages to the HST. Therefore accurate and timely fault diagnosis in the brake system is clearly an important issue.

Because of the increasing installation of monitoring systems on various industrial systems, data-driven fault diagnosis have became more popular [4]. The data-driven fault diagnosis consists of feature extraction, feature selection, and feature classification. Firstly, features are extracted from huge amounts of the monitor process data. Secondly, features are selected by the statistical methods. Finally, fault are classified based on the features. In [3], decision tree (DT) were used as classifiers for classifying brake faults using the statistical features extracted from the vibration signals. In [5], for fault diagnosis of brake system, descriptive statistical features were extracted from the vibration signals and the feature selection was carried out using the C4.5 DT algorithm. The selected features were classified into different faults by support vector machine (SVM). In [6], an automated fault detection method using SVM and AE parameters is proposed to diagnosis the valve condition in reciprocating compressor. It employs the SVM for AE parameters analysis in an attempt to reduce human intervention in the analysis process. However, these methods don't analyse the brake system deeply enough. The features lack of interpretability. It is bad for industrial fault diagnosis.

In this paper, a fault diagnosis model using the phased features and MLP for brake system is proposed. The phased features are extracted according to the structure of the brake system. Then the MLP is used for fault diagnosis. Section 2 analyses the brake system and proposes the phased features based on the analysis. Then the fault diagnosis model using the phased feature and MLP is introduced. In Section 3, the experiment platform is constructed to demonstrate the proposed fault diagnosis model. Finally, the conclusions are described in Section 4.

2. The proposed fault diagnosis model

A fault diagnosis model using the phased features and MLP for brake system is proposed in this paper. The proposed model is divided into two stages: offline and online stage. In the offline stage, the phased features is extracted according to the structure of brake system. Then the MLP model is trained for diagnosis. In the online stage, the model learned in the offline stage is used to diagnosis the fault.

2.1. The brake system

First, the structure of brake system is analysed in this subsection. In HST, the electro-pneumatic (EP) brake system is the most common brake system. The brake system applies air pressure to the train line to realizes the locomotive braking function based on driver's instruction. In this process, air pressure is regulated by air-filled and air-bled with control the valves opening and closing, such as relay valve, relief valve, air-filled valve and air-bled valve. During the braking process, the solenoid valve is in the state of switching frequently. Therefore, the solenoid valves become the main fault element of the brake system.

Table 1. The phased features of sensor signals

Driving current	Stage	phased features	
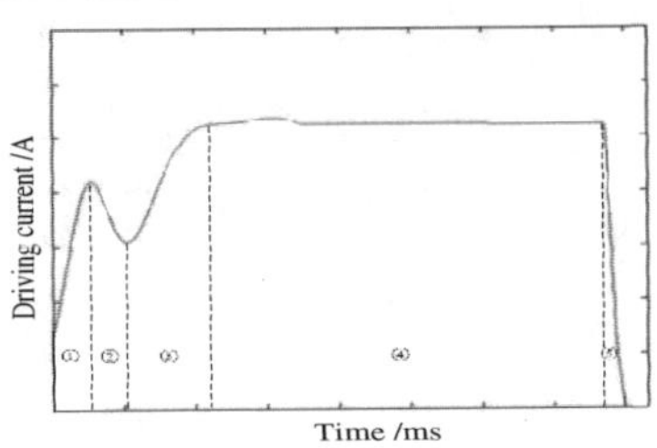	①Absorption touch	Peak current, maximum current change rate	Opening time
	②Actuation movement	Bottom current minimum current change rate	
	③Electrified holding	Maximum current change rate	
	④Release touch	Stable voltage	
	⑤Release motion	Minimum current change rate	

2.2. Data processing and the phased features extraction

The current and voltage of solenoid valve are changing in the process of braking. The trend of the sensor signals is studied in this paper to realize the fault diagnosis of valve.

During the process of data generation, the sensor data is mixed with noise which from various environmental factors. So the data processing is necessary to eliminate the noise. Median filtering as a non-linear noise processing method is used in the proposed model. It has better characteristic on edge preservation and depression of the noise. The basic principle of median filtering algorithm is to replace the value of any point in the digital sequence with the median value of the neighbourhood of each point.

Assuming the signal sequence is $x=\{x_1,x_2,\cdots x_n\}$ and sorting the N numbers by size, then the median filtering is used to smooth the sequence. After smoothing data, the phased features are extracted from sensor data to represent effective information for fault diagnosis. The working process of the solenoid valve is divided into five stages: the absorption touch stage, the actuation movement stage, the electrified holding stage, the release touch stage, and the release motion stage. Based on different stages, the phased features are extracted as Table 1.

Each phased feature contains the effective diagnosis information of sensor data. In order to enhanced generalization by reducing overfitting and reduce the computational complexity, wrapper algorithm is used to select feature in this paper.

In wrapper feature selection algorithm, the performance of the final learner is regard as the evaluation criterion of the feature subset as follows.

1) Setting the initial optimal error *E* to infinity. The best feature subset is the complete set of attributes *A*, and the number of repetitions *t* is 0.

2) A set of characteristic subset *A′* is generated randomly and calculating the error of the classifier is *E' when* using the subset of the feature.

3) If *E′* is smaller than *E*, then *A′* = *A*, *E′*= *E* and repeat 2), 3) step; otherwise *t*=*t*+*1* and jump out of the loop when *t* is greater than the stop control parameter *T*.

In this process, feature selection requires multiple training of the learner and select the best feature subset for the given learner.

2.3. Fault diagnosis based on MLP

Based on wrapper algorithm, the phased features are reduced for dimension reduction. After that, the Multi-layer Perceptron is used to diagnose the fault of brake system.

MLP is a supervised learning algorithm, and can learn a non-linear function approximator [7]. It is different from other algorithm, in that between the input and the output layer, there can be one or more non-linear layers, called hidden layers as Figure 2 shown.

At first, preparing the training dataset $D=\{(x_i,y_i)\}_{i=1}^n, x_i\in R^{m\times 1}, y_i\in R$. Where, n is the number of samples. $x_i(i=1,2,\cdots,n)$ is m-dimensional phased feature vector $I_i(i=1,2,\cdots,m)$ as the input of MLP. y_i is the label of fault. And the weighted input of *j* node in the hidden layer can be expressed as:

$$h_j=\sum_{i=1}^{m}W_{ij}\times I_i+b_j \tag{1}$$

where W_{ij} is the connection weight which from the input layer *i* node to the hidden layer *j* node, b_j is bias for the corresponding node, the output of the *j* node in the hidden layer is H_j.

$$H_j=\tanh(h_j) \tag{2}$$

After several iterations, the input o_k of output layer *k* note from hidden layers is

$$o_k=\sum_{j=1}^{J}W_{jk}\times H_j+b_k \tag{3}$$

where output layer contains *K* notes ($k=1,2,\cdots K$). The output O_k of the *k* node in the output layer corresponding to different activation functions.

Then, the mean square error minimization method is used to adjust the weights between layers and the mean square error E_k and gradient g_{ik} of the *n* training sample for the *k* node in the MLP output layer is

$$E_k = \frac{1}{N}\sum_{i=1}^{n}[y_k(i) - O_k(i)]^2 \tag{4}$$

$$g_{ik} = \frac{\partial \sum_{k=1}^{K} E_k}{\partial W_{jk}} \tag{5}$$

In the training process, it is necessary to judge that E_k is less than the given error or not. If E_k is less than the given error, the training of this node is completed, otherwise the next step weight is carried out as follows.

$$W_{jk} = W_{jk} + l \times g_{jk} \tag{6}$$

where l is the learning factor which determined by the experience value. Heavy gradient calculation and weight updating until network convergence. Based on this process, we can get the fault diagnosis model which learned from offline data.

3. Experimental study

3.1. Experiment platform

The BCU and PID control algorithms are used in the experimental platform. High-speed solenoid valves, sensors and equalizing reservoir are basically the same with the train loading components, which can simulate real vehicle condition very well. Among them, the model of high-speed solenoid valve (supply valve and exhaust valve) is MAC 35A-ACA-DDFA-1BA, and the model of pressure sensor is Keller PA-21Y (two-wire connection is adopted and the output signal is the current with 4-20MA) and the volume of equalizing reservoir is 1.2L. The pneumatic triplex contains desiccant, filter and pressure reducing valve to effectively filter the water in the wind source and achieve a reduced pressure output.

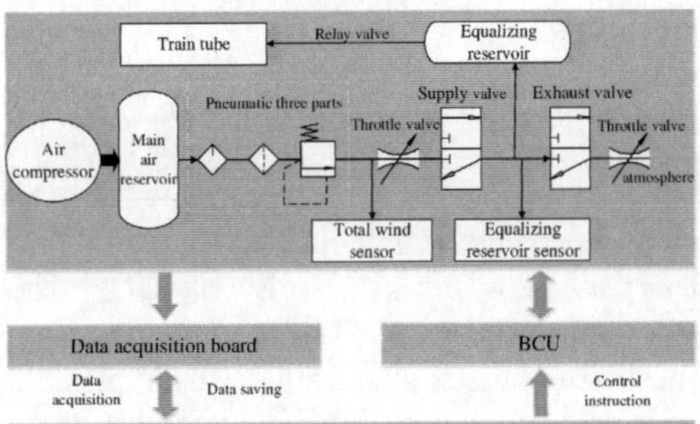

Figure 1. The schematic diagram of the experimental platform

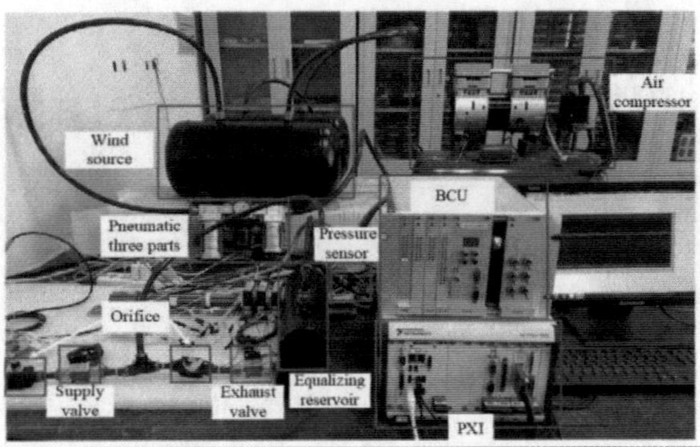

Figure 2. The experiment platform

In the closed-loop control mode, the brake control unit (BCU) receives the equalizing reservoir target value command issued by the upper computer, and compares the target value with the real-time pressure value of the equalizing reservoir fed back by the pressure sensor. If the pressure of equalizing reservoir is lower than the target pressure, then the main air reservoir inflates the equalizing reservoir through the supply valve. Conversely, the equalizing reservoir is exhausted to the atmosphere through the exhaust valve. The BCU issues a PWM control for the supply valve and exhaust valve. By precisely controlling the pressure of the equalizing reservoir, the train tube is controlled to achieve the braking of the train. Figure 1 is a schematic diagram of the experimental platform in closed loop control mode, Figure 2 is the experiment platform.

In the equalizing reservoir control system, according to the engineer's experience, in the normal case of the BCU, common faults include electric air valve failure, sensor failure, and air leakage. The existing fault diagnosis system is unrecognizable for the slight delay of the spool and the slight change of the return spring. Therefore, the stuck of the solenoid valve is studied in this paper.

3.2. Data acquisition and filtering

An external PXI box for data acquisition is used in this paper. PXI (PCI extensions for Instrumentation) is a rugged PC-based measurement and automation platform released by National Instruments. The collected signals include the voltage and current signals of the supply valve and the exhaust valve, the pressure of the equalizing reservoir, the pressure of main air reservoir. The sampling frequency is 10 000 Hz because the response time of the high-speed solenoid valve is generally within 10ms. The median filter window size used for the current is 7, for the voltage and the pressure of equalizing reservoir are 13, and for the pressure of total wind is 31.

3.3. Feature processing

The experiment used a total process of charge and exhaust to the equalizing reservoir as a sample, and the brake pressure was set to 300 KPa. In the experiment, the BCU uses the PI algorithm to control the switch of the high-speed valve in a closed loop, which leads to the process of multiple switching of the air charging and exhaust valve. Under normal circumstances, it can be stabilized by adjusting about 10 times. Due to noise interference, a precise extraction for the time domain characteristics of the current and voltage is challenging. Therefore, average and median processing are adopted to the valve's operating characteristics in a sample, reflecting the relevant characteristics of the valve as much as possible. Thus the proposed phased features mentioned before are used as input. In addition, the air filling time, maintenance time and exhaust time of the equalizing reservoir also used as input. The range of these features may be very different, which will have a great impact on the model training. Therefore, features need to be standardized.

3.4. Performance analysis

In this paper, the precision, recall and F1-score are used to evaluate the accuracy of the method, When the output of the algorithm is 1, indicates that the supply valve is stuck, while the output is 0, indicates that the supply valve is normal.

This section discusses the influence of the number of features on prediction accuracy and the comparison between different methods. The comparisons with other methods including Decision Tree (DT) and Support Vector Machines (SVM) are carried out to show the superiority of MLP. The parameters of other methods are optimized by grid search. A total of 28 sample data are obtained from the experimental platform. Among them, 18 samples are normal for the supply valve, 10 samples are stuck, and the exhaust valves used in the experiment are normal. Samples for training and prediction are randomly divided into portion of seven and three. The experimental results are taken the average of 10 trials to exclude the effects of random disturbances. The optimizer of MLP is Adam, the hidden layer is 2, the meurons in each layer is (64, 32), the learning rate is 0.05, the activation is sigmoid.

The Figure 3 shows the influence of the number of features selected by the Wrapper on the prediction precision. It can be seen that the pricision of the MLP is almost higher than that of the other two methods. That means that selected features can well reflect the fault characteristics. When the number of features is 12, 14, 18, 19, 20, 21, 23, 26, 29, 30, the precision reaches 1, indicating that the fault can be judged 100%. And DT does not have 100% precision. The precision of the SVM is lower than the other two methods. When 12 features are selected, the precision of the three methods is relatively high, under which the other evaluation methods are used for comparison. The experimental results are shown in the Table 2. As can be seen from the Table 2, MLP has the highest precision, recall, F-score, indicating that MLP can accurately determine faults and no faults. DT is superior to SVM in accuracy, recall and F1-score.

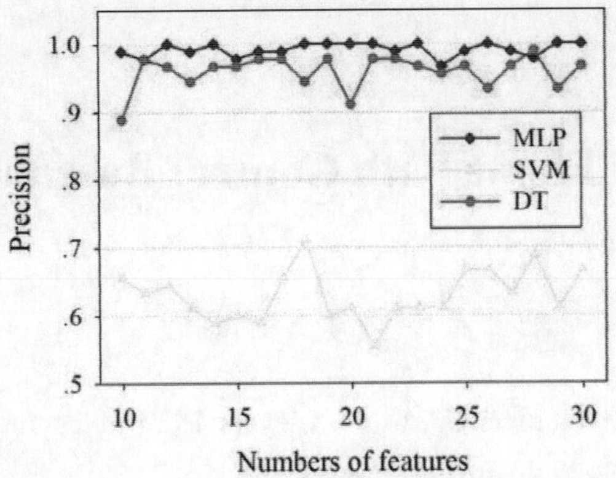

Figure 3. Prognostic performance of MLP, SVM and DT with different number of features chosen by Wrapper

Table 2 Prognostic performance of precision, recall and F1-score by different methods with 12 features chosen by Wrapper

Method	Precision	Recall	F1-score
MLP	1	1	1
DT	0.967	0.959	0.955
SVM	0.644	0.642	0.496

4. Conclusion

In this paper, for fault diagnosis in brake system, a model using the phased features and MLP is proposed. The phased features are proposed based on the analysis of brake system. The proposed phased feature is interpretable. The MLP is used as the classifier to classify the faults of brake system. We verified the effectiveness of this model by the experimental platform which is constructed for brake fault. The results shows the proposed model is able to diagnosis fault and has better performance compared with other diagnosis methods.

Acknowledgments

This work was partially supported by the National Key Research and Development Program of China (2016YFB1200401) and Hunan Science and Technology Plan Project (2016GK2003).

References

[1] He Z, Hu H, Zhang Y, Gao, S. Harmonic Resonance Assessment to Traction Power-Supply System Considering Train Model in China High-Speed Railway[J]. *IEEE Transactions on Power Delivery*, 2014, **29**(4), pp 1735-1743.

[2] NDRC. Mid-to-Long Term Railway Network Plan. Available online (accessed on 16 December 2017): http://www.nra.gov.cn/jgzf/flfg/ gfxwj/zt/other/201607/t20160721_26055.shtml .

[3] Jegadeeshwaran R, Sugumaran V. Comparative Study of Decision Tree Classifier and Best First Tree Classifier for Fault Diagnosis ff Automobile Hydraulic Brake System Using Statistical Features. *Measurement*, 2013, **46**(9), pp 3247-3260.

[4] Yin S, Ding S X, Xie X, Luo H. A Review on Basic Data-Driven Approaches for Industrial Process Monitoring. *IEEE Transactions on Industrial Electronics*, 2014, **61**(11), pp 6418-6428.

[5] Jegadeeshwaran R, Sugumaran V. Fault diagnosis of automobile hydraulic brake system using statistical features and support vector machines. *Mechanical Systems and Signal Processing*, 2015, **52**, pp 436-446.

[6] Ali S M, Hui K H, Hee L M, Leong M S. Automated Valve Fault Detection Based on Acoustic Emission Parameters and Support Vector Machine. *Alexandria Engineering Journal*, 2017, **57**(1), pp 491-498.

[7] Glorot X, Bengio Y. Understanding the Difficulty of Training Deep Feedforward Neural Networks. *International Conference on Artificial Intelligence and Statistics*, 2010, **9**, pp 249-256.

Unmanned Powered Paraglider Flight Path Control Based on PID Neutral Network

S T Qu[1]

[1]School of astronautics, Beihang University (BUAA), 37 Xueyuan Rd, Haidian District, Beijing 100191, P.R.China

qusite@buaa.edu.cn

Abstract. The unmanned powered paraglider (UPP) is a nonlinear time-varying dynamic system with high coupling degree. In view of how to make the UPP flying along arbitrary predetermined trajectory, the conventional control method is difficult to obtain the control trajectory with a better fitting degree of the presupposition trajectory. To solve this problem, the six degree of freedom (SDOF)dynamic model of the UPP is built and the control characteristics of the UPP is studied. In view of the control characteristics of the UPP, the controller suitable for the UPP is designed and compiled with the PID neural network method (PIDNN), and the simulation system for the flight process is built, which is used to simulate the flight path control. And the results show that the controller realizes the effective flight path control of the UPP.

1. Introduction

The unmanned powered paraglider (UPP) is a kind of flexible wing aircraft, which provides the lift needed by the parachute. It is mainly composed of four parts: parachute, control rope, propulsion system and payload cabin. Compared with other aircraft, the UPP has the advantages of lower cost, higher load capacity and less requirement for the site.[1][2]

However, due to the low flying speed of the UPP and the relatively high uncertainty of the complex flight environment, the magnitude of the disturbance is close to the flight speed, which makes it easier to be affected. In addition, because of the high coupling degree, nonlinear and time-varying dynamic characteristics of the UPP, the input control amount are only the control rope and the propeller thrust, and the dimension of the input is smaller when compared with the output. It also makes the self-adaptive of the control algorithm very necessary.[3]

In view of the above problems, to realize the autonomous flight according to a certain route, a more stable, reliable and adaptive control method is needed for the UPP. The PID neural network (PIDNN) method inherits the advantages of PID's simple structure and high reliability, and it involve the neural network algorithm. Although the traditional PID has simple structure and strong reliability [4][5], it is difficult to obtain good control result for the dynamic system with high coupling and nonlinear time-varying characteristics. To improve its self-adaptability[6][7], we consider the combination of neural network and PID control. There are two ways to combine the traditional neural network with the PID control. One is to determine the PID parameters by using the neural network.[8]-[10] Its structure is much more complex than the traditional PID controller and cannot avoid the weakness of slow convergence speed of the general neural network. It is also difficult to determine the number of neurons in hidden layer and the weight value between different layers. The other one is a single neuron structure PID controller[11], but it does not have the ability to approximate any function, so it is difficult to achieve good result in the complex control system such as UPP.

The PID neural network in this paper, instead of combining neural network and traditional PID control law, defines neurons with proportional, integral and differential functions, and uses neural network to adjust the connection weight directly. It inherits the advantages of traditional PID and neural network and avoids the shortcomings of the general method of combining PID with neural network.[12][13]

In this paper, the six degree of freedom (SDOF) model of the UPP is established. And the control characteristics is analysed based on the model. Then, the PID neural network method is used to design the controller for this system. Finally, according to the result of simulation, the UPP's flight path is effectively in accordance with the pre-set trajectory.

2. Six degree of freedom modeling (SDOF) of UPP

The SDOF model of the UPP is the basis of investigating the control problem. Considering the characteristics of the parafoil part, the force analysis requires the establishment of additional mass. And considering that the actual shape of the parafoil is curved, it should be differentiated into several same small parts in which the aerodynamic force and the aerodynamic torque should be considered.

2.1. The basic hypothesis and the selection of coordinate system

Some basic assumptions are as follows. First, the connections between parafoil and the load tank are considered as rigid connections, and the mass and the moment of inertia of the parafoil rope can be neglected. Second, the load compartment is regarded as a standard square with uniform mass distribution, and its lift and aerodynamic force, aerodynamic moment can be neglected, only its resistance is considered. Third: the quality center and pressure center of the parafoil are located at the 1/4 distance from the leading edge. Fourth: the hypothesis of the plane mechanism for earth.

Three coordinates are needed, which are the body coordinate system S_p, the geodetic coordinate system S_g and the air flow coordinate system S_a. The origin of S_g is O_g located at the center of mass of the model at the initial time. And the axis O_gZ_g points vertically to the earth. The O_gX_g points toward the leading edge along the horizontal direction, and O_gY_g forms the right-hand line with other two axes. As for the S_a, the origin O_a is located at the quarter of the string, and the O_aX_a is to the opposite direction of the large air flow. The O_aZ_a is in the vertical symmetry plane of the parachute, which is vertical to the O_aX_a, and points toward the lower wing surface of the wing. Similarly, the O_aY_a forms the right-hand line with other two axes. The three coordinate systems are shown in Figure 1 as follows:

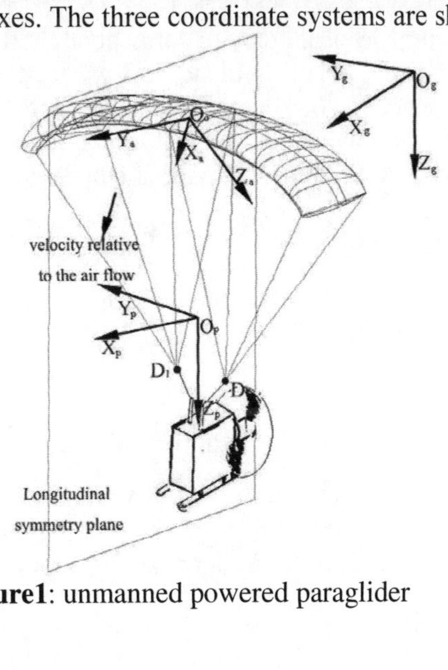

Figure1: unmanned powered paraglider

2.2. The force analysis and the establishment of dynamic equation
The force analysis of the UPP contains the force analysis of the parafoil and of the loading cabin. For the record, additional mass is needed and the force of the parafoil should be set up in different same section because of the inflatable structure and arc structure of the parafoil.

2.2.1. Additional mass and additional moment of inertia. The parafoil of UPP is inflated with air during the flight. Therefore, the density of the parafoil part of the UPP is similar to that of the surrounding fluid. The effect of the surrounding fluid on the whole motion can be added to the total mass and the moment of inertia in the form of additional mass and additional moment of inertia. The additional mass matrix is established by the method of Barrows 6 x 6 additional matrix[14], and the additional mass and additional moment of inertia is obtained as follows:

$$\boldsymbol{A}_{\mathrm{a,O}}=\begin{bmatrix}\boldsymbol{M}_{\mathrm{a}} & -\boldsymbol{M}_{\mathrm{a}}(\boldsymbol{L}_{\mathrm{O\text{-}R}}^{\times}+\boldsymbol{L}_{\mathrm{R\text{-}P}}^{\times}\boldsymbol{B}_2)\\ (\boldsymbol{B}_2\boldsymbol{L}_{\mathrm{R\text{-}P}}^{\times}+\boldsymbol{L}_{\mathrm{O\text{-}R}}^{\times})\boldsymbol{M}_{\mathrm{a}} & \boldsymbol{J}_{\mathrm{a,O}}\end{bmatrix},\ \boldsymbol{B}_2=\begin{bmatrix}0&0&0\\0&1&0\\0&0&0\end{bmatrix} \tag{1)(2}$$

The internal terms in the upper matrix are as follows:

$$\boldsymbol{J}_{\mathrm{a,O}}=\boldsymbol{J}_{\mathrm{a}}-\boldsymbol{L}_{\mathrm{O\text{-}R}}^{\times}\boldsymbol{M}_{\mathrm{a}}\boldsymbol{L}_{\mathrm{O\text{-}R}}^{\times}-\boldsymbol{L}_{\mathrm{R\text{-}P}}^{\times}\boldsymbol{M}_{\mathrm{a}}\boldsymbol{L}_{\mathrm{R\text{-}P}}^{\times}\boldsymbol{B}_2-\boldsymbol{Q}-\boldsymbol{Q}^{\mathrm{T}} \tag{3}$$

$$\boldsymbol{Q}=\boldsymbol{B}_2\boldsymbol{L}_{\mathrm{R\text{-}P}}^{\times}\boldsymbol{M}_{\mathrm{a}}\boldsymbol{L}_{\mathrm{O\text{-}R}}^{\times} \tag{4}$$

$\boldsymbol{L}_{\mathrm{O\text{-}R}}^{\times}$ and $\boldsymbol{L}_{\mathrm{R\text{-}P}}^{\times}$ are respectively the multiplication matrices of the vector pointing to the parafoil's roll center from the center of mass and the vector pointing from the parafoil's pitching center from its roll center. $\boldsymbol{M}_{\mathrm{a}}$ and $\boldsymbol{J}_{\mathrm{a}}$ are diagonal matrixes of which diagonal elements are the additional mass and the additional moment of inertia for the parafoil.

2.2.2. Aerodynamic force and aerodynamic torque. Considering the arc structure of the parachute, the parafoil is divided into eight segments on average, and the respective body coordinate system is set up in each part, and the lift and resistance are obtained as follows:

$$\boldsymbol{F}_{\mathrm{L}_i}=\frac{1}{2}\rho S_i\sqrt{u_{\mathrm{p}i}^2+w_{\mathrm{p}i}^2}k_iC_{\mathrm{L}_i}\begin{bmatrix}w_{\mathrm{p}i}\\0\\-u_{\mathrm{p}i}\end{bmatrix}\qquad \boldsymbol{F}_{\mathrm{D}_i}=-\frac{1}{2}\rho S_iV_{\mathrm{p}i}C_{\mathrm{D}_i}\begin{bmatrix}u_{\mathrm{p}i}\\v_{\mathrm{p}i}\\w_{\mathrm{p}i}\end{bmatrix} \tag{5)(6}$$

The subscript i represents the physical quantity corresponding to the section i; k is a product factor. S is the area of each segment of pterygoid; u_p, v_p and w_p are the projection of the velocity of the parafoil's each segment in respective coordinate system; V_p is the resultant velocity; C_L, C_D is the lift and drag coefficient. The total aerodynamics and aerodynamic torque expressions of the parafoil can be obtained by converting those of different section to the body coordinate system of the UPP.

$$\boldsymbol{F}_{\mathrm{aerop}}=\sum_{i=1}^{8}\boldsymbol{B}_{i\text{-}\mathrm{O}}(\boldsymbol{F}_{\mathrm{L}_i}+\boldsymbol{F}_{\mathrm{D}_i})\qquad \boldsymbol{M}_{\mathrm{aerop}}=\sum_{i=1}^{8}\boldsymbol{L}_{\mathrm{O}\text{-}i}^{\times}\boldsymbol{B}_{i\text{-}\mathrm{O}}(\boldsymbol{F}_{\mathrm{L}_i}+\boldsymbol{F}_{\mathrm{D}_i}) \tag{7)(8}$$

As for the the load tank, because it does not produce the lift, it is only necessary to analyse its drag. So, its aerodynamic force is as follows:

$$\boldsymbol{F}_{\mathrm{aerob}}=-\frac{1}{2}\rho S_{\mathrm{b}}V_{\mathrm{b}}C_{\mathrm{db}}\begin{bmatrix}u_{\mathrm{b}}\\v_{\mathrm{b}}\\w_{\mathrm{b}}\end{bmatrix} \tag{9}$$

Sb is the section area of the load tank. C_{db} is the drag coefficient. V_b is the resultant velocity of the load tank. u_b, v_b and w_b are the projection of the load tank's velocity in body coordinate system.

2.2.3. Dynamic model. Arranging the dynamic equations of the UPP in matrix form, the dynamic equations are as follows. G is the gravity component array. S_w is the multiplication matrix of the angular velocity, and Ft is the thrust provided by power system.

$$\begin{bmatrix} m_r \boldsymbol{I} + \boldsymbol{M}_a & -\boldsymbol{M}_a(\boldsymbol{L}^{\times}_{\text{O-R}} + \boldsymbol{L}^{\times}_{\text{R-P}}\boldsymbol{B}_2) \\ (\boldsymbol{B}_2\boldsymbol{L}^{\times}_{\text{R-P}} + \boldsymbol{L}^{\times}_{\text{O-R}})\boldsymbol{M}_a & \boldsymbol{J}_r + \boldsymbol{J}_{a,O} \end{bmatrix} \begin{bmatrix} \dot{\boldsymbol{V}}_O \\ \dot{\boldsymbol{W}} \end{bmatrix} = \begin{bmatrix} \boldsymbol{F}_{\text{total}} \\ \boldsymbol{M}_{\text{total}} \end{bmatrix} \tag{10}$$

The internal terms in the upper matrix are as follows. In the equation, the moment of inertia of the parachute to the centroid of the UPP is J_p, the moment of inertia of the load tank to the centroid of the system is J_b:

$$\boldsymbol{F}_{\text{total}} = \boldsymbol{F}_{\text{aerop}} + \boldsymbol{F}_{\text{aerob}} + \boldsymbol{F}_t + \boldsymbol{G} - \boldsymbol{S}^{\times}_w m_r \boldsymbol{V}_O - \boldsymbol{S}^{\times}_w \boldsymbol{M}_a \boldsymbol{V}_O + \boldsymbol{S}^{\times}_w \boldsymbol{M}_a \boldsymbol{D}\boldsymbol{W} \tag{11}$$

$$\begin{aligned} \boldsymbol{M}_{\text{total}} = {} & \boldsymbol{M}_{\text{aerop}} + \boldsymbol{L}^{\times}_{\text{O-p}}\boldsymbol{F}_{\text{aerop}} + \boldsymbol{L}^{\times}_{\text{O-b}}(\boldsymbol{F}_{\text{aerob}} + \boldsymbol{F}_t) - \boldsymbol{S}^{\times}_w(\boldsymbol{J}_r + \boldsymbol{J}_{a,O})\boldsymbol{W} \\ & - \boldsymbol{S}^{\times}_w \boldsymbol{D}^T \boldsymbol{M}_a \boldsymbol{V}_O + \boldsymbol{S}^{\times}_v \boldsymbol{M}_a \boldsymbol{D}\boldsymbol{W} \end{aligned} \tag{12}$$

$$\boldsymbol{D} = \boldsymbol{L}^{\times}_{\text{O-R}} + \boldsymbol{L}^{\times}_{\text{R-P}}\boldsymbol{B}_2 \qquad \boldsymbol{J}_r = \boldsymbol{J}_p + \boldsymbol{J}_b \qquad (13)(14)$$

3. Control characteristic of SDOF model of UPP

The flight path control of the UPP is mainly divided into longitudinal control and yaw control. The longitudinal control refers to controlling the flight speed and the flight height of the aircraft through the change of the propeller thrust. The yaw control is mainly through asymmetrically pulling down control rope, which causes the asymmetrical deflection of the rear edge of the parafoil, thus generating the torque to control the direction of the flight.

3.1. Longitudinal control

When given the different thrust, the UPP has different trajectories in the vertical plane which are shown as follows:

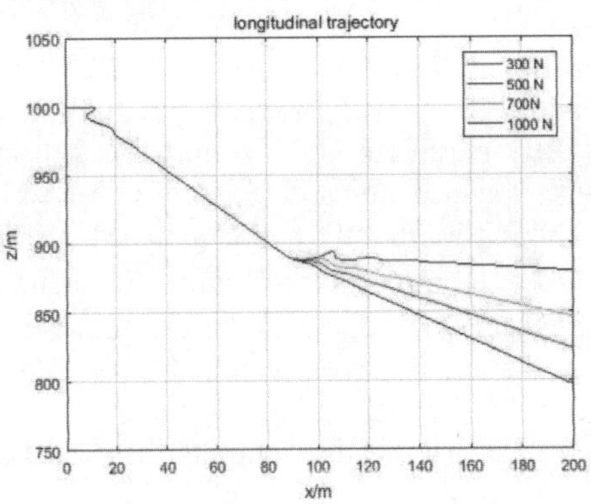

Figure2: longitude trajectory under different thrusts

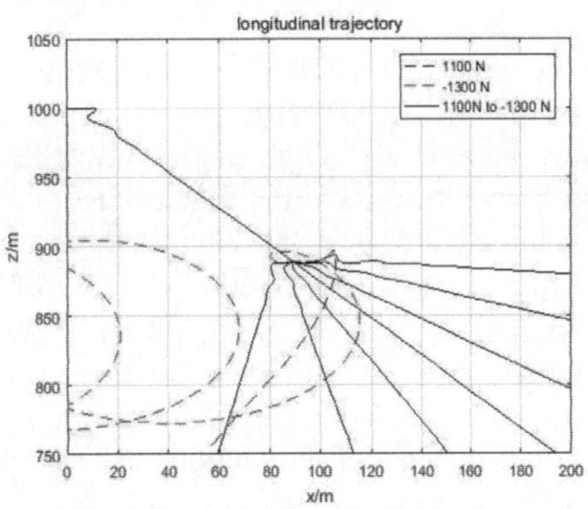

Figure3: Critical thrust to ensure no stall

It is known from Figure 2 that the different thrust is applied during the flight, and the trajectory is kept in the vertical plane because the device providing thrust is installed at the rear of the load tank, so there is no velocity component in the direction of the Y axis. After given thrust, the gliding ratio is improved. The difference is that the greater the thrust, the greater the glide ratio.

Besides, to prevent stall in the flight, the value of the thrust should be restrained in an effective range. According to the Figure 3, when the thrust is between 1100N and - 1300N, the flight trajectory of the UPP can be changed normally by the thrust. When the thrust exceeds this range, the stall happens. Therefore, when designing controller, the control volume should be controlled within this range.

3.2. Yaw control

When applied different pulling amount to the left control rope and the right control rope, the rear edge of the parachute produces an angle, which makes the rear edge of the parafoil asymmetrically deflected, and the trajectory response of the UPP based on the asymmetrical deflection of the following edge is shown as Figure 4.

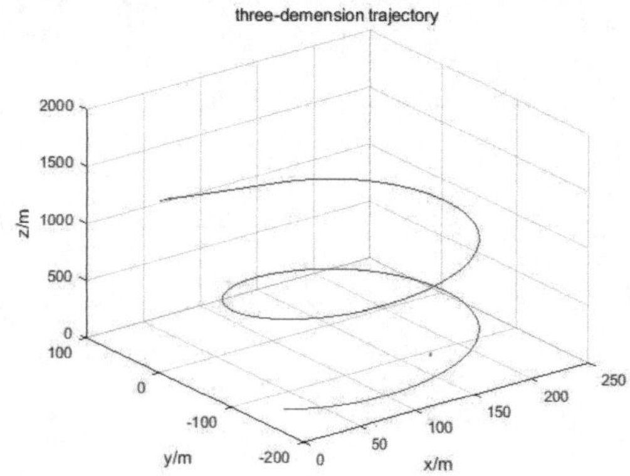

Figure 4: trajectory under the controlling of ropes

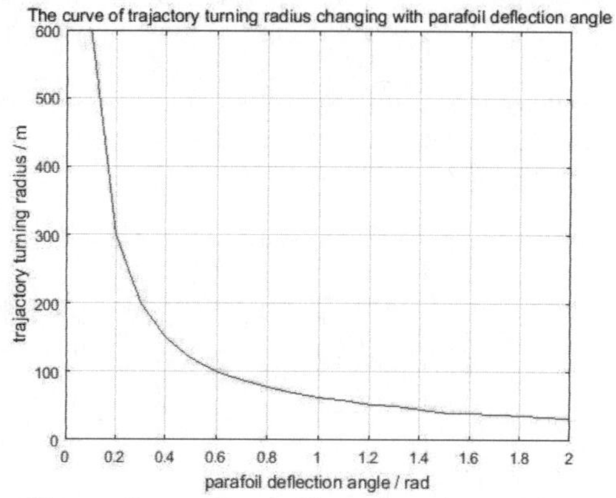

Figure 5: turning radius changing with parafoil deflection angle

It can be seen from Figure 4 that when the left control rope is pulled, the UPP begins to turn. Because of the difference of the parafoil's deflection angle created by the asymmetrical pulling amount of the left and right control ropes, the turning radius of the trajectory also changes. The relationship between these two variables can be obtained as Figure 5. It can be concluded that the turning radius of the UPP's trajectory decreases with the increase of the parafoil's rare edge deflection angle. And the minimum radius of the turning radius of the trajectory is about 30 meters. Therefore, the pre-set turning radius of the trajectory should not be set lower than this minimum according to the actual situation, otherwise the corresponding trajectory control cannot be achieved.

4. PID neutral network (PIDNN) structure and controller design

PID neural network (PIDNN) is not a simple combination of neural network and traditional PID control law. Instead of using neural network to set PID parameters, it defines the neuron with proportional, integral and differential function, and uses neural network to adjust the weight of connection directly[15].The most basic form of PIDNN is Single PID neural network (SPIDNN), which is a three level forward neural network of $2\times3\times1$. The input layer contains two neurons and receives external input information, which is the output feedback value and the pre-set value of the controlled object respectively. There are three neurons in the hidden layer, which have the function of the proportion, integral and differential operation respectively. The output layer has one neuron, which completes the synthesis of the outputs of hidden layer's neurons and then outputs the corresponding control volume. As is shown in Figure 6.

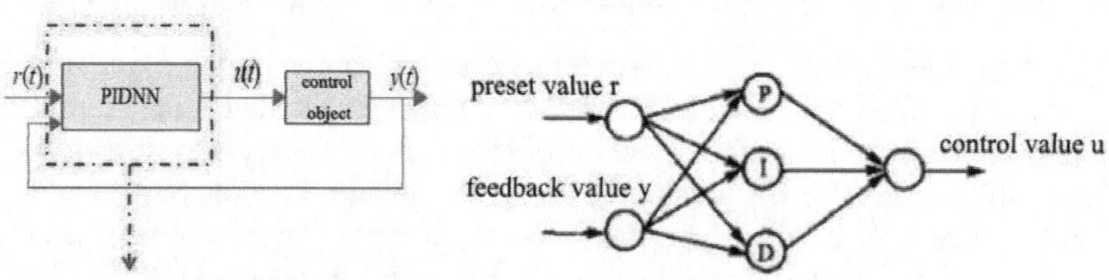

Figure 6: the most basic form of PIDNN

4.1. Network structure design

When the controlled object has m inputs and n outputs, MPIDNN needs 2n input units in the input layer, and there are 3n processing units in the hidden layer, including n proportional elements, n differential elements, n integral elements and there are m output units in the output layer. In the control process, the comparison between the actual position and the pre-set position is based on the position in the X direction, and the position in the Y direction and the Z direction is fed back to the controller. Figure 7 shows the control loop for UPP.

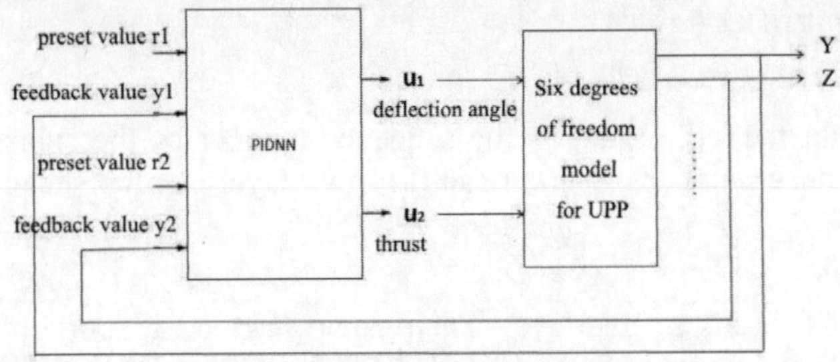

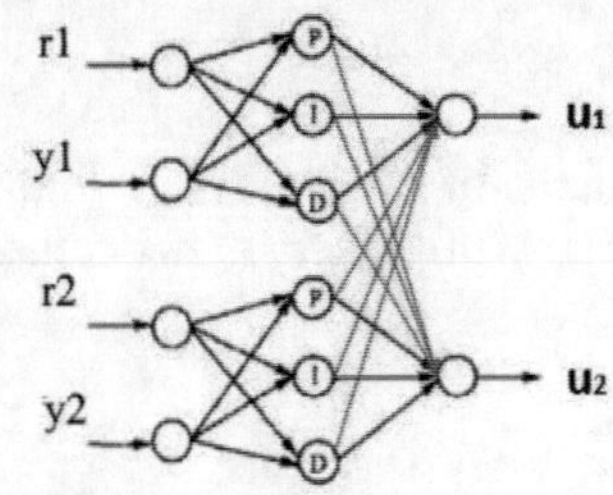

Figure 7: control loop for UPP **Figure 8**: the structure of MPIDNN

Therefore, the structure of the PIDNN controller which is shown in Figure 8 is a 4×6×2 three-layer forward neuron network, including two basic SPIDNN structures, the input amounts are the Y direction pre-set value and feedback value, the Z direction pre-set value and feedback value. The output amounts are obtained through the forward algorithm of PIDNN. The outputs are the values of the two control variables received by the UPP. PIDNN controller then adjusts the network weights between different layers according to the error value and prepares for the next calculation.

4.2. Design of control algorithm

The control algorithm is composed of two parts: forward algorithm and backpropagation algorithm. The forward algorithm is used to calculating the value of the control variables, and the backpropagation algorithm is used to adjust the network weights. The process is shown in Figure 9:

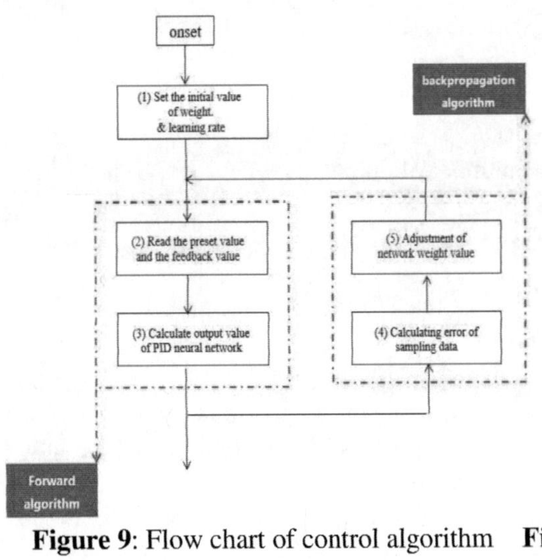

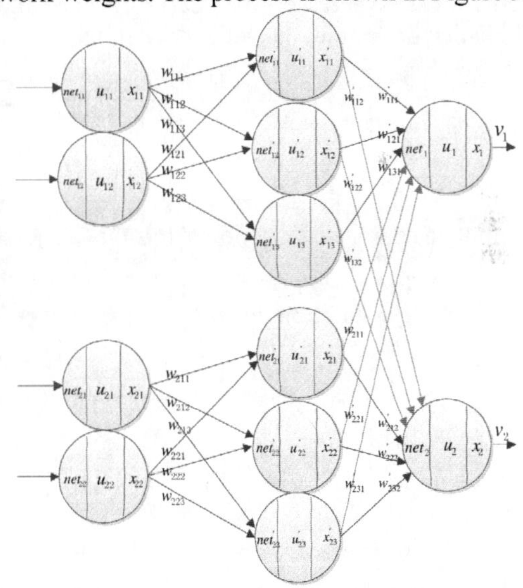

Figure 9: Flow chart of control algorithm **Figure 10**: internal structure of MPIDNN controller

4.2.1. forward algorithm. The forward algorithm is as follows. Each neuron is divided into the net (input of the neuron), u (the state function of the neuron), and x (the output of the neuron), as is shown in Figure 10 above:

Input layer: The input layer contains four identical neurons, which receive the Y direction's pre-set value r_1 and feedback y_1, Z direction's pre-set value r_2 and feedback y_2. The input of neurons is:

$$\begin{cases} net_{11}(k) = r_1(k) \\ net_{12}(k) = y_1(k) \\ net_{21}(k) = r_2(k) \\ net_{22}(k) = y_2(k) \end{cases} \tag{15}$$

The output of neurons is as follows and the subscript s is the sequence number of the subnet contained in MPIDNN,s=1, 2. i is the sequence number of neurons in the input layer of each subnet. i=1, 2

$$x_{si}(k)=u_{si}(k) = net_{si}(k) \tag{16}$$

Hidden layer: There are two groups of neurons in this layer. Each group has one proportional neuron, one differential neuron and one integral neuron. They are connected by the weight with the input layer's neurons. Therefore, the input of the hidden neurons is as follows: j is the sequence number of the hidden neurons in the subnet, j=1, 2, 3.

$$net'_{sj}(k) = \sum_{i=1}^{2} w_{sij} x_{si}(k) \tag{17}$$

Each group contains three neurons which respectively have different state function of proportion, integral and differential. Therefore, the output of neurons are as follows:

$$x'_{s1}(k)=u'_{s1}(k) = net'_{s1}(k) \tag{18}$$

$$x'_{s2}(k)=u'_{s2}(k) = u'_{s2}(k-1) + net'_{s2}(k) \tag{19}$$

$$x'_{s3}(k)=u'_{s3}(k) = net'_{s3}(k) - net'_{s3}(k-1) \tag{20}$$

Output layer: There are two neurons in the output layer, which are connected to the 6 neurons in the hidden layer through the weights The input of the neurons is:

$$\begin{cases} net_1(k) = \sum_{i=1}^{2}\sum_{j=1}^{3} w'_{sj1} x'_{sj}(k) \\ net_2(k) = \sum_{i=1}^{2}\sum_{j=1}^{3} w'_{sj2} x'_{sj}(k) \end{cases} \tag{21}$$

The output of neurons is as follows: h is the sequence number of output layer neurons, h=1, 2

$$x_h(k)=u_h(k) = net_h(k) \tag{22}$$

After the above calculation, the output of the entire MPIDNN controller can be obtained as follows:

$$\begin{cases} v_1(k)=x_1(k) \\ v_2(k)=x_2(k) \end{cases} \tag{23}$$

4.2.2. backpropagation algorithm. The aim of the backpropagation algorithm is to modify the connection weights between the two layers by the "error backpropagation method" with the "gradient descent method" as the core, so as to reduce the error. The error is:

$$J=\frac{1}{2}\sum E_h = \frac{1}{2}\sum_{h=1}^{2}\left[r_h(k) - y_h(k)\right]^2 \tag{24}$$

The modifying of weight values between hidden layer and output layer:
According to the weight iteration formula from hidden layer to output layer

$$w'_{sjh}(k+1) = w'_{sjh}(k) - \eta' \frac{\partial J}{\partial w'_{sjh}} \tag{25}$$

In which η' is the learning rate, $\frac{\partial J}{\partial w'_{sjh}}$ can be expressed as:

$$\frac{\partial J}{\partial w_{sjh}^{'}}=\sum_{p=1}^{2}(r_p(k)-y_p(k))\cdot \mathrm{sgn}\frac{y_p(k+1)-y_p(k)}{v_h(k)-v_h(k-1)}\cdot x_{sj}^{'}(k) \tag{26}$$

The modifying of weight values between input layer and hidden layer:
According to the weight iteration formula from input layer to hidden layer

$$w_{sij}(k+1)=w_{sij}(k)-\eta\frac{\partial J}{\partial w_{sij}} \tag{27}$$

According to the weight iteration formula from input layer to hidden layer:

In which η is the learning rate, $\frac{\partial J}{\partial w_{\mathrm{sij}}}$ can be expressed as:

$$\begin{aligned}\frac{\partial J}{\partial w_{sij}}=&\sum_{p=1}^{2}\sum_{h=1}^{2}(r_p(k)-y_p(k))\cdot \mathrm{sgn}\frac{y_p(k+1)-y_p(k)}{v_h(k)-v_h(k-1)}\cdot w_{sjh}^{'}(k)\\ &\cdot \mathrm{sgn}\frac{u_{sj}^{'}(k)-u_{sj}^{'}(k-1)}{net_{sj}^{'}(k)-net_{sj}^{'}(k-1)}\cdot x_{si}(k)\end{aligned} \tag{28}$$

5. Simulation result analysis

In the process of simulation, the predetermined three-dimensional trajectory is synthesized with two sinusoidal curves in the horizontal and vertical direction. The starting point coordinates of the simulation starting point are (0, 0, 800m), and the initial velocity is 10m/s. The pre-set trajectory is:

$$\begin{cases}x=x\\ y=500\cdot\sin(0.002\cdot x)\\ z=100\cdot\sin(0.002\cdot x)+800\end{cases} \tag{29}$$

The simulation results are shown in Figure 11 as follows:

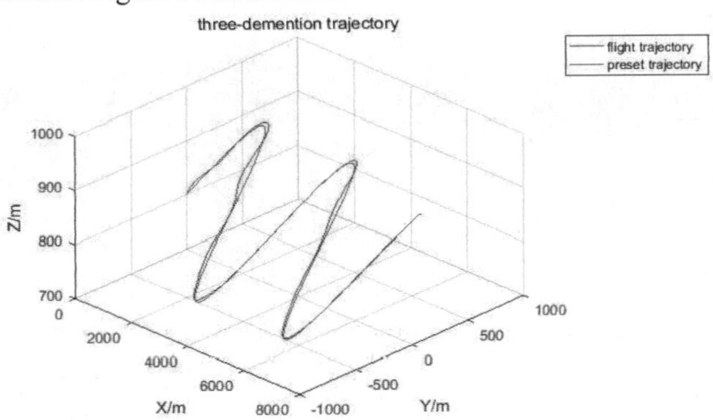

Figure 11: the comparation between flight trajectory and pre-set trajectory

According to the simulation results, during the flight, the UPP can fly according to the pre-set trajectory. The flight control of the trajectory is realized. To better analyse the result of yaw control and longitudinal control, the projections of the three-dimensional trajectory on the horizontal and vertical planes are analysed respectively.

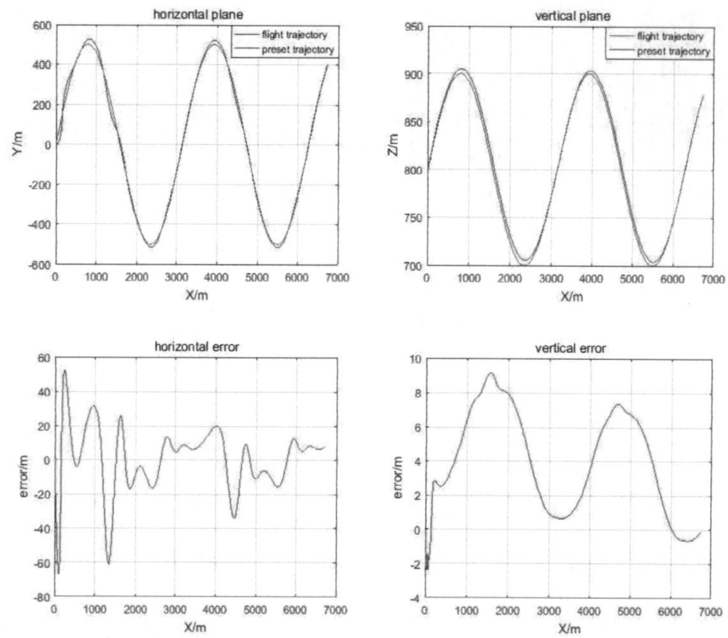

Figure 12: the comparation of trajectories and error in horizontal and vertical plane

As can be seen from the Figure 12 the maximum error is less than 25m when the UPP begins to stay in a stable flight state in the horizontal direction. In the vertical direction, the UPP flies well in accordance with the pre-set trajectory, and the maximum error can remain within 10 m.

Also, the values of the control variables of the PIDNN controller are recorded as follows:

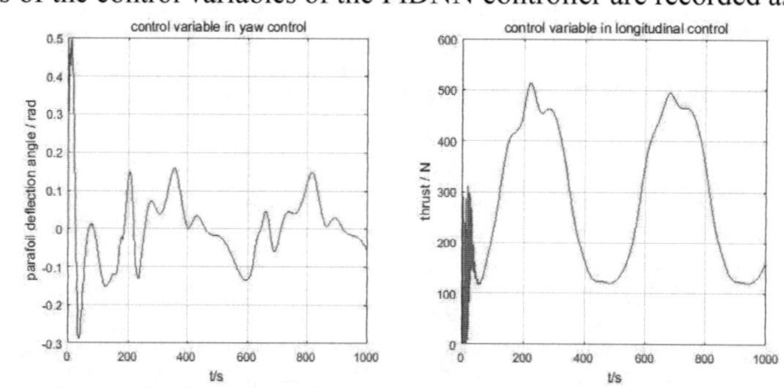

Figure 13: the value of control variable changing with time

As can be seen from Figure 13, the value of the yaw control variable is within 0.2 rad after being stable, and the turning radius is within the required range. The control range in the vertical direction is 0-500N, which is also within the effective thrust range of the propeller which can prevent the stall.

6. Conclusion

The key factors in the flight process are the 6 DOF model of the UPP, the structure of the PIDNN and the design of the control algorithm. From the simulation results, it can be seen that the PIDNN controller can output the control variables satisfying the control characteristics of the UPP, and control the UPP to fly according to the predetermined trajectory with a good fit.

Reference list

[1] Wise K A 2006 Dynamics of a UAV with Parafoil Under Powered Flight. AIAA-2006-6795.
[2] Guo H J, He W L and Sun Y 2017 Developing Prospect Analysis of Parawing UAV the 3rd China Aeronautical Science and Technology Conf. 19-20 Sep Beijing China

[3] Zhou X H, Chen Z L and Chen L P 2010 Study of Neutral Network PID in Heading Control of Unmanned Powered Parachute Aircraft **10** 58-59

[4] Hu W Z, Chen J P, Zhang H Y and Tong M B 2017 Design and Optimization in Multiphase Homing Trajectory of Parafoil System *Aeronautical Computing Technique* **47(6)** 0055-59

[5] Xiong J 2004 Optimal Design in Multiphase Trajectory of Parafoil System *Spacecraft Recovery & Remote Sensing* **25(3)** 0011-16

[6] Jiao L C, Yang S Y, Liu F, Wang S G and Feng Z X 2016 Seventy years Beyond Neural Networks: Retrospect and Prospect *Chinese Journal of Computers* **39(8)** 01697-01716

[7] Chen Z L, Zhang H, Qiu J G and Su L J 2018 Adaptive Backstepping Altitude Control for Unmanned Powered Parafoil *Control Engineering of China* **25(4)** 0554-58

[8] Zhang H, Chen Z L and Qi X H 2012 Unmanned Powered Parachute Aircraft Control Based on RBF Neural Network PID *Computer Technology and Development* **22(2)** 0206-08

[9] Li Y N, Li T L, Jiang Y and Fan J L 2016 Adaptive PID Control of Quadrotor Based on RBF Neural Network *Control Engineering of China* **23(3)** 0378-82

[10] Yang Y, Hu H D 2014 Simulink Simulation of BP Neural Network PID Controller Based on S-Function *Electronic Design Engineering* **22(4)** 0029-31

[11] Ma M, Zhang Y, Zhou M M and Li X J 2015 Study of Single Neuron PID Adaptive Flight Controller for Quadrotor *Computer Measurement & Control* **23(8)** 2723-25

[12] Shu H L 2000 The Combination of PID Control with Neural Network and the PID Neural Network Controller Used in Nonlinear Systems In: Proc. of the 19th Chinese Control Conf., 6-8 Dec 2000 Hong Kong China

[13] Chen Y M, He Y L, Kong L B and Zhou M F 2014 Decentralized PID Neural Network Control for a Quadrotor Helicopter *Journal of Chinese Inertial Technology* **22(2)** 0185-90

[14] Barrows T M 2013 Apparent Mass of Parafoil with Spanwise Camber *Journal of Aircraft* **39(3)** 445-451

[15] Shu H L 2006 *PID Neural Network and Control System* (Beijing: National Defense Industry Press) p 6-7

Effect of Pendulum Parameters on Performance of Electromagnetic Energy Harvester

Hongyan Wang[1,a], Fengjuan Miao[2], Liying Zou[2] and Tao Zhou[1]

[1] College of Mechatronic Engineering, Qiqihar University, Qiqihar 161006, China
[2] College of Communications and Electronics Engineering, Qiqihar University, Qiqihar 161006, China

[a] wanghongyan1993@163.com

Abstract. This paper presents an analysis of a vibration-based electromagnetic energy harvester with an attached pendulum. The pendulum is used to transform the horizontal harmonic motion of the complete system into the vertical motion of the magnetic oscillator. An electromechanical model of the coupled system is established for numerical simulation. Parametric studies are conducted subsequently on various pendulum mass and length configurations. The results indicate that the system with the natural frequency ratio of 1:2 has higher power outputs and wider frequency bandwidth as compared with other configurations. In the situation of natural frequency ratio of 1:2, higher power outputs can be achieved by an increase in the mass of the pendulum. The results demonstrate the feasibility of using the pendulum as a driver of vertical motion oscillator to harvest vibration energy from the horizontal excitation.

1. Introduction

In recent years, vibration energy harvesting has aroused increasing research interest due to its potential application as an alternative energy source for energizing low-power wireless electronic devices. Conventional vibration energy harvesters are designed as one-degree-of-freedom (1-DOF) oscillators with electrostatic, piezoelectric or electromagnetic transducers, which are efficient only when the motions of the oscillators are aligned with the direction of the excitation [1-4]. In reality, however, ambient excitations are often multi-directional, which demands the vibration energy harvesters to work in different excitation directions [5,6].

This paper proposes a type of vibration-based energy harvester combining an electromagnetic energy harvesting oscillator and a pendulum. The attached pendulum is used to transform the horizontal harmonic motion of the complete system into the vertical motion of the energy harvester. An electromechanical model of the coupled system is established. The dynamic behaviours of the system with various pendulum parameters configurations are investigated under horizontal harmonic excitation. The optimal power outputs versus the excitation frequencies and the load resistance for different pendulum parameters configurations are compared in the vicinity of auto-parametric region (1:2 resonance).

2. Electromechanical modeling

Figure 1 shows an electromagnetic energy harvester with an attached pendulum. The complete energy harvesting system consists of a 1-DOF oscillator, an attached pendulum and a fixed wound coil. The

1-DOF oscillator comprises magnetic mass m_1, spring k_1, and damper c_1. The pendulum has length l and mass m_2. As the complete system is excited by the horizontal harmonic motion, $a_x\cos\omega t$, the pendulum can swing around the pivot O with the pendulum angle θ, which induces the vertical motion of the magnetic mass m_1 in and out of the coil. Here, a_x is the excitation acceleration and ω is the excitation angular frequency. The current induced in the wire of the coil can provide power for the load resistance R.

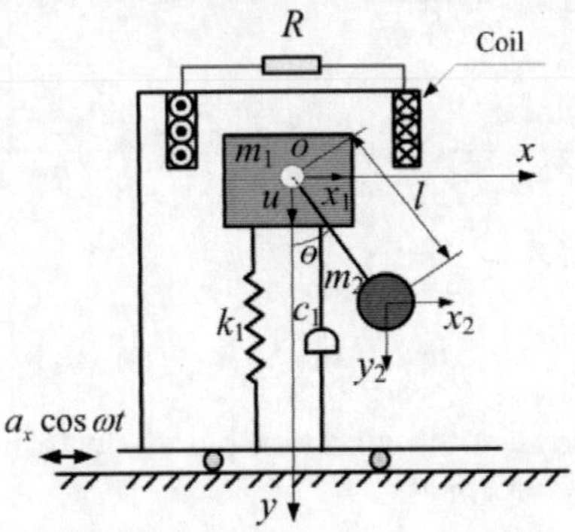

Figure 1. Electromagnetic energy harvester with an attached pendulum.

The position of the mass m_2 (x_2, y_2) depends on m_1 (x_1, u), which can be written as

$$x_2 = x_1 + l\sin\theta,\quad y_2 = u + l\cos\theta \tag{1}$$

where x_1 and u are the horizontal and vertical displacements of the mass m_1 relative to its original static equilibrium position, respectively.

The governing equations of the vibration-based electromagnetic energy harvester with an attached pendulum can be written as

$$(m_1 + m_2)\ddot{u} + c_1\dot{u} + k_1 u - m_2 l\dot{\theta}^2\cos\theta - m_2 l\ddot{\theta}\sin\theta + \theta_e I_R = 0 \tag{2}$$

$$m_2 l^2\ddot{\theta} + c_\theta\dot{\theta} - m_2 l\ddot{u}\sin\theta + m_2 g l\sin\theta + m_2 l a_x\cos\omega t\cos\theta = 0 \tag{3}$$

$$I_R = \theta_e\dot{u}/(R_{coil} + R) \tag{4}$$

where $(\bullet)$=d/dt (the overdot denotes differentiation with respect to time t). g is the acceleration due to the gravity. c_θ is the damping coefficient of the pendulum. θ_e is the electromagnetic coupling coefficient. I_R is the current generated in the coil. R_{coil} is the internal resistance of the coil. Assuming that the swinging of the pendulum is small, we can use Taylor series to expand the nonlinear functions $\sin\theta$ and $\cos\theta$ around the steady state for $\theta\approx 0$. Neglecting the terms of θ higher than the third power, we get

$$\sin\theta = \theta - \frac{\theta^3}{3!},\quad \cos\theta = 1 - \frac{\theta^2}{2!} \tag{5}$$

Substituting the terms given in equations (4) and (5) into equations (2) and (3), respectively, and rearranging them, we get two simplified equations as

$$\ddot{u} + 2(\zeta_1 + \zeta_e)\omega_1\dot{u} + \omega_1^2 u - rl\dot{\theta}^2\left(1 - \frac{\theta^2}{2}\right) - rl\ddot{\theta}\left(\theta - \frac{\theta^3}{6}\right) = 0 \tag{6}$$

$$l\ddot{\theta} + 2\zeta_\theta\omega_\theta l\dot{\theta} + (\omega_\theta^2 l - \ddot{u})\left(\theta - \frac{\theta^3}{6}\right) + \left(1 - \frac{\theta^2}{2}\right)a_x\cos\omega t = 0 \tag{7}$$

where ω_1 and ω_θ are the natural angular frequencies of the system and the pendulum, respectively. r is the mass ratio of m_1 to the sum of m_1 and m_2. ζ_1 and ζ_θ are the damping coefficients of the system and the pendulum, respectively. ζ_e is the electromagnetic damping coefficient. These parameters are given by

$$\omega_1 = \left(\frac{k_1}{m_1+m_2}\right)^{1/2},\ \omega_\theta = \left(\frac{g}{l}\right)^{1/2},\ r = \frac{m_2}{m_1+m_2} \tag{8}$$

$$\zeta_1 = \frac{c_1}{2(m_1+m_2)\omega_1},\ \zeta_\theta = \frac{c_\theta}{2m_2 l^2\omega_\theta},\ \zeta_e = \frac{\theta_e^2}{2(R_{\text{coil}}+R)(m_1+m_2)\omega_1} \tag{9}$$

Defining the state space vector as

$$\begin{bmatrix} u & \dot{u} & \theta & \dot{\theta}\end{bmatrix}^T = \begin{bmatrix} q_1 & q_2 & q_3 & q_4\end{bmatrix}^T \tag{10}$$

we can write the governing equations of the vibration-based electromagnetic energy harvester with an attached pendulum in the state space form as

$$\begin{bmatrix}\dot{u}\\ \ddot{u}\\ \dot{\theta}\\ \ddot{\theta}\end{bmatrix} = \begin{bmatrix}\dot{q}_1\\ \dot{q}_2\\ \dot{q}_3\\ \dot{q}_4\end{bmatrix} = \begin{bmatrix} q_2 \\ -2(\zeta_1+\zeta_e)\omega_1 q_2 - \omega_1^2 q_1 + rl\left(1-\frac{q_3^2}{2}\right)q_4^2 + rl\left(q_3 - \frac{q_3^3}{6}\right)\dot{q}_4 \\ q_4 \\ -2\zeta_\theta\omega_\theta q_4 - \left(\omega_\theta^2 - \frac{\dot{q}_2}{l}\right)\left(q_3 - \frac{q_3^3}{6}\right) - \frac{1}{l}\left(1-\frac{q_3^2}{2}\right)a_x\cos\omega t \end{bmatrix} \tag{11}$$

The average power delivered to the load resistor R is defined as

$$P = I^2 R \tag{12}$$

where I is the root mean square (RMS) current value for the current output I_R in the steady state.

3. Parametric study

In this section, a numerical calculation is performed to evaluate the effect of the pendulum parameters on the performance of the system. The parameters used to analyze the performance of the system are m_1=0.05 kg, k_1=130 Nm^{-1}, ζ_1=0.002, ζ_θ=0.01, g=9.8 m/s^2, θ_e=5 Tm, R_{coil}=100 ohm, and a_x=0.1 m/s^2. The excitation frequency is swept upward. In this research, we use the normalized natural frequency ratio $\alpha=1:(\omega_1(\omega_\theta)^{-1})$ to show the parameter variation. In section 3.2, the excitation angular frequency ω is normalized and represented as $\Omega=\omega(\omega_\theta)^{-1}$.

3.1. dynamics analysis

Figures 2(a) and 2(b) show the frequency response curves from θ and $\dot{u}$ for the electromagnetic energy harvesting systems with various pendulum mass configurations (m_2=0.083 kg, 0.25 kg and 1.15 kg), respectively. The coil is short-circuited. The length of the pendulum l=90.5 mm, which has ω_θ=10.4 rad/s. For m_2=0.083 kg, 0.25 kg and 1.15 kg, the natural angular frequencies of the systems are ω_1=31.2 rad/s, 20.8 rad/s, and 10.4 rad/s, respectively. Thus, we get the normalized natural frequency ratios α=1:3, 1:2 and 1:1 for the three different pendulum mass configurations. It is observed from figures 2(a) and 2(b) that with the increase of m_2, the shapes of the frequency response curves for θ and $\dot{u}$ can be converted from the soft spring type (m_2=0.083 kg, α=1:3) to both the soft and hard spring types (m_2=0.25 kg, α=1:2) and further to the hard spring type (m_2=1.15 kg, α=1:1). Especially, with the configuration of m_2=0.25 kg and α=1:2, the bifurcation appears in the vicinity of the excitation frequency ω=10.4 rad/s and the pendulum angle is reduced obviously. In figure 2(b), the configuration of m_2=0.25 kg and α=1:2 gives maximal $\dot{u}$ in the vicinity of ω=10.4 rad/s as compared with the two other configurations. It can be seen from figures 2(a) and 2(b) that larger θ and $\dot{u}$ for various m_2 always appear in the vicinity of ω_0. It is implied that the efficient excitation frequency for larger outputs depends on the length of the pendulum.

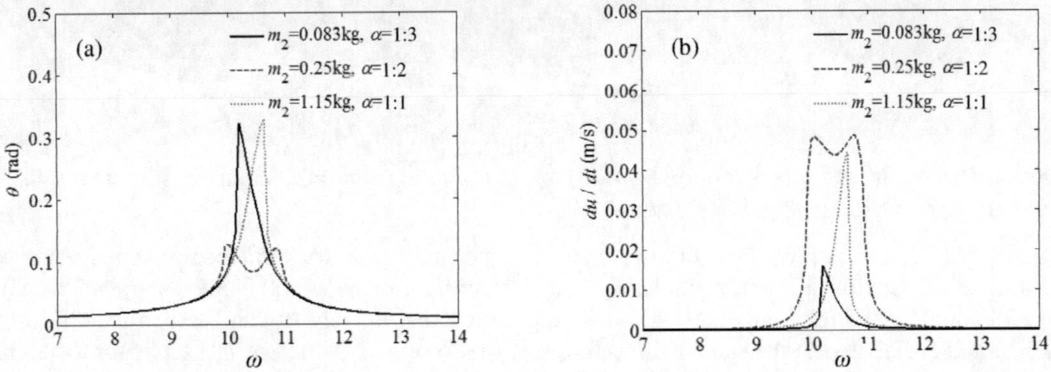

Figure 2. (a) Pendulum angle θ and (b) velocity $\dot{u}$ versus excitation frequencies for various m_2 configurations. l−90.5 mm. R=0.

Figures 3(a) and 3(b) show the frequency response curves from θ and $\dot{u}$ for the coupled system with various pendulum mass and length configurations (m_2=0.083 kg and l=22.6 mm; m_2=0.25 kg and l=90.5 mm; m_2=1.15 kg and l=203.5 mm), respectively. All three configurations achieve the same natural frequency ratio of α=1:2. It can be seen from figures 3(a) and 3(b) that with the increase of l, the response curve shifts from the right to the left. The configuration of m_2=1.15 kg and l=203.5 mm shows the maximal θ and $\dot{u}$.

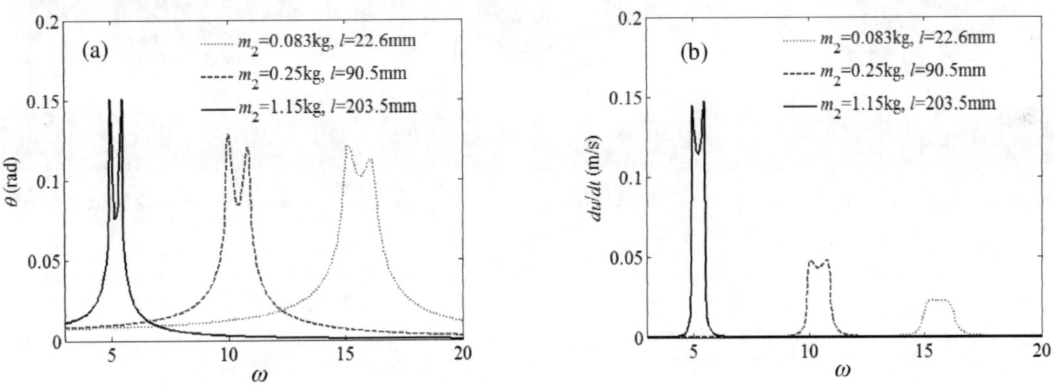

Figure 3. (a) Pendulum angle θ and (b) velocity $\dot{u}$ versus excitation frequencies for various m_2 and l configurations. α=1:2. R=0.

3.2. Power analysis

Figures 4(a), 4(b) and 4(c) show the power outputs per unit mass versus R and Ω for the same pendulum length (l=90.5 mm) with different pendulum mass, m_2=0.083 kg, 0.25 kg and 1.15 kg, respectively. It can be seen from figures 4(a), 4(b) and 4(c) that the electromagnetic energy harvesting system with the configuration of m_2=0.25 kg and α=1:2 has higher power outputs and wider frequency bandwidth as compared with other configurations.

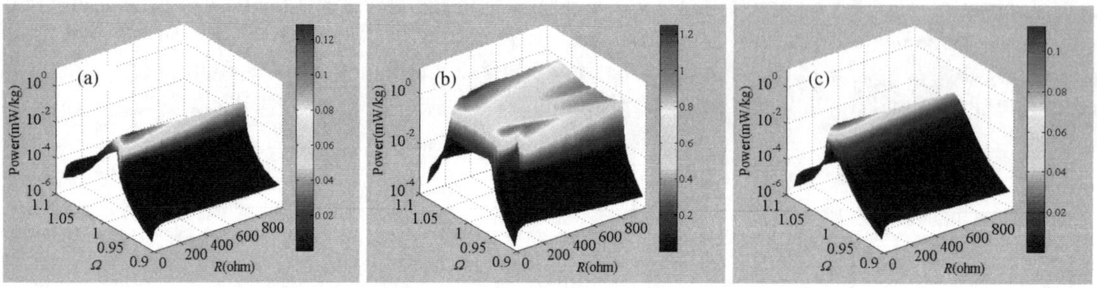

Figure 4. Power outputs versus R and Ω for different m_2 and α. (a) m_2=0.083 kg, α=1:3; (b) m_2=0.25 kg, α=1:2; (c) m_2=1.15 kg, α=1:1. l=90.5 mm.

Figures 5(a), 5(b) and 5(c) show the power outputs per unit mass for the electromagnetic energy harvesting systems with different pendulum parameter configurations (m_2=0.083 kg and l=22.6 mm; m_2=0.25 kg and l=90.5 mm; m_2=1.15 kg and l=203.5 mm), respectively. It can be seen from figures 5(a), 5(b) and 5(c) that with the same natural frequency ratio of α=1:2, the power graphs for different pendulum parameters configurations show the similar trends of double-peak responses for each constant-load resistance. Figures 6(a) and 6(b) plot the envelops of the maximal power outputs versus R and Ω based on the analytical results shown in figures 5(a), 5(b), and 5(c), respectively. It can be seen from figure 6(a) that the optimal power or near optimal power is achieved with a load resistance of 200 ohm. It can be seen from figure 6(b) that in the situation of the natural frequency ratio of α=1:2, the higher power outputs are achieved by an increase in the mass of the pendulum.

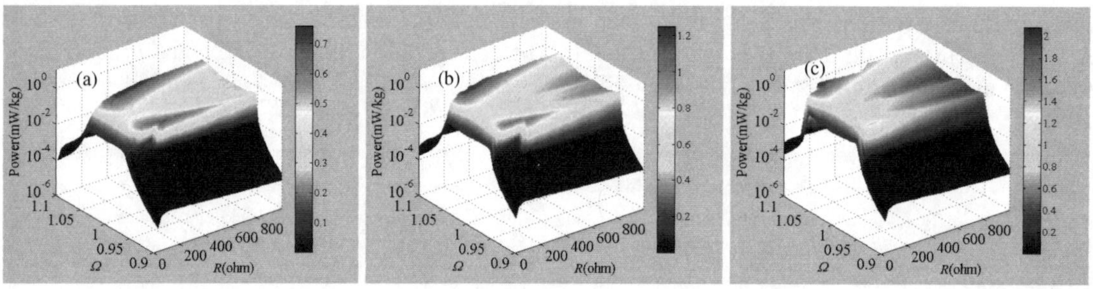

Figure 5. Power outputs versus R and Ω for different m_2 and same α. (a) m_2=0.083kg, l=22.6mm; (b) m_2=0.25 kg and l=90.5 mm; (c) m_2=1.15kg, l=203.5mm. α=1:2.

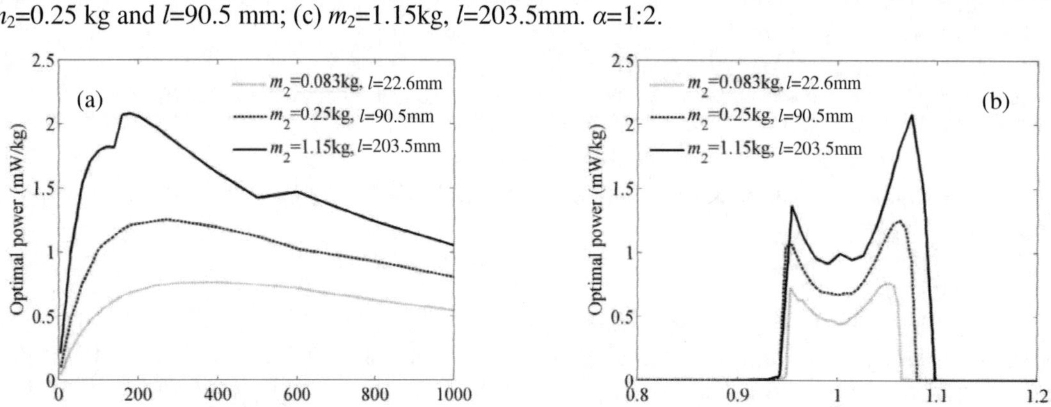

Figure 6. Optimal power versus (a) R and (b) Ω for different pendulum parameters configurations. α=1:2.

4. Conclusions

In this paper, we study the use of the vibration-based electromagnetic energy harvester with an attached pendulum for harvesting vibration energy from the horizontal excitation. An electromechanical model of the energy harvesting system is established. The dynamic performance and power generating capacity for various pendulum parameters configurations are investigated. The results show that the system with the natural frequency ratio of 1:2 has higher power outputs and wider frequency bandwidth as compared with other configurations. In the situation of natural frequency ratio of 1:2, higher power outputs are achieved by an increase in the mass of the pendulum.

Acknowledgments
This work is supported by Heilongjiang Provincial Natural Science Foundation for the Returned Overseas Scholars (No. LC2017028), China and also by the Fundamental Research Funds in Heilongjiang Provincial Universities (No.135209229), China.

References

[1] Tao K, Wu J, Tang L, Hu L, Lye S W and Miao J 2017 *J. Micromech. Microeng.* **27** 044002
[2] Tang L and Yang Y 2012 *J. Intel. Mat. Syst. Str.* **23** 1631-47
[3] Wang H, Tang L, Guo Y, Shan X and Xie T 2014 *J. Zhejiang Univ.-Sci A* **15** 711-22
[4] Challa V R, Prasad M G and Fisher F T 2009 *Smart Mater. Struct.* **18** 095029
[5] Xu J and Tang J 2015 *Appl. Phys. Lett.* **107** 213902
[6] Yang W and Towfighian S 2017 *Smart Mater. Struct.* **26** 095008

Effect of Surface Waviness on Aerostatic Bearing with Speed

Chou Fan[1], Shanghan Gao[1*] and Zhuang Ni[1]

[1]School of Mechanical Engineering, Guangxi University of Science and Technology, Liuzhou, China; 545006

*Corresponding author: Shanghan Gao.Tel: +86 13597026345; E-mail address: gxgshh@163.com

Abstract: In this paper, the Reynolds equation is solved by the different finite methods. On the basis of the ideal circle, the effect of surface ripple on the static pressure bearing is considered, and the performance parameters of the gas bearing are obtained by using the Newton iteration method. Considering the influence of different rotational speeds on the gas bearing, the influence of the bearing capacity, the air consumption and the friction coefficient on the static pressure bearing are analyzed, and the effect of the axial and circumferential waviness on the aerostatic bearing is analyzed. The results show that with the increase of the rotational speed and the error value, the bearing capacity becomes smaller and the friction coefficient becomes larger. In the specific error value, as the film thickness increases, the bearing capacity increases first and then becomes smaller.At the same time, the circumferential ripple error has a greater influence on the gas bearing than the axial direction.

Key words: Aerostatic bearing; Error; Capacity; Stiffness; Surface waviness

1. Introduction

Due to the low friction, high speed and pollution-free characteristics of gas bearings, they have been widely used in the production of precision instruments, such as aerospace, circuit board manufacturing, and missile technology.[1] Due to the relatively high compressibility of the gas,It is easy to cause instability when the gas bearing is running, and since the manufacturing error is unavoidable in the process of producing the gas bearing, it is necessary to carry out the gas static pressure bearing in the presence of an error. However, in the general analysis of the aerostatic bearing process, the influence of the rotational speed on the aerostatic bearing is often neglected. This paper will consider the influence of the aerostatic bearing with the waviness in the case of the rotating speed. Figure 1 is the structural diagram of the aerostatic bearing.

The discretization was derived by the nonlinear dimensionless Reynolds equation Newton method, and the iterative rate cutting method was solved by the improved Reynolds equation[2]. A new iterative method was proposed to solve the Reynolds method, and the method was applied to analyze the aerostatic bearing[3].Zhang[4] used the variable step size iterative method to solve the Reynolds equation, and this method had higher convergence efficiency. The surface error of the aerostatic bearing was analyzed and it was concluded that the large surface error may significantly weaken the bearing stiffness. Liu Wei used numerical analysis of the air hammer phenomenon of the aerostatic bearing[5]. In wang's [6]paper, the static characteristics of corrugated air bearing in the presence of corrugation in the presence of corrugated bearings were analyzed. Zhang Guanghui's influence on the gas static pressure bearing through different speeds and different number of air supply holes. The stability conditions of the aerostatic bearing motion were analyzed and the stability region was obtained.[7, 8]The effects of rotational speed, support stiffness and moment of inertia on the stability of the shafting motion are analyzed.Guo Liangbin analyzed the effects of supply pressure, film thickness and disturbance frequency on the dynamic characteristics of the bearing. Rao Heqing[9] and Sun Yazhou[10]simulated the porous aerostatic bearing and obtained the static performance of the porous bearing. At present, the analysis of gas compressibility in gas aerostatic bearings and changes in the internal flow field ignores the rotational speed. This paper will fully consider the effect of the rotational speed on the aerostatic bearing. The aerostatic bearing will also be analyzed for roundness and cylindricity in different modes of existence.

2. Mathematical model

The theoretical equations of roundness and cylindricity,and the dimensionless waviness error function are as follows:

$$\delta(x,z) = \frac{\left[\left(A\delta_A \sin\frac{\pi d}{\lambda_A}x\right)+\left(B\delta_B \sin\frac{z}{\lambda_B}\right)\right]}{(A+B)} \tag{1}$$

A, B are error coefficients, δ A, δ B are circumferential and axial error values, λ A, λ B are the wavelengths of the circumferential and axial corrugations, when there is roundness and cylindricity A =1, B=1, A=0, B=0 when there is no roundness and cylindricity.

Gas film thickness:

$$h = h_m * (1 + \varepsilon cos\theta) - \delta(x,z) \tag{2}$$

The Reynolds equation for the gas bearing watershed interval is as follows:

$$\frac{\partial}{\partial x}\left(h^3 p\frac{\partial p}{\partial x}\right)+\frac{\partial}{\partial z}\left(h^3 p\frac{\partial p}{\partial z}\right)+12\eta\frac{p_a}{\rho_a}\rho\tilde{v} = 12\eta\frac{\partial(ph)}{\partial t}+6\eta u\frac{\partial}{\partial x}(ph) \tag{3}$$

Due to the assumption that the entire bearing is a rigid body when analyzing the orifice static and dynamic pressure environment, the surface deformation is not considered, and the bearing is in a steady state during the entire movement process, so the bearing is stable throughout the process. State of motion, Derivation of dimensionless and elimination time terms by equation (3), we can get:

$$\frac{\partial}{\partial \bar{x}}\left(\bar{h}^3\frac{\partial \bar{p}^2}{\partial \bar{x}}\right)+\frac{\partial}{\partial \bar{z}}\left(\bar{h}^3\frac{\partial \bar{p}^2}{\partial \bar{z}}\right)+\bar{Q}\delta_K = \Lambda_x\frac{\partial(\bar{p}\bar{h})}{\partial \bar{x}}+\Lambda_z\frac{\partial(\bar{p}\bar{h})}{\partial \bar{z}} \tag{4}$$

Among them

$$\bar{Q} = 24\eta l^2 p_a \rho\tilde{v}/\left(h_m^{\ 3} p_r^{\ 2}\rho_a\right) \tag{5}$$

$$\Lambda_x = 12\eta l V_{x1}/\left(h_m^{\ 2} p_r\right) \tag{6}$$

$$\Lambda_z = 12\eta l V_{z1}/\left(h_m^{\ 2} p_r\right) \tag{7}$$

The dimensionless parameters are as follows:

$$\bar{h} = h/h_m,\quad \bar{p} = p/p_r,\quad \bar{x} = 2x/l,\quad \bar{z} = 2z/l,\quad \dot{m} = Ap_s\phi\sqrt{\frac{2\rho_0}{p_0}}\varphi \tag{8}$$

$$\varphi = \begin{cases} \dfrac{k}{k+1}\left[\left(\dfrac{p}{p_s}\right)^{2/k}-\left(\dfrac{p}{p_s}\right)^{k+1/k}\right]^{1/2} & \dfrac{p}{p_s} > \beta_k \\ \dfrac{k}{2}\left(\dfrac{k}{k+2}\right)^{(k+1)/(k-1)} & \dfrac{p}{p_s} \le \beta_k \end{cases} \tag{9a}$$

$$Q_i = \sum_i^n \bar{Q}_i h_m^{\ 3} p_r^{\ 2}\rho_a/\left(24\eta l^2 p_a\right) \tag{9b}$$

Solve the Reynolds equation and define the boundary conditions:
Atmospheric boundary conditions: p(z=0)=p(z=nz+1)
Symmetrical boundary conditions: p(x=0)=p(nx+1)

Table 1. Aerostatic bearing parameters

Name	Model parameter
Axis length $l(mm)$	30
Diameter $d(mm)$	30
Average radius gap thickness $hm(\mu m)$	20
Axial grid number nz	60
Radial grid number nx	60
Number of orifices in a single row	5
Throttle diameter $d0(mm)$	0.24
Number of orifices N	2
Air density $\rho(kg/m^3)$	1.204
Air viscosity coefficient $\eta(pa\bullet s)$	1.84×10^{-5}
Air specific heat ratio κ	1.404
Throttling coefficient ϕ	0.8
Atmospheric pressure $p_a(atm)$	1

Table 1 shows the structural parameters and lubrication parameters of the aerostatic bearing. Since the Reynolds equation in a compressible fluid is a nonlinear equation, therefore, the analytical solution cannot be directly obtained. Using Taylor's expanded Newton iteration method, It can be found that the Reynolds equation obtains an analytical solution. Then use the segmentation factor method to gradually reduce the pressure difference. The pressure calculation is gradually converge, and the super-relaxation iteration is mainly used as a method of accelerating iteration in the process of solving. Figure 1 shows the schematic diagram of static pressure bearing structure.

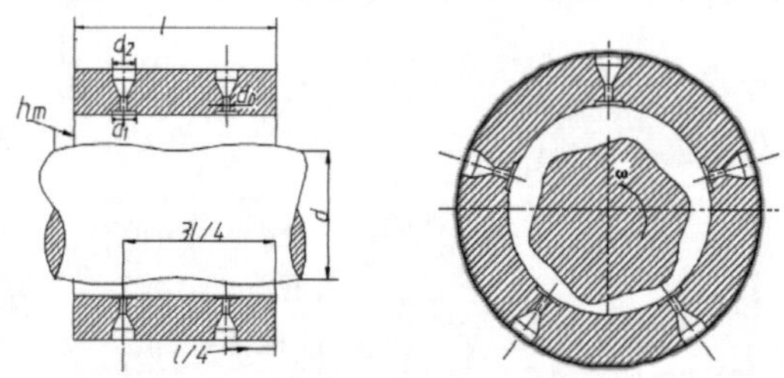

Figure 1. Schematic diagram of static pressure bearing structure

3.Characteristic analysis

The dimensionless bearing capacity is obtained by the static characteristics of the bearing:

$$\begin{cases} \overline{W}_x = \sum_i \sum_j \overline{p}_x \Delta\overline{x}\Delta\overline{z}\sin\left(\dfrac{(i-0.5)\Delta\overline{x}}{R}\right) \\ \overline{W}_y = \sum_i \sum_j \overline{p}_y \Delta\overline{x}\Delta\overline{z}\cos\left(\dfrac{(i-0.5)\Delta\overline{x}}{R}\right) \end{cases} \tag{10a}$$

$$\overline{W} = \sqrt{\overline{W}_x^{\,2} + \overline{W}_y^{\,2}} \tag{10b}$$

The frictional force in the circumferential direction and the axial direction is obtained by the flow resistance of the flow rate of the gas in the gas film section.

$$\begin{cases} f_x = \int_0^l \int_0^{2\pi R} \left(\frac{h_m}{2} \frac{\partial p}{\partial x} + \eta \frac{u_2}{h_m} \right) dxdz \\ f_y = \int_0^l \int_0^{2\pi R} \frac{h_m}{2} \frac{\partial p}{\partial x} dxdz \end{cases} \tag{11}$$

Using the bearing capacity and friction force, the coefficient of friction of the film area is obtained.

$$\mu = \frac{\sqrt{f_x^{\,2} + f_y^{\,2}}}{W} \tag{12}$$

The amount of gas consumed by the entire aerostatic bearing is calculated by using the amount of gas to be dissipated at each eccentricity of each orifice. The amount at its individual orifice:

$$Q_i = \sum_i^n \overline{Q}_i h_m^{\ 3} p_r^{\ 2} \rho_a \big/ \left(24\eta l^2 p_a\right) \tag{13}$$

4.Calculation results analysis

Figure2 is the key parameter: the eccentricity is 0, the rotational speed is 2000 rpm, and the axial and circumferential wavelengths are 10 mm, the dimensionless film thickness at the axial intermediate position and the distribution of dimensionless pressure are obtained.

Figure 2(a) is the circumferential film thickness from 0° to 360°. It can be seen that as the ripple amplitude value increases, the circumferential film thickness increases. Figure 2(b) shows the film thickness distribution in the axial direction. At the same wavelength, as the amplitude of the ripple increases, the axial amplitude also increases.

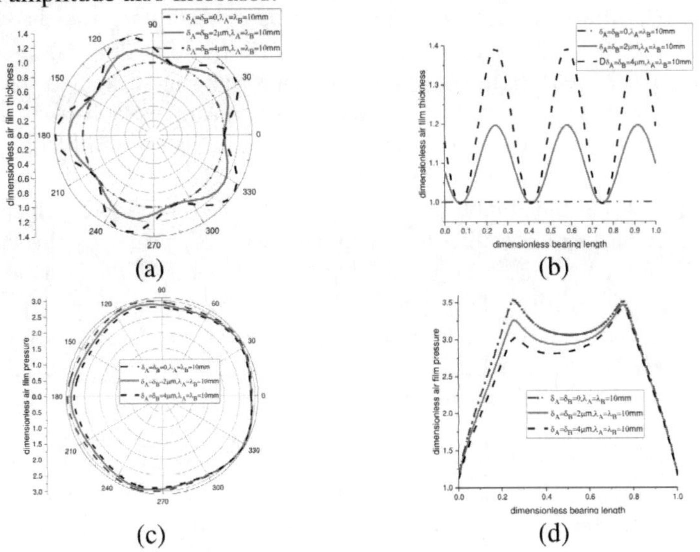

Figure 2. The dimensionless air film thickness and pressure distribution on the axial and circumferential middle section of the bearing(ε=0,n=2000rpm)

Figure 2(c) shows the distribution of the bearing in the circumferential direction. The pressure decreases with the increase of the amplitude value in the circumferential direction, and the pressure is asymmetrical in the axial distribution. This asymmetry may be caused by wavelengths and peaks. The location is related. Figure 2(d) shows the distribution of pressure in the axial direction. The pressure value decreases as the amplitude of the ripple increases, and the pressure distribution under various error conditions is more clearly distinguished.

Figure 3 is a dimensionless film thickness and pressure distribution of the aerostatic bearing with an eccentricity of 0.2, a rotational speed of 2000 rpm, and a wavelength of 5 mm in the axial and circumferential directions.

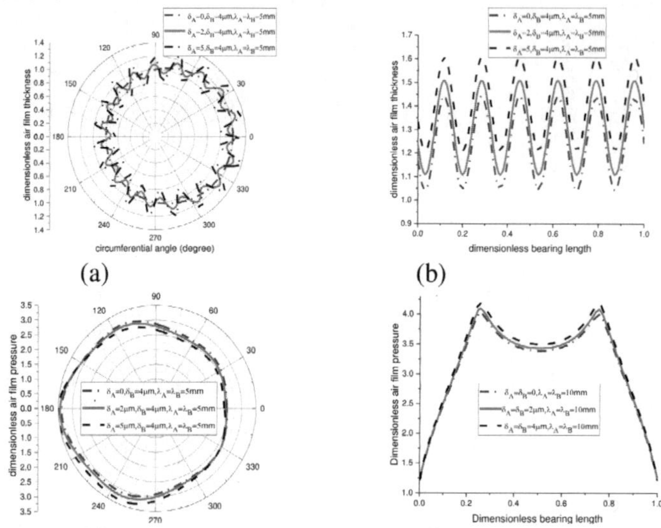

(a) (b)

(c) (d)

Figure 3. The dimensionless air film thickness and pressure distribution on the axial and circumferential middle section of the bearing(ε =0.2,n=2000rpm)

Figure 3(a) is the dimensionless gas film thickness distribution in the circumferential direction. When the circumferential and axial errors are not equal, the film thickness fluctuates back and forth around the ideal circular film thickness, and the amplitude of the fluctuation is The circumferential amplitude increases and becomes larger. Figure 3(b) shows the dimensionless film thickness distribution of the aerostatic bearing in the axial direction. The thickness distribution increases with the increase of the amplitude. Figure 3(c) shows the dimensionless pressure distribution in the circumferential direction. Pressure crossover occurred at approximately 150° and 330°. Figure 3(d) is the axial pressure distribution. The pressure distribution on the left and the right sides is symmetrically distributed, which decreases as the amplitude of the circumferential corrugation increases. At the same time, the pressure is reduced in the middle position between the orifices. In Figure 3, when the circumferential amplitude is continuously increased and the axial amplitude is constant, the pressure is equal in the circumferential direction of 150 to 330, because the rotational speed has a certain shearing force in the circumferential direction. There are extrusion and tensile forces on both sides of the minimum film thickness. Therefore, when considering a aerostatic bearing, the influence of error and rotation speed on the aerostatic bearing is not negligible.

4.1 Influence of circumferential ripple

Figure 4 is an analysis of the bearing capacity, stiffness, total gas consumption and friction coefficient of aerostatic bearings with different circumferential vortex amplitudes at 40,000 rpm under different eccentricities.

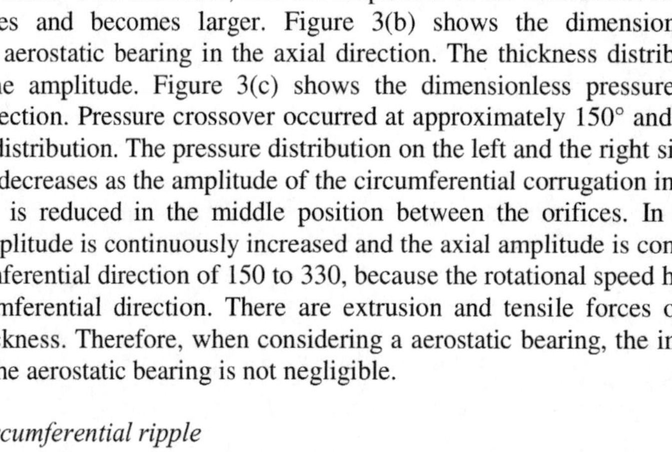

(a) (b)

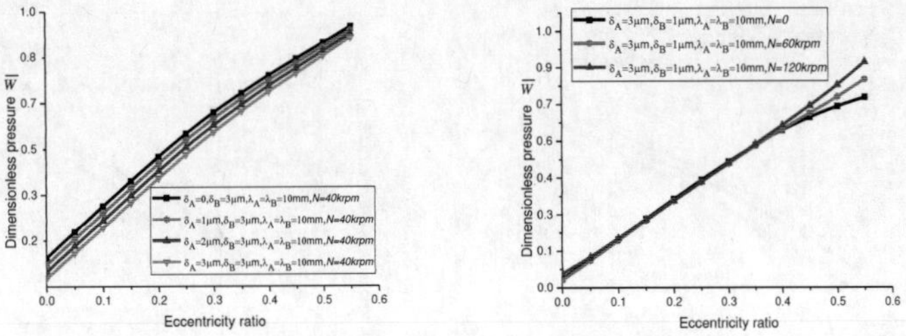

(c) (d)

Figure 4. Different circular amplitude and performance at different speeds in the case of performance changes(n=40000rpm)

Figure 4(a) is for a single row of 5 orifices. The distribution of the total gas consumption flow rate increases with the increase of the circumferential error amplitude. When the eccentricity increases, the total gas consumption decreases gradually. This may be due to the eccentric direction,The thickness of the gas film becomes smaller,Resulting in an increase in flow resistance in the gas flow area,Which leads to less air consumption. and an increase in the space of the gas film away from the eccentric position, which reduces the flow resistance and increases the gas supply amount. However, in the overall air supply amount, the increase in the air supply amount away from the eccentric position is smaller than the air supply amount in the eccentric direction. Figure 4(b) is a distribution diagram of bearing capacity. As the amplitude of the circumferential error increases, the bearing capacity of the bearing is gradually reduced. As the eccentricity increases, the bearing capacity increases. Figure 4(c) is a graph showing the variation of the friction coefficient of different circumferential ripple error amplitudes in the gas bearing film area. As the circumferential ripple error increases, the friction coefficient becomes larger, which indicates that the larger the error. The friction in the bearing area also increases with the frictional theory, which is consistent with the friction theory. Figure 4(d) is a graph showing changes in the bearing capacity of the gas bearing in the circumferential direction when the ripple amplitude is greater than the axial direction ripple amplitude, and the axial and circumferential wavelengths are equal, and the eccentricity changes. In the case of small eccentricity, the higher the bearing capacity is relatively larger, but as the eccentricity increases to the range of 0.35, the bearing capacity of the high and low speeds is almost equal, and as the eccentricity increases, the high speed The bearing capacity in the case will be significantly greater than the bearing capacity at low speeds, indicating that the effect of the speed and error on the aerostatic bearing is not negligible.

Figure 5 is the effect of circumferential wavelength on the aerostatic bearing when there are axial and circumferential ripple error amplitudes. Analysis of the aerostatic bearing using different circumferential wavelengths.

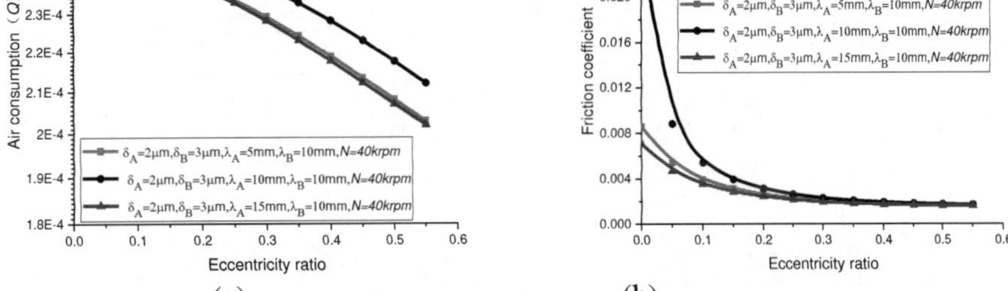

(a) (b)

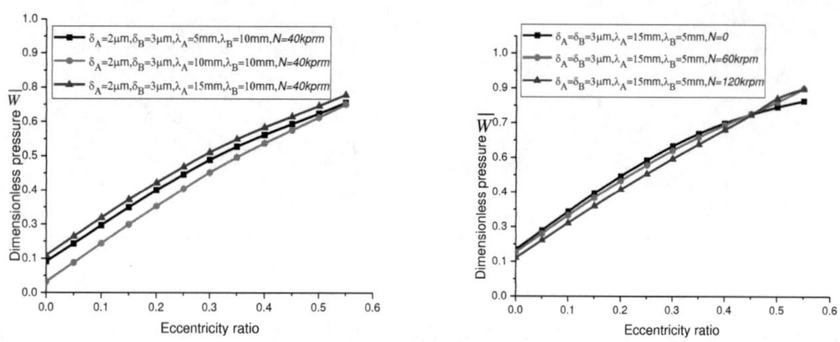

(c) (d)

Figure 5. Different circumferential wavelengths and performance changes at different speeds

Figure 5(a) is an analysis diagram of the total gas consumption of the bearing at three wavelengths. In the process of increasing the wavelength in the circumferential direction, the increase is more obvious when the wavelength is equal to 15 mm, and the change is smaller in 5 mm and 10 mm. There is a point of equal intersection, and the same phenomenon as in Figure 4(a) appears with the increase of the eccentricity. Figure 5(a) shows that the influence of the circumferential wavelength on the aerostatic bearing is also inconsequential. Figure 5(b) shows the bearing capacity becomes larger as the eccentricity increases. In Figure 5(c), the circumferential wavelength change has a great influence on the friction coefficient of the gas bearing, and the change between 0-15 mm is the first drop. Post-rise phenomenon, this phenomenon shows that the wavelength has a greater influence on the flow resistance in the gas region, Figure 5(d) is a graph showing changes in bearing capacity of different rotational speeds with eccentricity in the case where the amplitudes of the axial and circumferential corrugations are the same and the circumferential wavelength is greater than the axial wavelength.It can be seen from the figure that the low-speed gas bearing has a larger bearing capacity than the high-speed gas bearing in the case of a small eccentricity, but the bearing capacity of the gas bearing increases with the eccentricity increasing to 0.45. As the speed increases, it grows larger. It can be illustrated from Figure 5 that the wavelength of the circumferential corrugation is also important for the effect of the gas bearing on various performance parameters. Larger wavelengths have a greater impact on aerostatic bearings.

4.2 Influence of axial ripple

Figure 6 is the influence of the axial ripple error on the parameters of the aerostatic bearing.

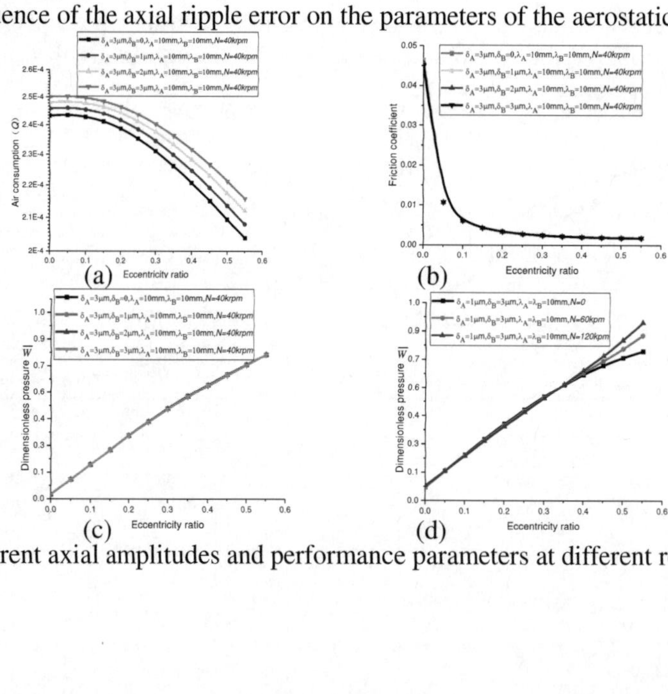

(a) (b) (c) (d)

Figure 6. Different axial amplitudes and performance parameters at different rotational speeds

IOP Conf. Series: Materials Science and Engineering **470** (2019) 012010 doi:10.1088/1757-899X/470/1/012010

Figure 6(a) shows the variation of the air consumption of different corrugation errors with the change of the eccentricity. The same variation and different circumferential ripple error with the eccentricity The change is the same, as shown in Figure 4(b).Figure 6(b) is the case where the different axial ripple errors change with the eccentricity. Figure 6(c) shows the variation of the bearing capacity of different axial ripple errors as the eccentricity changes. Figure 6(d) shows that the axial ripple amplitude is greater than the circumferential ripple amplitude.The change of the influence of the eccentricity on the bearing capacity of the gas bearing with the same axial and circumferential wavelengths and different rotational speeds. It can be seen that in the case of small eccentricity, the bearing capacity is almost the same, but in the case where the eccentricity is greater than 0.4, the influence of the rotational speed on the bearing capacity is not negligible, and the change is the same as the change in Figure 4(d). Increasingly, the bearing capacity becomes significantly larger. It can be seen from figure 6 that the variation of the axial ripple error value has little effect on the friction coefficient and bearing capacity of the gas bearing with the eccentricity change, but in the case of large eccentricity and different rotational speeds. The change of the bearing capacity is relatively obvious.

Figure 7 shows the influence of the axial ripple error on the performance parameters of the aerostatic bearing as the wavelength changes.

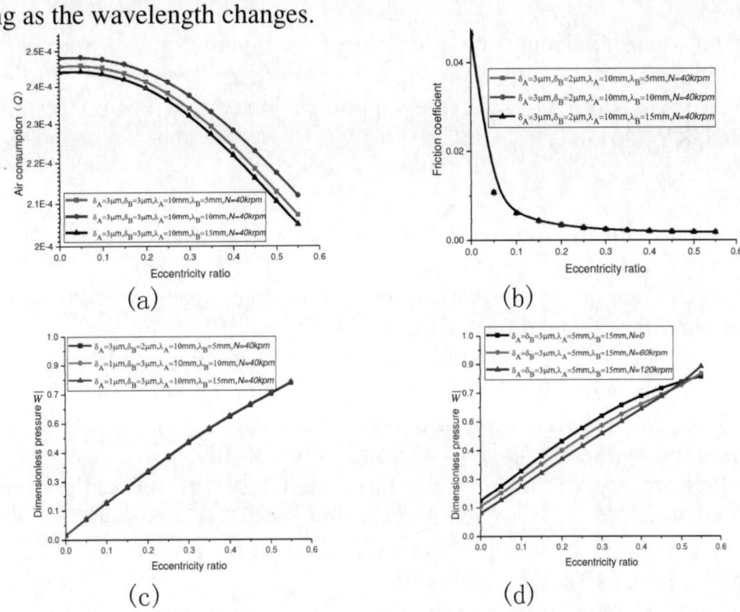

Figure 7. Different axial wavelengths and changes in performance parameters at different rotational speeds

Figure 7(a) shows the air consumption as a function of the eccentricity. Figure 7(b) shows the friction coefficient. In the case of eccentricity change, Figure 7(c) shows the variation of the bearing capacity of different axial wavelengths with the eccentricity. Figure 7(d) The axial wavelength is greater than the circumferential wavelength, and the different rotational speeds vary with the eccentricity. The curve of change is consistent with the distribution trend and trend of load capacity of Figure 5(d), indicating that in the case of a certain wavelength, the lower the rotation speed, the higher the bearing capacity, when a certain eccentricity is reached. In the case, the higher the rotational speed, the greater the bearing capacity. It can be seen from Figure 7 that the axial change in the axial direction can be found to have a small change in the performance parameters of the gas aerostatic bearing.

5.Conclusions

The numerical simulation of the orifice type static pressure gas bearing is carried out by using the nonlinear Reynolds equation coupled gas flow coefficient, and the characteristics of the gas aerostatic bearing are analyzed by the ripple error in the analysis. Through the calculation results, the conclusions of gas consumption, friction coefficient and bearing capacity change are obtained.

1. The increase of the circumferential ripple amplitude will lead to an increase in air consumption and friction coefficient, and the bearing capacity will gradually become smaller; for this reason, the circumferential ripple amplitude has a greater influence on the static pressure gas bearing.
2. The amplitude of circumferential ripple amplitude has a great influence on the static pressure gas bearing, but the variation law of the influence is not obvious.
3. Compared with the effect of circumferential corrugation on aerostatic bearings, the influence of axial ripple error is small.
4. As the eccentricity increases, the bearing capacity of the bearing increases continuously. The coefficient of friction in the entire film area gradually becomes 0 as the eccentricity increases, and the gas consumption gradually decreases as the eccentricity increases.
5. It is concluded that the magnitude of the axial ripple error has little effect on the performance of the gas bearing, and the axial ripple error can be appropriately ignored in the manufacturing.

Acknowledgments
The authors gratefully acknowledge the support of National Nature Science Foundation of China(Grant No.11562002), the support of the Science and Technology Project of Guangxi Province(Grant No.2015GXNSFAA139262),the support of Innovation Project of Guangxi Graduate Education(Grant No.GKYC201704).

References

[1] Jia C, Yang W and Qiu M. Research and development of gas bearings. 2014*The 7th China Bearing Forum Proceedings*(Luo Yang).

[2] Lo C,Wang C and Lee Y. Performance analysis of high-speed spindle aerostatic bearings. *Tribology International* 38(1)(2005), 5-14.

[3] Li Y, Zhou K and Zhang Z. A flow-difference feedback iteration method and its application to high-speed aerostatic journal bearings. *Tribology International* 84(2015), 132-141.

[4] Zheng S and Jiang S. Improved finite difference method for pressure distribution of aerostatic bearing. *Journal of Southeast University* 25(4)(2009), 501-505.

[5] N Bhat etc. Performance of inherently compensated flat pad aerostatic bearings subject to dynamic perturbation forces. *Precision Engineering* 36(3)(2012), 399-407.

[6] Wang X etc. Effect of surface waviness on the static performance of aerostatic journal bearings. *Tribology International* 103(2016), 394-405.

[7] Li S, Liu W and Li Q. Stability Analysis of Precision Centrifugal Static Pressure Gas Bearing Spindle System. *China Mechanical Engineering* 14(17)(2003), 1451-1453.

[8] Li S, Liu W and Zhang P. Effect of geometric error on the motion accuracy of gas static cylindrical bearings. *Lubrication and sealing* 03(2001), 45-48.

[9] Rao H. Simulation and Experimental Study of Porous Hydrostatic Bearings Based on FLUENT Software. *Harbin Institute of Technology* (2006), 70.

[10] Sun Y, Lu Z and Rao H. Simulation and experimental study on static characteristics of porous hydrostatic bearings based on FLUENT software. *Machine Tool & Hydraulics* 03(2007), 170-172.

IOP Conf. Series: Materials Science and Engineering **470** (2019) 012011 doi:10.1088/1757-899X/470/1/012011

Failure Strain Measurement of the Piping System of a Nuclear Power Plant Using Image Signals

S W Kim[1*], B G Jeon[1], D U Park[1] and D W Yun[1]

[1]Seismic Research and Test Center, Pusan National University, 49 Busandaehak-ro, Mulgeum, Yangsan, Kyungnam 50612, Republic of Korea

*corresponding author, Email: swkim09@pusan.ac.kr

Abstract. When seismic isolation devices are installed in piping systems, it is expected that such devices will be subjected to larger displacement than before their installation because they handle seismic loads, and it is likely that the seismic risks of some facilities will be increased by such displacement increase. In particular, it is highly likely that the seismic risks of piping systems, which connect seismic-isolated structures with conventional ones, will be increased. Previous studies have revealed that the vulnerable part of piping systems is the elbow, and that failures occur due to the low-cycle ratcheting fatigue in the hoop direction. Therefore, in this study, the ultimate state of the elbow was defined as a leakage, and an in-plane cyclic loading test was conducted. In addition, the hoop-direction strain of the steel pipe elbow was measured using an image measurement system, and the failure strain, which cannot be measured using the conventional sensors, was also measured.

1. Introduction

Seismic-isolated structures are likely to be subjected to larger displacement than conventional structures because seismic isolation devices deal with seismic loads. Therefore, in the case of piping systems that connect seismic-isolated structures with conventional ones, piping systems can be damaged due to plastic deformation concentrated on specific areas, such as the elbow and tee. As such, it is necessary to examine the seismic safety of seismic-isolated piping systems.

Studies to identify the vulnerable part of piping systems under seismic loads and to analyze the nonlinear behavior of such systems have been conducted using experimental and numerical methods, according to the development of the related industries. The dynamic behavior of piping systems with different materials under seismic loads was analyzed using a shaking table [1] particular, special attention was paid to the behavior of the pipe elbow. As a result, it was confirmed that plastic deformation occurred in each pipe elbow. Japan Nuclear Energy Safety Organization (JNES) and Nuclear Power Engineering Corporation (NUPEC) conducted an experimental study [2] to analyze the dynamic behavior of the representative piping systems of nuclear power plants under seismic loads. They conducted a cyclic loading test with piping systems and a shaking table test with the piping systems. They also performed tests to examine the design method and the limit state. As a result, it was found that the damage type of the piping systems under seismic loads was the fatigue failure at low frequencies [3]. In addition, the entire system was assessed to be safer compared to the results of the component test due to reasons such as the redistribution of loads [4]. A low-cycle ratcheting fatigue test [5] was conducted using a scaled model of the elbow, which previous studies have revealed to be the vulnerable part of piping systems. In the component test, it was observed that the elbow crack started from the inside and grew in the axial direction. The hoop-direction strain and the low-cycle ratcheting fatigue damage were also observed. As mentioned earlier, previous studies have revealed that the vulnerable part of piping systems is the elbow, and it was also revealed that failures occur due to the low-cycle ratcheting fatigue, but it is difficult to quantitatively define the limit state [6] of the elbow. As such, this study proposed a method of measuring the hoop-direction ratcheting strain using image signals to define a reliable limit state of the elbow. Towards this end, the ultimate state of the elbow was defined as a leakage, and an in-plane cyclic loading test was conducted.

2. Strain measurement using image signals

The image of an object acquired using the image measurement system was deformed by an external force, and the correlation between the reference and the deformed images was analyzed using the gray level values of the images. The deformation was measured by analyzing such correlation after separating the square-shaped images from the two images. In this case, the separated square-shaped area is called a "window," and the deformation was measured by analyzing the correlation between the window images separated by the gray level. Figure 1 shows the deformation measurement principle of the image correlation method. $f(x_i, y_i)$ is the reference window, which is the gray level pattern of the square area separated from the reference image before deformation, and $g(x_i', y_i')$ is the deformed window, which is the gray level pattern of the square separated from the deformed image. The size of the window was $(2M+1)\times(2M+1)$. M denotes the size of the target window. The deformation of a structure can be measured by analyzing the correlation between the reference window and the deformed window, and detecting the coordinates with the highest correlation. In this study, the zero-normalized sum of squared differences (ZNSSD) method of equation (1) was used to analyze the correlation between the two images.

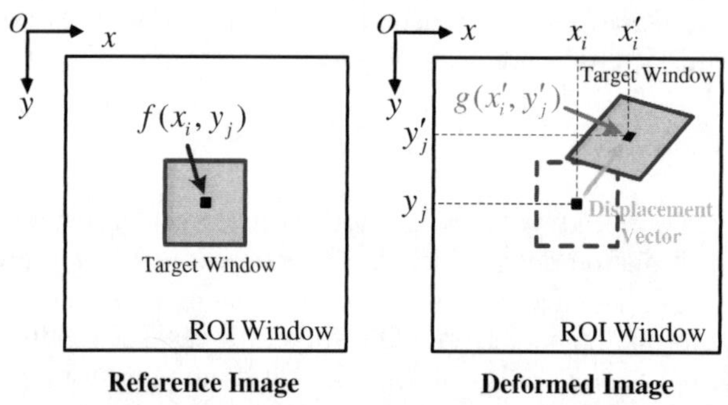

Figure 1. Deformation measurement using the image correlation method

$$C_{ZNSSD} = \sum_{i=-M}^{M} \sum_{j=-M}^{M} \left[\frac{f(x,y)-f_m}{\Delta f} - \frac{g(x',y')-g_m}{\Delta g} \right] \quad (1)$$

$$f_m = \frac{1}{(2M+1)^2} \sum_{i=-M}^{M} \sum_{j=-M}^{M} f(x_i, y_j), \quad g_m = \frac{1}{(2M+1)^2} \sum_{i=-M}^{M} \sum_{j=-M}^{M} g(x_i', y_j') \quad (2)$$

$$\Delta f = \sqrt{\sum_{i=-M}^{M} \sum_{j=-M}^{M} \left[f(x_i, y_j) - f_m \right]^2}, \quad \Delta g = \sqrt{\sum_{i=-M}^{M} \sum_{j=-M}^{M} \left[g(x_i', y_j') - g_m \right]^2} \quad (3)$$

As the object deformation caused by an external force occurs continuously, it is possible to estimate the deformation values of the surrounding points. To correct the geometric errors of the surroundings caused by deformation, subpixel was calculated using the shape function.

In equation (4), $\Delta x = x_i - x_0$ and $\Delta y = y_j - y_0$, u and v represent the displacement components of the reference window center in the x and y directions. u_x, u_y, v_x, and v_y denote the first-order displacement gradients of the reference window while u_{xx}, u_{xy}, u_{yy}, v_{xx}, v_{xy}, and v_{yy} represent the second-order displacement gradients. Considering the bending of the object caused by an external force as well as the influence of the nonlinear behavior, subpixel was calculated using the second-order shape function of equation (4).

$$\xi_2(x_i, y_j) = u + u_x\Delta x + u_y\Delta y + \frac{1}{2}u_{xx}\Delta x^2 + \frac{1}{2}u_{yy}\Delta y^2 + u_{xy}\Delta y^2 + u_{xy}\Delta x\Delta y$$

$$\eta_2(x_i, y_j) = v + v_x\Delta x + v_y\Delta y + \frac{1}{2}v_{xx}\Delta x^2 + \frac{1}{2}v_{yy}\Delta y^2 + v_{xy}\Delta y^2 + v_{xy}\Delta x\Delta y \quad (4)$$

The average strain measurement method [7] was applied to measure the failure strain of a structure using image signals. As shown in figure 2, an arbitrary area was divided by pixels, and the strain was measured for the divided area. For an arbitrary area, the hoop-direction strain was measured using the average strain. To measure the average strain in an arbitrary area, the slope was calculated first using the following linear function that plotted the hoop-direction relative displacement against the hoop-direction pixel position. The average strain was then measured using the slope calculated from each image over time. The ratcheting strain was calculated by using the curve fitting algorithm. Finally, it shows the contour analysis using the ratcheting strain measured at an arbitrary time.

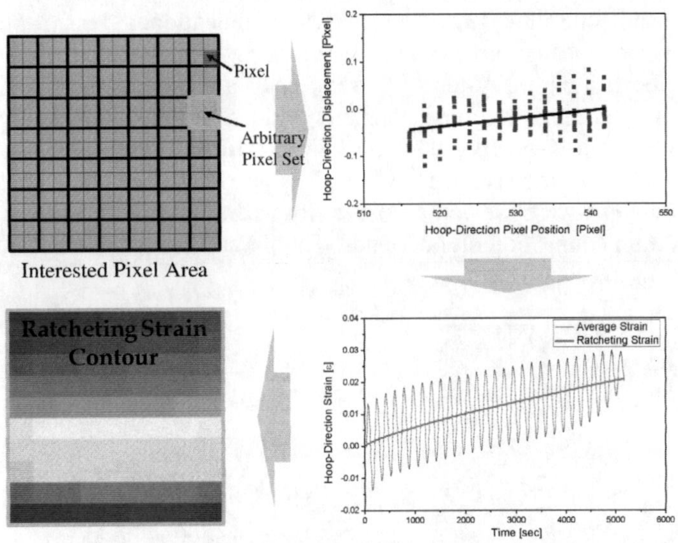

Figure 2. Method for estimating failure strain

Figure 3 shows the average strain measurement algorithm of piping systems using the image signal. The algorithm consists of six steps. In the acquired initial image, control points are designated on the points where the strain is to be estimated. The control points can be designated at various pixel intervals, and it is also possible to measure the strain in the separated area. In the area where the strain of the piping systems is to be estimated, ZNSSD is calculated, and the optimally matching coordinates are also calculated to analyze the correlation between the reference and deformed windows. At the matching coordinates, the subpixel is calculated using the second-order shape function, and the average strain is measured using the analyzed displacement.

In this study, the algorithm for measuring the hoop-direction strain of the piping system described above was utilized by creating an automated software program using MATLAB, as shown in figure 4.

In the figure, at the upper left part is the image acquired using the image measurement system at an arbitrary time, and at the upper right part is the pixel-based relative displacement of the steel pipe elbow. At the lower left part is the slope that was calculated for the upper left image, and at the lower right part is the strain that was measured using the calculated slope.

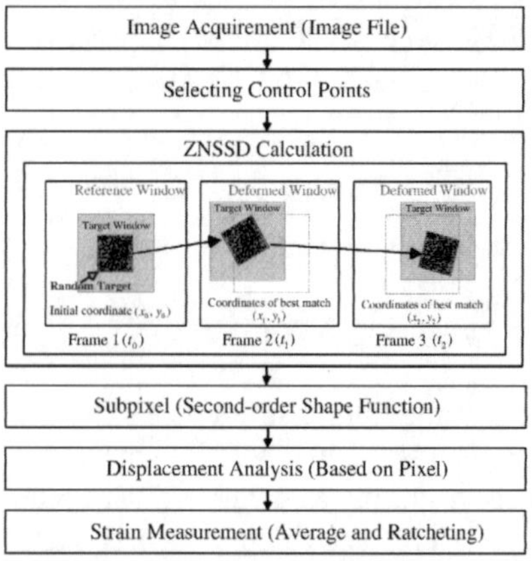

Figure 3. Summary of algorithm

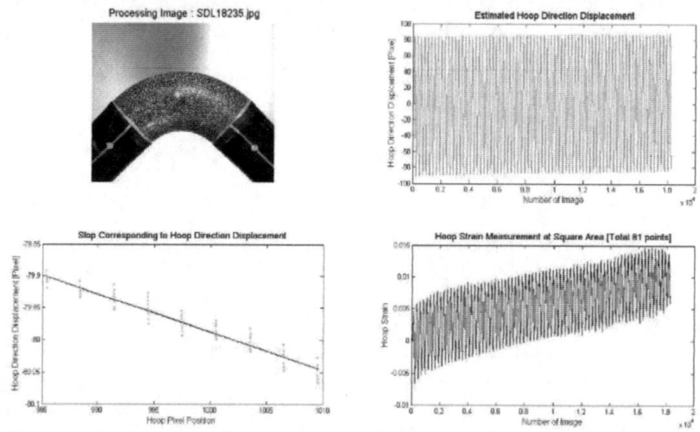

Figure 4. Program for estimating failure strain using MATLAB

3. Failure strain measurement of the steel pipe elbow

The image measurement system used a CMOS camera (IMB-7050G, 2448 x 2048 pixels) and two laptop computers considering portability and easy installation. The CMOS camera and laptop computers transferred and controlled data using LAN communication. A lens (M5018-MP2) was used in the experiment to measure the deformation of the remote piping systems.

In this study, the elbow of piping systems was selected as a vulnerable part in the case of a seismic load, and specimens with an 88.9 mm outer diameter and a 5.49 mm thickness were prepared using SA106 Grade B and SCH 40 (ASME, 2004) of ASME B36.10. To enable plastic behavior in the elbow, the straight length of each specimen was made more than three times longer than the outer diameter, and the specimens were welded onto the elbow. In addition, a jig was manufactured and installed to implement pin connection on both ends. The jig and the pin were manufactured precisely to ensure the accuracy of the experiment.

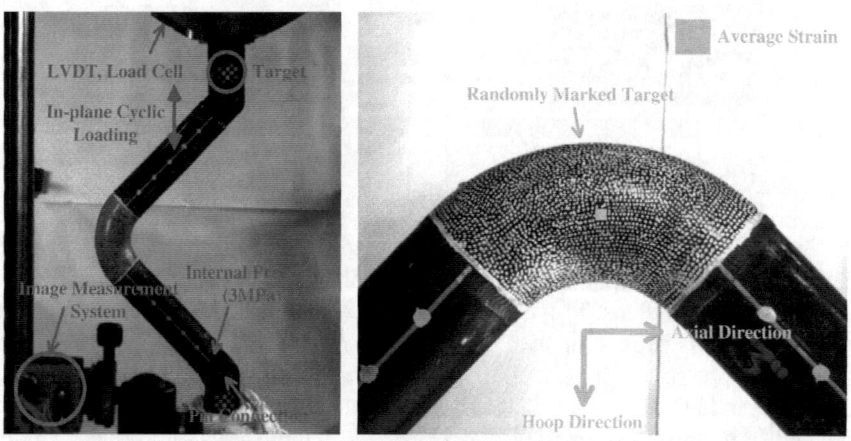

(a) Steel pipe elbow (b) Marked Target

Figure 5. Experimental setup

Figure 5(a) shows the steel pipe elbow installed at the universal testing machine (UTM). An in-plane cyclic loading test was conducted through displacement control, and a target was installed in the jig for the UTM connection, to compare its displacement with that of the linear variable differential transformer (LVDT) installed inside the UTM. Figure 5(b) shows the randomly marked targets that were installed to measure the hoop-direction strain of the piping systems. In the experiment, two 2448x2048 images were acquired at 2 frames per second using the image measurement system. The UTM performed measurement at the 1 Hz data acquisition speed, and the strain gage performed measurement at 0.25 Hz.

To verify the strain measurement algorithm using image signals, the in-plane cyclic loading test was conducted. Table 1 shows the load cases of the steel pipe elbow. The in-plane cyclic loading test was continued until a leakage occurred in the elbow. Figure 6 shows the steel pipe elbow in which a through crack occurred. It was confirmed that the crack occurred near the crown.

Table 1. Load cases

Load Case	Loading Amplitude (mm)	Internal Pressure (MPa)	Leakage Cycles N
1	±20		96
2	±40	3	15
3	±60		6

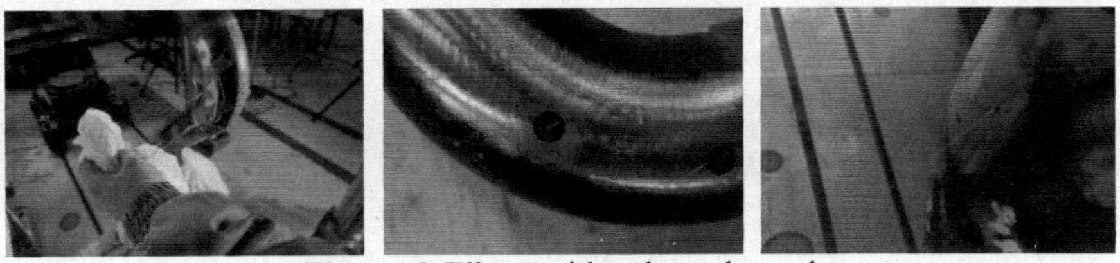

Figure 6. Elbow with a through crack

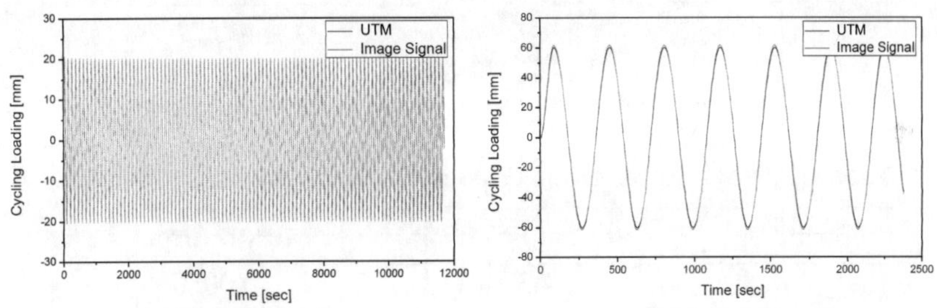

(a) Load case 1 (b) Load case 3

Figure 7. Comparison of the response measured at the LVDT of the UTM

Table 2. Error analysis of the response measured at the LVDT of the UTM

Load Case	Percent Error (%)	RMS Error (mm)
1	0.085	0.831
2	0.076	0.774
3	0.053	0.687

Figure 7 is a graph that compares the response measured by the LVDT of the UTM with those measured using image signals in the in-plane cyclic loading test. In table 2, the error rate of the percent error in each load case was within 1 %, and the RMS error was less than 0.5 mm, indicating that the error was very small. Therefore, it was found that the displacement response estimated using image signals exhibited reliability.

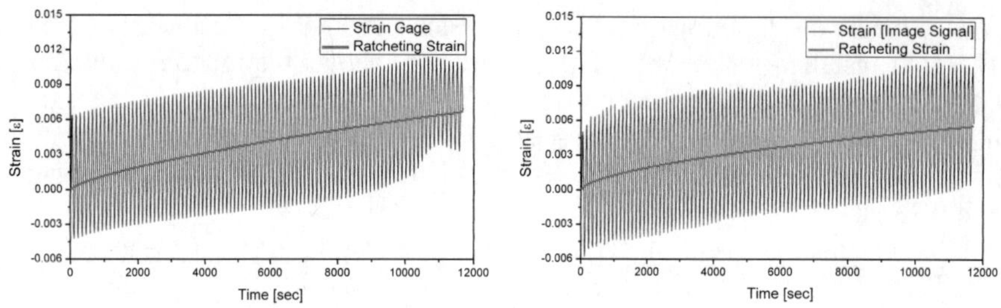

(a) Load case 1

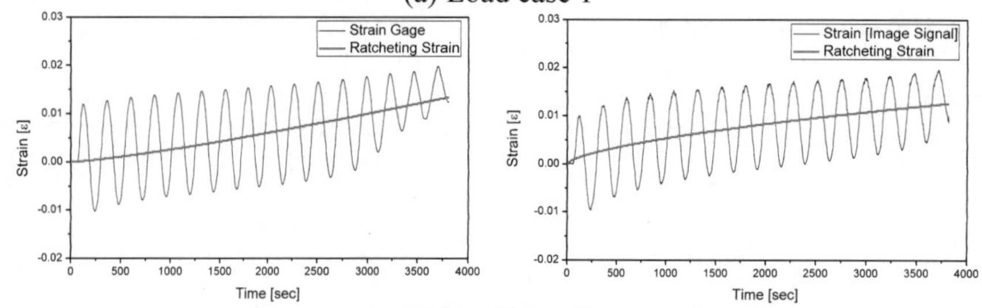

(b) Load case 2

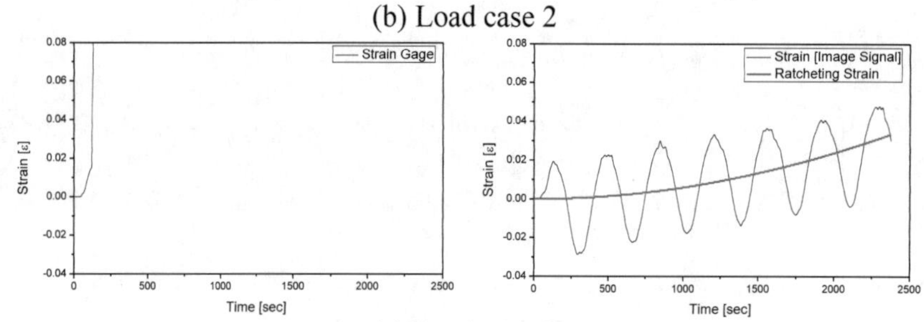

(c) Load case 3

Figure 8. Comparison of stain response

Table 3. Error analysis of the response measured by the strain gage

Load Case	Percent Error (%)	RMS Error
1	4.380	0.0026
2	5.645	0.0025
3	-	-

Table 4. Max. ratcheting strain

Load Case	Strain Gage	Image Signal	Difference (%)
1	0.0067	0.0061	8.955
2	0.0134	0.0126	5.970
3	-	0.0330	-

In figure 8, the strain in the hoop direction and the ratcheting strain measured using the image signals of the symmetry point of the area to which the strain gage had been attached was compared with the response measured by the strain gage. The black line shows the strain response and the red line shows the ratcheting strain. As shown in figure 8(c), it could not measure that the load case 3 exceeded the performance limit of the electric-resistance strain gage from the first cycle. This confirmed that the electric-resistance strain gage could not measure the strain until a leakage occurs in the elbow because its performance limit is exceeded soon after nonlinear behavior starts. In tables 3 and 4, error analysis was conducted for the measured data. In the table 3, the error was considered somewhat large because the percent error rate was within 6% and the RMS error rate was within 0.003. In table 4, the difference is also within 10%, which indicates that the difference is somewhat large. This appears to be due to the installation of the strain gage in the symmetry point as well as local problems. In table 4, it is considered that the limit state of a piping system can be quantitatively expressed in the future using the ratcheting strain.

4. Conclusions

In this study, for the case of conducting an in-plane cyclic loading test of a piping system, a non-contact measurement method through image signals based on an image measurement system is proposed as an appropriate method of measuring the hoop-direction strain of the elbow.

As the response measured using the LVDT of the UTM and the image signals obtained through the in-plane cyclic loading test exhibited small errors, the reliability of the image processing data was confirmed. In addition, although the response measured by the strain gage and that estimated using the image signals differed depending on the measurement location, the accuracy of the measured responses was confirmed by comparing them with the results of the electric-resistance strain gage.

Furthermore, it is expected that the measured strain response and the ratcheting strain can be used for estimating the failure criteria. It was confirmed that the strain could be measured from a remote distance without having to install the conventional sensors if the strain measurement algorithm that uses the image measurement system would be utilized. Therefore, the failure strain measured in this study is expected to be used in the future as an important factor in defining the failure criteria of seismic-isolated piping systems.

Acknowledgments

This work was supported by the Korea Institute of Energy Technology Evaluation and Planning (KETEP) and the Ministry of Trade, Industry & Energy(MOTIE) of the Republic of Korea (No. 20171510101910).

References

[1] Touboul F, Sollogoub P, Blay N 1999 Seismic behaviour of piping systems with and without detect: experimental and numerical evaluations *Nucl Eng Design* **192(2-3)** 243-260

[2] Zhang T, Brust FW, Wikowski G, Shim DJ, Nie J, Hofmayer CH, Ali SA 2010 Analysis of JNES seismic tests on degraded piping, *ASME 2010 Pressure Vessels and Piping Conference* Bellevue Washington USA PVP2010-25333

[3] Otoyo T, Otani A, Fukushima S, Jimbo M, Yamamoto T, Sakakida T, Onishi S 2014 Development of an evaluation method for seismic isolation system of nuclear power facilities (Part 4) failure behavior of crossover piping for seismic isolation system *ASME 2014 Pressure Vessels and Piping Conference* Anaheim California USA PVP2014-29011

[4] Park Y, DeGrassi G, Hofmayer C, Bezler P, Chokshi N 1997 Analysis of nuclear piping system seismic tests with conventional and energy absorbing supports *14th International Conference on Structural Mechanics in Reactor Technology* Lyon France BNL-NUREG-64173

[5] Mizuno K, Shimizu H, Jimbo M, Oritani N, Onishi S 2014 Development of an evaluation method for seismic isolation systems of nuclear power facilities (Part 5) fatigue test of the crossover piping *Proceedings of the ASME 2014 Pressure Vessels & Piping Conference* Anaheim California USA PVP2014-29032

[6] Jeon BG, Kim SW, Choi HS, Park DU, Kim NS 2017 A failure estimation method of steel pipe elbows under in-plane cyclic loading *Nucl Eng Technol* **49(1)** 245-253

[7] Kim SW, Choi HS, Jeon BG, Hahm DG, Kim MG 2018 Strain and deformation angle for a steel pipe elbow using image measurement system under in-plane cyclic loading *Nucl Eng Technol* **50(1)** 190-202

Finite Element Analysis for the Lifting Platform of a Self-Propelled Multi-Functional Machine Used in Orchard

Bai Zongchun[1,2], Lv Xiaolan[1] and Xia Liru[1]

[1] Institute of Agricultural Facilities and Equipment, Jiangsu Academy of Agricultural Sciences,50 Zhonglin St., Nanjing, 210014, P.R. China
[2] Key Laboratory of Urban Agriculture, Ministry of Agriculture, Shanghai, 200240, P.R. China

Email:vipmaple@126.com. Tel:+86-025-84390493, Fax:+86-025-84390435.

Abstract. Weight reduction is frequently concerned for general structure design as it can save energy in the life cycle of a machine. A small self-propelled multi-functional working machine for orchard use was designed. The lifting platform of the machine is analyzed using the finite element method for the purpose of structure optimization. By carrying out the finite element analysis, the inner stress for each component that forms the lifting platform was discussed under the designed maximum working load. Based on the relationship between the calculated von Mises stress in the structure and the yield strength of the material, the static strength of the component was verified and structure optimization suggestion was given. According to the FEA result, the overall structure of the lifting platform designed on the basis of theoretical calculation is statically safe, except for one of the long shaft sleeve. The dimension for most of the parts should be reduced since these components had an abundant static strength.

1. Introduction

For the past few years, the change in fruit tree cultivation mode in China requires more and more mechanized operations in orchard. However, the technical research in orchard working machine lags behind, which restricted the development of orchard greatly[1]. As the orchards in China are widely distributed in hilly and mountain area with complicated terrain, it is necessary to develop a small and multi-functional orchard working machine for the improvement of mechanization in orchard operation. Although some orchard machine products are readily available in western countries, e.g., the SOUIRREL lifting platform developed by an Australian company named CRENDON[2] and the Windegger Picking Platforms developed by a British company named N. P. SEYMOUR[3], their wheeled walking mechanism is not suitable for operations in complex mountains. To solve this problem, Zhu developed the first self-propelled multi-functional orchard working machine with tracks as its walking mechanism which can carry goods, trim branches of the tree and collect fruits[4]. Comparable machines with hydraulic lifting platform were also developed by Sun from agricultural university of Hebei and Liu from agricultural university of Hunan [5,6]. With abilities of trimming, weeding, picking and agricultural chemicals spraying, the moving platform designed by Wang can accept remote control[7]. To make the machine more accommodate complex landform, Wang designed a lifting platform with levelling mechanism included [8]. In addition of the works stated above, there are many other works reported about the orchard working machine. However, the general shortcoming of these machines is mono-functioning with large size of the structure. Besides, the operation of the machines is inconvenient.

The author tried to design a small self-propelled multi-functional working machine for orchard use, and the prototype machine is built. In this paper, the lifting platform of the machine is analyzed using the finite element method for the purpose of structure optimization.

2. Overall Structure of the Machine

Overall structure of the orchard working machine was illustrated in Fig. 1, including the crawler chassis, power unit, lifting platform, console and the heavy-duty platform. The power unit consists of an engine, storage battery, variable pump and a hydraulic motor.

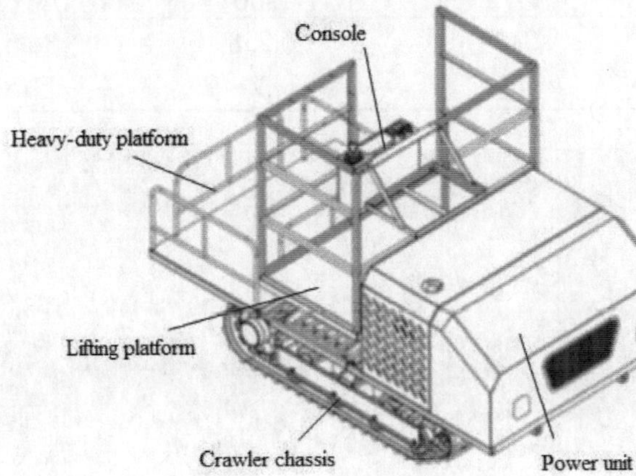

Figure 1. Overall structure of the orchard working machine

The principle diagram of the machine system was shown in Fig. 2. According to the function of each part, the machine can be divided into the control system, the power system, the hydraulic system and the mechanical structure. The power system provides power output for the machine's driving force, hydraulic pressure force and power for charging the battery. By control the hydraulic system and the power system, the control system is responsible for the move of the machine, rise and fall of the lifting platform and change in the hydraulic pressure.

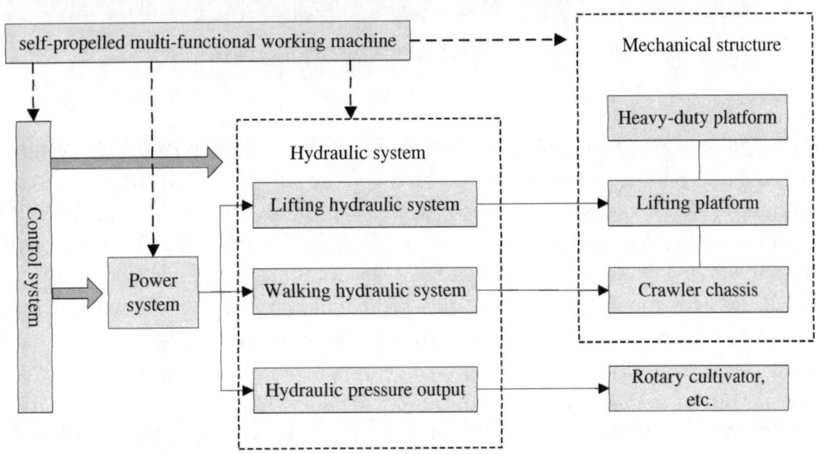

Figure 2. Principle diagram of the machine system

3. Finite Element Model of the Lifting Platform

3.1. Structure and Materials

Figure 3 showed the 3D model of the lifting platform when completely rose up, which consists of dozens of mechanical parts marked by Arabic numbers (hidden components not included). All the parts are made of Q235A steel (E235A steel in ISO 630 – 1995) except for the shaft parts and shaft sleeve parts which are made of 45 steel (ASTM 1045 steel), e.g., component 2, 6, 20, etc. Mechanical parameters of the two materials were shown in Table 1.

Table 1. Mechanical parameters of the materials used

Material No.	Elasticity modulus /GPa	Poisson's ratio	Yield strength /MPa
Q235A	212	0.238	235
45	209	0.269	515

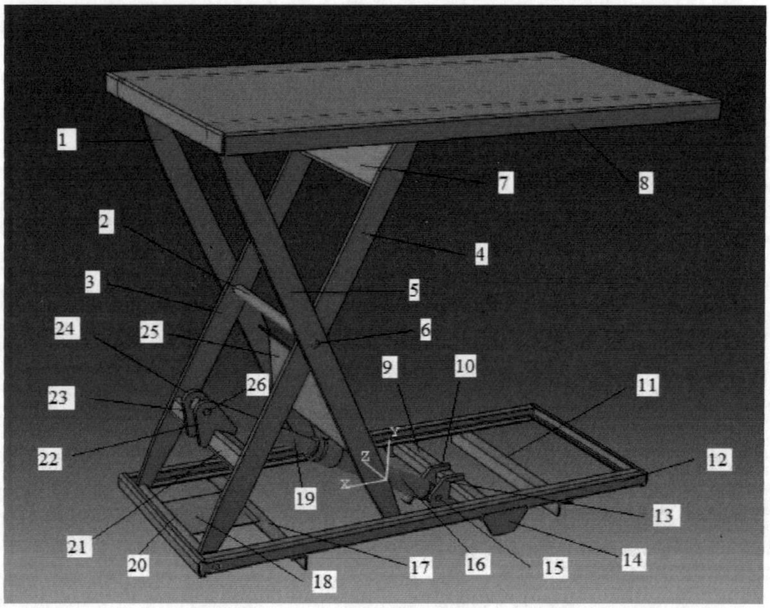

Figure 3. 3D model of the lifting platform

3.2. Finite Element Model

A general static analysis of the lifting platform was carried out employing the finite element code ABAQUS v6.13. As the stress characteristics in the pedestal of the lifting platform may be of no interest, the pedestal-related parts were removed to simplify the analysis, e.g., parts from number 9 (number convention is shown in Fig. 3) to number 19, 22, 23, etc. The finite element model of the assembly were shown in Fig. 4, which also showed the total size of the lifting platform. To mesh the parts, the C3D8R element (an 8-node linear brick with reduced integration) was prior used, otherwise the C3D10 element (a 10-node quadratic tetrahedron) was used. Allowing for the stress concentration, a finer mesh was used around holes. The contact between moving components were constrained by surface-to-surface contact with finite sliding, e.g. the contact between part 2 and part 6. Besides, the contacted surface was assumed to be frictionless. If there is no relative motion between two components with contact, e.g. contact between part 4 and part 7, then the "tie" constraint was used on the faces. The axis of the short shaft assembled in the hole of part 1 and in the hole of part 5 was constrained only allowed to rotate around z-axes, as well as the axis of part 20. As the design load lifting platform is 200 kg, to analyze severe loading situation, a concentrated force of F equals 2000N

IOP Conf. Series: Materials Science and Engineering **470** (2019) 012012 doi:10.1088/1757-899X/470/1/012012

was loaded near a corner of part 8, illustrated in Fig. 4. The distances between the loading position and the adjacent two edges are 150 mm.

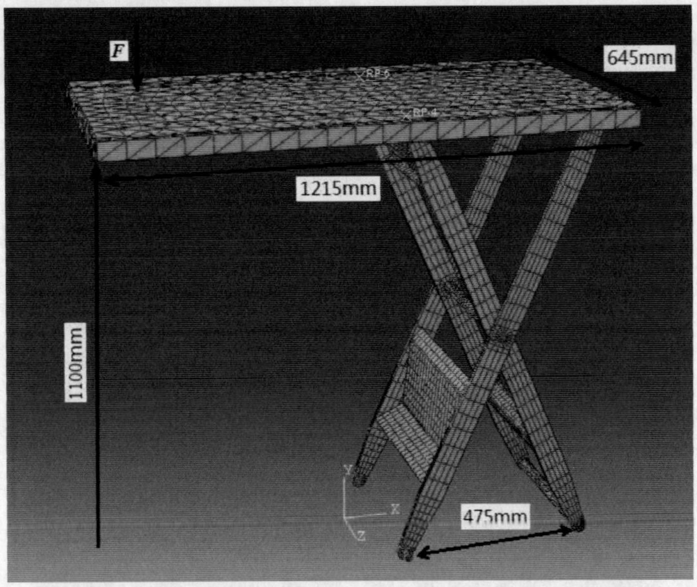

Figure 4. Finite element model of the lifting platform

4. Results and Discussion

4.1 General Result of Stress Characteristics

Figure 5 showed contour plots of von Mises stress distribution in the lifting platform. It can be seen that, generally, in the structure the stress may not exceed 100 MPa for most of the parts. However, for the shaft sleeve part 2, it can be seen that the stress level is significantly high, especially at the two ends, which may be over 400 MPa. Besides, the stress distribution in the supporting member, e.g. part 1, around the hole is also conspicuous. Therefore, it implies that these parts may be the mechanically weakness of the lifting platform which need to be further investigated.

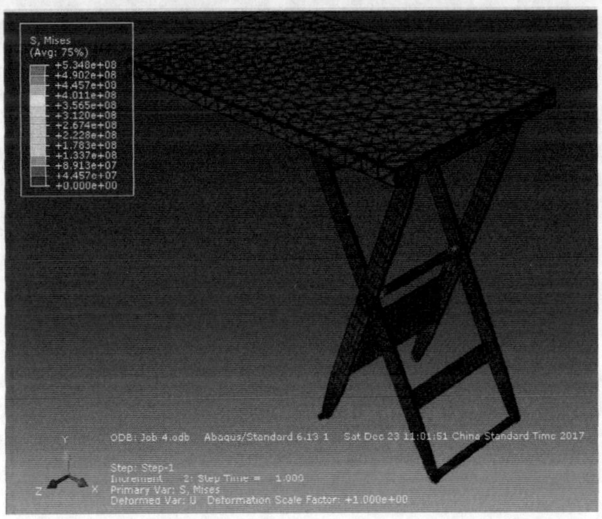

Figure 5. Contour plots of von Mises stress distribution in the lifting platform /Pa

IOP Conf. Series: Materials Science and Engineering **470** (2019) 012012 doi:10.1088/1757-899X/470/1/012012

4.2 Static Strength of Key Components

The von Mises stress distribution in the shaft sleeve *part 2* was shown in Fig. 6, which showed that the maximum stress existed in the shaft sleeve near the loading side. The stress level can be as high as 534.8 MPa and it is a little higher than 515 MPa, the yield strength of the material (refer to Table 1). Therefore, the shaft sleeve may need to be enhanced for further improvement of the platform. For instance, the static strength can be improved by appropriate heat treatment or thicken the wall of the sleeve. As a contact pair, the von Mises stress distribution in the shaft *part 6* was shown in Fig. 7. It can be seen that the stress level is pronounced than other parts showed in Fig. 5. Although the maximum stress could be 186.5 MPa, which appears in a different side in the shaft compared to the shaft sleeve, it is still safe for the structure allowing for the yield strength of the material. Moreover, the size of the shaft can be further reduced because the stress level is far below the yield strength.

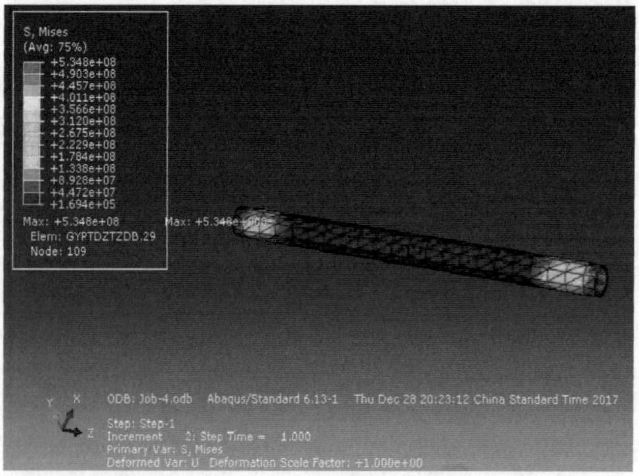

Figure 6. Von Mises stress distribution in *part 2* /Pa

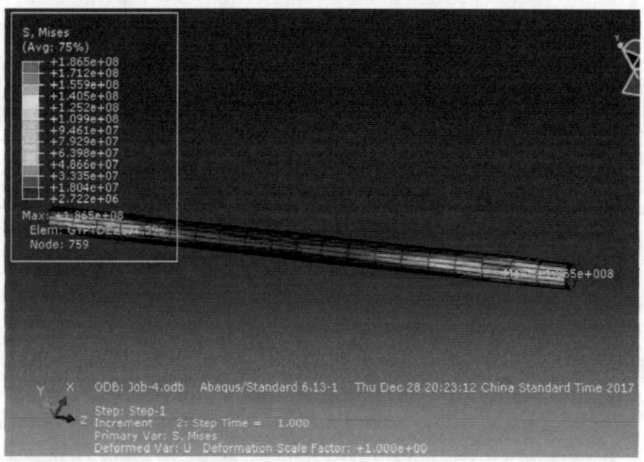

Figure 7. Von Mises stress distribution in *part 6* /Pa

As mentioned above, the stress distribution in the supporting member around the hole is also remarkable. A detailed stress characteristics in *part 1*, *part 3*, *part 4* and *part 5* were shown in Fig. 8. It can be seen that, in general, the higher stress level mainly exist in the surrounding material of the hole in the middle of these parts, whereas at the two ends the stress increase is not notable around the holes. Besides, a comparison between the four components showed that the highest stress level of approximately 138 MPa was found in *part 1*, which is far from the loading point compared to *part 5* in *z* direction. This may be attributed to a higher torque due to a bigger arm of force. In addition, the

IOP Conf. Series: Materials Science and Engineering **470** (2019) 012012 doi:10.1088/1757-899X/470/1/012012

comparison also showed that the stress is higher in *part 1* than that in *part 3*, and higher in *part 5* than that in *part 4*. This is because the stress concentration caused by holes is frequently discussed when the structure is subjected to a tension load, whilst under compression this phenomenon is not significant. In the present model, based on the knowledge of theoretical mechanics it may be found that generally *part 1* and *part 5* are under tension load, and *part 3* and *part 4* are under compression load. The static stress characteristics in the supporting members showed that in most region the stress level is far lower than the yield strength of the material (refer to Table 1), even for the maximum stress the value of which is just about half of the yield strength. Therefore, it may suggest that the size of these components could be further optimized for weight reduction of the machine, especially the cross-section size of the four components far from the middle hole, and the overall size of *part 3* and *part 4*.

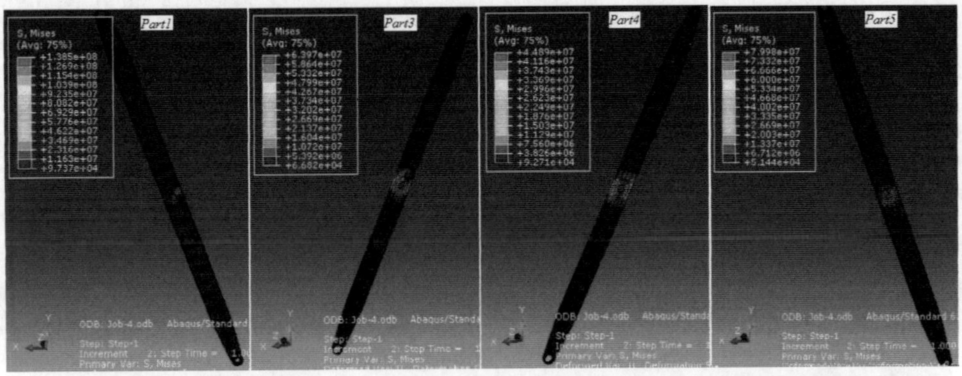

Figure 8. Von Mises stress distribution in the supporting members /Pa

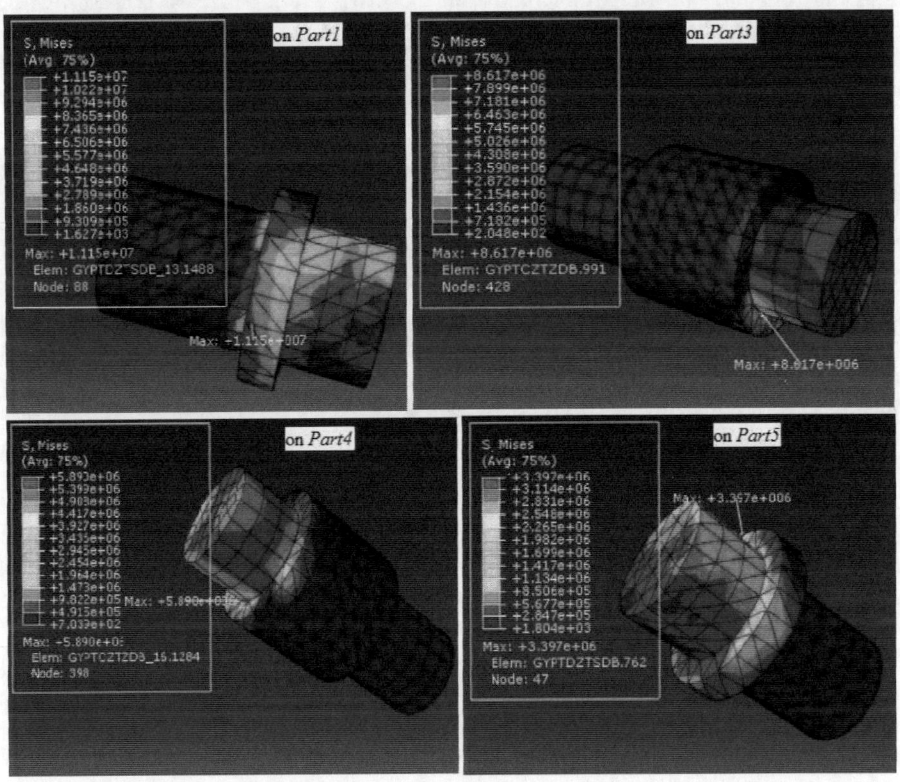

Figure 9. Von Mises stress distribution in the short shafts /Pa

Allowing for that the short shafts assembled at the bottom of *part 1* and *part 5*, as well as at the top of *part 3* and *part 4*, may suffer great shear forces, the von Mises stress distribution in the four components were shown in Fig. 9. It can be seen that, although these shafts are subjected to shear load, the stress distribution in the shafts is not as serious as expected. The result according to the finite element analysis (FEA) reveals that generally the maximum stress may not exceed 12 MPa under the design load of the lifting platform. Hence, the static strength of the short shaft is sufficient. On account that the shaft material has a yield strength of 515 MPa, the diameter of these shafts can be therefore further reduced to make the lifting platform lighter.

4.3 Structure Design Optimization

By carrying out the finite element analysis, the inner stress for each component that forms the lifting platform can be exhibited immediately under the overall working load. Based on the relationship between the calculated von Mises stress in the structure and the yield strength of the material, the static strength of the component can be verified and the dimension of the structure can be optimized accordingly. According to the FEA result, the overall structure of the lifting platform designed on the basis of theoretical calculation is statically safe. However, the FEA result also implies that the dimension for most of the parts should be reduced since these components had an abundant static strength.

For the shaft sleeve (*part 2*), the static strength should be improved by thickening the wall of the sleeve, if not by changing the heat treatment technology or replacing the material, whereas for the shaft (*part 6*) assembled in the sleeve, the diameter could be further reduced. For the supporting members (*part 1*, *part 3*, *part 4* and *part 5*), the size could also be further reduced for weight reduction of the machine, especially the cross-section size of the four components far from the middle hole, and the overall size of *part 3* and *part 4*. Besides, for the four short shafts that assembled in the supporting members and the rest parts of the lifting platform, the cross-section size in relation to the loading direction could also be reduced substantially.

5 .Conclusions

Weight reduction is frequently concerned for general structure design as it can save energy in the life cycle of a machine. Based on the knowledge of theoretical mechanics, the initial dimension of the lifting platform could be designed for each component and the prototype machine could be built. By carrying out the finite element analysis, the inner stress for each component that forms the lifting platform was exhibited under the design load. Based on the relationship between the calculated von Mises stress and the yield strength of the material, it showed that the weakest part is the long shaft sleeve used in the middle of the supporting members. The designed parameter according to theoretical calculation for the shaft sleeve may be destructive near the middle hole. On the contrary, the static strength for most of the components is sufficient. For the purpose of weight reduction, therefore, the dimension for most of the machine parts could be reduced.

Acknowledgement

This research was financially supported by the Jiangsu Agricultural Science and Technology Innovation Found (Grant No. CX(15)1045) and the Foundation of Key Laboratory of Urban Agriculture(No.UA201707), Ministry of Agriculture, P.R. China.

References

[1] J.L. Zhang, S.W. Chen, X.W. Zhuang, et al..2014. Research Status of Picking Machinery Used in Orchards on Southern Hillside, *Forestry Machinery & Woodworking Equipment*. 42 (9): 4-7. (In Chinese)

[2] CRENDON. 2017. Squirrel 536 LB SD. Online. http://www.crendon.com.au/squirrel656lb.htm. (Accessed August 2017)

[3] N. P. SEYMOUR. 2017. Windegger Picking Platforms. Online. http://www. npseymour.co.uk/ Windegger.htm. (Accessed August 2017)
[4] H.T. Zhu. 2011. Improved Design of LG-1—a Machine Used in Orchard, Xinjiang Agricultural Mechanization. 28 (5): 44-44.(In Chinese)
[5] Z.J. Sun. 2012. Design and Research of Crawler Multifunctional Orchard Operating Platform, Agricultural University of Hebei, Hebei. (In Chinese)
[6] D.W. Liu, F.P. Xie, X. Li, et al.. 2015. Design and experiment of small lifting platform in orchard, Transactions of the Chinese Society of Agricultural Engineering. 31 (3) :113-121.(In Chinese)
[7] Z. Wang. 2011. Small multi-function remote control power platform, Agricultural Equipment & Technology. 37 (3): 39-39. (In Chinese)
[8] J.C. Wang. 2013. Design on Hanging Operating Platform for Hilly Orchard, Agricultural University of Shandong, Shandong. (In Chinese)

IOP Conf. Series: Materials Science and Engineering **470** (2019) 012013 doi:10.1088/1757-899X/470/1/012013

Performance Differences of a Solid Attitude Control Thruster under Different Working Media

Zhong Yang [1,a], Hongwei Yan[1], Quanzhong Zhang[1] and Wubin Li[1]

[1] College of Transportation, Ludong University, Yantai, 264025, China

[a] ldu_yangzhong@163.com

Abstract. The solid attitude control thruster is widely applied in solid attitude control system of kinetic kill vehicle and other aircrafts. Its static and dynamic performances have significant effect on the control performance and accuracy. The thruster works under hot gas generated from the gas generator, so compressed air must be used for its detection. In this paper, a kind of solid attitude control thruster was proposed and its working principle was analysed. Theoretical and experimental research was done on the static and dynamic performances of the thruster. The results will provide a reference for the detection of the solid attitude control thruster's static and dynamic performances.

1. Introduction

The solid attitude control thruster is widely applied in various kinds of aircrafts, such as kinetic kill vehicle (KKV) and trajectory correction rocket projectile, etc. It received PWM (pulse width modulation) signal of certain frequency and regulates thrust to control the attitude according to the deviation of actual and ideal trajectory [1-5]. The static and dynamic performance indicators are very import which have a great influence on the control ability [6]. One of the static performance indicators is the value of thrust when the nozzle is wide open. It's an import manifestation of the thruster's control ability. And one of the most import dynamic performance indicators is dynamic delay time which is the time interval from the sending time of control signal to the generating time of thrust. It has a great influence on the control accuracy and dynamic performance of the attitude control system [7,8].

The thruster works under the hot gas from gas generator, so it can't cyclically utilize. Therefore, compressed air will be used to detect the thrusters instead of hot gas. During production compressed air will be used to detect the thrusters firstly and then hot gas will be used for spot check. Therefore, it's important and necessary to study the static and dynamic performances between compressed air and hot gas. In this paper, we proposed a kind of solid attitude control thruster and analysed its working principle. Then we did theoretical and experimental research on the static and dynamic performances of the thruster. The result will provide a reference for the detection of the solid attitude control thruster's performances.

2. Structure model

The solid attitude control thruster designed in this paper is shown in Figure 1. The hot gas flows into the thruster from the inlet and a small part goes into the control chamber through the fixed orifices f_1 and f_3 as control media. The adjustable orifices f_2 and f_4 are enclosed by the electromagnet and the outlet of the control chamber. As shown in Figure 1, when the adjustable orifice f_2 is closed and f_4 is open, the pressure is increased in control chamber 1 and decreased in control chamber 2. Then the pressure difference is formed and increased gradually. When it's large enough, the piston is driven to move right. The rocker arm is pulled to revolve around its axis and block the Laval nozzle in the left side. Eventually, the hot gas flows out through the right Laval nozzle to produce thrust.

The orifice model of the thruster is shown in Figure 2 [9]. The symbols P_s and P_b are the gas sourse pressure and backpressure separately. The orifices F_i (i =1,2,3,4) are the same with those shown in Figure 1. The symbol q_{mi} (i =1,2,3,4) means the mass flow of the four orifices and q_{ma}, q_{mb} are the mass flow of the control chamber 1, 2.

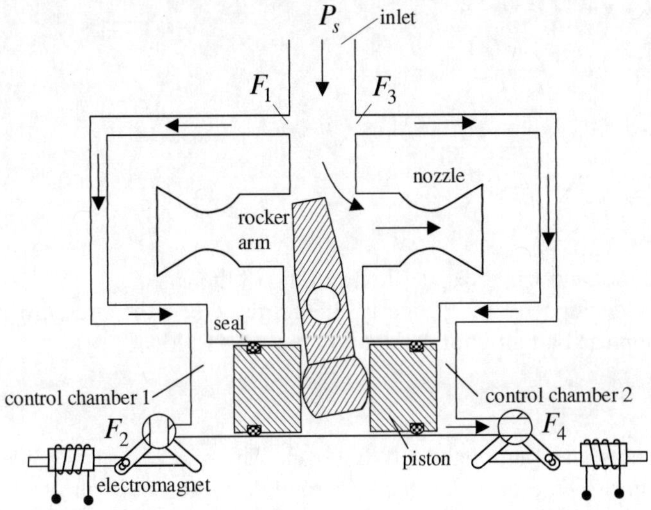

Figure 1. Schematic of the solid attitude control thruster

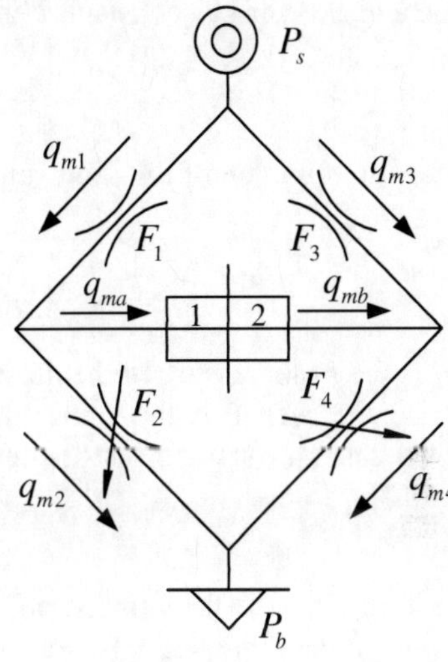

Figure 2. Orifice model diagram

The solid attitude control thruster works under hot gas as its working media which is destructive. Thus, compressed air must be used instead of hot gas to detect the thruster. Except temperature, compressed air and hot gas are also different in some aspects as follows:

- The specific heat ratio of the hot gas is lower than the compressed air. It's related to the gas component that the compressed air is composed of diatomic gas while the hot gas is polyatomic gas.
- The energy of the hot gas is much larger than the compressed air. It's because the temperature T and the gas constant R of the hot gas are larger than the compressed air.

The main parameters of the compressed air and hot gas in the experiment are listed in Table 1.

Table 1. The main parameters of the compressed air and hot gas

Designation	Specific heat ratio k	Gas constant R (J/kg·K)	Temperature T (K)	Pressure P (MPa)
Compressed air	1.4	287	300	7
Hot gas	1.25	360	1100	7

3. Research on static thrust

The static thrust is a significant indicator to evaluate the control ability of the attitude control thruster and also the primary indicator of design process. Therefore, it's important to study the static thrust. The difference between the static thrust of the two working media will decide whether the thruster can be tested by compressed air.

We assume that the thruster works under the optimum expansion condition, then the thrust can be calculated by the following equation:

$$F = \dot{m}V_e = \Gamma F_V P A_t \quad (1)$$

In equation (1), F means the nozzle's thrust and m, V_e are the mass flow rate and the velocity of exhaust, which can be given as:

$$\dot{m} = \Gamma P A_t / \sqrt{RT_0} \quad (2)$$

$$V_e = \sqrt{RT_0} \cdot F_V \quad (3)$$

In equation (2) and (3), P and T_0 mean the pressure and temperature of the gas in the inlet of the nozzle and R is the gas constant. Γ is a gas parameter and given as:

$$\Gamma = \sqrt{k}\left(\frac{2}{k+1}\right)^{\frac{k+1}{2(k-1)}} \quad (4)$$

Fv is the function of the flow velocity which can be calculated as follows:

$$F_V = \sqrt{\frac{2k}{k-1}\left[1-\left(\frac{P_b}{P}\right)^{\frac{k-1}{k}}\right]} \quad (5)$$

P_b is the backpressure of the nozzle's outlet and k is the specific heat ratio of the gas.

It can be seen that the static thrust is decided by the property of the gas when the pressure is definite. Then the static thrusts under compressed air and hot gas are compared as follows:

$$\frac{F^*}{F} = \frac{\Gamma^* F_V^*}{\Gamma F_V} = 1.03 \quad (6)$$

In formula (6), the superscript * means the parameters of hot gas. The static thrusts under compressed air and hot gas are consistent. Therefore, it's feasible to detect the static thrust by compressed air instead of hot gas in theory.

Then we do experiment on static thrust of the thruster. We fixed the piston to one side in order to generate thrust on one side. Then the thrust is tested by the force sensor and the experiment under hot gas is shown in Figure 3. The experimental results under compressed air and hot gas are shown in Figure 4. It's observed that the experimental result corresponded with the theoretical result. Therefore, the static thrust can be detected by compressed air.

Figure 3. Experiment of the static thrust under hot gas

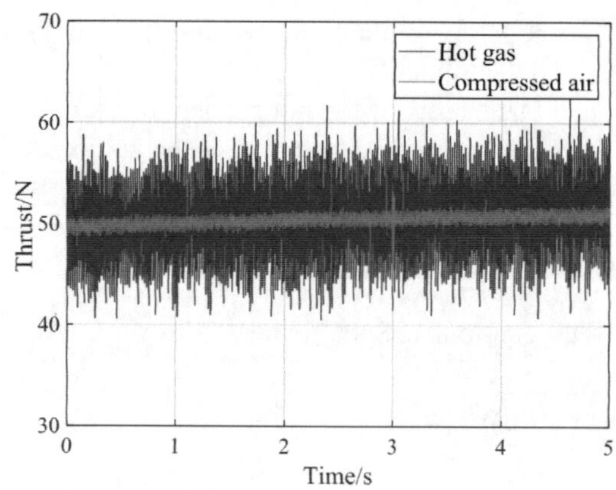

Figure 4. Static thrust of compressed air and hot gas

4. Research on dynamic delay time

The dynamic performance of the thruster is mainly characterized by dynamic delay time, which is the time interval from sending the control signal to the switching of the thrust. In order to test the delay time, two pressure sensors were applied to test the pressure in the two control chambers. Also, we assembled magnetic sheets on both sides of the piston and Hall sensors on the body correspondingly.

The output signals of the two Hall sensors were compared by the comparator circuit to obtain the piston's position signal. When the piston goes through the middle position, the piston's position signal goes to the opposite electric level. Therefore, the dynamic delay time can be obtained by contrasting the control signal with the control signal indirectly. The principle of the test system is shown in Figure 5 and the experiment devices in Figure 6.

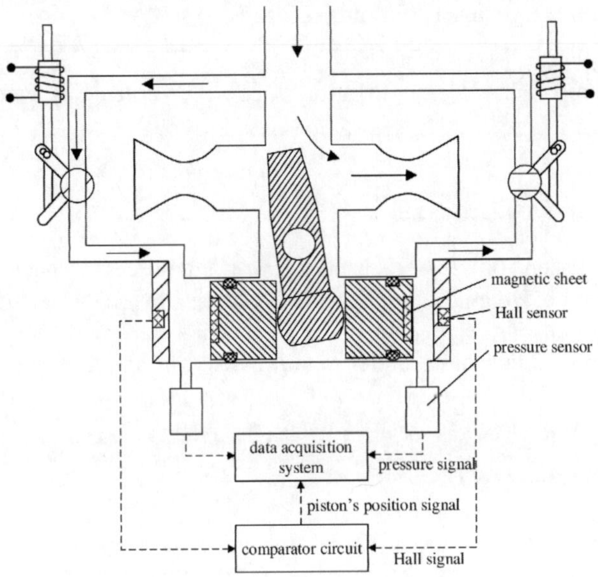

Figure 5. Schematic of the test system

Figure 6. Diagram of the experiment

The thruster was tested under compressed air and hot gas separately and the pressure of gas source was 7 MPa. The pressure curves in the two control chambers under compressed air and hot gas are shown in Figure 7 and the piston's position signal contrasted with the control signal under the two media are shown in Figure 8.

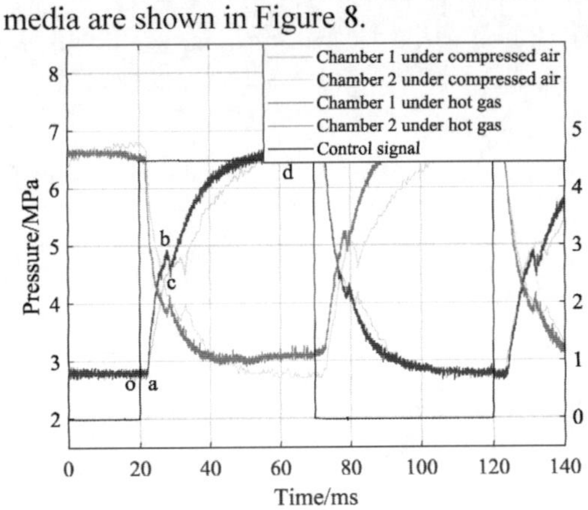

Figure 7. Pressure signal in the two control chambers under compressed air and hot gas

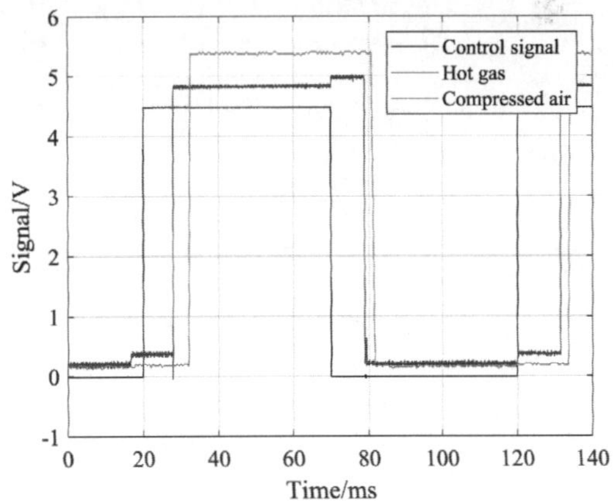

Figure 8. Piston's signal and control signal under compressed air and hot gas

Using pressure in chamber 1 under hot gas as an example, the working process can be divided into 4 parts obviously as shown in Figure 7. The first part is from point o to point a which is the delay time of the electromagnet. The second part is from point a to point b, which is called prepare phase. In this phase, the pressure in the two control are changing, but the pressure difference isn't large enough to drive the piston. The third part is from point b to point c, which is called motion phase. The piston moves to the other side under the pressure difference at this stage. The last part is from point c to point d, which is called stabilizing phase. The pressure becomes stable gradually until the next control signal comes. In Figure 8, the dynamic delay time of the thruster can be obtained by contrasting the control signal and the piston's position signal. The delay time of every stage can be obtained from Figure 7 and 8 and the results are shown in Table 2.

Table 2. The delay time of every stage (unit: millisecond)

Media	Delay time of the electromagnet	Prepare stage	Motion stage	Whole delay time
Compressed air	3.0	6.8	1.1	10.4
Hot gas	3.0	4.6	0.9	8.0

The pressure of hot gas varies more rapidly than compressed air during working process, especially in the prepare phase and stabilizing phase and the dynamic delay time of hot gas is obviously less than compressed air. Considering that the proportion of the prepare phase is the highest among all the phases, the difference between the dynamic delay time under compressed air and hot gas is concentrated in the prepare phase.

In the prepare phase, the pressure in the control chamber 1 is calculated as follows:

$$\dot{P}_1 = \frac{k}{V_1} Q_1 R T_s \tag{7}$$

In equation (7), P_1 and V_1 mean the pressure and volume of the control chamber 1 and Q_1 means the mass flow of the working media flowing into the control chamber 1. T_s is the temperature of source gas. The stream of the orifice F_1 is subcritical flow and the flow equation is given below:

$$Q_1 = \frac{\mu F_1 P_s}{\sqrt{T_s}} \varphi\left(\frac{P_1}{P_s}\right) \tag{8}$$

where

$$\varphi\left(\frac{P_1}{P_s}\right) = B\sqrt{\left(\frac{P_1}{P_s}\right)^{2/k} - \left(\frac{P_1}{P_s}\right)^{(k+1)/k}} \tag{9}$$

Where B is a parameter related to the gas and calculated as follows:

$$B = \sqrt{2k/R(k-1)} \tag{10}$$

The time of the prepare phase is obtained by integration:

$$t = \frac{V_1}{(k-1)\mu F_1 R B \sqrt{T_s}} \phi(k) \tag{11}$$

where

$$\phi(k) = \sqrt{1 - \left(\frac{P_b}{P_s}\right)^{\frac{k-1}{k}}} - \sqrt{1 - \left(\frac{P_1}{P_s}\right)^{\frac{k-1}{k}}} \tag{12}$$

The proportion of the delay time of prepare phase under compressed air and hot gas according to the equation (12) is calculated as 1.53 while the experimental result is 1.48 according to the data in Table 2. The experimental and calculation results are consistent basically. The delay time of the prepare phase under compressed air is about 50% more than hot gas while the delay time of motion phase under the two media has little effect on the dynamic delay time. Therefore, we can establish the evaluation system by the equation. When detecting the thruster, we should firstly estimate the performance of the thruster under hot gas by the experiment result of compressed air. Then carry out spot check on thrusters by hot gas in order to correct the evaluation system. Consequently, the detection of the thrusters can be realized by compressed air.

5. Conclusion

The solid attitude control thruster works under hot gas which is destructive. Therefore, it must be detected by compressed air. In this paper we proposed a kind of solid attitude thruster and analysed its working principle. Then we did theoretic and experimental research on the static thrust and dynamic delay time of the thruster under compressed air and hot gas. The two kinds of media had little effect on static thrust while a significant impact on dynamic delay time. The evaluation system of detecting the thruster was established preliminary. The result will provide feasible basis for the detection of the solid attitude control thrusters.

Acknowledgments

This work was supported by the Training Fund of Shandong Natural Science Foundation (Grant NO. ZR2017PEE016) and the Talent Introduction Project of Ludong University (Grant NO. LB2016002).

References

[1] P. Baker 1992 Army lightweight exo-atmospheric projectile (LEAP) *Aerospace Design Conf.* (Irvine, USA: AIAA) AIAA-92-1212

[2] J. G. Campbell 1993 Divert propulsion system for army LEAP kill vehicle *2nd Annual AIAA SDIO Interceptor Technology Conf.* (Albuquerque, USA: AIAA) AIAA-93-2637

[3] Stéphanie Joner and Isabelle Quinquis 2006 Control of an exoatmospheric Kill Vehicle with a solid propulsion Attitude Control System *AIAA Guidance, Navigation, and Control Conference and Exhibit.* (Keystone, USA: Colorado)

[4] HOU Xiao, FU Peng and WU Yuan 2017 Energy management technology of SRM and its development *J. Journal of Solid Rocket Technology* **40** 1-6+23

[5] Li Shipeng and Zhang Ping 1999 Analysis of solid propellant control thruster schemes for small KKV *J. Journal of Propulsion Technology* **20** 100-103

[6] CHANG Heng, Wang Yi-bai and LIU Yu 2013 Experimental investigation on solid attitude control system based on PWM principle *J. Journal of Solid Rocket Technology* **36** 608-612

[7] LIU Qing-hong, CHEN De-yuan and WANG Zi-cai 2003 Effect of system delay on guidance accuracy of interceptor *J. Journal of Astronautics* **24** 642-645

[8] DUAN Meijun and ZHOU Di 2017 Guidance law with finite time convergence considering thrust constraint and dynamics characteristic *J. Systems Engineering and Electronics* **39** 2068-207

[9] H. Keller and R. Isermann 1993 Model-based nonlinear adaptive control of a pneumatic actuator *Control Engineering Practice* **3** 505–511

IOP Conf. Series: Materials Science and Engineering **470** (2019) 012014 doi:10.1088/1757-899X/470/1/012014

Influence of Yield Stress Mismatch on the Creep Stress Field around Crack Tip of Stress Corrosion Cracking

Zhang Jianlong[1,2,a] and Xue He[1]

[1]School of Mechanical Engineering, Xi'an University of Science and Technology, Xi'an 710054, China
[2]Xi'an Special Equipment Inspection Institute, Xi'an710065, China

[a]zjl527449153@vip.qq.com

Abstract. To understand the influence of yield stress mismatch on the creep stressfield around crack tip in safe-end dissimilar metal welded joints at primary circuit, the influences of different mismatch coefficients on the creep stress field around the crack front were analyzed using ABAQUS software. Results show thatthe Mises stress distribution is not very different after the creep under different mismatch coefficients.The equivalent plastic strain(PEEQ) in the Nickel based 182 near crack tip is basically remain unchanged and the PEEQ of HAZ is decreasing with the increasing of the mismatch coefficient.The creep strain is mainly concentrated in the crack tip area and there is little change in creep strain under different mismatch coefficients.

1. Introduction

Stress corrosion cracking(SCC) is a failure mechanism which is caused by environment, susceptible material and tensile stress of nuclear power structural materials under high temperature and high pressure water environments[1-2].

As a key component of the primary circuit of nuclear power plant, the structural integrity of welded joints of dissimilar metal safety terminals has become a key issue affecting the long-term safe operation of pressurized water reactor power plants[3]. SCC also occurs in the welded joints of the primary circuit of pressurized water reactor[4]. The particularity of the welding process makes it easy to produce residual stresses in welded joints, and residual stress in welding process is one of the key factors leading to stress corrosion cracking in nuclear power plant[5]. There is a high degree of heterogeneity in mechanical properties of dissimilar metal welded joints, and the yield stress of the weld area varies very high[6]. At the same time, owing to the significant difference between the welded metal parts of the dissimilar metal welded joint components[7], a series of mechanical problems will arise after welding and after welding heat treatment[8]. Recently many researches have focused on the research of mechanical properties of welded joints. It is believed that the failure of the yield strength of the material at the welded joint will cause the change of the crack tip mechanical field and have a great influence on the SCC crack growth[9-10].

In order to provide some help for the safety evaluation of dissimilar metal welded joints, this paper takes the dissimilar metal welded joints at the safe end as the research object. The creep stress and strain field under different yield stress after creep was established using ABAQUS in this paper, and stress and strain field are analyzed under different yield stress.

2. Finite element modeling

2.1. Specimen model

The compact tensile(1T-CT) specimen was adopted in this paper using finite element software ABAQUS and its geometrical dimensions are confirmed to the ASTM399 standard [11], which as shown in Figure 1(a). Because the welded joints is including three parts, named as welded zone(WZ), heat affected zone(HAZ) and base metal(BM), respectively, and its material distribution is shown in Figure 1(b).

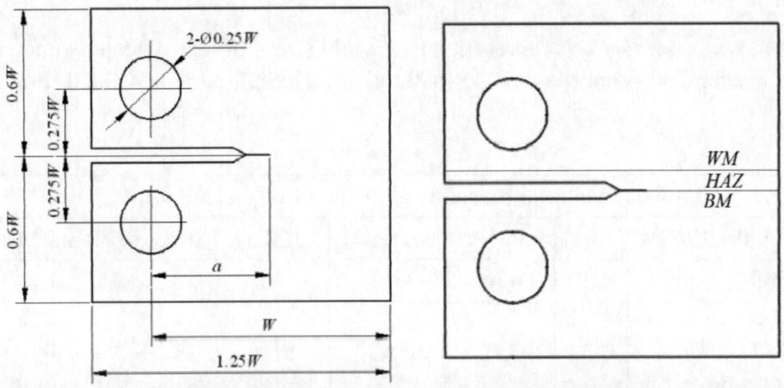

(a) Geometryand dimensions model (b) Sandwich model of welded joint

Figure 1. Geometryand dimensions of1T-CT model

To eliminate the stress singularity generated at the crack tip [12], the crack tip radius was taken as 0.5μm in the finite element simulation. The CPE8R was adopted in the global mesh type, as shown in Figure 2(a). In order to improve the accuracy of computation, sub-model technique is adopted. To obtain more detailed and accurate crack tip stress and creep strain, the mesh around crack tip was refined and its number was 15027, the sub-model mesh type is kept consistent with the global mesh model type, which is shown in Figure 2(b), and the total mesh number of sub-model is 3620.

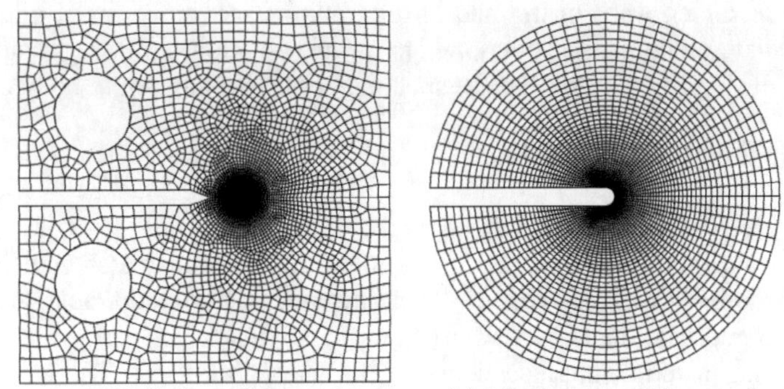

(a) Global mesh model (b) Mesh around the notch

Figure2. Finite element meshmodelofmodel of 1T-CT

2.2. Material model

The research object of this paper is the safe-end dissimilar metal welded joints in primary circuit of nuclear power plants, and which is composed of for different type materials, which is low carbon steel A508, heat affected zone material, welded zone material Alloy182 and austenitic stainless steel 316L, respectively.

To derive the creep property around crack tip of Alloy182, Creep constitutive relation is adopted power law model to describe the creep law in the simulation [13-14]. The constitutive relationship and its forms is as below.

$$\dot{\varepsilon}_{cr} = A\sigma^n t^m \tag{1}$$

Where $\dot{\varepsilon}_{cr}$ is the creep strain rate. The σ is the applied stress; A is the power law multiplier, and its value is 3.77×10^{-19}, n is the creep exponent, and the value is 6.71 [15].

The yield strength mismatch coefficient of the heat affected zone is defined as the following formula 2

$$M_Y=\sigma_{HAZ}/\sigma_{A182} \tag{2}$$

In the formula (2), σ_{HAZ} *and* σ_{A182} is represent as the yield stress of heat affected zone and Alloy182, respectively. The mechanical properties of the material in different part of welded joints is shown in Table 1 [16].

Table 1. Material mechanical property of welded joint

Material	E/GPa	ν	σ_0/MPa	Material	E/GPa	ν	σ_0/MPa	M_Y
A508	183	0.30	476	316L	176	0.30	227	1
A182	204	0.33	337	HAZ	204	0.33	674	2
							1011	3

2.3.Loading and boundary conditions

Two analysis steps are set in the finite element simulation analysis. The first one is the loading process, and loading is applied to the load hole by using the reference point of the rigid body in the Y direction. And the specific value of the concentrated force is applied to the reference point to ensure the crack tip stress intensity factor K_I is $30MPa.m^{0.5}$ in the tensile direction [17]. The second step is the creep process, and creep time is set as 1000h.

In order to eliminate the rigid body displacement during the experiment period, two points in the right side of the CT specimen was constrained in rotation and movement of X direction, so that they can only move and rotate along the Y direction. The centre position on the right side of the specimen was also fixed along the Y direction, so that the specimen does not have rigid body displacement.

3. Results and Discussion

3.1.Effect of yield stress mismatch on the Mises stress at crack front

The Mises stress distribution around the crack front under different mismatch coefficient after creep is shown in Figure 3, which shows that when the welding material around crack tip equal-matches the yield strength of the material in the heat affected zone, the crack tip Mises stress distributes symmetrically along the material interface between HAZ and Alloy182. The Mises stress gradually decreases with the increasing distance from the crack front under any mismatch coefficient. And when the welded material occurs mismatch, the Mises stress in the HAZ is less than the Mises stress in the Alloy182, which indicating that the yield stress of the HAZ will increase, which also will offset the high stress zone to the welding material. And the Mises stress distribution under different mismatch coefficients change very little after the creep.

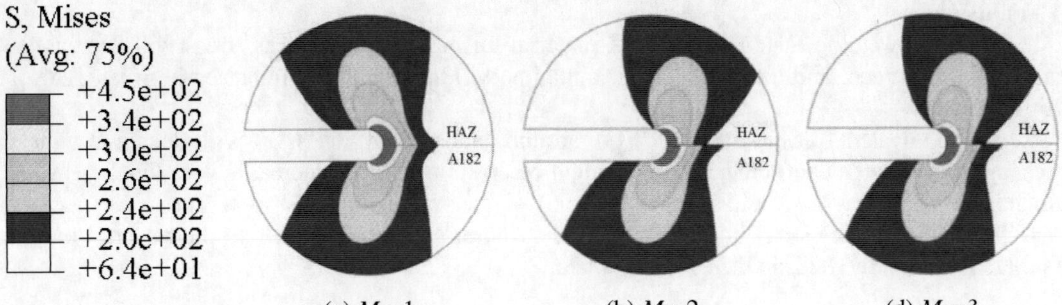

(a) M_Y=1 (b) M_Y=2 (d) M_Y=3

Figure 3 Mises stress distribution in front of notch front under differentmismatch coefficient after creep(MPa)

3.2.Effect of yield stress mismatch on the PEEQ at crack front

The equivalent plastic strain(PEEQ) distribution around notch front under different mismatch coefficient after creep is shown in Figure 4, which shows that the PEEQ also distributes symmetrically along the material interface under the equal mismatch between HAZ and Alloy182. Equivalent plastic strain in the Nickel based 182 near crack front is basically remain unchanged under different mismatch coefficient after creep, but the *PEEQ* of HAZ is gradually decreasing with the increasing of the mismatch coefficient.

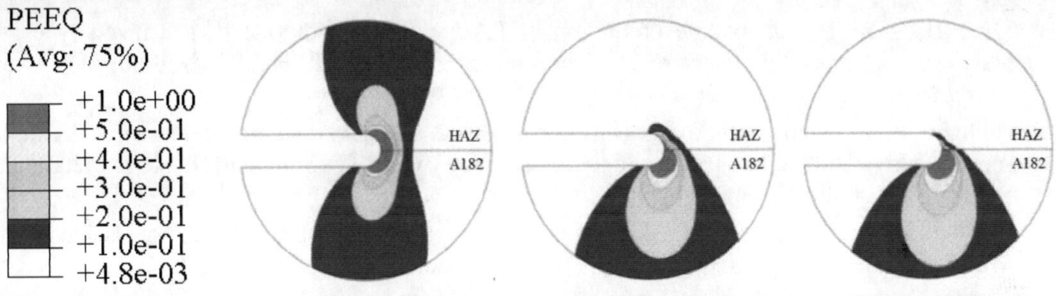

(a) M_Y=1 (b) M_Y=2 (d) M_Y=3

Figure 4 PEEQ distribution in front of notch front under differentmismatch coefficient after creep

3.3.Effect of yield stress mismatchcoefficienton the CEEQ at crack front

The equivalent creep strain(CEEQ) is also an important crack growth driving force parameter in the stress corrosion cracking tip, and its distribution around the crack front notch under different mismatch coefficient after creep is shown in Figure 5. Figure 5 shows that the creep strain is mainly distributed around the crack tip region, and its value is becoming less and less with the increasing distance from crack tip. The value of CEEQ is becoming larger and larger when the mismatch coefficient of HAZ is increasing.

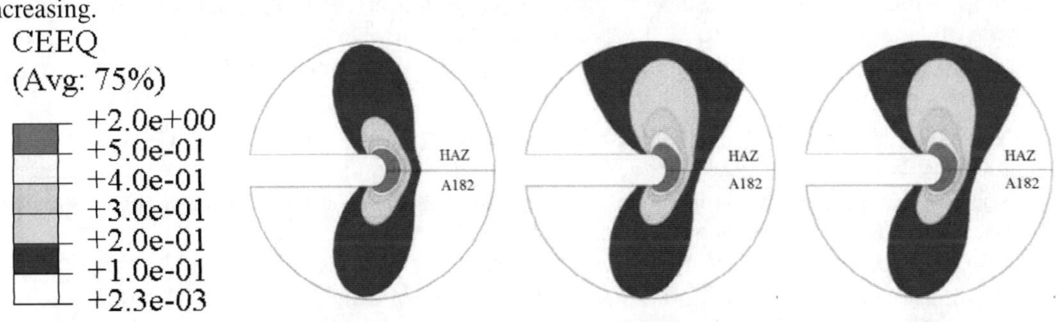

(a) M_Y=1 (b) M_Y=2 (d) M_Y=3

Figure 5 PEEQ distribution in front of notch front under differentmismatch coefficient

4. Conclusions

(1) When the welding material and HAZ material mismatches, the Mises stress will decrease to a same level after creep, and the Mises stress mutation will appear on the interface between HAZ and Alloy182.

(2) The equivalent plastic strain(PEEQ) around crack front in HAZ will decrease with the increasing of mismatch coefficient, but equivalent creep strain(CEEQ) increases with the increasing of mismatch coefficient.

(3) The PEEQ, Mises stress and CEEQ also distributes symmetrically along the material interface when HAZ and Alloy182 is in the equal mismatch.

Acknowledgements
This workis financially supported by the National Natural Science Foundation of China(Grant Nos. are 51475362 and 11502195).

References

[1] ZHAO Lingyan, CUI Yinghao, XUE He. Effect of crack propagation initiation angle on the stress corrosion cracking behavior in welding transition zone of Ni-based alloy[J]. CHINA SCIENCEPAPER, 2016, 11(16):1864-1866.

[2] Andresen P L, Karen G, Lawrence N J. Stress Corrosion Cracking of Sensitized Type 304 Stainless Steel in 288C Water: A Five Laboratory Round Robin[M]. John Wiley & Sons, Inc. 2013.

[3] Xue H, Shoji T. Quantitative prediction of EAC crack growth rate of sensitized type 304 stainless steel in boiling water reactor environments based on EPFEM[J]. Journal of pressure vessel technology, 2007, 129(3): 460-467.

[4] LI Guang-fu, LI Guan-jun, FANG Ke-wei, et al. Stress corrosion cracking behavior of dissimilar metal weld A5008/52M/316L in high temperature water environment[J]. Acta Metallugica Sinica, 2011, 47(7):797-803

[5] XIONG Linyu. Effect of Strength Mismatching on Elastic Plastic Deformation Behaviors of Welded Joints with Cracks. China Mechanical Engineering, 2012,23(6):733-738.

[6] LUO Li-wei, FU Xiao-chao. Generation and erase method of residual stress by welding[J]. Shanxi Architecture, 2008, 34(20) : 134-135.

[7] GONG Xiao-yan, JIAO Kang, ZHAO Ling-yan, et al. Effect of welded mechanical heterogeneity on fractureparameters of pipeline cracks[J]. Journal of Xi'an University of Science and Technology, 2013, 33(2) : 211-215

[8] Lundin C D. Dissimilar metal welds-transition joints literature review[J]. Welding Journal, 1982.61(2) : 58-62

[9] Zhao L, Xue H, Tang W, et al. Effect of welded mechanical heterogeneity on local stress and strain ahead ofgrowing crack tips in the piping welds[C]. ASME2011 Pressure Vessels and Piping Conference.AmericanSociety of Mechanical Engineers, 2011: 223-232.

[10] HOU Juan, PENG Qun-jia, Shoji T, et al. Study of microstructure and stress corrosion cracking behavior in welding transition zone of Ni-Base Alloys[J]. Acta Metallurgica Sinica, 2010, 46(10) : 1258-1266.

[11] ASTM Standard E399-90. Annual book of ASTM standards[M]. USA: ASTM International, 2002.

[12] Xue He, Guo Rui, Chen Zheng. Effect of oxide film thickness on creep at tip of stress corrosion cracking[J]. China Sciencepaper, 2016, 11(22).

[13] Zhu Zhi, Zhang Liwen, Gu Sendong. Stress relaxation test of HastelloyC-276 alloy and its creep constitutive equation[J] .The Chinese Journal of Nonferrous Metals, 2012, 22(4):1063-1067.

[14] HAN Ningning, GU Jianfeng. Creep behavior of Inconel 600 alloy used for nuclear power[J]. HEAT TREATMENT OF METALS, 2016, 41(6):1-3.

[15] Yatomi M, Tabuchi M. Issues relating to numerical modelling of creep crack growth[J]. Engineering Fracture Mechanics, 2010, 77(15):3043-3052.
[16] Wang H, Wang G, Xuan F. Numerical Research of Ductile Crack Growth Behavior forDissimilar Metal Welded Joint of Nuclear Pressure Vessel[J]. Nuclear Power Engineering, 2012, 33(5): 36-40.
[17] Xue He, Guo Rui, Chen Zheng. Effect of oxide film thickness on creep at tip of stress corrosion cracking[J]. China Sciencepaper, 2016, 11(22).

The Study of Imperfect Maintenance Strategy based on Improved Age Reduction Factor

Duan Junhang, Zhu Ling and Cao Li

Department of Automation, Tsinghua University, Beijing, 10084

duan_junhang@163.com

Abstract. This paper is going to study the maintenance strategy optimization of repairable equipment in the imperfect maintenance mode. By introducing age reduction factor, repairable equipment can effectively describe the effects of imperfect maintenance. For the current age reduction factor is usually given by industry experts without a clear quantitative calculation formula, this paper offers an improved formula on the basis of its definition. The new formula takes not only traditional factors like repair times,costs, etc. but also subjective factors like equipment importance into account. Based on this and NSGA II(the second generation of non-dominated sorting genetic algorithm), the traditional scheduling maintenance strategy tries to solve its model whose objective function includes the maximum reliability, maximum availability, minimum total costs. Finally, by taking the radio relay equipment of mobile communication vehicle as an example to calculate and demonstrate, its validity was proved.

1. Introduction

when building a maintenance model for some equipment, it is usually assumed that after a preventive maintenance period the equipment will be repaired "as good as new" or "as good as old". These two states were defined according to the equipment's failure rate. "As good as new" is also called perfect maintenance, which refers to restoring the equipment's failure rate to default setting(λ_0);"as good as old" is also known as the minimum maintenance, which means restoring its failure rate to pre-maintained state(λ_i). Obviously, these two assumptions are designed for describing and calculating the maintenance model. But in actual situation, most equipment's failure rate(λ_k) is between the two after maintenance[1], that is $\lambda_i > \lambda_k > \lambda_0$, and this is called imperfect maintenance. It is more reasonable, because after i-1 times of preventive maintenance, the equipment will be whole-update or restore the failure rate to default setting at high cost at the i-th time.

In the early, Brown et al[2] proposed a model to describe imperfect maintenance,that is to say, the failure rate can attain the perfect maintenance state according to the probability of P, therefore, the minimum maintenance can be achieved at the probability of (1-P); Pham and Wang[3] summarized various types of imperfect maintenance model such as the (p,q) rule, the improved factor method, the (α,β) rule and the virtual age method,etc.; F.G.Bad and M.D.Berrade discussed the costs of signal imperfect maintenance[4]; Wee Meng Yeo, Xue and Ming Yuan discussed the relationship between costs and maintenance effects[5]; Qu Yuxiang stated the role of reduction factor[6]; Tzong-Ru Tsai et al putted forward that there is a liner relationship between costs and maintenance effects[7]; Ye Penfan used Markov Chain to describe deterioration process relying on the multi-stage deterioration model[8]. In this essay, starts from the maintenance effect, introduces the parameters, the author tries to describe achieved degree after each imperfect maintenance.

2. Improved age reduction factor

At present, the method of improved factor and age reduction factor is the main method which introduced parameters to describe the effect of imperfect maintenance. Age reduction factor was proposed by Dedopoulos and Smeers[9], and its basic idea is still to describe the maintenance effect by introducing parameters. Differ from improved factor, reduction factor describes the state after maintenance at t point according to (t-α). It is more intuitive by the "regression" on time dimension. The quantitative reduction factor is generally given by industry experts after calculating and combined with experience. The method is simple, but exists subjective effects and numeric value differs widely from different experts. Supposing age reduction factor $\alpha \in [0,1)$,the equipment undergoes N-1 periods of preventive maintenance (the last period's node is not counted), and its failure rate at each maintenance node can be described as equation(2-1):

$$\lambda_1 = \lambda(t)$$

$$\cdots\cdots$$

$$\lambda_N = \lambda\left(t - \textstyle\sum_{j=1}^{N-1} \alpha_j T\right) \tag{2-1}$$

the corresponding node's reliability degree R_n can be described as equation(2-2):

$$R_N = R\left(t - \textstyle\sum_{j=1}^{N-1} \alpha_j T\right) \tag{2-2}$$

After getting equations of the failure rate and degree of reliability at each node of preventative maintenance period, what matters is to confirm the age reduction factor α[10].

Combined with the preventative maintenance costs, times and equipment importance, this paper presents the equation of age reduction factor as follow:

$$\alpha_j = \begin{cases} \beta(kC_{p_i})^{b \cdot i}, i = 1,2,\dots,N-1 \\ 1 \qquad\qquad\quad , i = N \end{cases} \tag{2-3}$$

The meaning of each parameter is: C_{p_i} is the cost of preventative maintenance at the i-th time, k is the cost regulatory factor and satisfies the condition $0 < k < 1/C_{p_i}$; b is the time regulatory factor and satisfies the condition $0 < b < 1$;β is the equipment importance factor and satisfies the condition $0 < \beta < 1$. At each maintenance, the more cost is putted in, the bigger α_j is, the better the maintenance effect is, therefore α_j is proportional to C_{p_i}. At the same time, the more times the equipment is maintained, reflecting equipment's degradation accelerate to the end of working life, therefore α_j is inversely proportional to i. Each participant in the maintenance dominates the process and should not be ignored, hence taking β as the equipment importance factor. For different equipment, the value of β reflects which equipment is more important; for one equipment, the value of β reflects which stage is more important. Taking weaponry as an example, its importance degree is obviously different in peacetime and in combat readiness. The bigger β is, the equipment or the stage, the more valued. The maintenance effect is obviously better under the condition that cost and times remain the same, so β reflects the subjective factor of maintenance participants. Meanwhile, the equipment will be replaced instead of being maintained at the last node of design lifetime.

From equation(2-3), calculating equations of reliability degree R_n and failure rate λ_n are as follows:

$$R_N = \begin{cases} R\left[\left(i - \sum_{j=1}^{i} \alpha_j\right)T\right], i = 1,2,\dots,N-1 \\ R\left[\left(N - \sum_{j=1}^{N-1} \alpha_j\right)T\right], i = N \end{cases} \tag{2-4}$$

$$\lambda_N = \begin{cases} \lambda[(i - \sum_{j=1}^{i} \alpha_j)T], i = 1,2, \ldots, N-1 \\ \lambda[(N - \sum_{j=1}^{N-1} \alpha_j)T], i = N \end{cases} \tag{2-5}$$

Since maintenance nodes are made according to the planing period,the degradation will accelerate and the cost will increase over time . Dedopoulos and Smeers proposed that the maintenance cost, time and times satisfy the equation: $C_{p_i} = C_{f_i} + C_{v_i}(T_x)$; YCh.R. H and Lo.H.C also proved the rationality of the preventative maintenance cost as a liner function. Therefore, this paper uses the the equation (2-6) to describe the preventative maintenance cost. And in it, i is the maintenance times, C_f is the fixed cost of each maintenance, C_{v_i} is the variable cost. According to equations(2-3), (2-4), (2-5) and (2-6), the failure rate and the reliability degree of each node can be calculated.

$$C_{p_i} = C_f + iC_{v_i}, i = 1,2 \ldots, N \tag{2-6}$$

3. The regular planned-maintenance strategy and construction of its multi-objective optimization model

Taking maintenance activities according to regular planned-maintenance strategy as in figure 1[11]: according to planning period time T, 2T, ...(N-1)T, taking T_y hours' preventative maintenance each time, breakdown maintenance will be used if failure occurs within the period; when failure occurred after T_1 point, after the maintenance of T_x hours, the equipment will continue to run for T_2 hours until the coming of the next planned maintenance node. At this period, the maintenance at each node is imperfect maintenance. When arriving at the NT-th period node, the equipment will be replaced directly rather than taking preventative maintenance. The most important point of regular planned-maintenance strategy is the determination of planned-period T. This paper takes the period T as the decision variable basing on the regular planned-maintenance strategy, constructs the multi-objective preventative maintenance optimization model whose objective function is equipment availability, total cost and reliability according to the imperfect maintenance method.

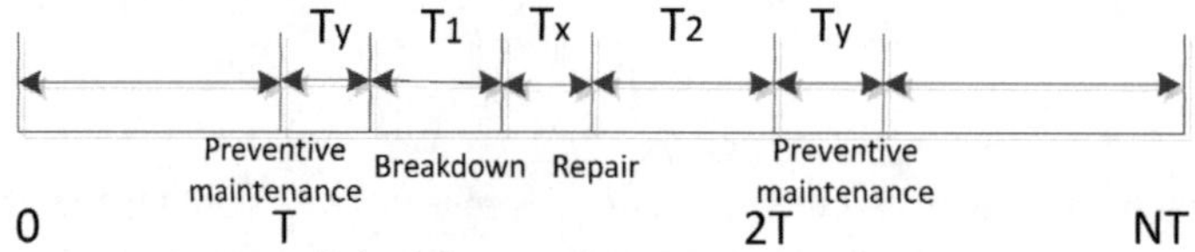

Figure 1 The regular planned-maintenance strategy

3.1. Taking availability degree as objective function

As shown in Figure 1, T_y is the average time of preventative maintenance, T_x is the average maintenance time after failure, T_r is the replacing time, T_0 means that the equipment newly puts into use at zero point, and has no preventative maintenance. MDF(Mean Down Failure) in N periods T can be described as:

$$MDT = (N-1)T_y + \sum_{i=1}^{N} T_x \int_{T_{i-1}}^{T_i} \lambda_i(t)\,dt + T_r \tag{3-1}$$

MUT(Mean Up Failure) can be described as:

$$MUT = NT - \sum_{i=1}^{N} T_x \int_{T_{i-1}}^{T_i} \lambda_i(t)\,dt \tag{3-2}$$

Supposing A is availability degree, with its definition, A in N periods can be described as:

$$A = \frac{MUT}{MUT+MDT} = \frac{NT-\sum_{i=1}^{N} T_x \int_{T_{i-1}}^{T_i} \lambda_i(t)dt}{NT+(N-1)T_y+T_r} \tag{3-3}$$

3.2. Taking the total cost as objective function

Supposing $C(T)_i$ as the i-th period cost, which represents the cost from the ending of T_{i-1} to the end of T_i; supposing C_{p_i} as the cost of preventative maintenance while C_s the cost of breakdown maintenance, which includes practical maintenance cost and downtime lose, hence coming the equation:

$$C(T)_i = C_{p_i} + C_s \int_{T_{i-1}}^{T_i} \lambda_i(t)\,dt \tag{3-4}$$

3.3. Taking the average reliability degree as objective function

From equations (2-5) and (2-6), the reliability degree function R_i of each planned-maintenance node can be calculated, therefore the average reliability degree can be calculated according to equation (3-5):

$$\overline{R} = (\textstyle\sum_{i=1}^{N} R_i)/N \tag{3-5}$$

Taking the regular planned-maintenance period T as the decision variable, basing on equations (2-5),(2-6),(2-7),(2-8),(3-3),(3-4),(3-5),under the regular planned-maintenance strategy, constructing the optimization model with the objective function which has the maximum availability, the lowest average cost and the maximum average reliability degree as in equation (3-6):

$$\begin{cases} A_{max} = \dfrac{NT - \sum_{i=1}^{N} T_x \int_{T_{i-1}}^{T_i} \lambda_i(t)\,dt}{NT + (N-1)T_y + T_r} \\ C_{min} = \displaystyle\sum_{i=1}^{N-1} (C_f + iC_{v_i}) + \sum_{i=1}^{N} C_s \int_{T_{i-1}}^{T_i} \lambda_i(t)\,dt \\ \overline{R}_{max} = \dfrac{\sum_{i=1}^{N-1} R\{[i - \beta \sum_{j=1}^{i} k^{b\cdot j}(C_f + jC_{v_i})^{b\cdot j}]T\} + R\{[N - \beta \sum_{j=1}^{N-1} k^{b\cdot j}(C_f + jC_{v_i})^{b\cdot j}]T\}}{N} \end{cases}$$

Converted into minimum form, the model above equals to:

$$\begin{cases} A_{min} = 1 - \dfrac{NT - \sum_{i=1}^{N} T_x \int_{T_{i-1}}^{T_i} \lambda_i(t)dt}{NT + (N-1)T_y + T_r} \\ C_{min} = \sum_{i=1}^{N-1} (C_f + iC_{v_i}) + \sum_{i=1}^{N} C_s \int_{T_{i-1}}^{T_i} \lambda_i(t)\,dt \\ \overline{R}_{min} = 1 - \dfrac{\sum_{i=1}^{N-1} R\{[i - \beta \sum_{j=1}^{i} k^{b\cdot j}(C_f + jC_{v_i})^{b\cdot j}]T\} + R\{[N - \beta \sum_{j=1}^{N-1} k^{b\cdot j}(C_f + jC_{v_i})^{b\cdot j}]T\}}{N} \end{cases} \tag{3-6}$$

4. Solving multi-objective optimization problem by NSGAII

Genetic algorithm can generate multiple points, which search towards multiple directions at the same time, and one search can obtain several Pareto solution sets. Therefore,genetic algorithm was seen as an important method to solve the multi-objective optimization problem. In 2002, Deb el ta improved non-dominated sorting genetic algorithm (NSGA) proposed by them and got the second generation of NSGA, which was called NSGAII. Compared to the first generation, the second has following improvements: ① dominating sorting quickly. The process of classifying N individuals will continue until all the fronts have completed (deleting individual F1, then classifying solutions'number dominated by these individuals in order that minus one each time), recycling degree is $O(N^2)$ at the worst time. ② using "congestion distance" to distinguish diversity. In NSGAII, calculating the congestion distance by ranking the individual and neighboring individual, and this replaced the first generation of σ_{share} parameters. The smaller the individual distance D_i is , the greater the congestion degree is. ③ using elitist strategy. In the selection of offspring, if the individual group is smaller than the parent group, select them in order ($F_i < F_j$) and fulfill them; if the individual group is greater than the parent group, rank according to the congestion degree and select ($D_i > D_j$), that is:

if (Fi <Fj) or (Fi=Fj) and (Di >Dj)

then i ≺ j//there i,j are selecting orders

5. Example Analysis

Now studying on the signal transmission system of node communication vehicle, as supposed in the third part, related parameters of current transmission system are shown in Table 1. this system includes switches, relays,power supplies, multiplex switching equipment, routers, etc. It mainly switch and transmit the radio-frequency signal. The relay is the most important, and other equipment's analysis process is similar, so the following study takes the relay as the object.

5.1. Analysis of an incomplete maintenance strategy for a life cycle

Firstly, the relay machine is analyzed using the above-mentioned incomplete maintenance strategy model within a lifetime. This section mainly analyzes the failure rate, reliability, and the average reliability, availability, and cost of a service life.

Part parameters of the relay are shown in Table 2.

Table 1. Part parameters of transmission system

System life	Preventative maintenance Period	The average time of preventative maintenance	Downtime lose	Regulator factor k	Regulatory factor b	Importance factor β
30 (year)	1.9 (year)	0.1 (year)	95 (ten thousand)	1	0.005	0.5

Table 2. Part parameters of the relay

System Life	Fixed cost of preventative maintenance	Variable cost of preventative maintenance	Average replacing Time	Replacing cost	Actual maintenance costs	m	t
30(year)	3 (ten thousand)	0.2 (ten thousand)	0.1(year)	50 (ten thousand)	5 (ten thousand)	38	1.2

M is the proportional parameter and t is the shape parameter. For Weibull distribution failure rate, when m is greater, the value of the failure rate is smaller and faster, and when t is greater, the failure rate rises faster. When t is greater than 1, the failure rate increases with time. When t is less than 1, the failure rate decreases with time. For Weibull Distribution Reliability, the smaller the m is, the faster the reliability declines. When t is greater, the reliability function is more fluctuated and the fall is faster. The Weibull distribution simulates the reliability and failure rate of actual weapon equipment. The value of M,t reflects the performance of different equipments, where m takes 38 and t takes 1.2. Set a preventive maintenance period of 2 years.

The system life is 30 years, the preventative maintenance period is 2 years, so the number of preventative maintenance is at most 15 times. Observe changes in age-related retirement factors in Table 3.

Table 3. The changing of age reduction factors

N	1	2	3	4	5	6	7	8
C_{p_i}	3.2	3.4	3.6	3.8	4.0	4.2	4.4	4.6
α_j	0.503	0.506	0.510	0.514	0.518	0.522	0.527	0.531
N	9	10	11	12	13	14	15	
C_{p_i}	4.8	5	5.2	5.4	5.6	5.8	0	
α_j	0.537	0.542	0.547	0.553	0.559	0.565	15	

The fallback factor is 15 in the 15th cycle, that is, the equipment is replaced in the 15th cycle so that the equipment reaches the latest non-destructive state. The site where the equipment was installed was inspected for 40 years, and the failure rate and reliability were observed as the number of cycles changed.

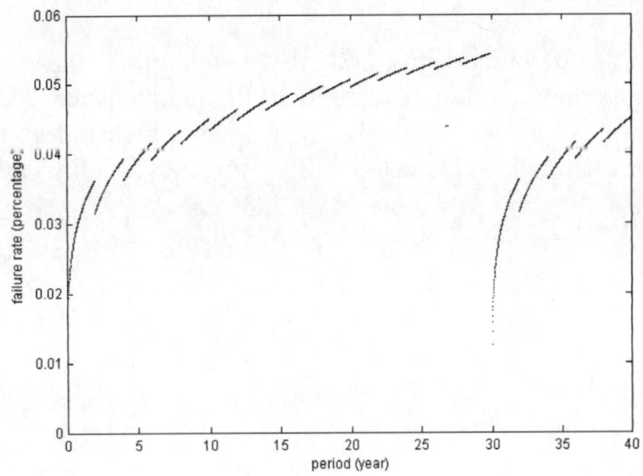

Figure 2. The changing of failure rate along with the period

By observing the year-to-year changes in failure rate in Figure 2, it can be seen that every two years, that is, the maintenance cycle, will return to a previous state after maintenance, and this retreat effort depends on the improved retirement age factor proposed in this paper. When the equipment life is reached and the equipment is replaced, the failure rate returns to the initial state. The overall trend of failure rate change depends on Weibull distribution parameters, and parameters are set by experience based on different equipment.

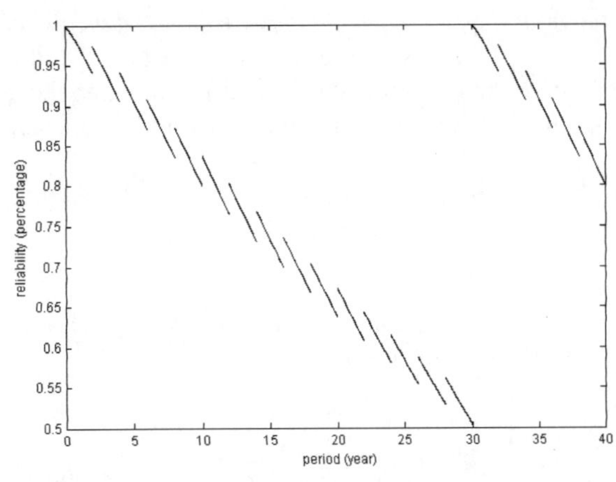

Figure 3. The changing of reliability degree along with the period

Similarly, in Figure 3, it can be seen that the reliability changes with the year, and the reliability returns to a certain state in the previous cycle after maintenance at the maintenance node. Here, when the reliability drops to 50 percent, replace the new parts.

The changing of availability degree and average reliability degree in Figure 4 and the changing of total cost in Figure 5.

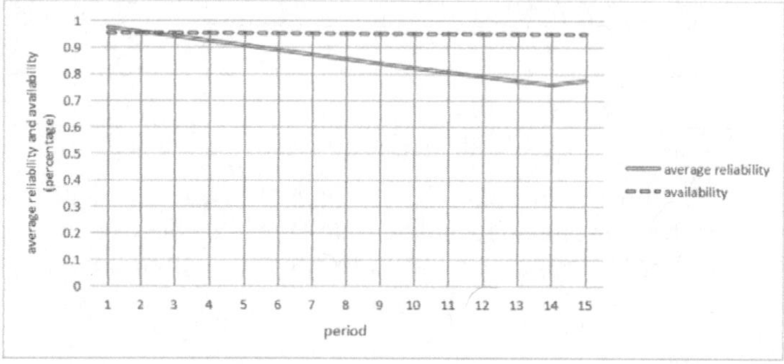

Figure 4. The changing of availability degree and average reliability degree along with the period

By observing the average reliability and availability at the maintenance cycle nodes in Figure 4, it can be seen that the availability remains basically unchanged, which means that the equipment can guarantee a higher utilization rate. The average reliability decreases steadily with time.

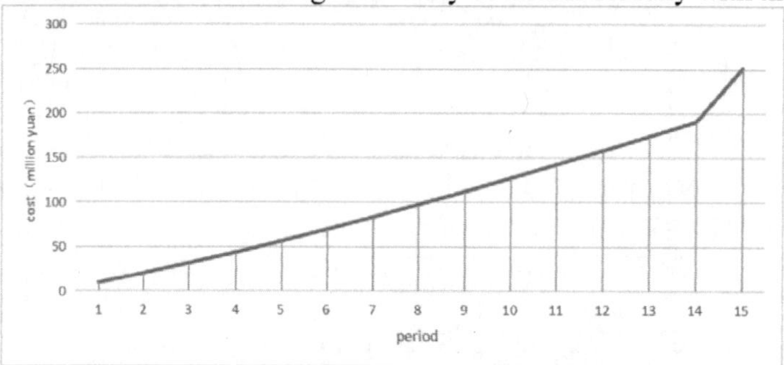

Figure 5. The changing of cost along with the period

Similarly, observing the change in the total cost with the number of cycles in Figure 5, we can see that the total cost has risen steadily, and there has been a jump in the cost of replacement at the end of its life.

5.2. Analysis of optimal maintenance cycle under incomplete maintenance strategy

Setting different maintenance cycles will result in different failure rates and reliability curves, so that the average reliability, availability, and cost indicators are different when the equipment is used up. So we need to find the optimal maintenance cycle so that the above indicators achieve optimal at the same time.

since the availability degree almost unchanged along with the changing of period, the optimal maintenance period should be seek aiming at cost and reliability.

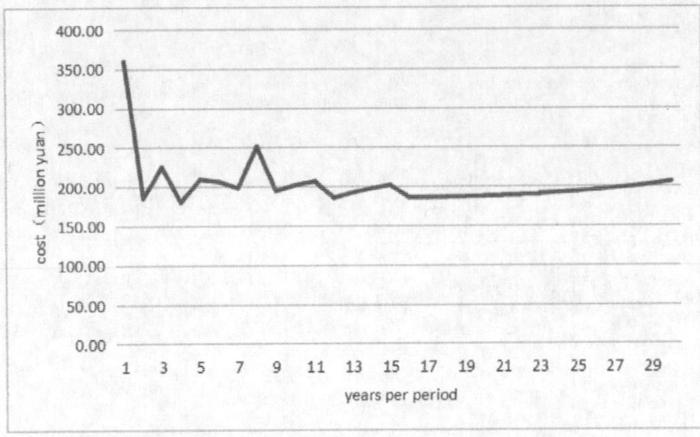

Figure 6. The changing of total cost along with the period

In Figure 6, the occasional increase in cost was due to the replacement of the equipment, which resulted in the increase of the total cost. Meanwhile, when the maintenance period lengthen to 15 years, the average reliability degree rises at the maintenance node, because the maintenance year is the replacement year.

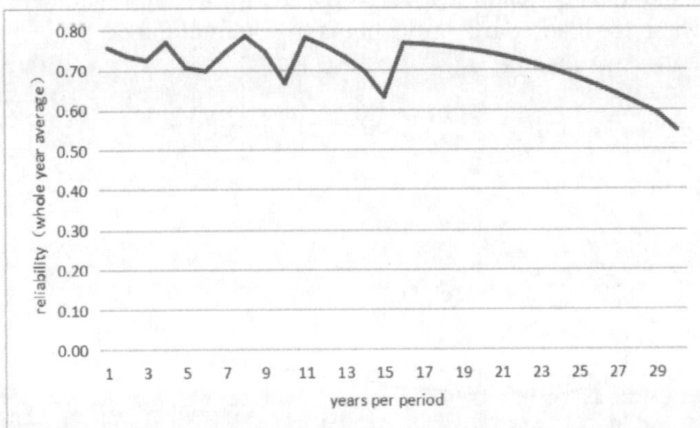

Figure 7. The changing of the whole year's average reliability degree along the period

The mean of the maintenance node can not reflect the real reliability degree of equipment, so it is necessary to calculate the whole year's average reliability degree as shown in Figure 7. The occasional increase in reliability degree is due to the replacement of equipment during the lifetime when the maintenance period can not be divided exactly by the lifetime, hence the average reliability degree increased.

In order to observe the reliability and cost change trend with the maintenance cycle better , we first normalize the data . X is the data set, and x is the original data. The normalized result is as (5-1).

$$\bar{x} = \frac{x - \min(X)}{\max(x - \min(X))} \tag{5-1}$$

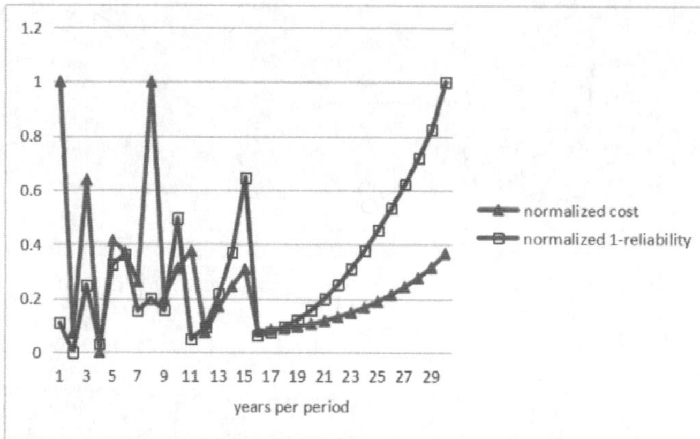

Figure 8. The changing of the normalized 1-reliability degree and normalized cost along with the period.

The reliability and cost are stretched over the maintenance cycle curve so that the data is tiled at the beginning and end of the axis as shown in Figure 8.

The scatter chart of the average reliability degree R, the availability degree A and the total cost C are as in Figure 9, in which the marked triangle point is the optimal Pareto set.

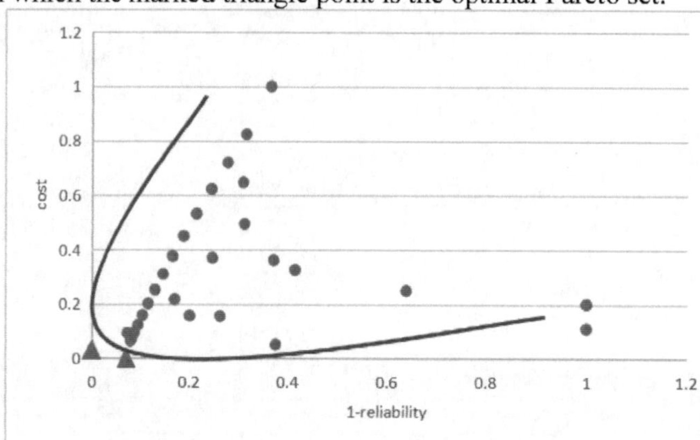

Figure 9. The value of the average reliability degree R, the availability degree A and the total cost C

From the scatter chart of the average reliability and cost, it can be seen that the point at the lower left is the desired scheme, that the cost and (1-reliability) are simultaneously minimized when the scheme is on the convex hull and both of the cost and (1-reliability) have the advantages. , that is,the pareto set, the same as the optimal set.

Base on the actual decision,weighing the importance degree between the cost and reliability degree, selecting appropriate maintenance period as shown in Table 4.

Table 4. The optimal Pareto set

Maintenance period	Cost	The Average Reliability
2	185.12	0.784
4	180.00	0.780

Through experiments, we have found that the optimal maintenance time of the relay under a given parameter is 2 or 4 years in Table 4. The final solution is determined by weighing the importance of actual cost and average reliability.

6. Conclusion

As seen from the verification results, the introducing of improved age reduction factor can better present the changes of equipment failure rate and reliability degree along with the changing times of the period during the imperfect maintenance process, and verify the feasibility of its degradation mechanism. For the enumerable set, the method of fast dominating sorting in NSGA II is used to find the front of Pareto set and then determine its optimal solution, which have certain guiding significance for the formulation of equipment decision-making.

References

[1] Bad'ıa F G, Berrade M D. Optimum Maintenance Policy of a Periodically Inspected System under Imperfect Repair. Advances in Operations Research, 2009,691203:1-13.
[2] M Brown, F Proschan. Imperfect Repair. Application Probability, 1983(20):851-859.
[3] Pham H, Wang H Z. Imperfect maintenance. European Journal of Operational Research, 1996,94(3):425-438.
[4] Bad'ıa F G, Berrade M D. Optimum Maintenance Policy of a Periodically Inspected System under Imperfect Repair. Advances in Operations Research, 2009,691203:1-13.
[5] Yeo W M, Yuan X. Optimal warranty policies for systems with imperfect repair. European Journal of Operational Research, 2009,199(1):187-197.
[6] Qu Yuxiang Research on Maintenance Decision of Deteriorating System Based on Incomplete Maintenance Theory[D], Tsinghua University, 2011.
[7] Tsai T, Liu P, Lio Y L. Optimal maintenance time for imperfect maintenance actions on repairable product. Computers & Industrial Engineering, 2011,60(4):744-749.
[8] Ye Peifan.Study on optimization model of state maintenance strategy based on incomplete maintenance[D].Tsinghua University 2011.
[9] Dedopoulos I T, Smeers Y. An age reduction approach for finite horizon optimization of preventive maintenance for single units subject to random failures. [J].Computer &Industrial engineering, 1998,34(3):643-654.
[10] Zhao Yongqiang,Liang Gongqian.Research of Periodic Preventive Maintenance under Reliability and Dynamic Maintenance Cost[J].Aeronautical Manufacturing Technology,2013(7).
[11] Wang Shiping, Zhu Minbo. Reliability and maintainability of electronic machinery. Beijing: Tsinghua University Press, 2000.100~120..

Multi-objective Optimization of the Rapidly Assembled Helipad Deck

Dan Wei[1,2], Xudong Deng[1], Lixun Zhang[1] and Jiru Gong[1]

[1] College of Mechanical and Electrical Engineering, Harbin Engineering University, China

[2] Haifeng General Aviation Technology CO., LTD, China

Email: zhanglixun@hrbeu.edu.cn

Abstract. Aiming at the potential edge bearing problem of the helipad deck, as well as the contradiction between the deck weight and the bearing capacity, a multi-objective optimization design method is proposed in this paper. On the condition of avoiding the most dangerous taking off and landing situations, carrying out the layout of the plane size ranges and the static analysis of the current size by establishing plane coordinate system and deck size constraint inequality. The deck design variables are confirmed based on the deck weight, processing capacity and other constrains, so as to establish the quadratic polynomial response surface model(RSM) of the objective function, and experimental design sample points which uniformly distributed in the design space are obtained through Latin square hypercube sampling(LHS) method. The multi-objective optimization of the deck is performed using the Screening method to obtain the pareto feasible solutions and the response relationship between the design variables and the objective function, which provides the basis for the design of deck size and connector.

1. Introduction

With the rapid development of aviation industry in our country, helicopters are increasingly used in daily life. Especially in all kinds of emergencies and natural disasters, helicopters are widely used in passenger and cargo transportation, emergency rescue, logistics and so on. It plays an increasingly important role in all these areas. However, limited by complex terrain and other environments, helicopters are often unable to land directly which causes serious loss of life and property because of lack of enough time for rescue. To solve this problem, a kind of helipad can be carried easily and paved rapidly is needed.

The rapidly assembled helipad can be divided into two categories according to whether it has landing legs: 1) One of them is rapidly assembled helipad with legs which is mainly applied to forest cutting, mineral exploration and other fields. It is suitable to be used in mountainous areas, steep slopes and other rough terrain; 2) The other is deck type rapidly assembled helipad[1-3] which is mainly applied to medical rescue, military and other fields.It is suitable to be used in grassland, farmland, desert and gravel environment.

In the aviation emergency rescue area, different types of helicopters are used according to different rescue missions, the number of airborne, land topography and some other conditions. Medium and small-scale skid helicopter is also applied to family rescue and most health assistance. It requires smaller landing area and low requirement for rapidly assembled helipad. Wheeled helicopter is mainly

IOP Conf. Series: Materials Science and Engineering **470** (2019) 012016 doi:10.1088/1757-899X/470/1/012016

applied to major industrial accidents and natural disasters which has great carrying capacity and bigger take-off weight. It requires bigger landing area and high requirement for rapidly assembled helipad.

This paper focuses on helicopter emergency rescue. A size optimization design method of deck type rapidly assembled helipad is put forward in this paper by taking a kind of three-point helicopter for example. The main material of the deck is high density polyethylene. The deck unit can be connected by stainless steel connectors. When we plan the size of the deck unit, we should firstly confirm the size range according to actual rescue conditions, transportation conditions, etc. And then we should conduct dimensional optimization analysis according to the type of support aircraft to confirm the size of the deck unit.

2. Size range of the deck

2.1. Size planning

We should consider the transportation, assembly, loading and disassembly, etc when we plan the size of the deck. The main modes of transportation of the helipad mainly include hoisting and airborne, so the size of the deck unit should meet the size requirement of hoisting and the size of the cabin door. The size of the deck unit should not be too large, otherwise it is inconvenient to carry and install.

The most important thing is to minimize the marginal load after assembling the decks. The relief ground is usually soft grass, fields, sand and so on while high density polyethylene is a kind of elastic material. It is inevitable to have large deformation caused by marginal load[4]. So, we should plan the size of the deck for the helicopter wheel's imprint size to reduce the possibility of marginal load.

2.2. Size range planning

The situation that the wheels land on the deck junction is the first thing to consider and unavoidable. We should try to avoid multiple wheels landing on the deck junction at the same time in the design. We suppose a wheel has landed on the deck junction. Then we should try to avoid the second wheel landing on the deck junction in this case. Only three situations should be analysed because of the symmetrical structure of wheels. The layout of the wheel areas is shown in figure 1. Twelve typical points are chosen: A1 (x_{a1}, y_{a1}), A2 (x_{a2}, y_{a2}), B1 (x_{b1}, y_{b1}), B2 (x_{b2}, y_{b2}), C1 (x_{c1}, y_{c1}), C2 (x_{c2}, y_{c2}), D1(x_{d1}, y_{d1}), D2 (x_{d2}, y_{d2}), E1 (x_{e1}, y_{e1}), E2 (x_{e2}, y_{e2}), F1 (x_{f1}, y_{f1}), F2 (x_{f2}, y_{f2}).

Let the transverse dimension of the deck unit be l_x and the longitudinal dimension be l_y. The junction point of the deck is L (x_L, y_L).

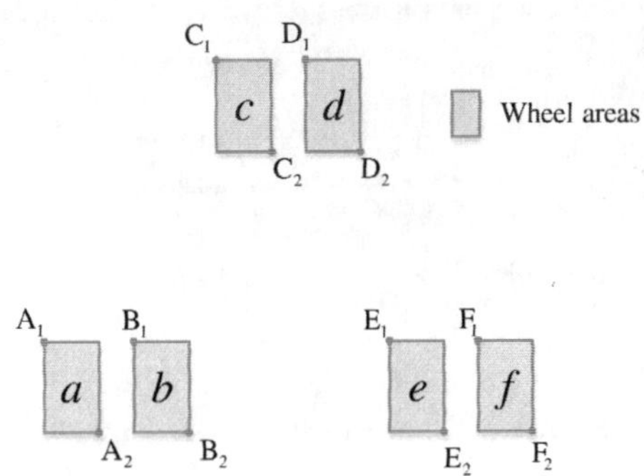

Figure 1. The layout of the wheel areas.

As shown in figure 2, coordinate system O_a-xy can be set up when the wheel a lies on the junction point of the deck.

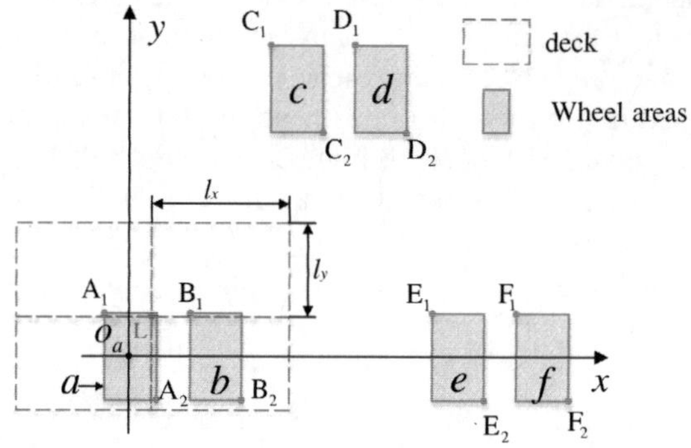

Figure 2. The wheel a lies on the junction point of the deck.

The following conditions need to be met if other wheels do not lie on the edge of the deck unit:

$$
\begin{aligned}
&(n_x l_x + x_{\mathrm{L}} - x_{i1})(n_x l_x + x_{\mathrm{L}} - x_{i2}) > 0 \\
&(n_y l_y + y_{\mathrm{L}} - y_{j1})(n_y l_y + y_{\mathrm{L}} - y_{j2}) > 0 \\
&n_x, n_y = 1,2,3,4......
\end{aligned}
\qquad (1)
$$

where *i=b,c,d,e,f,j=c,d.*

Coordinate system can be set up at the geometric center of the wheel b when the wheel b lies on the junction point of the deck. In this case, *i=a,c,d,e,f,j=c,d.*

Coordinate system can be set up at the geometric center of the wheel c when the wheel c lies on the junction point of the deck.In this case, *i=a,b,d,e,f,j=a,b,e,f.*

2.3. Planar size range planning

In practice, planar size of the deck was restrained by the time of splicing and transportation. So in the following calculation, we set $n_x \leqq 4$, $n_y \leqq 4$, 1000mm $\leqq l_x \leqq$ 2400mm, 1000mm $\leqq l_y \leqq$ 2400mm. The transverse size range and the longitudinal size range of the deck can be shown in figure 3 and figure 4 by solving the restraint conditions in the previous section.

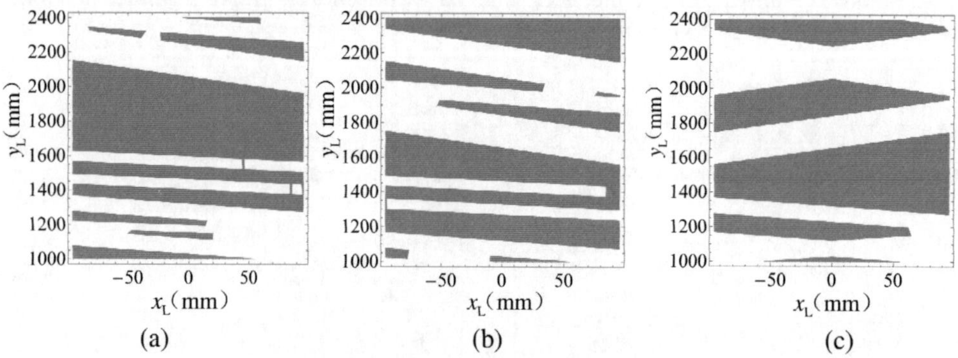

Figure 3. Range of transverse size.

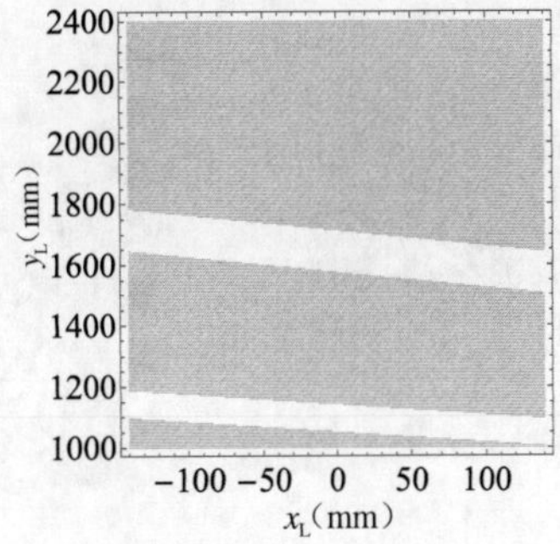

Figure 4. Range of longitudinal size.

As shown in figure 3 and figure 4, the transverse size of the deck unit is sensitive to the distance between wheels, and it's value range is small and discrete. It is because the number of wheels is so large that there are too many restraint conditions on size. The transverse size can be determined to 1500 mm by confirm the size range: $1496.7<l_x<1503.3$. The restraint conditions on longitudinal size are much less so the size range is wider. The process of optimization is shown in figure 5.

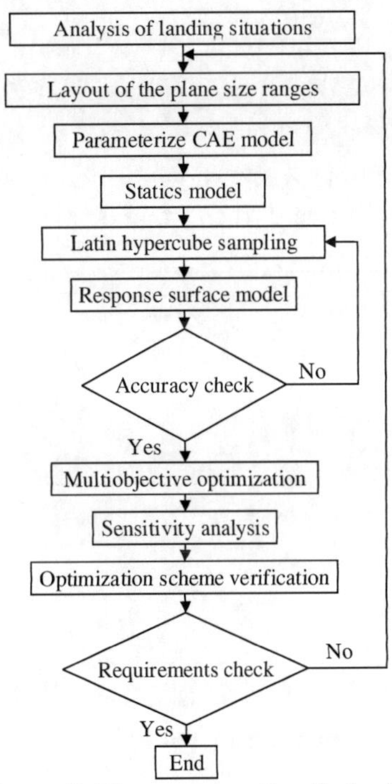

Figure 5. The process of optimization.

3. Static analysis of the deck

3.1. Buildup of model

Size range of the deck thickness can be confirmed according to the planar size range in section 2.3, requirement of deck unit's dead-weight and processing craft. On this basis, the three-dimensional model of the deck can be built as shown in figure 6. l_x stands for the transverse size, l_y stands for the longitudinal size, h represents for the thickness of the deck, d represents for the the diameter of the linking hole. As shown in figure 7, the finite element mesh can be refined partly according to the bearing condition and the contact surface.

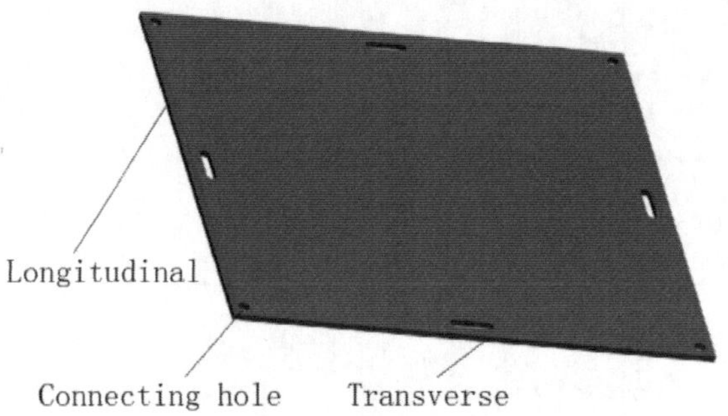

Figure 6. 3D model of the deck.

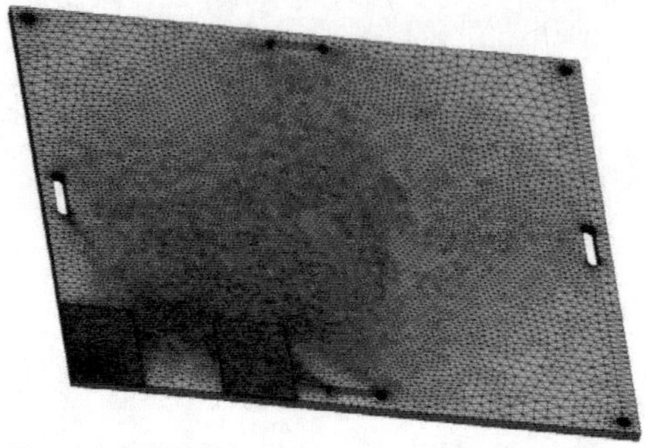

Figure 7. Finite element model of the deck.

3.2. Static analysis

High density polyethylene is chosen in this paper of which the density is 952Kg/m^2 and the elasticity modulus is 1.07×10^9N/m^2. The relational expression is shown as follows according to the constitutive model[5-6] researched by Kown, etc.

$$Y_{(\varepsilon)} = \begin{cases} E\varepsilon & (\varepsilon \le \varepsilon_y) \\ d\left[a(\varepsilon+b)^{(c-1)} - a(\varepsilon+b)^{-c}\right]+e & (\varepsilon_y \le \varepsilon \le \varepsilon_n) \\ \alpha k\varepsilon^N & (\varepsilon_n \le \varepsilon \le \varepsilon_t) \\ k\exp(M\varepsilon^\beta) & (\varepsilon \ge \varepsilon_t) \end{cases} \tag{2}$$

The stress-strain curve fitted in MATLAB is shown in figure 8. As we can see from the picture, the relation between stress and strain is linear when strain is less than ε_y.

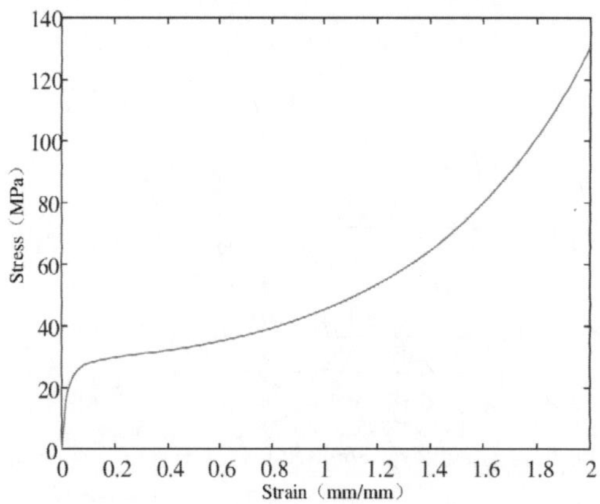

Figure 8. Stress-strain curve.

High density polyethylene is a kind of high-molecular polymer. Some filler and auxiliaries would be added in the deck during the course of fabrication. And it's mechanical property could be affected by temperature and density greatly. So, we calculate the allowable stress of the material according to 1.5 times the safety factor.

The helicopter has a certain speed while landing vertically which causes a dynamic load on the deck[7-8]. Therefore, load factor should be taken into consideration when conduct static analysis. The vertical force F_z between the wheel and the landing surface is:

$$F_z = (1+C_\delta\dot{\delta})C_\delta\delta^n \tag{3}$$

In the formula, C_δ stands for tire force coefficient, $\dot{\delta}$ stands for the compression of the tire, n stands for the tire force index. The relation between the magnitude of F_z and the draft gear travel could be described by an appropriate clothoid of which the magnitude of the peak is related to weight, buffer performance and landing speed. When the helicopter drops at a large speed for hard landing, we set the load factor to 3.5 and consider the safety factor for load of 1.5. The deck is loaded with a constant while the linking hole is subjected to horizontal force. The static analysis result of the deck is shown in Table 1.

Table 1. Static analysis result of the deck.

Mass of the deck unit M (kg)	Maximum equivalent stress Σ(MPa)	Maximum equivalent strain E(mm/mm)
13.748	13.748	0.129

4. Buildup of response surface model

4.1. Confirming the design variables

For the rapidly assembled parking apron, the larger the deck unit is, the shorter the assembly time will be, the main influence factors are the transverse length l_x and the longitudinal length l_y. The larger the deck thickness is, the stronger the bearing capacity will be. But the deck would be too heavy to carry if the transverse length l_x, the longitudinal length l_y and the deck thickness are too large. Though the diameter of the linking hole D has little effect on weight and maximum equivalent stress, a nonnegligible transverse force would generate which would operate on the linking hole and the connector. In this respect, the diameter of the linking hole and the edge distance become important factors of transverse force. l_y, h, D are chosen as the design variables while the transverse length has been determined in section 2.3 and the value range is shown in Table 2.

Table 2. Value range of the design variables.

Design variables	Value range(mm)
l_y	(1200, 1500)
h	(20, 30)
D	(28, 38)

4.2. Latin Hypercube Sampling

The choice of test points is very important when constructing the response surface while choosing the test points randomly could influence the construction of response surface. There are three design variables in the optimization of the deck and take samples by LHS. LHS is helpful to select sample points of experiment design more evenly by optimizing the distribution of the input values. 15 sample points are selected and the design space with two factors is shown in figure 9 in which the sample points are equally distributed.

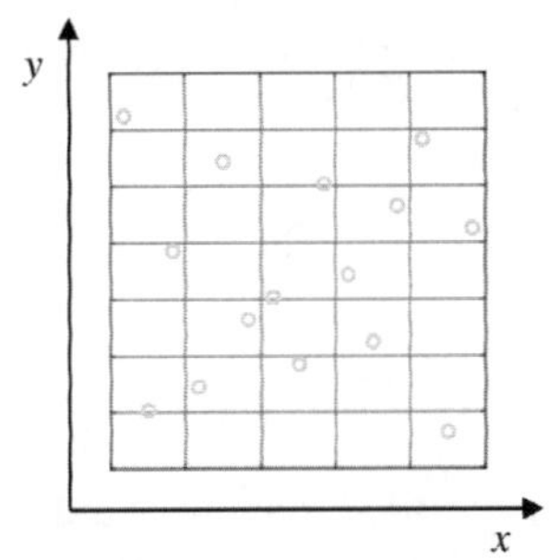

Figure 9. Distribution of test points.

4.3. Quadratic polynomial response surface

Considering that the design variables and the variable to be optimized are not completely linear, the quadratic polynomial response surface model is adopted to predict non-experimental points response. The mathematical model expression is:

$$\hat{y}_{(x)} = a_0 + \sum_{i=1}^{n} b_i x_i + \sum_{i=1}^{n} c_{ii} x_i^2 + \sum_{1 \le i < j \le n}^{n} d_{ij} x_i x_j \tag{4}$$

In the formula, $\hat{y}_i$ stands for the response surface prediction, a_0, b_i, c_{ii}, d_{ij} stand for the constant term and the undetermined coefficients of the primary, secondary and cross terms respectively, x_i, x_j are design variables and n is the number of design variables. Experimental design points should be subjected to polynomial regression analysis to verify the fitting accuracy of the response surface model.

The undetermined coefficient vector A(a_0, a_1..., a_s) could be acquired by least squares method[9-10]. And that:

$$\boldsymbol{A} = (\boldsymbol{D}^T \boldsymbol{D})^{-1} \boldsymbol{D}^T \boldsymbol{Y} \tag{5}$$

$$s = (n+1)(n+2)/2 \tag{6}$$

$$\boldsymbol{D} = \begin{bmatrix} D_{11} & \cdots & D_{s1} \\ \vdots & D_{ij} & \vdots \\ D_{1L} & \cdots & D_{sL} \end{bmatrix} \tag{7}$$

In the formula, s stands for the number of the undetermined coefficients, D stands for the basis function matrix, Y is the response vector at L-th test point which could obtained through AWB DX, D_{ij} is the value of the design variables' derivation function corresponding to the i-th regression coefficient at the j-th test point.

The quadratic non-liner polynomial is converted to the linear regression to verify the fitting situation of the sample points and use the multi-correlation index[10] R^2 to evaluate the fitting situation :

$$R^2 = 1 - \frac{\sum_{i=1}^{N} (y_i - \hat{y}_i)^2}{\sum_{i=1}^{N} (y_i - \overline{y}_i)^2} \tag{8}$$

In the formula, N is the number of the test points, y_i is the actual response value, $\overline{y}_i$ is the average value of the actual response values. The closer R^2 is to 1, the higher the fitting accuracy of the response surface model. The multi-correlation index of the equivalent stress σ in the quadratic response model is 0.9887, the relation between the objective function m and the design variables contains only constants and cross terms theoretically, The multi-correlation index of mass m is infinitely close to 1, indicating that the quadratic response model fits well and can be used for optimization calculation instead of the actual model.

5. Multi-objective optimization

5.1. Optimization problems and mathematical models

The unit deck area of the rapidly helipad should be as large as possible, so as to reduce the assembling time and workload. As the horizontal length of the deck has been insured, its longitudinal length should be as large as possible. At the same time, the mass and equivalent stress of deck should be as small as possible. The mathematical model of the deck multi-objective optimization can be described as follows:

$$\begin{aligned} &\min(m,\sigma) \\ &\max(l_y) \\ s.t.\quad & m_{(X)} \le m_U \\ &\sigma_{(X)} \le \sigma_{von} \\ &X_L \le X \le X_U \end{aligned} \tag{9}$$

Where, $m(X)$ is the mass of unit deck, $\sigma(X)$ is the equivalent stress of deck, m_U is the upper mass limit of unit deck, σ_{von} is the upper equivalent stress limit of deck, $X=(l_y, h, D)^T$, X_L is the lower limit of design variables, X_U is the upper limit of design variables.

5.2. Multi-objective optimization

In multi-objective optimization, objective functions are constrained with each other. Thus, the optimization results are usually some non-inferior solutions which also named as Pareto sets. The optimization is just a initialy size design for the deck. Therefore, it's carried out by Screen in AWB DX, and pareto feasible solutions[11] are shown in figure 10. In order to obtain a large domain, select 6 candidate points in 500 sample points.

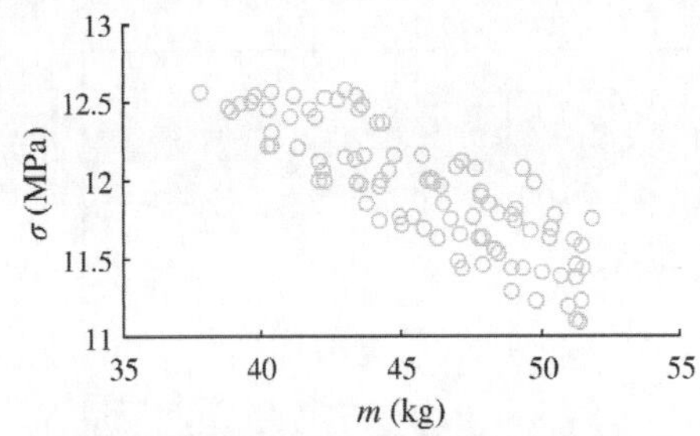

Figure 10. Pareto feasible solutions.

Table 3 gives 6 design scheme corresponding to the feasible solutions, which can be selected for different design conditions and requirements.

Table 3. Design scheme of feasible solutions.

Serial number	Unit deck mass M (kg)	Equivalent stress Σ (MPa)	longitudinal length l_y(m)
1	43.41	12.54	1.44
2	50.19	12.41	1.49
3	41.32	12.20	1.36
4	48.46	11.79	1.41
5	38.78	12.47	1.34
6	49.83	12.07	1.43

To further illustrate the effect of design parameters on objective function, partial sensitivity analysis charts are presented this paper. Figure 11 shows the effect of different design parameters on the maximum equivalent stress on the deck, as can be seen from the figure, the diameter of the connecting holes and the thickness of the deck have negative effects on the maximum equivalent stress, and the diameter of the connecting holes has the greatest influence on it.

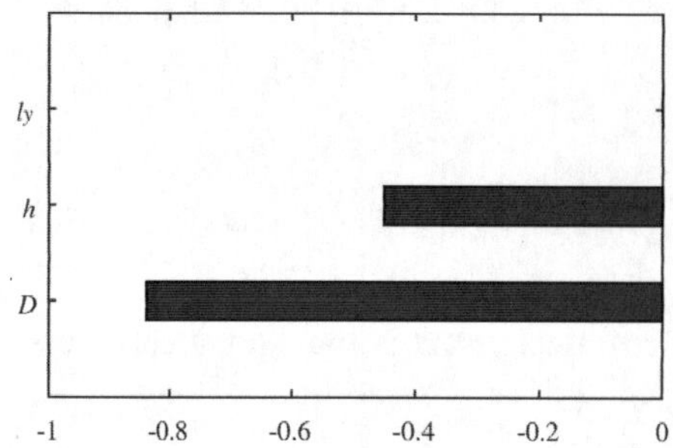

Figure 11. Sensitivity of maximum equivalent stress.

In order to further study the effect of the diameter of the connecting holes on the maximum equivalent stress, the effect of the connecting holes' diameter on the maximum equivalent stress is presented in figure 12. It can be seen from figure 12, as the diameter of the connecting holes increases gradually, the decrease of the maximum equivalent stress on deck gradually slows down. There is a roughly negative square correlation between the maximum equivalent stress and diameter of the connecting holes, which theoretical supports for the design of connecting holes.

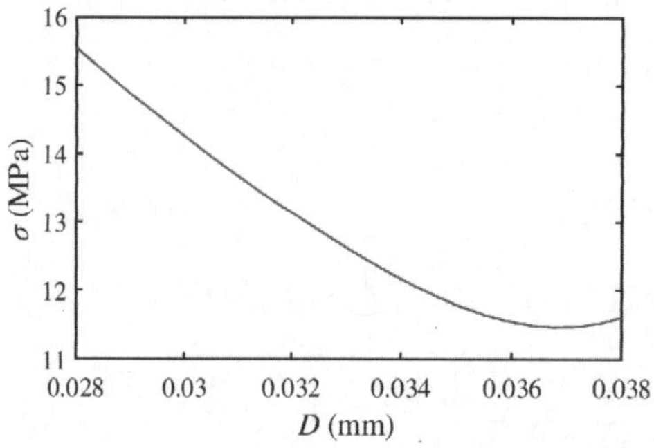

Figure 12. Effect of diameter on maximum equivalent stress.

Figure 13 shows the effect of different design parameters on the deck mass. Due to the existence of the connecting holes and the long holes, the longitudinal length and thickness do not have the same effect on the mass of deck. It is consistent with the theoretical results. Besides, the diameter of connecting holes has little effect on the mass of deck.

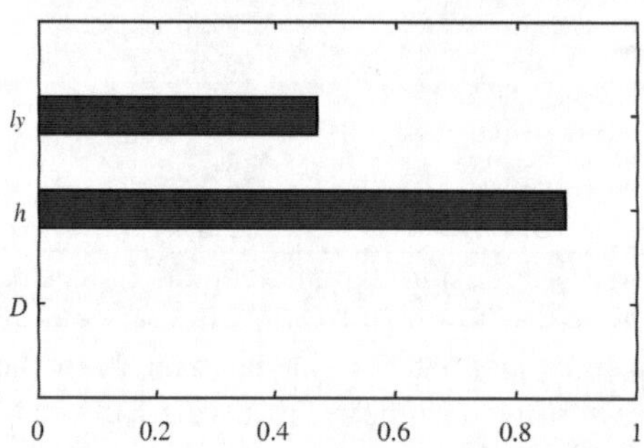

Figure 13. Sensitivity of the mass.

This paper chooses number 4 design scheme of feasible solutions, making round the design variables, take D=36.0mm, l_y=1.4m, h=24.4mm, and the comparison with the result of pre-optimization is shown as Table 4. After optimization, on the basis of the original plane size, the mass of unit deck decreased by 2.46%, and the maximum equivalent stress decreased by 14.17%. Obviously, it's comprehensive performance was greatly improved, which could be used as a preliminary size design.

Table 4. Parameters comparison of the optimization.

Parameters name	Before optimization	After optimization
Mass of unit deck m(kg)	49.421	48.206
Maximum equivalent stress σ(MPa)	13.748	11.8
Longitudinal length l_y(m)	1.4	1.4

6. Conclusions

(1) In view of the most dangerous conditions of deck, the plane coordinate system and inequality group are established, which determines the plane size range of the deck. For two-wheeled helicopters, horizontal length of the deck has a small desirable range while longitudinal length has a wide desirable range.

(2) Deck made of High-density polyethylene (HDPE) has a good wheeled-bearing property, and equivalent strain could be controlled under the range of linear elastic strain while determining the thickness of the deck.

(3) For wheel-type helicopters, longitudinal length and thickness are the main factors affecting the mass of unit deck, and the thickness has a greater effect.

(4) Horizontal force has great effect on the connection of deck and designers can never neglect it. There is a roughly negative square correlation between the maximum equivalent stress and diameter of the connecting holes. In general design conditions, the diameter of the connecting holes has enough range of adjustment under the condition of allowable stress, which provides reference for design and adjustment of connectors.

(5) The design route and multi-objective optimization method adopted in this paper can identified the global pareto feasible solutions, which provides design reference for helicopter decks' preliminary design of different design requirements and applications, and the design of high density polyethylene sheets.

References

[1] Timothy W. Rushing and Isaac L. Howard 2015 *J. Terramechanics* 58
[2] Fan H.L., Liu F.J., and Long, H. 2010 *J. Terramechanics* 48
[3] Doyle, J.D., Howard, I.L., Gartrell, C.A., Anderton, G.L., Newman,J.K., Berney IV and E.S. 2014 *Int. J. Geomech* 161–70
[4] Rushing, T. W., and Howard, I. L. 2011 *Transp. Res. Rec* 92–101
[5] H.J. Kwon and P.-Y.B. 2008 *Int. J. Solids. Struct* **45** 3526
[6] S. Muhammad and P.-Y.B. Jar. 2013 *Finite. Elem. Anal. Des* 70
[7] Zhang, W., Zhang Z., Zhu Q.D. and Xu S.Y., 2009 *Chinese. J. Aeronaut* 374
[8] C. Howcroft, M. Lowenberg, S. Neild and B. Krauskopf 2013 *J. Aircraft* 1908–22
[9] Sun, G. Y., Li, G. Y., Zhou, S. W., Xu, W., Yang, X. J and Li,Q. 2011 *Struct. Multidiscip. Optim* **44** 111–24
[10] Zheng J., Shao, X.Y., Gao, Liang., Jiang, Ping. and Qiu, H.B. 2014 *Expert. Syst. Appl* 2117
[11] Sun, G. Y., Li, G. Y., Zhou, S. W., Xu, W., Yang, X. J., and Li, Q. 2011 *Struct. Multidiscip. Optim* **44** 107

On-Policy Learning for the Swing Process Control of a Cutter Suction Dredger

Changyun Wei, Xiujing Chen and Fusheng Ni

College of Mechanical and Electrical Engineering, Hohai University, China

E-mail: c.wei@hhu.edu.cn

Abstract. It is hard to describe the swing process of a cutter suction dredger with accurate mathematical models because the dynamic characteristics of the swing process are complex, and the relationship between the parameters is not clear. Currently, the swing process control depends entirely on the operators, and it is sometimes difficult to obtain a high production efficiency and construction accuracy. In this paper, an approach that combines SARSA-Lambda with a linear neural network is proposed to solve the intelligent control of the swing process. The dynamic model of the swing process is built using the generalization capability of the linear neural network that solves the output problem of the continuous state space. SARSA-Lambda is used to realize the adaptive control of the swing process by means of learning. The simulation results show that the proposed approach can quickly and effectively learn and achieve goals in uncertain environmental conditions and achieve better control consequents.

1. Introduction

Dredging equipment is widely used in the national economic infrastructure such as the construction and maintenance of port and waterways,flood control and dredging as well as environmental renovation [1-3]. Among them, the cutter suction dredger is a kind of dredging equipment that is widely used in dredging projects (see Figure 1). It mainly cuts the underwater soil by means of a rotary cutter, and then uses a centrifugal pump to carry out long-distance transportation in the slurry mixture. It can be used in river navigation, rivers, lakes and offshore operations. In the dredging operation of cutter suction dredgers, various factors must be considered to adjust the lateral movement speed in real time to ensure safe and efficient operation of dredgers [4].

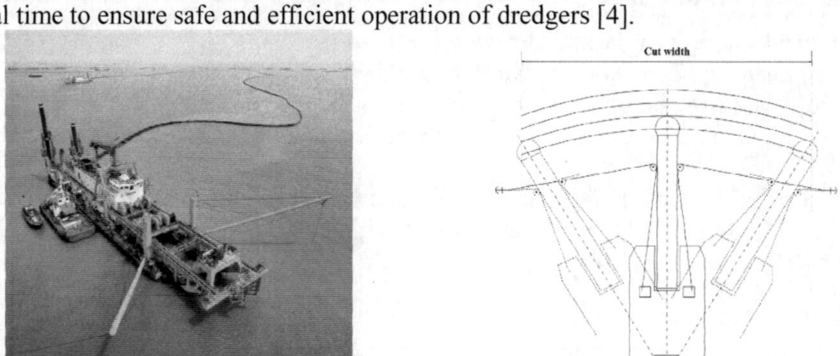

Figure 1. A cutter suction dredger and its swing process.

At present, the control of the swing speed is still in the manual operation stage. Due to the dredging area environment, in order to make the dredger production relatively stable, operators need to continuously adjust the swing speed by operating the traverse handle according to their long-term dredging experience [5]. However, the dredging operation environment is not always constant. Different environments and soil quality have a great impact on dredging. Therefore, there are certain limitations that the operator uses the previous construction experience to handle the current dredging construction. In addition, the production efficiency and completion of dredging operations are affected by the operator's physical fatigue and personal work ability [6,7]. In order to reduce the dependence of the dredging construction process on personnel, it is necessary to develop a swing process control system that is highly efficient, accurate, and highly automated.

Reinforcement learning (RL) is an important machine learning method, which was first proposed by Minsky, emphasizing the environment-based action to maximize the expected benefit [8]. In RL algorithms, SARSA-Lambda is an on-policy learning algorithm that can perceive the environment and choose the optimal action through self-learning [9]. The learning agent has strong robustness and adaptability and this it is suitable for dealing with multi-dimensional nonlinear systems. Because of the need to connect to the environment in SARSA-Lambda, this paper uses the generalization ability of neural network in the simulation experiment. The linear neural network is applied to SARSA-Lambda and realizes the continuous state space value within a certain range. In this paper, SARSA-Lambda is applied to control the swing system of a cutter suction dredger, which is more efficient and more automatic than conventional control.

2. SARSA-Lambda Learning for Swing Process Control

SARSA-Lambda is a value-based on-policy method for learning the optimal policy [10]. It mainly obtains corresponding rewards in the environment through the action of agent output, and then adjusts the agent behaviour policy according to the reward [11]. SARSA-Lambda establishes the Q table and eligibility traces E table, and selects actions based on the strategy where all Q values corresponds to the current state of the Q table. After that, the Q table will be updated according to the E table and the E table will also be updated at the same time.

2.1. SARSA-Lambda learning model

The learning agent in SARSA-Lambda chooses actions according to Q value in Q table and action space under an unknown environment. The basic process can be described as follows: at the current time t, the agent receives the environmental feedback state s_t, reward r_t and action space A_t, and selects the action a_t in the Q table through the strategy π. Afterwards, the action a_t is feed back to the environment, and then the state is transferred to the next moment, in addition to determining its action space and rewards. Finally, according to the E table, the Q value is updated, and the E table is updated at the same time. Based on the given strategy π, $Q(s,a)$ is a function of the value of the state s and the action a ,then:

$$Q^{\pi}(s,a)=E^{\pi}\left(\sum_{t=0}^{\infty}\gamma^{t}r_{t}\mid s_{0}=s,a_{0}=a\right) \tag{1}$$

where s_0 is initial state, a_0 is the initial action. According to the Bellman equation [11], $Q(s,a)$ satisfy:

$$Q^{\pi}(s_t,a_t)=\sum_{s_{t+1}}\left[p(s_t,a_t,s_{t+1})r(s_t,a_t,s_{t+1})\right]+\gamma\sum_{s_{t+1},a_{t+1}}\left[p(s_t,a_t,s_{t+1})Q^{\pi}(s_{t+1},a_{t+1})\right] \tag{2}$$

where $p(s_t,a_t,s_{t+1})$ is the probability of transferring it to the next state s_{t+1} in the agent's current state s_t and action a_t; $r(s_t,a_t,s_{t+1})$ is reward which is obtained when transferring it to the next state

s_{t+1} in the agent's current state s_t and action a_t; γ is a discount factor which weights immediate rewards relative to future rewards, and its value is between [0,1].
After each action, the Q value needs to be updated. The Q table is updated according to the following rule:

$$Q_{t+1}(s,a)=Q_t(s,a)+\alpha\delta_t e_t(s,a) \tag{3}$$

where α is the learning rate. Then the δ_t and $e_t(s,a)$ can be updated as follows:

$$\delta_t = r_{t+1}+\gamma Q_t(s_{t+1},a_{t+1})-Q_t(s_t,a_t) \tag{4}$$

$$e_t(s,a)=\begin{cases}\gamma\lambda e_{t\text{-}1}(s,a)+1, s=s_t, a=a_t\\ \gamma\lambda e_{t\text{-}1}(s,a), otherwise\end{cases} \tag{5}$$

Where $e_t(s,a)$ is the current eligibility traces, λ is the trace decay parameter. Decrease the $E(s,a)$ value with time, the farther away from the reward, the smaller the "indispensability". The E table is updated for all s, a on the basis of the following rule:

$$E=E*\gamma*\lambda \tag{6}$$

Because SARSA-Lambda emphasizes action based on environment, the environment is essential. Thus, the environment setup of the swing process is introduced in the next section.

2.2. Swing Process Analysis

To analyse the swing process of a cutter suction dredger, the state and action parameters need to select so as to build the environment model. In dredging operations, the performance of the dredger is directly reflected on the output to some extent, and also directly determines the project benefits [5-7]. The production of a cutter suction dredger is calculated by:

$$W=QC_w t=(\pi r^2 v)C_w t \tag{7}$$

where W is the production; Q is the flow in the pipe; C_w is the mud concentration; r is the radius of the mud pipe; v is the flow rate in the pipe; t is the time. The flow rate of the pipeline changes little during the dredging process, so the mud concentration directly reflects the production. The formation of mud within the pipeline goes through three main processes:

Cutter cutting soil: The output of the dredger is determined by the quality of the soil that is cut by the cutter. The volume of the soil V is calculated as follows:

$$\begin{aligned}A_c &= kb_c h_c\\ V &= A_c V_S\end{aligned} \tag{8}$$

In the above equation, A_c is the cutting cross-sectional area perpendicular to the swing speed; k is the scale factor; h_c is the cutting depth; b_c is the cutting width; V_S is the swing speed. After cutting, small particles of the soil will be sucked up through the suction pipe to form a slurry. The particle size of silt particles is related to the thickness d_d :

$$d_d=\frac{60V_S}{Zn} \tag{9}$$

In the form, Z is the number of the cutter teeth, and the n is the cutter speed. In the pipeline, the percentage of soil volume is f , the volume of sediment V_{in} entering the pipeline per unit time can be:

$$V_{in}=fV_m \tag{10}$$

In the formula, V_m is the volume of sediments. However, there are missing losses η in this process, which can be characterized by the slippage rate ($0\leq\eta\leq 1$). The final dredging output is:

$$W = (1-\eta)A_c V_s \tag{11}$$

Combing Equation (7) and (11), we can get:

$$C_w = f(\lambda, \eta, A_c, v)V_s \tag{12}$$

Where λ is the deposition loss rate during the slurry pipeline transportation. The motor-driven cutter performance directly reflects its cutting ability. By analysing the data collected at the construction site, the cutter motor current is selected. In the dredging operation, the suction vacuum is installed on the upper part of the cutter and is very sensitive to the change of mud concentration in the pipeline. Once the cutter sucks the suction, the suction vacuum will change. Therefore, the suction vacuum number reflects the production rate of the dredger in real time. In actual operation, the operator adjusts the swing speed to ensure that the slurry concentration is stable within a certain range. In summary, the action in reinforcement learning is the swing speed V_S, and the environmental status parameters include the cutter motor current I_d, suction vacuum p_v and mud concentration C_v.

2.3. Learning Model for Swing Process Control

The SARSA-Lambda learning model for swing process control is depicted in Figure 2.

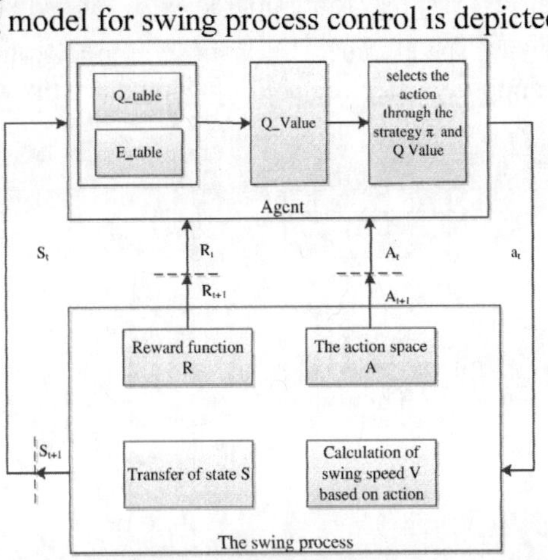

Figure 2. SARSA-Lambda learning model for swing process control.

We use S=[I_d, p_v, C_v] to define the state space of the swing process, and the reward function r is expressed by the following formula:

$$r = \begin{cases} -10 & I_d \in (0,500) \cup (950,960), b \in (-10,-20) \cup (-70,-100) \\ C_v & I_d \in (500,950), b \in (-20,-70), C_v \in (0,45) \\ 10 * C_v & I_d \in (500,950), b \in (-20,-70), C_v \in (45,60) \end{cases} \tag{13}$$

Before performing intelligent control, it is necessary to establish and improve the Q and E tables in the agent. In the SARSA(λ) process, the action a_t is selected according to the state s_t and the action space A_t under the strategy π. Afterwards, the Q table and E table are updated using the reward value generated by the action a_t.

Due to too many states in the process of swing process, it is impossible to set the Q table and E table directly, so the Q table and the E table need to be explored by agents themselves. The setting of the strategy π directly affects the exploration ability of the agent,and its value is between [0 , 1], which is expressed as the random selection action in the current state. And the probability of 1-π chooses the behaviour according to the optimal Q value of the Q table. If π is too small, the agent will not be able to fully explore the environment; otherwise, it will cause the agent to be unstable in system control. Therefore, this paper sets that as the number of explorations continues to increase, strategy π gradually decreases in the Q table exploration process. The strategy π can be calculated as:

$$\pi = \pi_0 - \frac{e}{e_{\max}} \pi_0 \tag{14}$$

In the formula, π_0 is the initial strategy; e is the current exploration round number; $e_{\max}$ is the maximum number of exploration rounds. When the Q table is established and perfected, set the initial strategy $\pi_0 = 0.5$, the maximum number of exploration rounds $e_{\max} = 500$, the agent constantly explores the state space, obtain and update the Q value. After the Q table and E table is complete, a swing process intelligent control simulation test can be performed.

3. Simulation and Results

In the simulation experiment, we set the SARSA(λ) parameters as follows: $\alpha = 0.01$, $\gamma = 0.9, \pi = 0.1$. The initial state for the swing process simulation is s_0=[939.08,-54.09,30.91], and the swing speed is V_S=0. The simulation test results are compared with the actual manual operation shown in Figure 3. We can see that the swing speed under the proposed model control fluctuates greatly because there is a 10% probability of randomly selecting actions. However, the maximum change range of the swing speed is [-2,2].

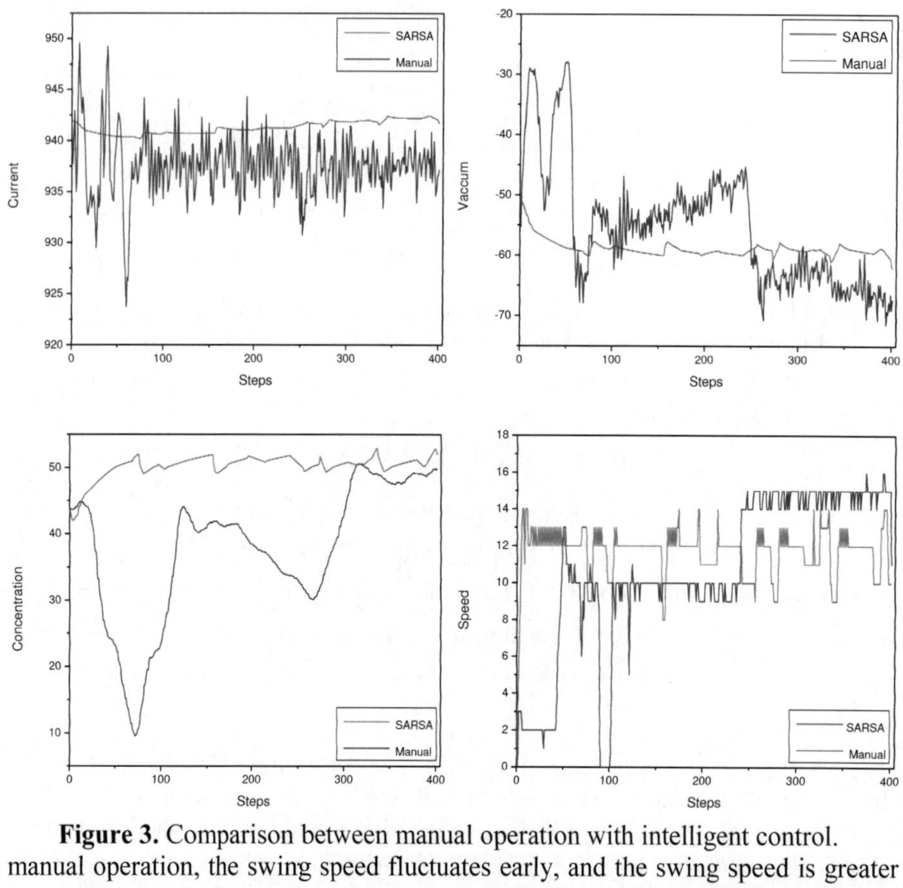

Figure 3. Comparison between manual operation with intelligent control.

In the manual operation, the swing speed fluctuates early, and the swing speed is greater at lower concentrations. However, under the control of the SARSA(λ), the swing speed fluctuates less, and the swing speed is lower than the manual operation. In addition, the cutter motor current is smaller than manual operation and fluctuates without overload. At the same time, suction vacuum fluctuations are small, so it is not easy to have suction or over-absorption. In the same way, the mud concentration also reaches relatively fast and stabilizes at high concentrations. In summary, we can claim that the overall SARSA(λ) operation is more stable and efficient, and the total production volume can be higher than manual operation.

4. Conclusions

The operation of a cutter suction dredger is affected by various uncertainties such as soil quality, environment, and ship status. It is difficult to describe the complex and dynamic characteristics of the swing process with an ideal model. Therefore, it is often not possible to maintain optimal control during actual construction. In this paper, the proposed approach is adaptive to the swing process control model of a cutter suction dredger based on SARSA(λ). When the swing process environment model is changed, the swing speed can be adjusted independently to achieve the best efficiency.

The comparison between the simulation results of the SARSA(λ) model with the actual operational results shows that the control effect of this model is obviously better than the manual operation, and the optimization speed is fast and relatively stable. This model connects the SARSA(λ) algorithm theory with the actual control object, and thus it has certain significance and application value.

Acknowledgments

This research is supported by the National Natural Science Foundation of China (Grant No. 61703138), and Natural Science Foundation of Jiangsu Province (Grant No. BK20170307).

References

[1] Li W. Analysis of China Dredging Market. *Port and Economy*, 2006(1):38-39.
[2] NI F. Review of dredging equipment development at home and abroad. *Journal of Hohai University Changzhou Branch*, 2004(1):1-9.
[3] Wang Y, Xie H and Wu Y. Present State and Development of Domestic and International Dredging Markets. *Journal of SSSRI*, 2004(2):109-113.
[4] Zhu W,Ni F, Wei C,and Wang S. Modelling and analysis of dredging control system based on state space method. *Journal of Computer Applications*, 2016,36(S2): 120-124.
[5] Wei C, Ni F, Dai W. The predictive control for the swing process of a cutter suction dredger. *Machine Design and Manufacturing Engineering*, 2018(1):70-74.
[6] Jiang S, Ni F, Wei C, and Li H. Research on the Multi-loop Control for Traverse System of Cutter Suction Dredger. *Process Automation Instrumentation*, 2016(12):62-66.
[7] Zhang D, Jiang S, Li Z. Research of an Improved Smith Production Control Method for Cutter Suction Dredger. *Machinery and Electronics*, 2015(7):57-60.
[8] Ge J, Zhou L. Real-time traffic signal control model based on SARSA. *Computer and Engineering and Appliccations*, 2015, 51(24):244-248.
[9] Xue L, Yin C, and Li L. Fused Sarsa-lambda learning algorithm based on multi-agent. *Computer Engineering and Application*, 2008, 44(4):182-184.
[10] Lin L, Wang H, and Wang Y. Sarsa Reinforcement Learning Algorithm based on Neural Networks. *Computer Technology and Development*, 2006(1):30-32.
[11] Eamat P, Mohsen P, Maziar P. Multi-criteria expertness based cooperative method for SARSA and eligibility trace algorithms. *Business Media New York*, 2015(43):487-49

IOP Conf. Series: Materials Science and Engineering **470** (2019) 012018 doi:10.1088/1757-899X/470/1/012018

A Path Planning Method for Robotic Wire and Arc Additive Manufacturing of Thin-Walled Structures with Varying Thickness

Guocai Ma, Gang Zhao, Zhihao Li and Wenlei Xiao*

School of Mechanical Engineering & Automation, Beihang University, Beijing, China

*xiaowenlei@buaa.edu.cn

Abstract. Wire and arc additive manufacturing (WAAM) is a potential technology for manufacturing components with a shorter period of time and by a lower cost. Path planning is a key research interest in WAAM. This paper presents an adaptive path planning method for manufacturing thin-walled structures with varying thickness. Firstly, the straight skeleton, which represents the skeleton of a given geometry, is extracted to express the geometry. Then the deposition with weaving technology is applied to obtain required varied width of the geometry. The resulting adaptive paths are able to fill thin-walled structures with varying thickness less than twice of the single-bead width.

1. Introduction

Wire and arc additive manufacturing (WAAM), an additive manufacturing (AM) technology, has great prospects and potential due to its high deposition rate, high energy efficiency and low cost [1]. It is a deposition technology that employs either Gas Metal Arc Welding (GMAW), Gas Tungsten Arc Welding (GTAW) or Plasma Arc Welding (PAW) [1]. It has great advantages in manufacturing large scale components that are often seen in the fields of aviation and shipbuilding.

One of the essential steps in AM is the generation of paths, which determines the material accumulating process of the filling layers and has a dominating influence to the quality of the fabricated part. So far, many path patterns have been developed for AM to generate tool paths, such as raster, zigzag, contour, continuous, hybrid, etc. Generally, these path patterns are designed for universal applications with no considering structural speciality of the built part. As a result, it is often less satisfactory to apply these patterns in manufacturing thin-walled structures by WAAM because of its large width of the single bead [2]. Therefore, the path generation should be specially considered for manufacture of complex structures using WAAM.

To minimise the number of stopping-ending points in paths for WAAM, Ding D. etc. proposed a continuous path planning algorithm that can generate a closed tool path using a combination of zigzag and contour pattern strategies [3]. But it is not applicable for thin-walled structures. Based on medial axis transformation (MAT) which can represent the skeleton of a give geometry, Ding D. etc. proposed a practical path planning algorithm for WAAM to build thin-walled structures [2]. First, the skeleton is obtained through computing the bisector of each pair of segments. Then, branch loops are generated based on the skeleton considering the step-over distance. After that, each loop is offset from the inner part to the outer boundary of the geometry. The algorithm generates paths from inner to outside can avoid inner gaps that will occur using contour path patterns. However, this approach generates too many discontinuous paths that are not desirable for WAAM. To solve this problem, an improved adaptive MAT algorithm [4] is proposed. The improved algorithm offsets the path loop with varying step-over distance. To fulfil the varying step-over distance, parameters need to be changed gradually to generate the desired bead width during the material accumulating process. Both of the above methods are designed for thin-walled structures having multi-bead width. However, when the thickness of the thin-walled structure is less than twice of the single-bead width, these methods are incapable of generating satisfactory paths.

This paper presents an adaptive path planning method for wire and arc additive manufacturing of thin-walled structures with varying thickness. The use of deposition with weaving technology in WAAM is explored. Based on the model of deposition with weaving, an adaptive path planning method for WAAM is proposed. Experiments validate the effectiveness of the method.

2. Deposition with weaving

Deposition with weaving is a technology used in welding applications to obtain wide weld beads. In related studies, it is proved to be of efficiency [5] and beneficial for the solute to diffuse [6]. To solve the problems in path planning of thin-walled structures with varying thickness, the deposition with weaving technology is studied in this paper.

2.1. Modeling of deposition with weaving

The applications of deposition with weaving is often carried out by industrial robots. In robot programs, there exists mainly three parameters to control the weaving movement of welding tools, namely, travel speed along a guiding line (v_t), amplitude (A) and weaving length (W_l), as shown in figure 1. Preliminary studies [4,7] show that the section profile of a weld bead without weaving can be fitted by a parabola curve. Thus, a simplified weaving bead model, mainly generated by scanning a parabola curve along the weaving trajectory, is presented, as shown in figure 2, to understand the bead generated by deposition with weaving. Given the travel speed (v_t) along the leading line, the nominal moving speed (v) along the weaving trajectory can be calculated as equation (1). Suppose the width of the bead generated with the calculated speed v is w_v, then the width of the bead deposited with weaving is formed by equation (2). And the average bead height can be calculated as equation (3).

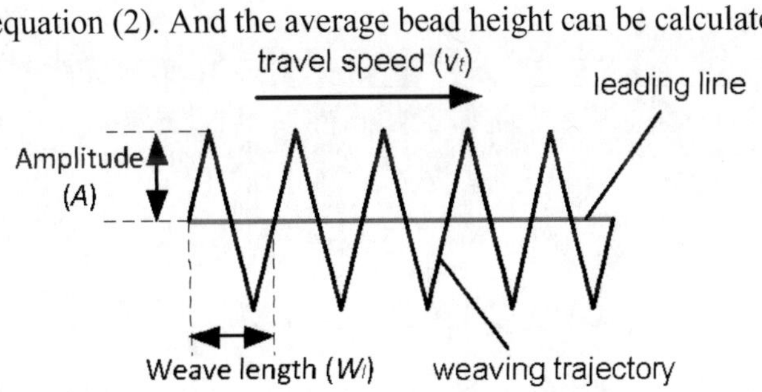

Figure 1. Parameters for the control of deposition with weaving.

$$v = \frac{4(A^2 + (\frac{W_l}{4})^2)^{1/2}}{W_l} \cdot v_t \tag{1}$$

$$width = 2A + w_v \tag{2}$$

$$height = \frac{\pi \cdot r^2 \cdot v_w}{width \cdot v_t} \tag{3}$$

where r is the radius of the wire material and v_w is the wire feed rate. Obviously, the amplitude has a great influence on the width of the weld bead.

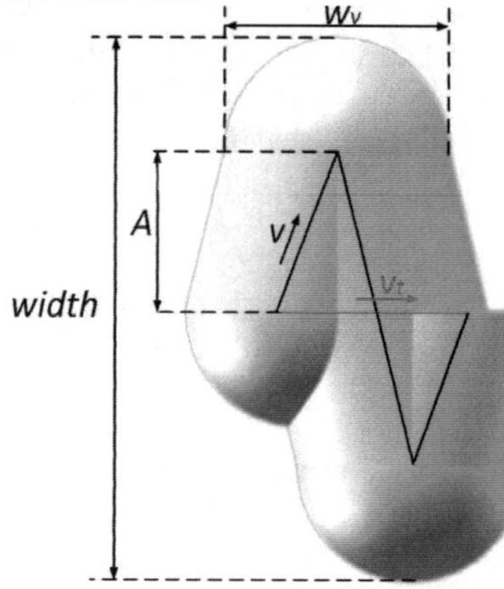

Figure 2. Weaving bead model.

2.2. Regression model of process parameters

Practically, it is not easy to obtain the width of the bead (w_v) in figure 2. Therefore, a second degree regression model is adopted to establish the relationship between process parameters and bead geometry as equation (4),

$$Y = CP \tag{4}$$

where Y is the bead width or height, P the combination of process parameters as shown in equation (5), and C the coefficient as shown in equation (6).

$$P = \begin{bmatrix} 1 & v_w & v_t & W_l & A & v_w^2 & v_t^2 & W_l^2 & A^2 & v_w v_t & v_w W_l & v_w A & v_t W_l & v_t A & W_l A \end{bmatrix}^T \tag{5}$$

$$C = \begin{bmatrix} c_0 & c_1 & c_2 & c_3 & c_4 & c_5 & c_6 & c_7 & c_8 & c_9 & c_{10} & c_{11} & c_{12} & c_{13} & c_{14} \end{bmatrix} \tag{6}$$

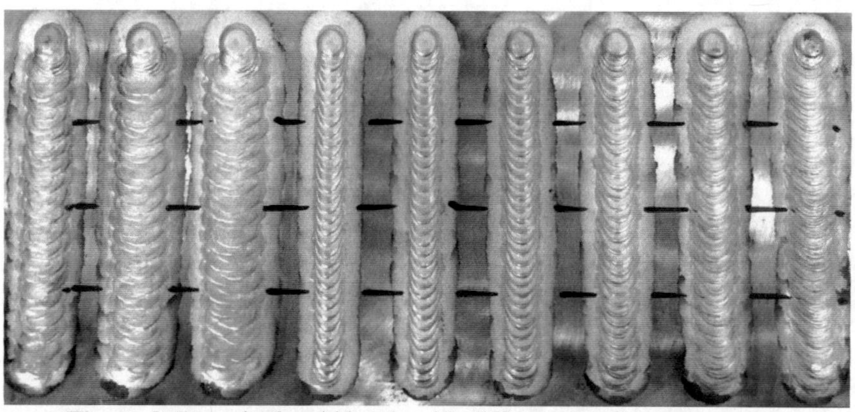

Figure 3. Deposited weld beads with different weaving parameters.

To determine these coefficients, totally 168 weld beads are deposited with different parameters (weave length from 2mm to 5mm and amplitude from 2mm to 8mm). Part of them are shown in figure 3. Then the widths of these beads are measured and the heights are calculated by equation (3). Part of the results are shown in table 1 (due to space limitations, only part of the data are listed). By using these data, the coefficients of the regression model are obtained as equation (7) and equation (8)

Table 1. Widths and heights of beads deposited with different parameters.

No.	v_w(m/min)	v_t(m/min)	W_l(mm)	A(mm)	width(mm)	height(mm)
1	8	0.3	2	3	12.63	2.39
2	8	0.3	2	6	17.70	1.70
3	8	0.3	3	5	16.35	1.84
4	8	0.3	4	4	15.97	1.89
5	8	0.3	5	4	13.70	1.32
6	8	0.3	5	5	15.18	1.19
7	8	0.5	2	4	13.35	1.36
8	8	0.5	3	3	10.23	1.77
9	8	0.5	3	6	15.77	1.15
10	8	0.5	4	8	18.63	0.97
11	8	0.5	5	5	15.18	1.19
12	7	0.3	2	2	11.46	2.30
13	7	0.3	2	5	14.20	1.86
14	7	0.3	3	7	18.36	1.44
15	7	0.3	4	3	13.95	1.89
16	7	0.3	4	6	16.01	1.65
17	7	0.3	5	5	14.96	1.76
18	7	0.5	2	4	11.84	1.34
19	7	0.5	3	6	12.66	1.25
20	7	0.5	4	2	9.78	1.62
21	7	0.5	4	5	12.34	1.28
22	7	0.5	5	4	11.83	1.34
23	7	0.5	5	8	17.01	0.93
24	6	0.3	2	5	12.55	1.80
25	6	0.3	3	7	17.11	1.32
26	6	0.3	4	3	10.69	2.12
27	6	0.3	5	2	10.22	2.21
38	6	0.3	5	8	18.43	1.23
39	6	0.5	2	5	11.18	1.21
30	6	0.5	3	4	9.87	1.38
31	6	0.5	3	7	13.72	0.99
32	6	0.5	4	6	12.27	1.11
33	6	0.5	5	5	11.14	1.22
34	6	0.5	5	8	15.83	0.86

$$C_w = [4.21 \quad 0.66 \quad -8.69 \quad 0.53 \quad 0.96 \quad 0.06 \quad 0 \quad -0.12 \quad 0.07 \quad -0.28 \quad 0.07 \quad -0.07 \quad -0.08 \quad -0.16 \quad 0.02] \tag{7}$$

$$C_h = [4.21 \quad 0.66 \quad -8.69 \quad 0.53 \quad 0.96 \quad 0.06 \quad 0 \quad -0.12 \quad 0.07 \quad -0.28 \quad 0.07 \quad -0.07 \quad -0.08 \quad -0.16 \quad 0.02] \tag{8}$$

where C_w and C_h are calculated using the width and height data, respectively. The mean errors for the width and height model are 4.99% and 5.57%, respectively. By combination of equations (4) to (8), regression models for width and height can be obtained as equation (9) and equation (10). Using the regression models, the width and height of the weld bead can be predicted directly from process parameters.

$$width = C_w P \tag{9}$$

$$height = C_h P \tag{10}$$

3. Path planning method

Generally, in path generation for AM, CAD models in STL format are sliced, obtaining 2D polygons. Then within each polygon, paths are generated to fill the inner area of the polygon. The adaptive path planning method presented in this paper is based on straight skeleton and the deposition with weaving technology discussed above.

3.1. Straight skeleton

The straight skeleton is first defined for simple polygons by Aichholzer et al. [8,9] to describe a geometry by a topological skeleton which is defined by a continuous shrinking process in which the edges of the polygon are moved inwards parallel to themselves at a constant speed. As the edges move in this way, the vertices where pairs of edges meet also move. Finally, the straight skeleton is the set of straight lines traced out by the moving vertices. The computation of the straight skeleton can be performed by simultaneously parallel offset the polygon edges. In a planar polygon, as shown in figure 4a, each skeleton edge (the red line) is a bisector of two polygon edges. For each endpoint of a skeleton edge, the distance between the endpoint and one of the corresponding two polygon edges is computed. Then the width of the polygon at the endpoint position is twice the distance. The centerline of the polygon can be roughly defined by comprising all skeleton edges that are not incident to a vertex of the polygon. It is an approximation of the medial axis of the polygon. For each free-end edge in the centerline, it is extended to intersect with the polygon edge, as shown in figure 4b.

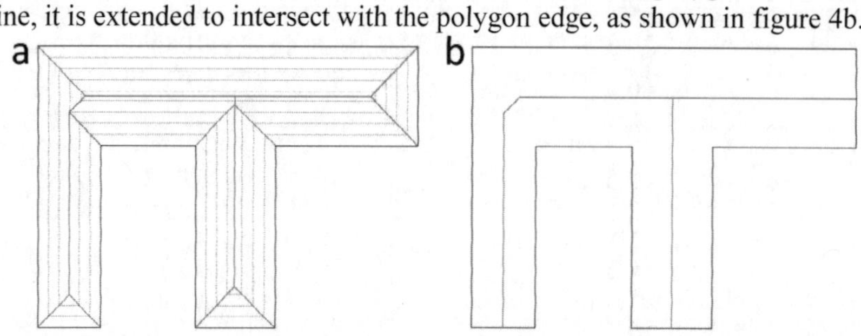

Figure 4. Straight skeleton and centerline (a) straight skeleton (b) centerline of the polygon.

3.2. Parameter determination

In the process of WAAM, the welding torch moves along the centerline of the polygon. The task of the path generation process is to determine the parameters adopted in the torch movement process. To complete the task, first, the layer height is artificially specified in the slicing procedure. Then, for each edge of the centerline, it is divided into several line segments. The number of the line segments, n, is determined by equation (11)

$$2 \le W_l = \frac{L_e}{n} \le 8 \tag{11}$$

where L_e is the length of the edge. Suppose the endpoints of an edge are P_s and P_e, and the widths of the polygon at the endpoints of the edge are w_s and w_e. After division, there will be $n-1$ interpolation points on the edge, as shown in figure 5. The interpolation points and corresponding widths are calculated as equation (12) and equation (13). For each line segment, if the widths of the polygon at the two endpoints are not the same, the large one is selected as the width of the line segment. In this way, the polygon is approximated by a series of rectangles, as shown in figure 6.

$$P_i = P_s + \frac{i}{n}(P_e - P_s) \tag{12}$$

$$w_i = w_s + \frac{i}{n}(w_e - w_s) \tag{13}$$

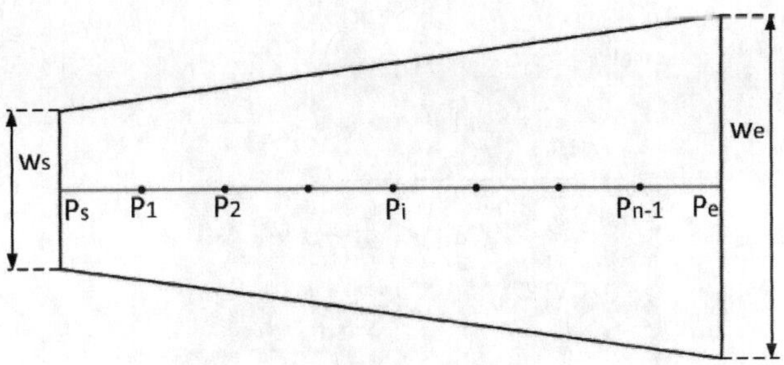

Figure 5. Division of an edge.

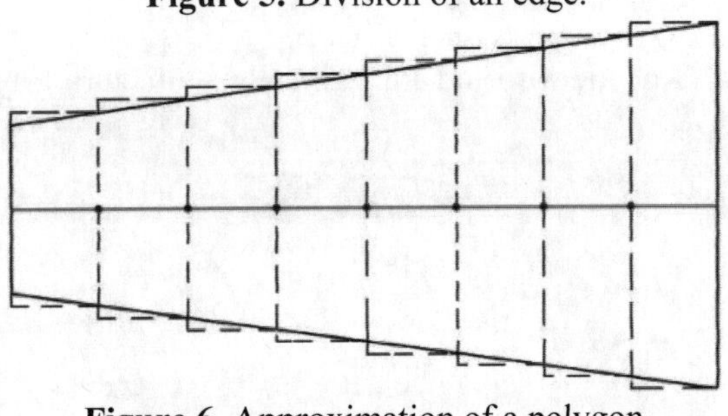

Figure 6. Approximation of a polygon.

From an robot-moving viewpoint, if the nominal moving speed is constant during the movement of the torch, the robot runs much smoothly. Thus, another limitation can be obtained as equation (14)

$$\frac{4(A^2 + (\frac{W_l}{4})^2)^{1/2}}{W_l} \cdot v_t = const \tag{14}$$

where $const$ is a constant value which is artificially specified. According to the specified layer height and the computed width using equation (13), bringing equations (11) and (14) into equations (9) and (10), the parameters for each line segment can be calculated.

The line segments and corresponding process parameters form the adaptive paths for wire and arc additive manufacturing of thin-walled structures.

4. Experiments and discussions

To validate the effectiveness of the proposed path planning method, experiments are designed and executed by a robotic wire and arc additive manufacturing system as shown in figure 7. A KUKA robot controls the movement of a welding torch and a welding machine control cabinet controls the electrical arc and wire feeding process. A 2-axis tilt and rotatory positioning platform serves as the workbench.

Two structures are deposited to demonstrate the ability of the proposed method. One is a polygon structure with a gradually varying thickness from 10mm to 20mm as shown in figure 8. Though the polygon is approximated by a series of rectangles, no obvious staircase shapes, as shown in figure 6, are observed in the manufactured part. The other is a structure with a sharp changing thickness, from 9mm to 18mm, as shown in figure 9. It can be seen that the sharp change of the thickness is well reflected in the manufactured part. The experimental results show that the proposed method is an effective method for robotic wire and arc additive manufacturing of thin-walled structures with varying thickness.

Figure 7. Robotic wire and arc additive manufacturing system.

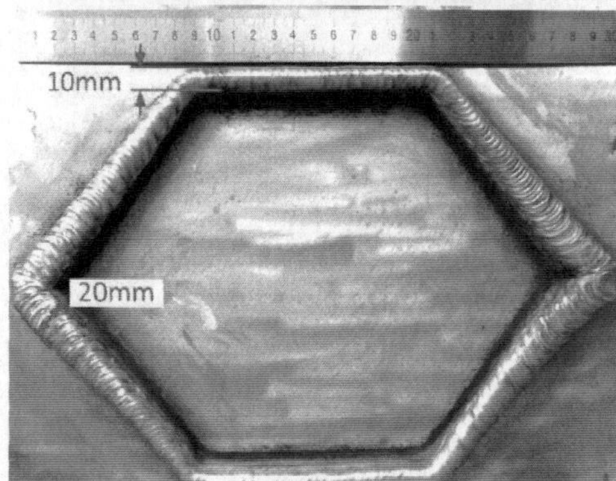

Figure 8. A polygon structure with a gradually varying thickness.

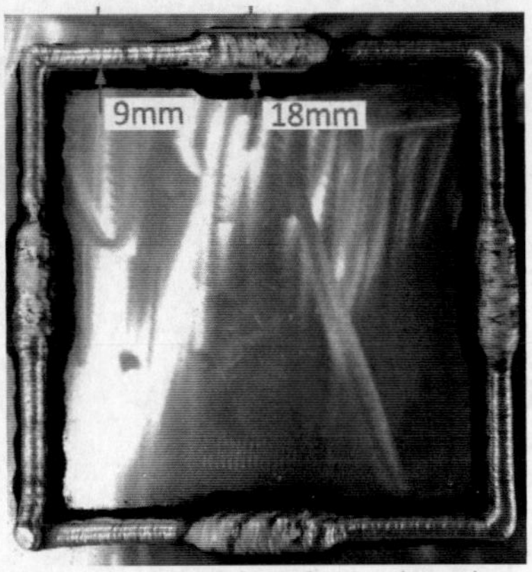

Figure 9. A structure with a sharp changing thickness.

5. Conclusion

This paper proposes a path planning method for building thin-walled structures with varying thickness by robotic WAAM. The deposition with weaving technology is detailed studied. A second degree regression model is established to describe the relationship between process parameters and bead geometry. The centerline of a polygon and varied process parameters generate the adaptive paths for robotic WAAM. Manufacturing examples demonstrate that the proposed method is an effective strategy for building thin-walled structures with varying thickness.

Acknowledgments

The research was supported by the Beijing Municipal Project of Science and Technology (NO. Z161100001516005).

6. References

[1] Panchagnula J S and Simhambhatla S. "Manufacture of complex thin-walled metallic objects using weld-deposition based additive manufacturing." *Robotics and Computer-Integrated Manufacturing* **49**(2018):194-203.

[2] Ding D, et al. "A practical path planning methodology for wire and arc additive manufacturing of thin-walled structures." *Robotics & Computer Integrated Manufacturing* **34**.C(2015):8-19.

[3] Ding D, et al. "A tool-path generation strategy for wire and arc additive manufacturing." *International Journal of Advanced Manufacturing Technology* **73**.1-4(2014):173-183.

[4] Ding D, et al. "A multi-bead overlapping model for robotic wire and arc additive manufacturing (WAAM)." *Robotics & Computer Integrated Manufacturing* **31**.C(2015):101-110.

[5] Chen Y, et al. "Effect of weave frequency and amplitude on temperature field in weaving welding process." *International Journal of Advanced Manufacturing Technology* **75**.5-8(2014):803-813.

[6] Zhan X, et al. "Microstructure and property characteristics of thick Invar alloy plate joints using weave bead welding." *Journal of Materials Processing Technology* **244** (2017): 97-105.

[7] Suryakumar S, et al. "Weld bead modeling and process optimization in hybrid layered manufacturing." *Computer-Aided Design* **43**.4 (2011): 331-344.

[8] Aichholzer O, Aurenhammer F and Alberts D. "A novel type of skeleton for polygons." *J. UCS The Journal of Universal Computer Science*. Springer, Berlin, Heidelberg, 1996. 752-761.

[9] Aichholzer O and Aurenhammer F. "Straight skeletons for general polygonal figures in the plane." *International computing and combinatorics conference*. Springer, Berlin, Heidelberg, 1996:117-126.

Optimization of Key Parameters of Two-way Inner and Outer Tube Differential Pressure Flowmeter

FAN Bao-tong[1,2] and MENG Jiang[2,1]

[1]School of Mechanical Engineering, North University of China, Taiyuan,China
[2]Advanced Manufacturing Technology Key Laboratory of Shanxi Province , North University of China,Taiyuan,China

1837985971@qq.com

Abstract. In view of the existing insufficient research on the parameters of the key structure of the two-way inner and outer tube differential pressure flowmeter, on the basis of the structure of the two-way inner and outer tube differential pressure flowmeter, the front and rear cone angles are respectively set. The study of the length of the thin tube for the bidirectional inner and outer tube differential pressure flowmeter was added. For the three structural parameters of the front cone angle α, the rear cone angle β, and the length of the thin tube l, nine sets of orthogonal tests were designed, which is different from the previous ones. The pressure loss ratio is proposed as the main evaluation index, and the simulation is performed by using FLUENT. By analyzing the results, the optimal combination is selected. Through simulation, the optimal model has good performance in measurement accuracy and pressure loss.

1. Introduction

At present, among the flow measurement of various industrial productions, the pressure difference type flowmeter has a low price and has a lot of experience. After a long period of research, the accuracy is getting higher and higher, so the application is very extensive [1]. The two-way inner and outer tube pressure difference flowmeter has the advantages of small disturbance to the fluid and large pressure difference signal[2]. In this paper, a new evaluation index of pressure loss difference is proposed. On these basis, the front and rear cone angles and the length of the thin tube are optimized to make the measurement more accurate and the disturbance to the fluid smaller.

2. Model structure and theoretical basis

The pipe for the throttle is selected to select the pipe diameter of the DN32. According to previous studies, the equivalent diameter ratio is optimal at around 0.7, and the diameter of the tube of the throttle member is 18 mm and 12.6 mm, respectively. The basic structure and three-dimensional model of the inner and outer tube differential pressure flowmeter are shown in Figs. 1 and 2.

IOP Conf. Series: Materials Science and Engineering **470** (2019) 012019 doi:10.1088/1757-899X/470/1/012019

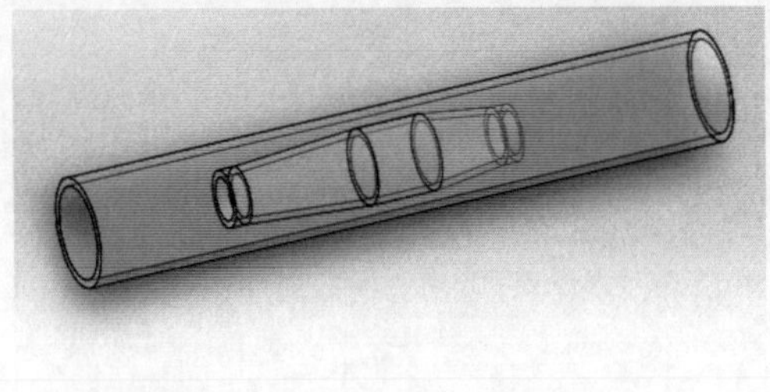

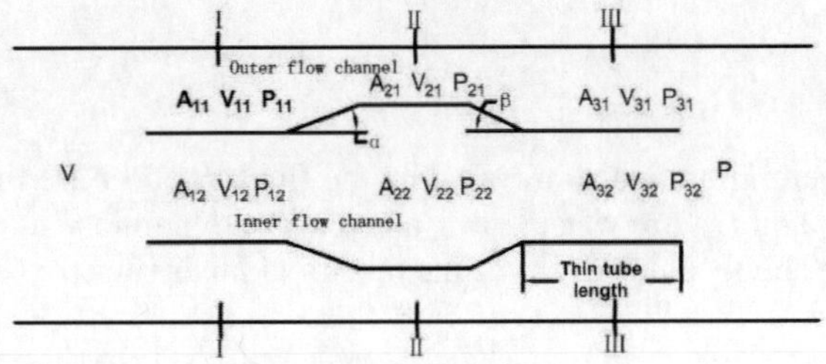

Figure 1. Flowmeter three-dimensional structure **Figure 2.** Flow chart structure diagram

The theoretical basis of asymmetric bidirectional internal and external tube differential pressure flowmeters is derived from the continuous equation and the Bernoulli Equation [3]. When the fluid passes through the throttle, in the I-I plane, the external flow channel compresses the throttle, the flow rate increases, the pressure decreases, the internal flow fluid diffuses, the flow velocity decreases, and the pressure increases; in the II-II plane, The inner and outer runners are relatively stable, and the flow velocity and pressure tend to be stable. In the III-III plane, contrary to the I-I plane, the outer runner diffuses, the flow rate decreases, the pressure increases, and the inner runner fluid compresses and the flow rate increases. The pressure is reduced. As a result, the largest pressure difference will be formed inside and outside the II-II section:

Similar to the measurement formula of the traditional differential pressure flowmeter [4]:

$$Q=\frac{C\lambda^2 A}{\sqrt{1-\lambda^4}}\sqrt{2\frac{\Delta P}{\rho}} \tag{1}$$

where C is the outflow coefficient, A is the cross-sectional area of the flow channel, and λ is the equivalent diameter ratio.ρ is the density of the measured fluid, and ΔP is the pressure difference between the internal and external measuring points.

As shown in Fig. 1, in the figure, A is the cross-sectional area of the flow channel, V is the average flow velocity of the inlet, and P is the outlet pressure. When m is 1 in $A_{mn}V_{mn}P_{mn}$, it represents the parameter of I-I section. When m is 2, it represents the parameter of II-II section. When m is 3, it represents the section III-III parameter. When n is 1, it represents the external channel parameter, n When it is 2, it represents the internal flow channel parameter. α and β are the front and rear cone angles, respectively. According to the Bernoulli equation, the outer flow path and the inner flow path are respectively:

$$Z_1+\frac{P_{11}}{\gamma}+\frac{V_{11}^2}{2g}=Z_2+\frac{P_{21}}{\gamma}+\frac{V_{21}^2}{2g}+h_{11}=Z_3+\frac{P_{31}}{\gamma}+\frac{V_{31}^2}{2g}+h_{21} \tag{2}$$

$$Z_1+\frac{P_{12}}{\gamma}+\frac{V_{12}^2}{2g}=Z_2+\frac{P_{22}}{\gamma}+\frac{V_{22}^2}{2g}+h_{12}=Z_3+\frac{P_{32}}{\gamma}+\frac{V_{32}^2}{2g}+h_{22} \tag{3}$$

where Z_1, Z_2 and Z_3 are the positional potentials, and h_{11}, h_{21}, h_{12}, h_{22} are the head losses. A head loss includes the pipeline loss h_λ and local loss h_ζ[5], The losses along the outer and inner runners are:

$$h_{\lambda 1}=\frac{8\mu L}{\pi\left[R^4-r^4-\frac{(R^2-r^2)}{\ln(R/r)}\right]\rho g}Q_{v1} \tag{4}$$

$$h_{\lambda 2} = \frac{8\mu L}{\pi r^4 \rho g} Q_{v2} \tag{5}$$

where L is the path traveled by the fluid, R and r are the hydraulic radii of the inner and outer runners, Q_{v1} and Q_{v2} are water flows, and μ is the dynamic viscosity coefficient.

The local head loss of the outer and inner runners is:

$$h_{\zeta 1} = \zeta_1 \frac{V_{21}^2}{2g} + \zeta_2 \frac{V_{31}^2}{2g} = \frac{\lambda_1 \left(K_1^2 - 1\right) V^2}{2\gamma g \sin\alpha} + \frac{\lambda_2 \left(K_1^2 - 1\right) V^2}{2\gamma g K_1^2 \sin\beta} + k_1 \left(1 - \frac{1}{K_1}\right) \frac{V^2}{2g} \tag{6}$$

$$h_{\zeta 2} = \zeta_3 \frac{V_{22}^2}{2g} + \zeta_4 \frac{V_{32}^2}{2g} = \frac{\lambda_3 \left(1 - K_2^2\right) V^2}{2\gamma g \sin\alpha} + \frac{\lambda_4 \left(1 - K_1^2\right) K_2^2 V^2}{2\gamma g \sin\beta} + k_2 \left(1 - K_2\right) \frac{K_2^2 V^2}{2g} \tag{7}$$

where λ_1, λ_2, λ_3, λ_4 are the resistance coefficients along the path, k_1 and k_2 are the coefficients related to the front and rear cone angles, K_1 and K_2 are the compression ratios of the water flow channel. In addition to the aspect ratio associated with the head loss and local head loss along the path, there are also the length of the throttle and the front and rear cone angles α and β.This is the optimal level combination to look for in this article.

3. Experimental Design and Flow Field Simulation

3.1. Experimental design

According to the analysis, the main parameters affecting the pressure loss difference are the length l of the throttle tube, the front cone angle α, and the rear cone angle β. Three parameters are selected for each of the three levels. The length of the thin tube is divided into three levels of 0mm, 5mm and 10mm. The front cone angle α is selected from three levels of 5°, 25° and 45°, and the rear cone angle β is also selected from three levels of 5°, 25° and 45°. The test is a three-factor three-level, so the selected orthogonal table $L_9(3^4)$ is suitable [6]. There are nine groups of tests, and the specific design of the test design is shown in Table 1.

Table 1. Orthogonal test group test parameters

Test number	Front cone angle α/°	Rear cone angle β/°	Thin tube length/mm
1	5	5	0
2	5	25	5
3	5	45	10
4	25	5	10
5	25	25	0
6	25	45	5
7	45	5	5
8	45	25	10
9	45	45	0

3.2. Flow field simulation

The three groups of experiments were modelled by three-dimensional software, and the flow field simulation was simulated by ANSYS FLUENT with the RNG k-ε model. The unstructured tetrahedral meshes were used [7], and the number of grids was about 13000. The simulation medium was selected from liquid water at room temperature of 20℃ and the incompressible fluid. The diameter D of the pipe was 27.2 mm for the inner diameter of the DN32 PVC pipe, the simulated inlet flow rate was 2 m/s, and the outlet condition was outflow. Two pressure taps are respectively set at a distance of 1D before and after the throttle member, and two other pressure tapping points are set at the center of the throttle member and the center of the outer flow passage. When modeling, set the symmetry plane to facilitate observation of the pressure cloud map. The symmetry plane pressure cloud diagram of Experiment 2 is shown in Figure 3.

4. Simulation results and data analysis

Each group of experiments was simulated, the relative pressures of the four pressure points were recorded, and the pressure difference between the front and back points was calculated. The pressure difference between the internal and external points was measured. The pressure loss ratio of the traditional evaluation index was the internal and external pressure difference signal and the front and rear pressure. The ratio of losses[8]. The original intention of using the pressure loss ratio as the evaluation index is to select the structural parameters with the inner and outer pressure difference as large as possible and the pressure loss before and after as small as possible [9], but the internal and external pressure difference and the front and rear pressure loss cannot be avoided, but their ratio is small. But it is a big situation. Therefore, the pressure loss has certain loopholes in this evaluation index. The pressure loss difference is the difference between the internal and external pressure difference signals and the pressure loss before and after. This index is more scientific and reasonable than the pressure loss ratio. All the measurement parameters and calculation results are shown in Table 2.

Table 2 . Parameters and calculation results of each test point

Test ID	Pressure of front point P_f/Pa	Pressure of rear point P_r/Pa	Pressure of inner point P_i/Pa	Pressure of outer point P_o/Pa	$\Delta P_{fr}=P_f-P_r$/Pa	$\Delta P_{io}=P_i-P_o$/Pa	$\Delta P=\Delta P_{fr}-\Delta P_{io}$/Pa
1	60.0666	-2110.894	-13.5472	-7860.6333	2170.9606	7847.0861	5676.1255
2	487.7056	-2098.6714	-4.7488	-6958.9761	2586.377	6954.2273	4367.8503
3	761.4759	-2646.1289	-15.0528	-6181.8521	3407.6048	6166.7993	2759.1945
4	828.038	-2189.9231	-2.6676	-7071.1963	3017.9611	7068.5287	4050.5675
5	2754.9976	-280.3122	120.1933	-4863.2441	3035.3098	4983.4374	1948.1276
6	2101.8513	-1494.4583	36.4988	-4966.9932	3596.3096	5003.492	1407.1824
7	1701.3268	-1353.7277	-50.1978	-6480.0381	3055.0545	6429.8403	3374.7858
8	1970.1716	-1332.1194	-43.9494	-6176.3115	3302.291	6132.3621	2830.0711
9	5734.3848	1266.8462	183.0813	-2134.8748	4467.5386	2317.9561	-2149.5825

According to the data in Table 2, the most important parameters affecting the pressure loss difference are analyzed. The range calculation table of each factor is shown in Table 3, and the range R of each factor can be calculated [10].

It can be easy to learn that the magnitude of each factor is ranked as $R(\beta) > R(\alpha) > R(L)$. Therefore, the factor that has the greatest influence on the pressure loss difference is the post-cone angle β. According to Table 2, it can be analyzed that the optimal horizontal combination is 5° for the front cone angle, 5° for the rear cone angle, and 0mm for the length of the thin tube. The optimal level combination obtained is exactly the model of Test 1, which is a symmetrical structure, and there is no difference between forward measurement and reverse measurement. The pressure cloud image obtained by FLUENT simulation is shown in Fig. 4.

Table 3 . Calculation table of each factor

	Front cone angle	Rear cone angle	Thin tube length
T_1	12803.1703	13101.4788	5474.6706
T_2	7405.8775	9146.049	9149.8185
T_3	4055.2744	2016.7944	9639.833
$\overline{T}_1$	4267.7234	4367.1596	1824.8902
$\overline{T}_2$	2468.6258	3048.683	3049.9395
$\overline{T}_3$	1351.7581	672.2648	3213.2777
R	2915.9653	3694.8943	1388.3875

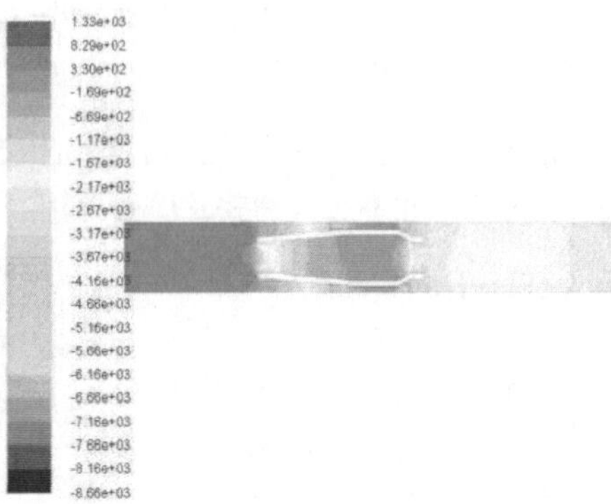

Figure 3. Pressure cloud diagram of test 2

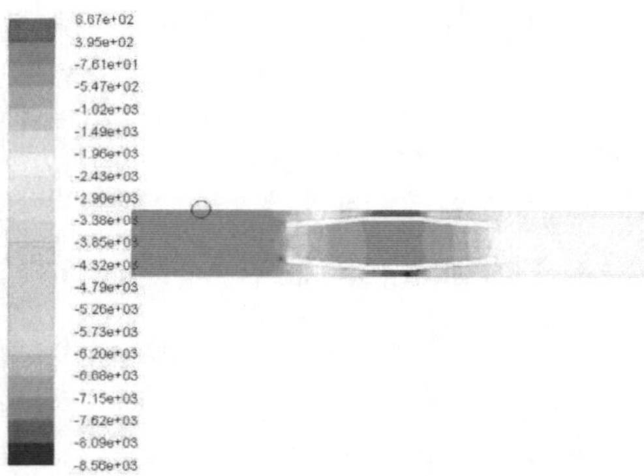

Figure 4. Optimal structure simulation

According to the data in Table 2, it can be analyzed that the optimal horizontal combination is 5° for the front cone angle, 5° for the rear cone angle, and 0 for the length of the thin tube. The optimal level combination obtained is exactly the model of test 1. Through simulation verification, it is not difficult to find that the optimal horizontal combination of the inner runner and the outer runner has the largest pressure difference, and the pressure loss before and after the throttle is the smallest. The pressure loss difference is the largest and the expected effect is achieved. Moreover, the optimal structure is a front-back symmetrical structure, and there is no case where the forward measurement and the reverse measurement are different. The pressure cloud image obtained by FLUENT simulation is shown in Fig. 4.

5. Conclusion

This paper simply analyzes the loopholes of the pressure-loss ratio of the evaluation index of the internal and external tube pressure difference flowmeters, and proposes a more reasonable evaluation index-pressure loss difference. According to the experimental design and simulation, the optimal horizontal combination was selected. The front and rear cone angles of the optimal model are both 5° and the length of the thin tube is 0 mm. Through verification, the optimal level combination sought is in agreement with the simulation results. Compared with the previous flowmeter, it achieves more accurate measurement and less effect on fluid disturbance, and can adapt to more industrial occasions.

Acknowledgment

The authors would like to acknowledge the support from the Advanced Manufacturing Technology Key Laboratory of Shanxi Province under Grant no. XJZZ201705.

References

[1] Lu Jiamin, Zhang Qiang, Zhang Dayuan, et al. Overview of Differential Pressure Flowmeters[J]. Metrology Technology, 2018(1).

[2] Zhai X, Shen X, Wang G, et al. Performance Study on the New Types of Differential Pressure Flowmeters[J]. Process Automation Instrumentation, 2017.

[3] Liao Y. Analysis on Solving Methods and Application of Bernoulli Equation[J]. Journal of Anshun University, 2015.

[4] Li Xiaoting, Wang Xiaojie, Fang Lide, et al. Characteristics of a new type of internal and external tube differential pressure flowmeter[J]. Chinese Journal of Scientific Instrument, 2012, 33(10): 2371-2379.

[5] Zhang Jun, ZHANG Xiaoting. Calculation of Fluid Resistance and Head Loss in Fluid Transmission[J]. Fluid Power Transmission and Control, 2011(4): 24-27.

[6] Mao Shisong. Experimental Design [M]. China Statistics Press, 2012.

[7] Chen M, Jiang W, Liu W, et al. Research on V cone flowmeter used for jet fuel flowrate measurement[J]. Journal of Electronic Measurement & Instrumentation, 2016.

[8] Fan Yulang, Meng Jiang, Zhao Wei, et al. Pressure loss of internal and external tube differential pressure flowmeters[J]. Mechanical Design and Research, 2014, 30(1): 98-100.

[9] Meng J, Liu Z, An K, et al. Simulation and Optimization of Throttle Flowmeter with Inner-Outer Tube Element[J]. Measurement Science Review, 2017, 17(2):68-75.

[10] Tang Jiali, Xu Zhangwei. Range Analysis and Variance Analysis in Orthogonal Experiments[J]. Middle School Mathematics, 2017(9): 31-34.

IOP Conf. Series: Materials Science and Engineering **470** (2019) 012020 doi:10.1088/1757-899X/470/1/012020

Applications of Spray Cooling Technology in Aerospace Field

Ran Liu, Lei Zhang and Xian Zhang

Beijing Institute of Spacecraft Environment Engineering, Beijing 100094, China

liuran_cast511@126.com

Abstract. Spray cooling is an effective heat transfer technology that uses liquid atomization and boiling to to take away the heat from the surface of a cooled object. It has the characteristics of large heat transfer coefficient, good temperature uniformity, small degree of superheat, high critical flux and small circulation flow. It has become a research hotspot of efficient heat and mass transfer technology. The electronic components in the aerospace field are generally in the harsh environment of high heat flux density, high vacuum, microgravity and high radiation. With the increasing use of high-power components in spacecraft, the heat dissipation of spacecraft has become the first problem to be solved in spacecraft design. In response to this problem, the author has studied a large number of spray cooling technologies for aerospace applications, and has certain application value for the research of heat dissipation of high heat flux components in aerospace field.

1. Introduction

With the improvement of satellite functions, the development of the space industry and the continuous development of deep space exploration, the compact design of the spacecraft and the miniaturized design concept enable the integration of numerous electronic components in smaller and smaller areas. The heat flux density of electronic devices has increased dramatically. Coupled with the harsh environment of high-vacuum, micro-gravity and super-radiation, the spacecraft's heat dissipation problem has become the first problem to be solved in spacecraft design.

Spray cooling technology is a phase change cooling technique that uses atomized droplets to be sprayed onto a heat exchange surface to remove heat through phase change evaporation of the medium. Its characteristics are: high heat transfer coefficient, good temperature uniformity, low superheat, high critical heat flux density and small circulation flow, which is the most competitive high heat flux density thermal control technology. At present, NASA has listed closed spray cooling circuit system technology as one of the research priorities of future thermal control systems. The US Navy and the Department of Energy have set up special funds for the research of spray technology in the past ten years, and successfully achieved the heat dissipation target of heat flux density of nearly 1000W/cm^2 [1-2]. In recent years, ACT has successfully realized the commercialization of atmospheric gravity spray technology[3].Domestic research on spray cooling technology has also been carried out in many aspects and has achieved certain results.

2. Spacecraft thermal control system

In order to apply the spray cooling technology to the spacecraft thermal control system, relevant research under microgravity conditions should be carried out. A large number of experimental studies have been carried out in the United States to study the heat transfer effect of spray cooling by using aircraft parabolic flight or free fall to simulate the microgravity environment. NASA and the US Air Force have conducted a lot of research on the effects of spray cooling and heat transfer under microgravity by providing historical flight data, falling towers, gravity-reduction flight tests and laboratories, and many achievements [4 -10]. In NASA's CEV (Crew Exploration Vehicle) project, the heat dissipation of the spacecraft thermal control system usually adopts the evaporative cooling method of heat sink sublimation to take away heat. The spray cooling can achieve high heat dissipation under relatively low working fluid flow, which is applied to flash evaporator. The open spray cooling system has been tested and verified [11-13].

The use of water as a flash evaporator working fluid takes away excess heat from the spacecraft, and the vaporized water is directly discharged into space. This type of flash evaporator is called compact flash evaporator. The spray flash evaporator developed by NASA is a cylindrical heat exchanger. The surface of the heat exchanger is covered with fins to enhance the heat transfer during spray cooling. Figure 1 shows the spray nozzle array of the flash evaporator, and Figure 2 shows the simulation calculation model for the liquid flowing through the flash evaporator.

Figure 1. Spray nozzle array of flash evaporator

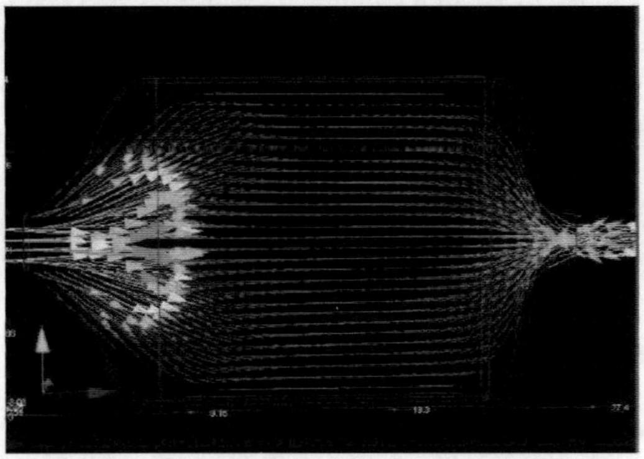

Figure 2. Simulation calculation model for liquid flowing through flash evaporator

To simulate the spray cooling effect in a vacuum environment in space, NASA tested a single nozzle in a vacuum chamber with a vacuum of 10^{-3} torr[13],as shown in Figure 3.Test proved that the compact flash evaporator system with open spray cooling is lighter and more compact than the current heat sink sublimation system, and is more suitable for future spacecraft needs. The original intention of this project was to develop open spray cooling technology, but the closed spray cooling system also carried out a lot of research for the thermal control of future space missions.

Figure 3. A single nozzle is tested in a vacuum chamber

Many ground test studies have been carried out on closed spray cooling technology at domestic and foreign. Schematic diagram of spray cooling closed loop system is shown in Figure 4. The system includes micro pumps, spray nozzles, spray chamber, condensers, heaters, etc. The micro pump provides drive power to the system. The liquid comes out of the pump port, passes through the heater to adjust the temperature and then enters the spray chamber. The liquid is atomized and sprayed on the heat source through the nozzle. Two-phase flow returns to the condenser to cool to a liquid state, completing a cycle.

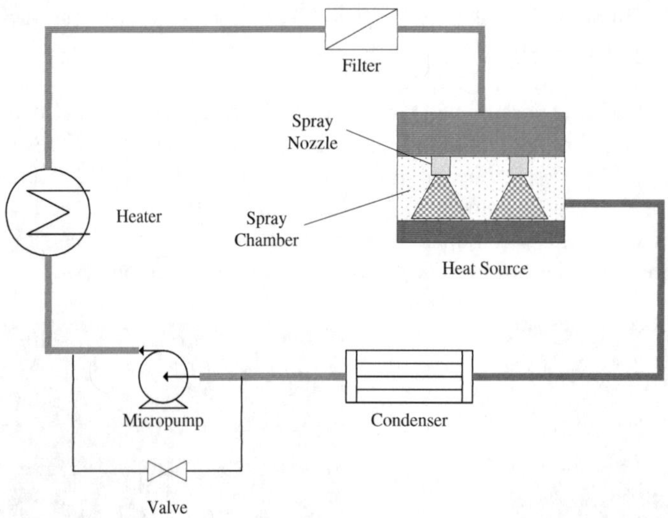

Figure 4. Schematic diagram of spray cooling closed loop system

NASA's Glenn Research Center conducted a flight test on spray cooling. The KC-135 microgravity test bed was used for flight test to verify and study its operation ability, heat transfer performance and characteristics under microgravity conditions.FC-72 is used as the working fluid [14].China academy of space technology also developed a prototype of the spray cooling system with an ultimate heat flux density exceeding 300 W/cm^2[15]. However, the closed spray cooling system is still only in the experimental research, and it is not really applied to the spacecraft thermal control system.

3. High power laser cooling system

In the past decade, high-power diode lasers have developed rapidly. Laser-diode arrays (LDAs) technology can be applied to Mars satellites, earth science, and Mercury exploration. However, the heat generated by LDAs is very high, which requires a thermal control method with high heat dissipation capability, and spray cooling is currently the hottest research focus.

In 2003, Universal Energy researched spray cooling technology for high-power semiconductor laser arrays with funding from the US Air Force Laboratory[16]. The device has eight nozzle arrays with a closed system, the nozzle is 8.8mm away from the hot surface, the spray chamber size is 28.5mm (length) × 17.0mm (width) × 8.8mm (height), and the working medium includes FC-87, FC - 72, methanol and water, the power of the circulatory system is provided by a micro pump. Prototype of laser spray cooling is shown in Figure 5.The super-cooled liquid is sprayed through the nozzle on the hot surface, and most of the droplets form a thin liquid film on the hot surface, and a small portion of the droplets evaporate to remove heat. The steam leaves the spray chamber together with the liquid, enters the condenser through the two-phase flow passage and re-condenses into liquid, participating in the next stage of the cycle.

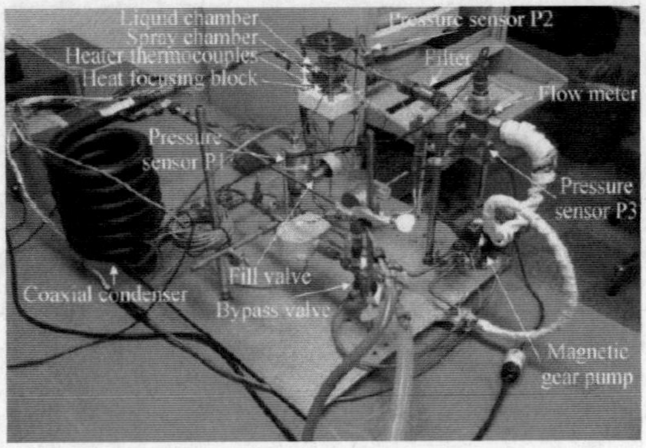

Figure 5. Prototype of laser spray cooling

4. Thermal Protection systems for Hypersonic Vehicle

Thermal protection in ultra-high temperature areas is a key issue for hypersonic vehicles. NASA proposed a solution for the design of a liquid spray cooling thermal protection system, as shown in Figure 6 [17]. The liquid spray cooling thermal protection system uses a sandwich structure. The skin material is directly heated, using a corrugated sheet as the core element of the sandwich structure and secured to the heated skin side. Spray tubes is installed inside the corrugated sheet as a passage for the coolant. Each of spray tubes is provided on one surface thereof with rows of closely spaced holes. The coolant is sprayed from the hole to the heated area, the coolant is heated to a gaseous state, and the gas is discharged to the adjacent side by the passage on the side wall of the corrugated sheet. Inside the corrugated sheet, the pipe is used for exhausting the gas. During operation, the temperature of the spray cooled sheet is the same as the coolant evaporation temperature. For large-area cooling, the system's cooling capacity is 57kW/m^2; for a small area of 5cm^2, the system's cooling capacity can reach 230kW/m^2, and the internal structure temperature can be maintained at 330K.

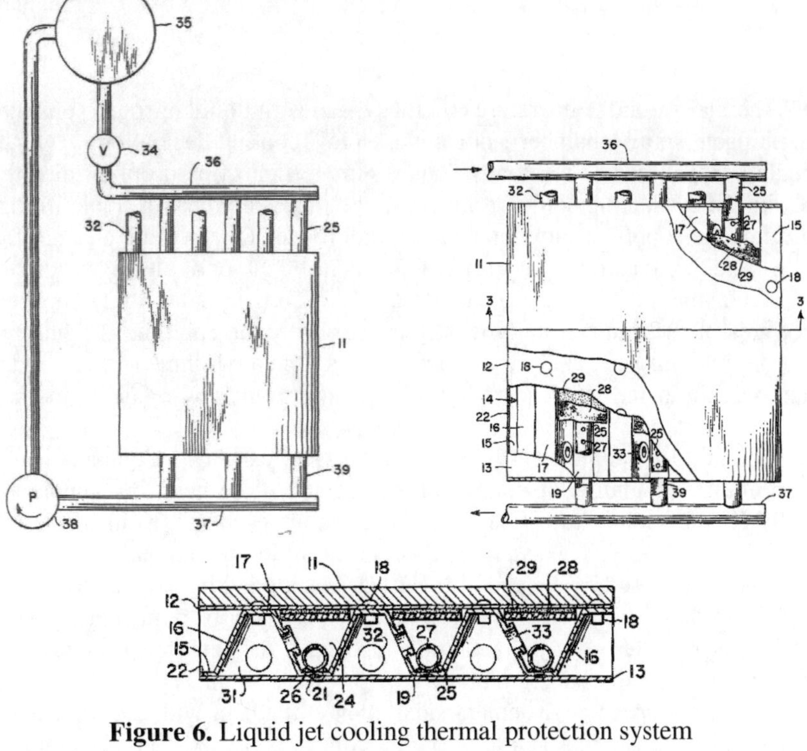

Figure 6. Liquid jet cooling thermal protection system

In China, Liu Shuang[18] proposed a new method for applying spray cooling to metal thermal protection systems for the head cone and wing leading edge of hypersonic vehicles. Designed and manufactured a sample of metal thermal protection system using spray cooling, built an oxyacetylene experimental platform, and tested the overall performance of the spray cooling thermal protection system sample. Test results show that the spray cooling can dissipate 800-1000 kW/m^2 of heat.

5. Liquid nitrogen spray chamber in space environment simulation equipment

In the current space cryogenic environment simulation in the aerospace field, the main method of using liquid nitrogen is to adopt a structure of a liquid nitrogen coil and a heat exchanger. However, when the ambient temperature outside the satellite changes significantly with time, a dynamic orbital thermal environment simulation is required. At this time, a variable temperature gas nitrogen system is required to change the heat sink temperature. One of them is the heat sink temperature control system with a liquid nitrogen spray chamber, as shown in Figure 7.

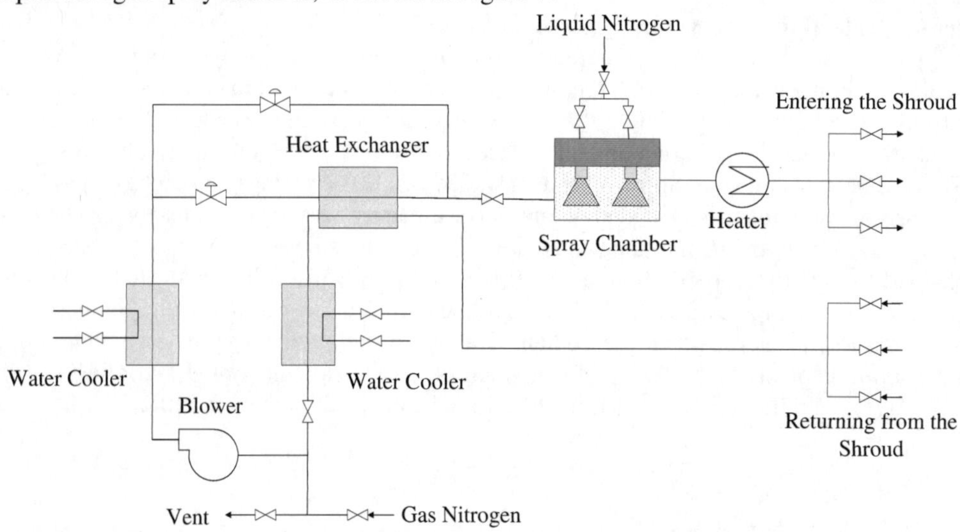

Figure 7. Thermal shroud temperature control system with liquid nitrogen spray chamber

The liquid nitrogen spray chamber adopts the spray cooling technology, which uses a low temperature liquid nitrogen nozzle to atomize liquid nitrogen into fine droplets, directly contacts the heat exchange with the circulating cold nitrogen, and uses the latent heat of liquid nitrogen to achieve cooling, by adjusting the supply of nitrogen and liquid nitrogen. The pressure changes the flow rate of liquid nitrogen to achieve dynamic and precise adjustment of the outlet nitrogen temperature, which can control the inlet temperature of the heat sink to ±1 °C. Moreover, after liquid nitrogen atomization, the total surface area of the liquid is significantly increased by the continuous liquid breaking into a large number of discrete droplets, and the contact area with the environment is remarkably improved, and the heat and mass transfer efficiency is greatly increased, thereby reducing the consumption of liquid nitrogen.

The vacuum thermal test equipment [19] produced by American PDM Company is equipped with a gas-nitrogen temperature-regulating heat sink, which can adjust the heat sink temperature within the range of 117-394 K, the temperature rise and fall rate can reach 1.1 K/min, and the system can withstand the heat load of 50kW. The external process of gas-nitrogen temperature regulation and heat sinking adopts a one-way closed cycle, and nitrogen gas is supplied by two vaporizers, each of which can generate 1870 m^3/h of nitrogen gas for 8 hours. Nitrogen is regulated by six high-density vaporization units and then fed into the heat sink to control the nitrogen flow and temperature by controlling the output of the vaporization unit.

In 2013, Turkey built a space environment simulation equipment with a gas-nitrogen temperature control system [20], and the gas-nitrogen temperature control unit used a liquid nitrogen shower. The liquid nitrogen flows out of the storage tank through the vaporizer and is vaporized into nitrogen. The nitrogen gas passes through the turbo compressor and the regenerative heat exchanger, enters the liquid nitrogen spray chamber, and is thoroughly mixed with the liquid nitrogen, and the heater power after the shower is adjusted. Nitrogen is obtained at a temperature of -165 °C to +110 °C and enters the heat sink.

In 2016, Beijing Institute of Spacecraft Environment Engineering successfully developed a gas-nitrogen temperature control system using liquid nitrogen spray cooling. The surface temperature range of the heat sink surface is -170.0°C~+150.0°C, and the temperature rise and fall rate reaches 3.5°C/ Min, heat sink uniformity is better than ±3.0 °C.

6. Conclusion

In spacecraft environmental testing, spray cooling has been widely used in engineering applications, and it is easier to implement engineering applications on the ground than the difficulty of using spray cooling in space environments. The use of spacecraft, especially the closed-loop spray cooling technology, is still in the laboratory stage and is not widely used. In addition to the lag of theoretical research, the control method of controlling the temperature of the cooled object reduces the volume and weight of the test prototype, which is also the direction for future research on spacecraft.

References

[1] LC Chow, MS Sehmbey, MR Pais. High Heat Flux Spray Cooling[J]. Heat Transfer. 1997, 129(8):291-318
[2] Chunlin Xia. Spray/jet Cooling for Heat Flux High to 1 kW/cm2[J]. Annual IEEE Semiconductor Thermal Measurement and Management Symposium. 2002:159-163.
[3] http://www.1-act.com[EB/OL]
[4] K Sone, K Yoshida, T Oka, Y Abe. Spray Cooling Characteristics of Water and FC-72 under Reduced and Evaluated Gravity for Space Shuttles[C].Energy Conversion Engineering Conference.1996, 2(2):1500-1505
[5] K Yoshida, Y Abe, T Oka.et a1.Spray Cooling under Reduced Gravity Condition[J].Journal of Heat Transfer.2001,123(2):309-318
[6] Y Ma, Chung J.N. A Study of Bubble Dynamics in Reduced Gravity Forced-convection Boiling [J].International Journal of Heat and Mass Transfer.2001, 44(2):399-415
[7] Y Ma, Chung J.N. An Experimental Study of Critical Heat Flux (CHF) in Microgravity Forced-convection Boiling[J].International Journal of Multiphase.2001,27(10):1753-1767
[8] P Stephan,C Sodtke, J Kern, N Schweizer,.High Resolution Measurements of Wall Temperature Distribution underneath a Single Vapors Bubble under Low Gravity Conditions[J].International Journal of Heat and Mass Transfer. 2006, 49(5):1100-1106
[9] Levi J. Elston, Kirk L. Yerkes, Scott K. Thomas, and John McQuillen. The Effect of Variable Gravity on the Cooling Performance of a 16-Nozzle Spray Array[R].AD:AFRL-RZ-WP-TP-2008-2199
[10] Kirk L. Yerkes, Travis E. Michalak, et al. Variable-gravity Effects on a Single-phase Partially Confined Spray Cooling System [J]. Journal of Thermophysics and Heat Transfer,2006,20(3):361-370
[11] E Golliher, J,Romanin , H Kacher,et al. Development of the compact flash evaporator system for exploration [C].//37th International Conference on Environmental Systems. 2007
[12] D M Althausen, E Golliher. Testing of a R134a spray evaporative heat sink[C].//38th International conference on Environmental Systems. 2008
[13] E Golliher, A Licari, T Jin. Testing of a compact flash evaporator system for exploration[C].//38th International Conference on Environmental Systems.2008
[14] Silk E A. Spray cooling and the next generation of NASA space flight. [R] .NASA Report,No.20050182030
[15] MIAO Jianyin, et al. Development of Heat Transfer Technologies for Spacecraft [J], Spacecraft

Engineering,2010.3,Vol:19 NO.2:106-112
[16] Lin L C. Closed-loop spray cooling of high-power semiconductor lasersi[R].AD Report. No.ADA413471
[17] R. A. Anderson, R. T. Swann. Sandwich Panel Construction. US Patent. NASA Report,No.70N37979
[18] Liu Shuang. Active Cooling Mechanism and Cooling Capacity Evaluation of Thermal Protection systems for Hypersonic Vehicle[D]. Harbin Institute of Technology.2010.6
[19] Zhang Lei, Liu Min, Wang Zijuan, He Chao. A Review of Technologies of Temperature-adjustable Heat Sink[J]. Spacecraft Environment Engineering. 2012,29(2):179-184.
[20] H. Gürgüç Işık, A. Bilge Uygur, Cem Ömür, Değer Akın, Murat Sancak. A New Thermal Vacuum Test Chamber Built In Turkey[R].IEF.2013,6581267

Simulation Study on the Influence Law of Two-body Wear in Sliding Guide Ways

Shengfang Zhang, Qiang Hao, Zhihua Sha, Yu Liu*, Jian Yin, Fujian Ma and Dapeng Yang

School of Mechanical Engineering, Dalian Jiaotong University, Dalian, China 116028

*E-mail: liuyu_ly12@126.com

Abstract: In view of the typical two-body wear issue in sliding guide ways, this paper established the two-body wear simulation model of the guide ways by the finite element software. Based on the friction and wear mechanism, and the influence of friction conditions on the two-body wear of the guide ways was considered. The wear process of the guide ways was analysed, and the influence of load and sliding speed on the wear of the guide ways was researched. The simulation results showed that large sliding speed with the wear would accelerate and a large load would increase the wear of the guide ways.
Keywords: sliding guide ways, friction factor, two-body wear, finite element simulation

1. Introduction

The sliding resistance and energy consumption cause the two-body friction by the interaction of the two contact surfaces, and the two-body wear is a process in that the surface material changed the shape, size, structure and properties under two-body friction. Friction causes wear and friction occurs in the wear. Therefore, for the two-body friction and wear mechanism, the tribology community has conducted a lot of researches for many years.

Based on the idealized classical model, Rabinowicz laid the theoretical foundation for the qualitative analysis of wear[1]. Lee et al obtained an empirical formula for estimating the tool wear model by Archard model analysis[2]. Soerberg et al used the Archard wear model to describe the relationship between friction material wear rate and braking distance[3]. And the methods for wear research were not limited to theoretical analysis, but also focused on experimental research. Peterseim et al established a model for numerical simulation of pin-disc friction and wear tests, and its self-developed program is used to calculate wear volume and wear surface appearance parameters[4]. Mukras et al used a wear processor to simulate the dry friction process of the swing pin on the pivot[5]. However, the experimental research method consumes a lot of manpower and material resources, and the complexity of the system and different working conditions will cause some simulation test results to be not well used in practical engineering applications. Numerical simulation techniques can make up for these deficiencies. Bortoleto et al used finite element modelling to numerically simulate the stress field distribution and analysed the variation of the surface profile of the disc[6]. Shen et al used the Archard model to analyse the dynamic wear process of the bearing and learned the wear between the contact surfaces by finite element simulation is that the discrete quasi-static model can be used to simulate the nonlinear wear process[7]. Ripoll et al used finite element analysis to analyse the effect of reciprocating sliding friction on the plastic deformation of the textured surface under different groove morphology and loading conditions[8].

The above researches mainly aim at the change of the friction performance of the friction pair, but the material wear analysis and research are not deep enough. In this paper, the typical two-body wear of sliding guide ways is studied by Deform-3D. The two-body wear simulation model of the guide ways is to carry out finite element analysis, and the influence of friction condition on the two-body wear is explored, which provides a theoretical basis for practical application.

2. Mechanism analysis of two-body wear

Wear is a process in that the surface materials continue removal in the relative motion of the objects in contact with each other, which is an inevitable result of friction. The interaction of the two friction surfaces can be either mechanical or molecular. Mechanical effects include elastic deformation, plastic deformation, and furrowing effects. It can be caused by the direct mesh of the rough peaks between the two surfaces, or it also can be caused by external abrasive particles. The surface molecular action includes two kinds of attraction and adhesion effects, the former has a small force and the latter's force is large. When the friction pair slides, the rough peak of the soft surface is easily deformed, and a smoother surface formed by the soft rough peak first breaks under load. Thus, the contact state is no longer a rough peak versus a rough peak, but the rough peak of the hard surface slides on a relatively smooth soft surface, and when the hard surface rough peak slides on the soft surface, the points on the soft surface are subjected to a load, the surface shear deformation will occur and cause wear.

The relationship between various factors in the friction and wear process is very complicated. The surface layer will undergo mechanical, structural, physical and chemical changes under the interaction of the friction surface. The plastic deformation of the surface layer makes the metal cold. It hardens and becomes brittle. If the surface is subjected to repeated elastic deformation, fatigue damage and wear will occur. The high temperature of the surface contact caused by frictional heat can soften the surface metal, and rapid cooling after contact will cause recrystallization or solid solution decomposition, these will cause wear.

3. Establishment of finite element model for two-body wear of guide ways

Based on friction and wear mechanism, a simplified model of the two-body wear of the rectangle guide ways is established by using the finite element software Deform. The model is shown in Figure 1. In the simplified model, the sliding guide way length is L_1=10mm, the width is l_1=20mm, the height is h_1=5mm, the static guide way length is L_2=100mm, the width is l_2=20mm, and the height is h_2=5mm. At the same time, the material properties are added to the sliding guide way and the static guide way, as shown in Table 1.

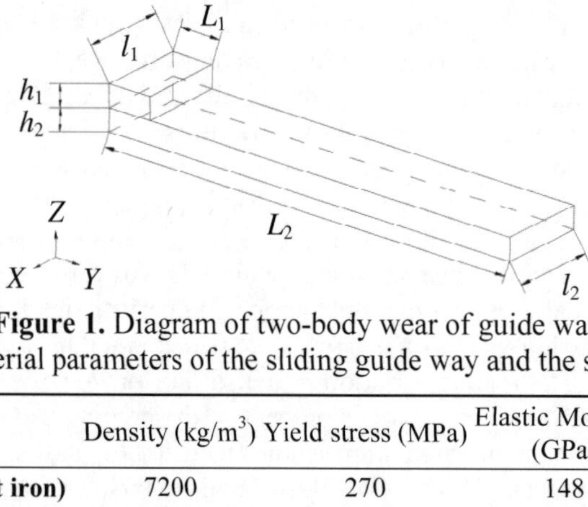

Figure 1. Diagram of two-body wear of guide ways.

Table 1. Material parameters of the sliding guide way and the static guide way.

Material	Density (kg/m^3)	Yield stress (MPa)	Elastic Modulus (GPa)	Poisson's ratio	Hardness
the static guide way **(cast iron)**	7200	270	148	0.3	HRC17
the sliding guide way **(45 # steel)**	7800	355	210	0.3	HRC26

In the simulation process, the initial mesh of the model will be distorted and degraded. In order to avoid the problem of reduced accuracy or non-convergence caused by severe distortion of the mesh, the adaptive mesh re-division technique to Deform is adopted. Encryption and automatic compensation of the lost mesh volume not only ensures the accuracy of the local deformation, but also saves the solution time and memory consumption.

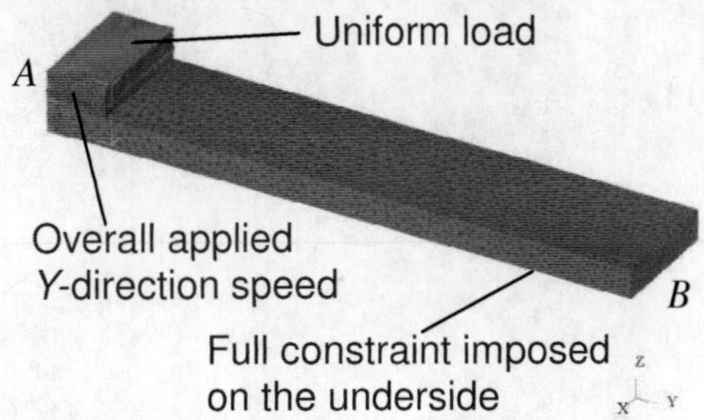

Figure 2. Finite element model of the guide ways wear.

The boundary conditions of the sliding guide way and the static guide way are set separately. The wear process is achieved by relative sliding, so that the sliding guide way needs to apply a speed in the *Y* direction while the upper surface of the sliding guide way applies a load in the negative direction of the *Z* axis. In the actual working condition, the static guide way is fixed, and full constraint is applied to the bottom surface of the static guide way to control the movement of the static guide way. The model is shown in Figure 2. It is assumed that the heat exchange coefficient between the friction surfaces is 45N/sec/mm/C, the friction coefficient is 0.1, and the friction coefficient does not change with temperature. Set the environment temperature to 20°C. In the simulation process, the wear calculation is based on the Archard wear calculation formula, which is:

$$W = \int K \cdot \frac{p^a \cdot v^b}{H^c} \mathrm{d}t \tag{1}$$

Where, W is wear depth, p is the interface pressure, v is sliding velocity, H is material hardness, K, a, b, and c are experimentally calibrated coefficients.

The simulation control increment is 0.01s/step, and the reading is saved every 20 steps. The time step setting cannot be too large, otherwise the solution accuracy will be reduced, and will result in mesh distortion or even non-convergence. The heat transfer and deformation are selected in the simulation model, and the international unit system will be chosen. The conjugate gradient method is used to increase the solution accuracy. The wear process under different parameters is simulated by a single variable method, and the influence of different friction conditions on the wear of the two-body of the guide ways is analysed.

4. Simulation results and analysis

4.1. Analysis of two-body wear process of guide ways

When the load is 1000N and the sliding speed is 5mm/s, the sliding guide way slides from point *A* to point *B* on the static guide way through the simulation of the finite element software Deform, the nephogram of wear can be obtained, as shown in Figure 3.

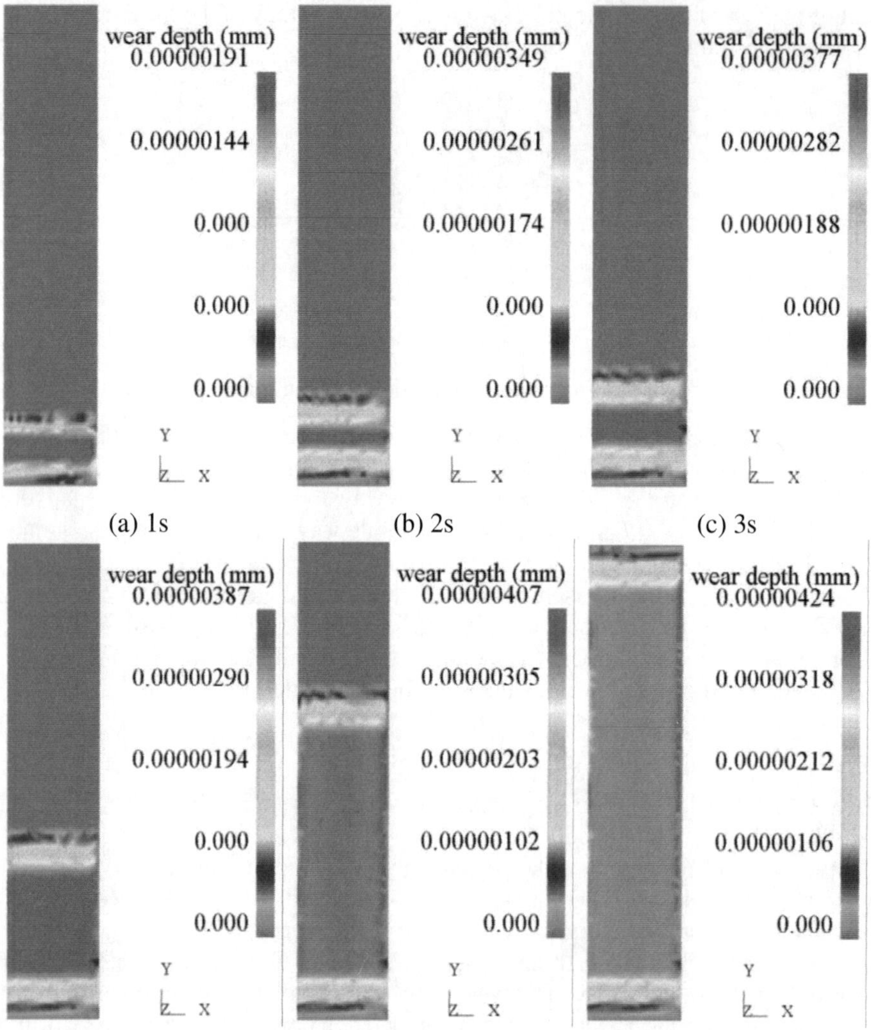

(a) 1s (b) 2s (c) 3s

(d) 6s (e) 12s (f) 18s

Figure 3. Nephogram of the static guide way wear in different time.

It can be seen from Fig.3(a) that the wear value of the contact area between the static guide way and the sliding guide way is generally 1.89×10^{-6}mm when the sliding guide way starts to slide, and the wear value on both sides is below 8.78×10^{-7}mm. It shows the tendency that the middle area wear is large and various, and wear of both ends are small. As the sliding guide way continues to slide, the middle area wear of the static guide way is continuously expanded, and the area with small wear at both ends remains basically unchanged. In Fig.3(f), the middle portion of the static guide way wears relatively uniformly, and the both ends area wears very small. And the material of the edge region is not supported, and the stress concentration caused by the extrusion, and the material strength is low and easily broken. The maximum wear values at both ends of the static guide way are 1.84×10^{-6}mm, and the maximum wear value in the middle is about 3.58×10^{-6}mm. This is because the sliding guide way are in the initial sliding and ending sliding state, the contact friction area is small. When it is at both ends of the static guide way, the wear of the static guide way is small.

4.2. Effect of load on the wear of the static guide way

The static guide way is subjected to different loads, and the degree of wear will change accordingly. In order to study the relationship between wear and load, the simulation loads are selected to be 1000N,

1300N and 1600N respectively when the sliding speed is 5mm/s. By extracting 100 nodes data uniformly distributed from the point *A* to point *B* on the static guide way, the curves of the single-wear surface appearance of the static guide way under different loads are plotted, as shown in Fig.4.

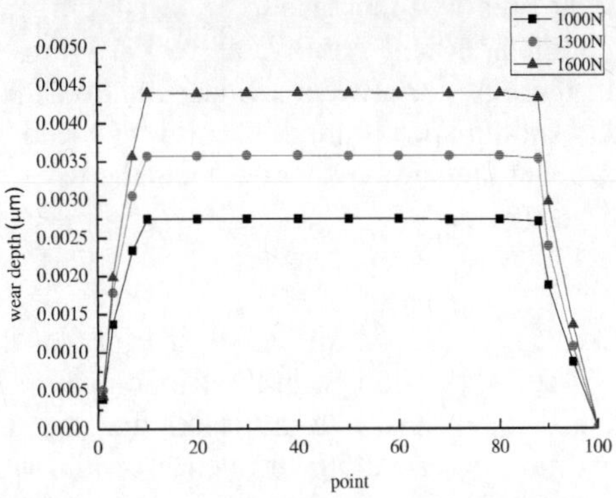

Figure 4. Chart of parameters with different loads.

It can be seen from Fig.4 that the static guide way wear tends to increase at the node 1-10, and the wear value at the node 10 reaches a maximum of 2.75×10^{-3}μm, and the wear value of the middle portion is substantially unchanged. At the node 88, the static guide way wear value begins to decrease, and the wear value decreases to zero when the sliding guide stops. When the load is 1000N, the wear depth in the middle of the static guide way is 2.75×10^{-3}μm. When the load is increased to 1300N, the wear depth in the middle of the static guide way increases to 3.58×10^{-3}μm. When the load is 1600N, the middle of the static guide way wear depth is 4.40×10^{-3}μm. When the hardness, sliding speed and friction coefficient of the guide way material are constant, the increase of the load will increase the wear depth of the static guide way. The increase in wear at both ends of the static guide way also increases as the load changes. The larger the load, is the more inclined the curves, is which means that the degree of material wear is greater.

4.3. Effect sliding speed on the wear of static guide way

The sliding speed of the sliding guide way is one of the factors that affect the wear of the static guide way. In order to study the influence of different sliding speeds on the wear of the static guide way, other friction parameters were assumed unchanged, and the sliding speeds were selected as 2.5mm/s, 5mm/s and 7.5mm/s respectively for finite element simulation. The post-simulation processing results are extracted, and the curves of the single-wear surface appearance of the static guide way at different sliding speeds are shown in Fig.5.

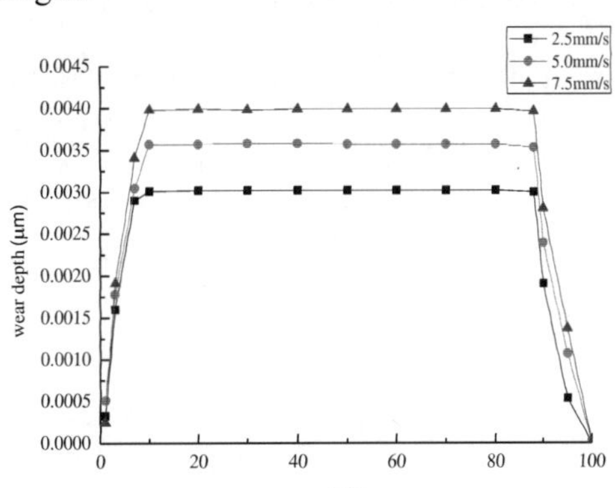

Figure 5. Chart of parameters with different sliding speed.

It can be seen from Fig.5 that the trend of the surface appearance wear depth contrast curves of the static guide way is similar to Fig.4. When the sliding speed of the sliding guide way is 2.5mm/s, the wear value of the middle portion of the static guide way is maintained at 3.02×10^{-3}μm. At a speed of 5mm/s, the wear depth in the middle area will increase to 3.58×10^{-3}μm, and when the sliding speed is 7.5mm/s, the wear depth in the middle area will stabilize at 3.99×10^{-3}μm. When the hardness, friction coefficient and load are constant, there is not always a linear relationship between the sliding speed and the wear depth. In the specific sliding speed range, the sliding speed increases, while the wear of the static guide way becomes large, but the sliding speed effects the wear of the sliding guide way not obviously.

5. Conclusions

In this paper, based on the Archard wear calculation model, the load of the static guide way and the sliding speed of the sliding guide way are considered in the friction pair. The simplified finite element model of the two-body wear of the guide ways is established, and the wear process is studied. The results show that the friction condition has a certain influence on the wear of the guide ways. The middle area of the surface of the static guide way wears uniformly, and the wear at both ends is light. Within a certain range, the wear of the static guide way is severe with the increase of the load, and when the sliding speed of the sliding guide way becomes large, the wear of the static guide way aggravates.

Acknowledgments

The research work is supported by National Natural Science Foundation of China under Grant No. 51475066 and No. 51675075 and Natural Science Foundation of Liaoning Province under Grant No. 2015020114.

References

[1] Rabinowicz E, Tanner R I 1995 Friction and Wear of Materials *J. Journal of Applied Mechanics* **33(2)** pp 606-611
[2] Lee H C, Lee Y and Lee S Y 2008 Tool life prediction for the bolt forming process based on high-cycle fatigue and wear *J. Journal of Materials Processing Tech* **201(1)** pp 348-353 (in Chinses)
[3] Söderberg A and Andersson S 2009 Simulation of wear and contact pressure distribution at the pad-to-rotor interface in a disc brake using general purpose finite element analysis software *J. Wear* **267(12)** pp 2243-2251
[4] Peterseim J, Elsing R and Deuerler F 1998 Simulation of sliding wear of hard-phase-containing metallic compound materials J. Metall 52(10-11) pp 643-651
[5] Mukras S, Kim N H and Sawyer W G 2009 Numerical integration schemes and parallel computation for wear prediction using finite element method *J. Wear* **266(7–8)** pp 822-831
[6] Bortoleto E M, Rovani A C and Seriacopi V 2013 Experimental and numerical analysis of dry contact in the pin on disc test *J. Wear* **301(1–2)** pp 19-26
[7] Shen X, Liu Y and Cao L 2012 Numerical Simulation of Sliding Wear for Self-lubricating Spherical Plain Bearings *J. Journal of Materials Research & Technology*, **1(1)** pp 8-12 (in Chinses)
[8] Ripoll M R, Podgornik B and Vižintin J 2011 Finite element analysis of textured surfaces under reciprocating sliding *J. Wear* **271(5)** pp 952-959

Backlash Analysis of Compound Planetary Gear Set in Rotary Actuators

Y Wang and F Yang*

School of Mechanical Engineering, Shandong University, Jinan 250061, China

*Corresponding author e-mail: yfc26@163.com

Abstract. This paper proposes a model for backlash analysis of compound planetary gear set in rotary actuators. Three main types of errors in the compound planetary gear set are analysed, that is, manufacturing errors, assembly errors and deformation errors. Due to the randomness of the errors, the backlash model of the gear set was established based on probability theory to improve the backlash estimation for the practical applications. Finally, the backlash of a compound planetary gear set was calculated and the influences of main errors to the backlash were also investigated.

1. Introduction

Aviation actuator is an important transmission mechanism of automatic flight control system. It receives the control instructions of the control system to drive the deflection of plane rudder, and then completes the control of the flight attitude and trajectory. Its performance affects the flight quality of the aircraft, such as maneuverability, stability, safety and reliability. The structure size of the actuator is limited by the aircraft space, and its weight should be as lighter as possible. Meanwhile, sufficient precision and load requirements should be ensured. Therefore, the design of actuator needs to meet these requirements of the special working environment of the aircraft [1-2].

One of the main factors influencing the performance of the actuator is the gear backlash, which is defined as the lag of the output shaft at the rotation angle when the input shaft changes the direction of rotation in the working state [3]. So in the design stage of the actuator, it is necessary to estimate the numerical range of the backlash in order to evaluate whether the design scheme is reliable.

In the previous studies, the extremum method [4-5] is often used to calculate the gear backlash. However, all the influencing factors are not simultaneously at the maximum. So the system synthetic error obtained from the extremum method exceeds the actual value, thus unreasonably improving the manufacturing precision and resulting in the uneconomic consequences [6].

In the design stage, the tolerances of each part are given, and the errors of these parts are usually distributed in certain statistical rules. Therefore, it is more reasonable to use probability method to establish the statistical synthesis formula of transmission chain backlash [7-8].

2. Design

According to the working conditions of the aircraft, basic structure and transmission technical requirements of the actuator are transmission ratio over 100, output torque over 200Nm, output angular velocity about 30r/min, structure diameter within 150mm, and backlash angle within 0.5°.

In the design process of the actuator, the new structure adopted is compound planetary gear set and the diagram of compound planetary gear set is shown in Figure 1.

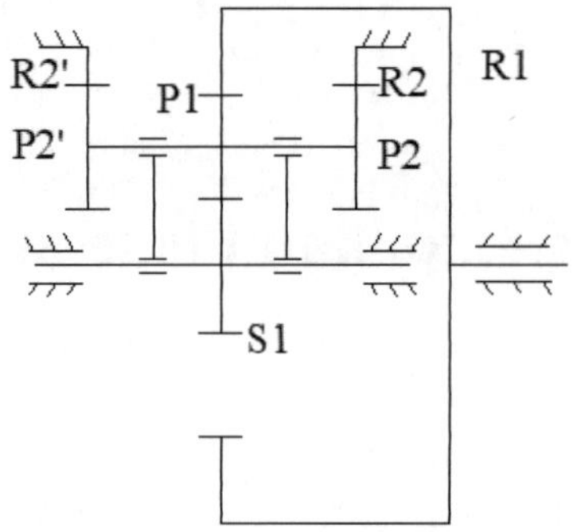

Figure 1. Diagram of compound planetary gear set.

Three planetary gears on the same shaft form a planetary gear set and there are four meshes on each planet gear set. The power system controls the sun gear S1 which engages with gear P1 to drive the planetary gear set. The planetary gear P1 drives the ring gear R1 as output. The fixed ring gear R2 and R2' adopt symmetrical design in order to reduce the wear of the tooth surface caused by bending moment of the large output torque. In order to realize power distribution, the whole mechanism uses three planetary gear sets. The gear ratio i_{S1R1} can be expressed as:

$$i_{S1R1} = \left| \frac{Z_{R1}(Z_{P1}Z_{R2} + Z_{S1}Z_{P2})}{Z_{S1}(Z_{R1}Z_{P2} - Z_{P1}Z_{R2})} \right| \tag{1}$$

In this scheme, the gear module is 1mm and the pressure Angle is 20 degrees. The gear geometry parameter and precision are shown in Table 1. According to the equation (1), the ratio is 178. And the transmission efficiency is 0.631.

Table 1. Gear geometry of the compound planetary gear set.

Gear	S1	P1	R1	P2	R2
Tooth number	18	45	108	42	105
Tooth Width\mm	15	15	15	12	15
Precision grade (GB2363-90)	7e	7e	7e	7e	7e

3. Modelling

3.1. Error sources

The main factors affecting the backlash mainly cover three categories including manufacturing error caused by the machine tool error, assembly error caused by errors of gearbox assembly and deformation error caused by the load. Due to these errors, there is a backlash between the gear pair in the process of meshing, and the circumferential backlash is defined as j_t.

3.2. Manufacturing error

The influence of manufacturing error on the backlash is reflected in two aspects, tooth thickness deviation and geometric eccentricity.

3.2.1. Tooth thickness deviation. Tooth thickness deviation is an important error in gear manufacturing. The tooth thickness deviation can be estimated according to the deviation ΔE_{wm} of the common normal line length. ΔE_{wm} is a negative value, so the circumferential backlash can be expressed as:

$$j_{tEwm} = -\frac{\Delta E_{wm}}{cos\alpha_n} \tag{2}$$

3.2.2. Geometric eccentricity. Due to the positioning error of the machine tool worktable, there is an eccentricity between the base axis of the gear blank and the rotary axis of the worktable. In the probability method, the influence of the eccentricity is analyzed by the total composite radial deviation $\Delta F''_i$ of the gear. $\Delta F''_i$ represents the difference between the maximum and minimum center distance, so the value of the eccentricity is approximately equivalent to $\frac{1}{2}\Delta F''_i$. The eccentricity e_g follows Rayleigh distribution and the numerical characteristics are:

$$E_{eg} = \sqrt{\frac{\pi}{2}} \times \frac{1}{2} F''_i \approx 0.62 F''_i \tag{3}$$

The circumferential backlash can be expressed as:

$$j_{teg} = 2e_g \tan\alpha_n \approx 1.24\Delta F''_i \tan\alpha_n \tag{4}$$

3.3. Assembly error

3.3.1. Parallelism error of bearing mounting holes. The error of gearbox during the assembly results in the parallelism error of gear shaft axis. In this paper, a rectangular coordinate system of gear is established, with the shaft axis as X axis, the center line direction of meshing gear pair as Z axis, and the direction perpendicular to X axis and Z axis as Y axis. The deflection Δf_{xy} of gear X axis in XY plane is shown in Figure 2. When the deflection occurs, Δf_{xy} increases the actual tooth thickness in the meshing plane, and the change of effective tooth thickness is:

$$\Delta S = S(\cos\lambda_{xy} - 1) + b\sin\lambda_{xy} \approx b\lambda_{xy} = b\frac{\Delta f_{xy}}{b} = \Delta f_{xy} \tag{5}$$

In this equation: S -The original tooth thickness; b-the tooth width; λ_{xy} -A tiny angular displacement.

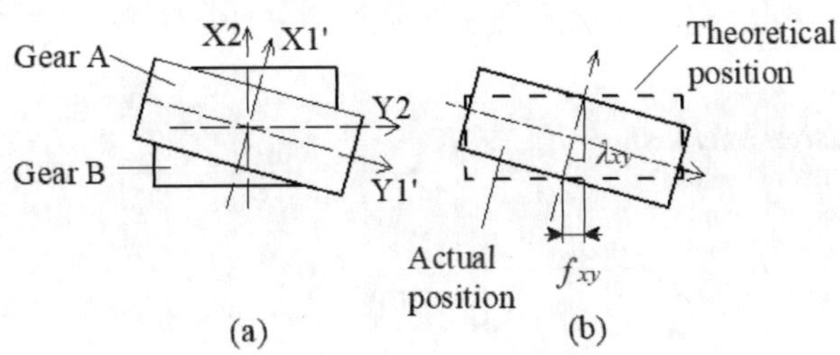

Figure 2. The deflection of gear X axis in XY plane.

The circumferential backlash can be expressed as:

$$j_{txy} = -\Delta S \approx -\Delta f_{xy} \tag{6}$$

The deflection Δf_{xz} of gear X axis in XZ plane is shown in Figure 3.

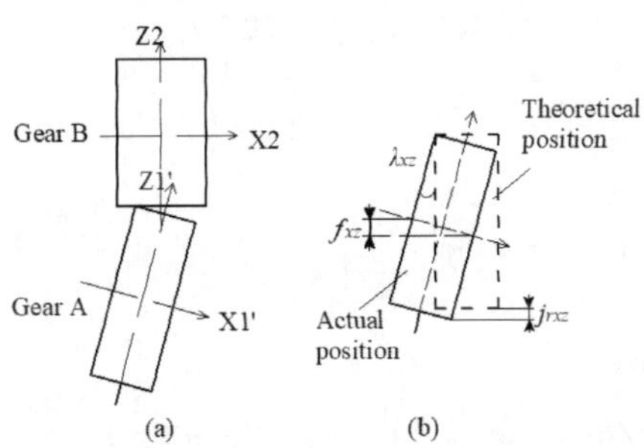

Figure 3. The deflection of gear X axis in XZ plane.

The radial displacement at any point of the involute is:

$$j_{rxz} = \frac{b}{2}\sin\lambda_{xz} - r(1-\cos\lambda_{xz}) = \frac{b}{2}\frac{\Delta f_{xz}}{b} - r(1-\cos\lambda_{xz}) = \frac{\Delta f_{xz}}{2} - r(1-\cos\lambda_{xz}) \approx \frac{1}{2}\Delta f_{xz} \tag{7}$$

In this equation: b-the tooth width; λ_{xz} -A tiny angular displacement.

The circumferential backlash can be expressed as:

$$j_{txz} = -2 j_{rxz} \tan\alpha_n \approx -\Delta f_{yz} \tan\alpha_n \tag{8}$$

3.3.2. Error of bearing. The clearance errors introduced by bearing include radial clearance Δ_r of bearing, assembly gap ΔC_{bi} of bearing inner ring and shaft, assembly gap ΔC_{bo} of bearing outer ring and mounting hole. The bearing clearance causes gear movement in the radial direction increasing the center distance of gear pair. The circumferential backlash can be expressed as:

$$j_{t\Delta r} = \Delta_r \tan\alpha_n ; j_{tCbi} = \Delta C_{bi} \tan\alpha_n ; j_{tCbo} = \Delta C_{bo} \tan\alpha_n \tag{9}$$

3.4. Deformation error

3.4.1. Torsional deformation of the shaft. When the rotating shaft is under the load of output torque T, the torsion deformation of the driving shaft is:

$$\varphi_n = \frac{180\times 60}{\pi}\frac{TL}{GI_n} \tag{10}$$

3.4.2. Bending deformation of the shaft. In the process of gear transmission, under the action of normal force F_n on the tooth surface, the shaft will bend and deform. The bending deformation f_c can be divided into tangential component f_t and radial component f_r. According to the equation (6) and (8), the following relationship is derived:

$$j_{tFn} = 2f_t + 2f_r \tan\alpha_n = 2f_c \cos\alpha_n + 2f_c \sin\alpha_n \tan\alpha_n = \frac{2f_c}{\cos\alpha_n} \tag{11}$$

3.5. The calculation model of backlash

The backlash angle caused by j_t is:

$$B_{jt} = \frac{j_t}{\frac{1}{2}\times m_n z\times 1000}\times\frac{180}{\pi}\times 60 = 6.876\times\frac{j_t}{m_n z} \tag{12}$$

So the backlash angle of gear A caused by the errors is:

$$B_A = 6.876\times\frac{1}{m_n Z_A}(2j_{tEwmA} + j_{tegA} + j_{txyA} + j_{txzA} + j_{t\Delta rA} + j_{tCbiA} + j_{tCboA} + j_{tFnA}) + 2\varphi_{nA} \tag{13}$$

The backlash angle of gear A caused by gear pair A and B is:

$$B_{AB} = B_A + \frac{Z_B}{Z_A}B_B \tag{14}$$

For the compound planetary gear set in this paper, the backlash converted to the output gear R1 is:

$$B_{output} = \frac{1}{i_{S1R1}}B_{S1P1} + B_{R1P1} + \frac{Z_{R2}Z_{P1}}{Z_{P2}Z_{R1}}B_{R2P2} \tag{15}$$

4. Results

4.1. The backlash angle of the compound planetary gear set

On the basis of accuracy standards, Table 2 shows the main tolerances of the error. According to the equation (13), (14) and (15), the backlash angle of each gear pair and output gear is shown in Table 3.

Table 2. The main tolerance of the error.

Gear		S1	P1	R1	P2(P2')	R2(R2')
Manufacturing error	$\Delta E_{wm}\,/\,\mu m$	-20	-20	-27.5	-20	-27.5
	$\Delta F''_i\,/\,\mu m$	25	32	39	32	39
Parallelism error	$\Delta f_{xz}\,/\,\mu m$	5				
	$\Delta f_{xy}\,/\,\mu m$	3				
Bearing error	$\Delta_r\,/\,\mu m$	10.5				
	$\Delta C_{bi}\,/\,\mu m$	8				
	$\Delta C_{bo}\,/\,\mu m$	19				
Deformation	$\varphi_n\,/\,arc\min$	0	3.05	0	1.53	0
	$f_c\,/\,\mu m$	0	0.08	0	0.41	0

Table 3. The backlash angle of the compound planetary gear set.

The backlash between gear pair		Backlash angle/*arc*min
Gear pair S1 and P1	B_{S1P1}	8.4
Gear pair P1 and R1	B_{P1R1}	13.6
Gear pair P2 and R2	B_{P2R2}	10.8
Sum(Convert to R1)	B_{Output}	24.8

4.2. The error analysis

Figure 4 shows the influence of tooth thickness deviation, composite radial error and parallelism deflection of each gear on the backlash. According to Figure 4, the errors of gear P2 and R2 have the greatest influence and the gear P1 and R1 play an important role on the backlash. Because of the large radio, the backlash angle caused by the errors of gear S1 is much smaller than the errors of other gears.

From Figure 4, it can be seen from the results that manufacturing errors are fundamental and the main source of the backlash. The errors of assembly have little impact on the backlash. Because the value of parallelism deflection is usually less than other errors and the actual backlash angle by the bearing will decrease with three sets of evenly distributed gears.

Deformation error is a dynamic error, varying according to the load. In order to meet the strength requirements, the deformation error should be kept within a reasonable range.

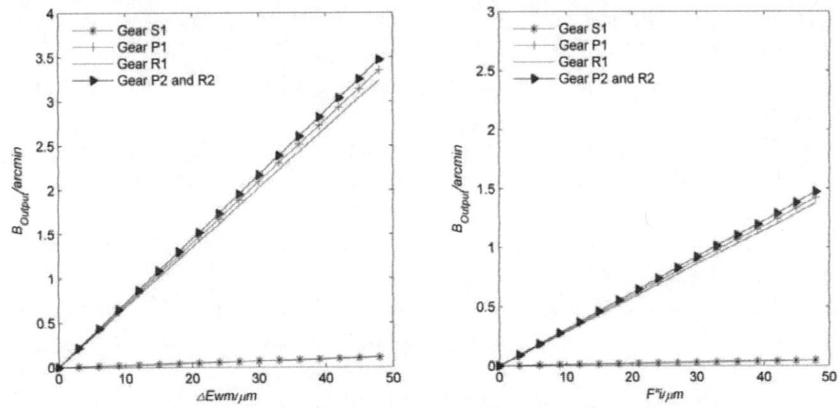

(a) (b)

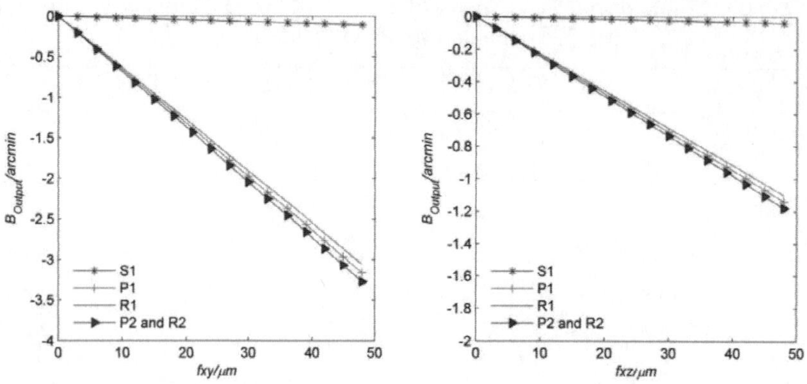

(c) (d)

Figure 4. The effect of the main errors on backlash.

5. Conclusions

In this paper, the actuator uses a new structure of compound planetary gear set and realizes the transmission of the actuator with large transmission ratio, high torque, small size and high precision. For the compound planetary gear set, this paper systematically analyzed the main factors of the gear set including manufacturing error, assembly error and deformation error, and the system backlash analysis model is established by probability method.

According to the model, the backlash angle is 24.8 *arc*min which meets the accuracy requirements of actuator. The influences of the errors on the backlash can be obtained from the model. The main factors affecting backlash is the manufacturing error which mainly caused by the tooth thickness deviation. It should be noted that the influence of each gear error on output backlash angle can also be achieved by the model. This model can be a reference to the error allocation.

Acknowledgements

This work was supported by National Natural Science Foundation of China (No.51775309) and Young Scholars Program of Shandong University.

References

[1] X Yao, D Qin and R Li 2008 Study on contact teeth and load distribution of planetary gearing with small tooth number difference *Int. Conf. on Frontiers of Design and Manufacturing(Hong Kong)* pp 545-551

[2] Y Zhang 2016 Return Difference Analysis of Two-input Planetary Gear Reducer *J. Mech. Transm.* **40** pp 132-135

[3] X Guo 2004 *Firing accuracy analysis for long range gun weapon systems* (Beijing: National Defense Industry Press) PP 1-4

[4] W A Tuplin 1962 *Gear design* (London: The Machinery Publishing Co. Ltd)

[5] P J Sweeney 1996 Gear Transmission Error Measurement Using Phase Demodulation *J. Mech. Eng. Sci.* **210** 201-213

[6] Q Ye 2003 Calculating the Return Stagnant Angle of the Gearing by Use of Probability Statistic *J. Mech. Elec. Eqp.* **01** pp 16-19

[7] D R Slatter and I R Degen 2005 Miniature zero-backlash gears and actuators for precision position applications *Proc. of the 11th European Space Mechanisms and Tribology Symposium* pp 1-5

[8] E Chen and D Walton 1990 The Optimum Design of KHV Planetary Gear with Small Tooth Differences *Int. J. Mach. Tools and Manu.* **30** pp 99-110

Simulation Calculation of Flow Characteristics of Air Separation System Based on Flowmaster

Yu Zeng[1], Bing Sun[2]and Haoyang Sun[1]

[1] Beihang University
[2] Aviation Industry Corporation of China

zengyu@buaa.edu.cn; 13810545207
sunbing0720@163.com; 18618405545
sunhaoyang@buaa.edu.cn; 15652909619

Abstract. The air separation system is an important part of aircraft fuel inerting system, which is responsible for the important task of supplying nitrogen-enriched air to the aircraft fuel tank. The flow characteristic of air separation device is the key problem of its design. In this paper, based on one-dimensional fluid simulation software Flowmaster, we did some research on the simulation model of membrane modules, established the simulation model of air separation system, obtained the simulation results of the flow characteristics and analyzed the results in detail. The results showed that under low pressure and flow conditions, compared with the experimental values, the relative error of nitrogen-enriched air and oxygen-enriched air flow rate was about 5%.In the case of high inlet pressure and flow, the pressure loss of the junctions and the long pipeline was larger, and the inlet pressure, outlet pressure and pressure loss of the 12 membrane modules were significantly different. The research work was of reference value for the design optimization of air separation system in aircraft fuel inerting system.

Keywords—Fuel inerting system, air separation system, pressure loss, flow characteristics, numerical simulation

1. Introduction

For modern military aircraft, fuel system fire disaster is the most common and the most serious potential danger, and the survival of fuel tank is of great significance to improve the survival of the whole machine. The fuel system mainly includes storage system, distribution system, measurement system, inertial system and many other complex subsystems[1]. The inerting system takes the high-pressure gas from the engine compressor and processes it into nitrogen-enriched air through the air separation system which is suitable for concentration of the fuel tank. Its function is to ensure that the oxygen concentration of the oil free space above the tank level is below 9%, to ensure the safe state of the oil tank. On the other hand, as the air pressure increases in the external environment during the landing, the inert gas is also used to increase the pressure in the tank to prevent excessive negative pressure.

The traditional aircraft fuel system research and design process often involves large workload, complicated tasks, inconvenient data usage, low efficiency, poor fault-tolerance, and difficulty in retrieving. With the continuous development of computational simulation technology, the designers usually choose the simulation calculation method to verify the feasibility of the complex system[2].

This paper mainly studies the flow characteristics of air separation system in fuel system based on fluid simulation software Flowmaster. As the world's leading one-dimensional fluid system simulation software facing the engineering, Flowmaster includes one-dimensional hydrodynamic system solver and fluid system simulation software package which is mature and complete, has received more and more application from aerospace, automobiles, ships, energy and chemical industry, hydraulic pump station and other industrial field. Flowmaster provides a set of simple and visual development environment to build elements and system. When building the simulation system, designers can call multiple pipelines and components more than 400 species, such as pipe, bend, T-branch tube, valve, pump, fan, heat exchanger, voltage regulator, fuel tank, spring, thermal bridge, control element and so on. By using Flowmaster, we can carry out accurate pressure, flow, temperature and flow velocity analysis of each node in the system, to help engineers complete and optimize the system design quickly. It can reduce production cost, improve product performance, shorten product development time and test times. [3-5]

2. Air separation system

Air separation system is loaded in aircraft inerting system, using hollow fiber technology, to separate oxygen and nitrogen in the air which is requested in pressure, flow, temperature and cleanliness, drain the enriched oxygen, and collect the enriched nitrogen for fuel tank inerting, to prevent serious disaster caused by aircraft fuel tank explosion and improve the survivability of the aircraft.

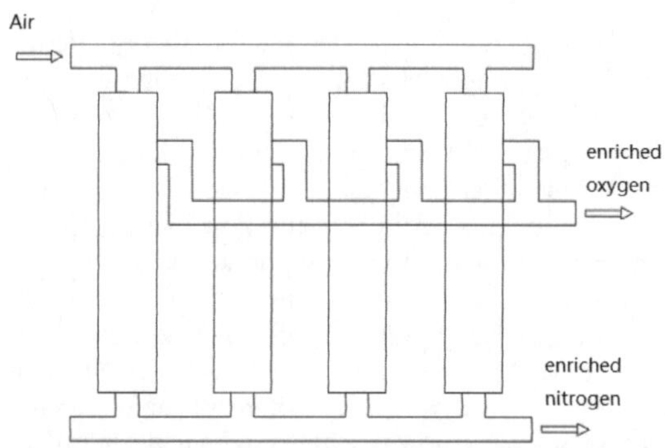

Figure 1. The upper layer of air separation system

The upper layer of air separation system is showed in Fig. 1, which is mainly composed of membrane modules, frames, pipelines, pipe supports, flexible clamp components and so on. In the air separation system, the air is bled from the environmental control system, entered into 12 membrane modules in the upper, middle and lower layers, where the separation of oxygen and nitrogen is achieved. The three -layer membrane modules are similar. The enriched oxygen bled from each membrane module is sent through the enriched oxygen pipe to the back of the aircraft cabin, the enriched nitrogen bled from each membrane module is collected into the enriched nitrogen pipe, adjusted into Nitrogen gas with different oxygen concentration and flow by a pressure regulator, and transported to the corresponding parts in fuel tank according to the system requirements.

3. Simulation model construction

3.1. Modeling purposes

The modeling purposes are formatted as follows:

- In view of the air separation system, how to utilize flow characteristics components in Flowmaster to build reasonable simulation calculation model of the air separation system and guide the subsequent simulation calculation and experiment of the air separation system is necessary.

- In view of the existing air separation system, there is a large pressure loss between the gas source and the entrance of 12 membrane modules. This paper will focus on two T-branch tubes and a 1.3 meters long and straight pipeline, analyze the pressure loss by the simulation calculation and put forward the reasonable optimization in order to improve process performance of the structure.

3.2. Construction of membrane module model

Membrane module is an important part of air separation system, which pipeline structure is composed of air inlet, enriched nitrogen outlet and enriched oxygen outlet. In practical application, experimenters usually control the flow of enriched nitrogen by adjusting valve opening at the nitrogen gas outlet to control the oxygen concentration of enriched nitrogen, enriched oxygen is bled directly into the atmosphere. oxygen concentration, gas flow ratio and flow characteristics of membrane module are all related to inlet pressure and enriched nitrogen pipeline pressure.

In order to study membrane module's flow characteristics of enriched oxygen pipe and enriched nitrogen pipe, two flow resistance components, the pressure loss coefficient of which respectively were known from the experimental data of the existing single membrane module, were adopted to match membrane module.

The pressure loss coefficient is related to inlet pressure and pressure loss (the difference between inlet pressure and enriched nitrogen pressure), The relation of them is:

$$\Delta p = \frac{KQ^2}{2A\rho} \tag{1}$$

Q is mass flow, A is the area of pipe section, ρ is the density and Δp is the pressure loss.

According to the experimental data of single membrane module provided by Hefei jianghang aircraft equipment company, there are 8 groups of flow resistance coefficient of Enriched nitrogen pipeline K1 and flow resistance coefficient of Enriched oxygen pipeline K2 under different inlet pressure and pressure loss shown in Table 1.

Table 1. K1 and K2 under different inlet pressure and pressure loss

Inlet pressure/MPa	Pressure loss/ MPa	K1	K2
0.4	0.01	10675.5	10124.1
0.5	0.015	3375.9	2863.3
0.6	0.015	1102.9	1669.2
0.4	0.015	3306.4	11269.0
0.5	0.025	1182.5	3008.9
0.6	0.025	410.9	1492.7
0.4	0.035	822.0	4844.1
0.5	0.05	271.2	2317.1

3.3. Simulation model

In view of the actual situation of air separation system, this paper builds the simulation model based on Flowmaster—one-dimensional thermal flow simulation platform. The simulation model is shown in Fig. 2.

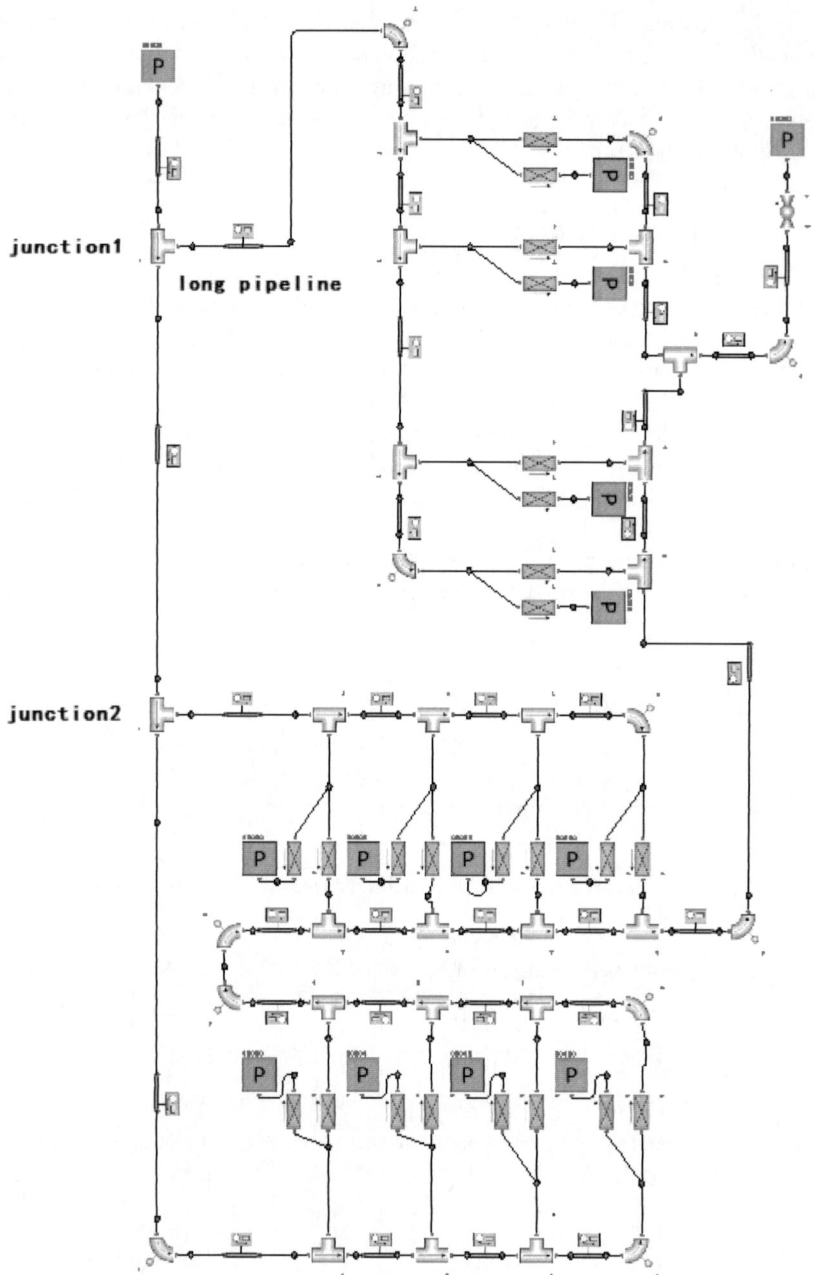

Figure 2. Simulation model of air separation system

The components in the model included pipe, bend, valve, source, T-branch tube, loss and so on. In the simulation model of air separation system, pressure source was adopted to simulate the air source of inlet and outlet, global valve was adopted to simulate flow control valve at the enriched nitrogen outlet. The 12 membrane modules were numbered 1~12, as shown in Fig. 2. The variable of simulation calculation was the opening of inlet pressure and the opening of flow control valve.

3.4. Component model selection and parameter setting

Table 2 introduces the model and the parameters to be set in the simulation calculation model according to the actual structure and components.

Table 2. The model and the parameters to be set

Components	Model	Parameters to be set
pipe	Pipe cylindrical	Length,diameter, Friction data
Constant pressure air supply	Source:Pressure	liquid type, total pressure
Flow resistance element	Loss:discrete	cross-sectional area, forward loss coefficient, forward loss coefficient
Valve	Valve:globe	Diameter,valve opening
Bend	Bend:circular	deflection angle, radius/diameter, roughness, diameter
T-branch tube	Junction:T(90°)	through pipe diameter, branch pipe diameter
medium	Dry air as real gas	No set up

4. Result and analysis

According to 8 groups of the experimental data of single membrane module, 8 groups of simulation calculations were performed.

In the simulation calculation, in the different air pressure conditions, flow resistance coefficient of membrane module was assigned in accordance with experimental data of single membrane module, the average pressure loss of the membrane modules was consistent with that of the single membrane module by changing the opening of the flow control valve, which ensured that the flow characteristics of the membrane modules were more consistent with the actual situation. The key parameters of simulation calculation of air separation system are shown in Table 3.

Table 3. Key parameters of simulation

Constant pressure air supply /MPa	The opening of flow control valve /%	Average pressure loss of the membrane modules /MPa	Average inlet pressure of the membrane modules /MPa
0.4	3.6	0.00999	0.398
0.5	6.33	0.01492	0.492
0.6	9.42	0.01501	0.583
0.4	7.1	0.01489	0.397
0.5	12.57	0.02496	0.489
0.6	19.72	0.02501	0.572
0.4	21.4	0.03505	0.390
0.5	59.3	0.04999	0.468

4.1. Analysis of pressure loss of T-branch tube and the long pipeline

IOP Conf. Series: Materials Science and Engineering **470** (2019) 012023 doi:10.1088/1757-899X/470/1/012023

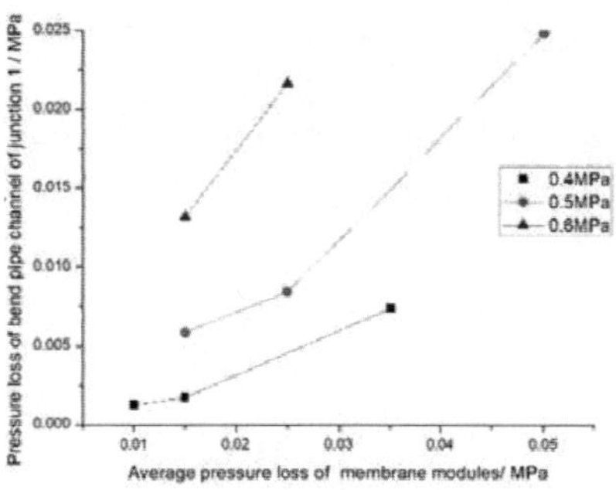

Figure 3. Pressure loss of bend pipe channel of junction 1

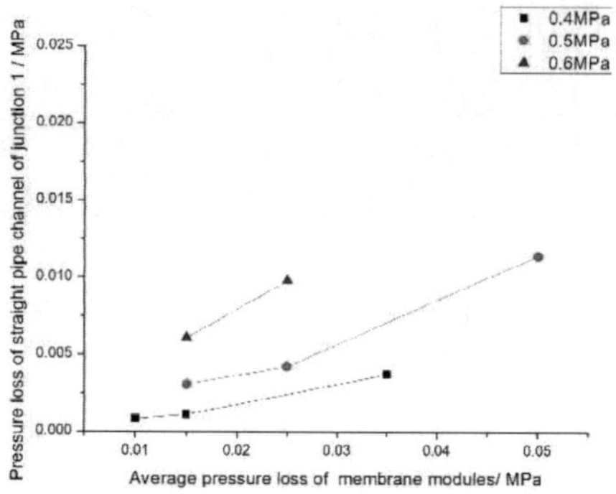

Figure 4. Pressure loss of straight pipe channel of junction 1

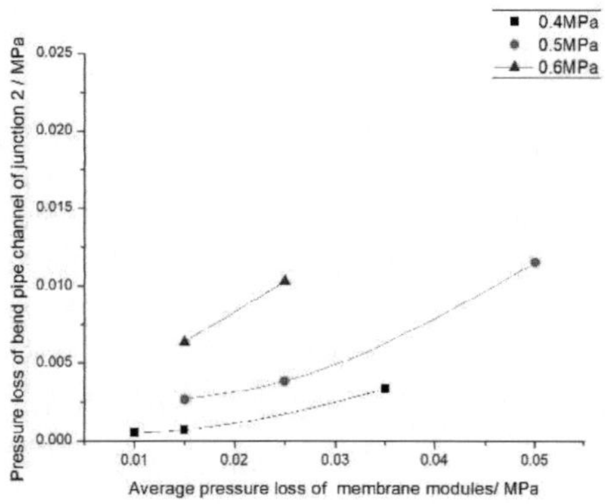

Figure 5. Pressure loss of bend pipe channel of junction 2

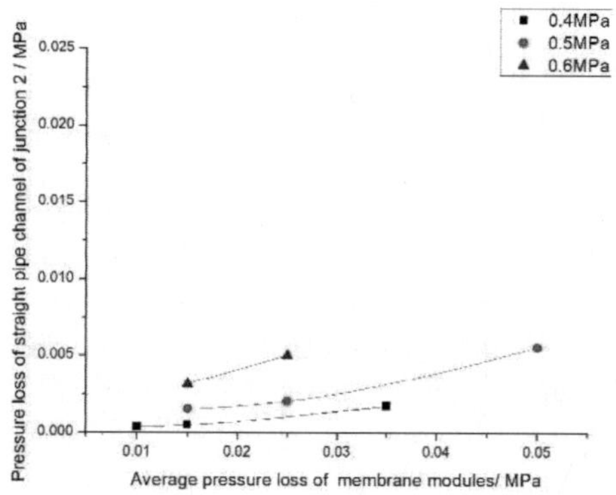

Figure 6. Pressure loss of straight pipe channel of junction 2

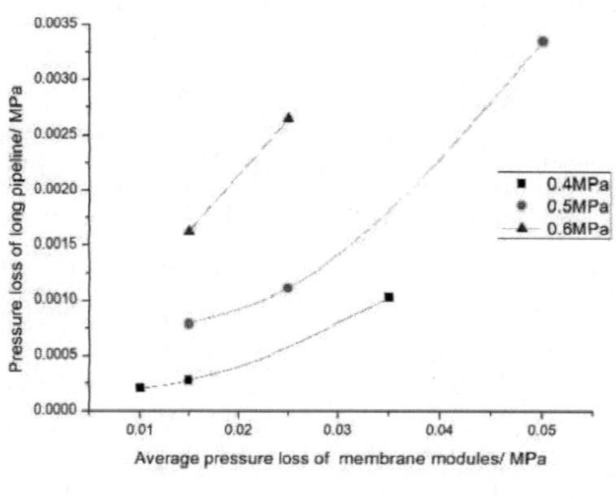

Figure 7. Pressure loss of the long pipeline

As can be seen from figure 3-7:

- The Pressure loss of bend pipe channel was greater than that of straight pipe channel;
- The junction 1 was in the front of the junction 2, and the flow rate of junction 1 was larger than that of the junction 2, so the pressure loss of junction 1 was greater.
- With the increase of air source pressure and flow, the pressure loss also increased.

- The pressure loss of junction was an order of magnitude greater than that of the long pipeline of 1.3 meters.

4.2. Analysis of simulation 8

The air source pressure and flow rate of simulation 8 were the largest in the 8 simulation experiments, and the differences between the 12 membrane modules were more obvious, which was convenient for analysis.

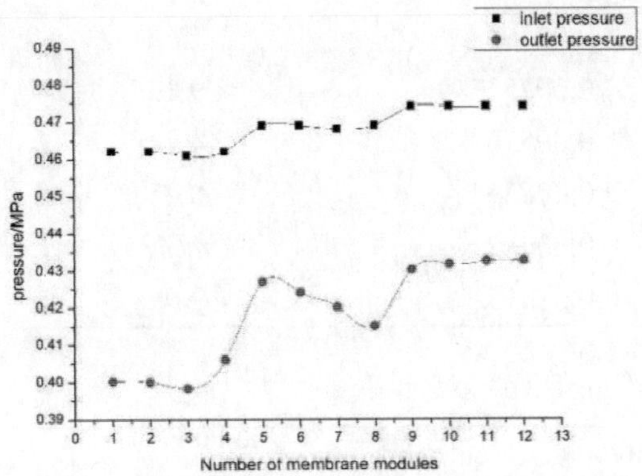

Figure 8. Inlet and outlet pressure of membrane modules in simulation 8

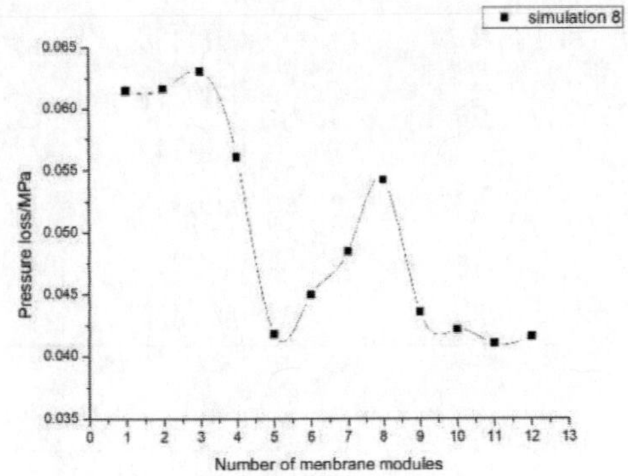

Figure 9. Pressure loss of membrane modules in simulation 8

It could be seen that when the inlet pressure and flow rate were larger, the inlet pressure, outlet pressure and pressure loss of each membrane module were very different. The following conclusions could be obtained from Fig. 8-9.

- The front pipe structure of membrane module in air separation system for had certain influence for inlet pressure of 12 membrane modules. As a result of the existence of two T-branch tubes, pressure loss of larger flow of T-branch tube was bigger, pressure loss of the bend pipe channel was bigger, so the inlet pressure of membrane module 1-4 was the minimum, the inlet pressure of membrane module 5-8 was secondary, and the inlet pressure of membrane module 9-12 was the maximum.
- The back pipe structure of membrane module in air separation system for had certain influence for outlet pressure of 12 membrane modules. Pressure loss of back pipe of membrane module1-4 was less than that of membrane module5-8, Pressure loss of back pipe of membrane module5-8 was less than that of membrane module9-12, so the outlet pressure of membrane module 1-4 was the minimum, the outlet pressure of membrane module 5-8 was secondary, and the outlet pressure of membrane module 9-12 was the maximum.
- The pressure loss of membrane module in air separation system was related with front pipe structure and back pipe structure. In a word, the pressure loss of membrane module 1-4 was the maximum, the pressure loss of membrane module 5-8 was secondary, and the pressure loss pressure of membrane module 9-12 was the minimum.

4.3. Simulation error analysis

The simulation values of the average flow of enriched nitrogen and enriched oxygen were compared with the experimental values of the single membrane module, and the error analysis was showed in Table 4-5.

Table 4. Error analysis of enriched nitrogen flow

Simulation number	Experimental values of enriched nitrogen flow /m³ ·s^-1	The average calculation value of enriched nitrogen flow /m³ ·s^-1	Relative error /%
1	0.00125	0.00122	2.25
2	0.00244	0.00240	1.94
3	0.00389	0.00393	1.13
4	0.00278	0.00252	9.13
5	0.00542	0.00516	4.74
6	0.00833	0.00837	0.45
7	0.00886	0.00782	11.78
8	0.01667	0.01515	9.09

Table 5. Error analysis of enriched oxygen flow

Simulation number	Experimental values of enriched oxygen flow /m³ ·s^-1	The average calculation value of enriched oxygen flow /m³ ·s^-1	Relative error /%
1	0.00523	0.00545	4.18
2	0.01001	0.01042	4.09
3	0.01312	0.01395	6.28
4	0.00502	0.00539	7.30
5	0.00997	0.01002	0.55
6	0.01412	0.01477	4.66
7	0.00808	0.00751	7.03
8	0.01199	0.01127	5.95

Table 4-5 showed that, under lower pressure and flow rate, the simulation error is within 5%; under higher pressure and flow, the maximum simulation error was 11.8%. In addition to the model error of the simulation calculation and the measurement error of the test, the main error was method error, the cause of which was that inlet pressure and pressure loss of each membrane module in simulation and experiment with single membrane were not the same, and the resistance coefficient of the membrane module K1 and K2 assigned different from the actual condition. A new idea was proposed that adopting the method of iterative calculation—according to the inlet pressure and pressure loss of membrane module in the last simulation calculation, assigning resistance coefficient of the membrane module K1 and K2 again and repeating the above operation until convergence.

5. Conclusion and optimization

The conclusion and optimization are formatted as follows:

- In the T-branch tube, the pressure loss of bend pipe channel was much greater than the pressure loss of straight pipe channel. As the pressure increased, it appeared a large pressure loss even influenced the experiment. Therefore, it was necessary to adjust T-branch tube to reduce the pressure loss.
- There was a 1.3-meter-long pipeline in experiment system, the pressure loss caused by which increase as the air source pressure increased. It is necessary that minimizing the length of the

pipeline or increase the diameter of it to increase its circulation area and reduce the pressure loss in the subsequent experiments.

- It is necessary to properly allocate the intake and exhaust mode of the membrane module, otherwise, the inlet pressure, outlet pressure and pressure loss of each membrane module will be different when the air source pressure and flow are high.
- The flow characteristics of membrane module were decided by the inlet pressure and the pressure loss of enrich nitrogen pipe. Therefore, we need a lot of experimental data and use interpolation to get the corresponding resistance coefficient K1 and K2. When sufficient experimental data are available for support, an iterative method can be used to analyze the flow characteristics of air separation system, which can obtain the accurate inlet pressure, pressure loss and resistance coefficient of each membrane module in the certain conditions of air source pressure and flow control valve opening, to avoid situations where the single assignment of resistance coefficient to all membrane modules is different from the actual working conditions.

References

[1] Langton R, Clark C and Hewitt M M 2009 *Aircraft Fuel Systems*(New York: Wiley–Interscience)

[2] Guangtuo G 2014 *J. Platform construction research for fluid numerical simulation of an airplane fuel system basing on flowmaster*. *Equipment Manufacturing Technology* **9** 42-44.

[3] Yi T, Guiping L and Guodong L 2009 *J Computer analysis of transport aircraft oxygen system based on flowmaster*. *Journal of Beijing University of Aeronautics and stronautics* **35** 1306-10

[4] Kexue Y 2008 *J Flowmaster Software Based on a Certain Type of Helicopter Fuel System Simulation*. *Helicopter technique* **4** 15-18

[5] Chao L, Zhihong L, Hong Shi and Yanlong J 2012 *J Simulation calculation of aircraft anti-ice system catheter based on Flowmaster software*. *Science & Technology Information* **32** 699-702

Modal Analysis of Double-Bolted Connection under Different Operating Conditions

Chao Cao[1], Xueyan Zhao[1], Weikang Du[1], Shourong Liu[1,*] and Dehao Wu[1]

[1]College of Engineering, China Agricultural University, Beijing, 100083, P. R. China.

* Corresponding author: lsr@cau.edu.cn

Abstract: Based on the finite element analysis and modal analysis, the vibration characteristic parameters and the law of behaviour changes in different bolted connection structures were studied. According to the different contact status of each component and bolt, the 16 typical working conditions were designed and selected, and the subspace iteration method was used to simulate the experiment with the ANSYS Workbench software. The simulation results showed that the natural frequency of the double-bolted connection structure tends to be stable under different working conditions. A certain degree of fluctuation of the natural frequency sometimes occurred, but the fluctuation range was not large. This research thus laid the foundation for subsequent structural health monitoring and diagnosis.

1. Introduction

The bolted connection structure is simple in form, convenient in disassembly and diverse in form, thus it is widely used in industry,such as docking of separation surface between spaceflight structure cabins and main shaft structure of wind generator. According to the distinct operational principle, the bolt connection is mainly divided into bolt connection and reaming bolt connection, with notable differences in the gap between the hole wall and the bolt rod. Although the bolt connection structure can provide a certain connection stiffness and strength for the structure, however, because of the geometric catastrophe and the destruction of the structure to the continuity, the local deformation of the joint will be considerable and the distribution of stress and deformation is uneven. The relaxation of bolt pre-load and the fatigue as well as creep at the joint will impinge on the reliability of the connection. More importantly, due to the contact and friction between the adjacent interfaces, it is vulnerable to rendering structure damping, nonlinearity, uncertainty in the analysis process of bolted connection and even catastrophe [1-5].

In the past decades, scholars across the world had studied the bolted joint structure mainly from the dynamic performs of the bolted joint stiffness of the bolted joint structure and the loosening mechanism. Yao et al. studied the effects of aero-engine bolting loads and structural parameters on joint stiffness [6]. Wang established the corresponding models and verified the nonlinear dynamics of bolted joint structures [7]. Liu analyzed the loosening mechanism of bolted joints under axial load excitation [8]. Klok studied mechanical behavior of bolted joints under impact rates of loading [9]. Tan researched the effect of bolt spacing on the tightness behavior of bolted flange joints [10].

Recent years, structure health monitor had increasingly become prevail. He explored vibration-based structural damage detection with applications to structures with bolted joints [11]. Ma et al. investigated nonlinear vibration characteristics of a rotor system with pedestal looseness fault under different loading conditions [12]. Zhang et al. researched application of sub-harmonic resonance for the detection of bolted joint looseness [13].

Besides, many scholars took modal analysis as a helpful tool to explore the dynamic performs of the system. Barad et al. found that natural frequency could help crack detection in cantilever beam though experimental study [14]. Tse proposed modal analysis for simplified model of a linked building system [15]. Huang et al built a simplified mode of dynamometer to achieve the theoretical natural frequency and the results indicated that the dimension of the key structure had huge impact on the natural frequency [16].Wang et al carried out an optimal design of the dynamic characteristics on the accessory driving system and the results revealed that the first-order natural frequency was sensitive to different tensioner dimensions [17].

The connection status of double-bolted joint has a notable effect on the reliability of the entire structure. Hence, it is indispensable to develop an effective method for health monitoring of bolted joints to detect the loss of the pre-load or the looseness of the bolts, which can trigger failure of bolted joint, causing possible engineering disasters, if not promptly detected. In this paper, based on the finite element analysis and subspace iteration method in the modal analysis, the characteristic parameters of vibration and the law of behavior changes in different double-bolted connection structures were investigated. According to the distinctive contact status of each part and bolt, the 16 typical working conditions were designed, and the ANSYS Workbench software was served as the main simulation platform to support all of these simulation experiments.

2. Calculation of modal analysis

The finite element method is an effective numerical method for solving partial differential equations. Its theoretical basis is calculus of variations. In essence, it is almost a weighted residual method for piecewise interpolation [18]. And it is an effective approximate calculation method, thus the ANSYS Workbench platform is chosen to conduct the corresponding simulation experiments.

Modal analysis is used to determine the vibration characteristics of the design structure or machine parts, namely natural frequencies and modes of structures, which are critical parameters in the design of dynamic load. Meanwhile it is also a necessary procedure for other dynamic analysis problems such as transient dynamic analysis, harmonic response analysis and spectral analysis [19].

The approximate calculation methods of modal analysis are mainly divided into Ritz method, matrix iteration method and subspace iteration method. The matrix iteration method does not require the hypothetical modal vectors, but every iteration of the first order inherent characteristic is required. However, the original high order eigenvalue problem is converted to a lower order eigenvalue problem via Ritz method, greatly reducing the degrees of freedom, and some low order eigenvalues and its eigenvectors can be obtained at the same time, but the precision depends mainly on the accuracy of the hypothetical mode. So the subspace iteration method ingeniously combines the above two methods. On the one hand, Ritz method is applied to shrinking the complexity of degree of freedom, and to determine the corresponding natural frequencies and natural modes. On the other hand, the matrix iteration method is employed to perform iterative operation, which can obtain sufficient accurate hypothesis mode of vibration and improve the accuracy of calculation [20].

The typical process for modal analysis as follows:

Establish system vibration differential equation, shown in Equation (1).

$$\boldsymbol{M}\ddot{\boldsymbol{x}} + \boldsymbol{C}\dot{\boldsymbol{x}} + \boldsymbol{K}\boldsymbol{x} = \boldsymbol{f}(t) \tag{1}$$

Where M is the mass matrix, C is the damping matrix, K is the stiffness matrix, $f(t)$ is the exciting force vector and x is the displacement vector.

Then neglect the damping item and let $f(t)$=0, then obtain the frequency characteristic equation of the system, shown in Equation (2).

$$([K] - \omega_{\mathrm{n}}^2[M])\{u\} = \mathbf{0} \tag{2}$$

The above equations were solved by subspace iterative method.

Firstly, Ritz method was used to reduce the degree of freedom of the problem. For a system with n degrees of freedom, in order to calculate the first P-order inherent properties, S had to be assume a certain mode larger than P.

$$[u]_0 = [\{\varphi\}_1, \{\varphi\}_2, ..., \{\varphi\}_{n1}] \tag{3}$$

In order to approximate the real vibration mode, the first iteration mode matrix $[\varphi]_0$ assumed above had to be calculated by the matrix iteration method.

$$[\varphi]_1 = [D][u]_0 = [\delta][M][u]_0 \tag{4}$$

Before the next iteration, $[\varphi]_1$ was performed orthogonalization so that its column vectors converge to the dominant modes of different orders after iterations. And with $[\varphi]_1$ served as a hypothetical mode, the generalized mass matrix and generalized stiffness matrix of reduced dimension subspace for the first iteration was reckoned by using Ritz method.

$$\begin{cases} [\tilde{M}]_1 = [\varphi]_1^T [M][\varphi]_1 \\ [\tilde{K}]_1 = [\varphi]_1^T [K][\varphi]_1 \end{cases} \tag{5}$$

Then the eigenvalue equation was obtained, shown in Equation (6).

$$([\tilde{K}]_1 - \tilde{\omega}_{ni}^2 [\tilde{M}]_1)[Y]_1 = \{0\} \tag{6}$$

The dimension- reduced subspace, i.e., the first iteration result of the S-orders natural frequencies and the corresponding modal vectors in the original system can be calculated by Equation (6), shown in Equation (7).

$$\begin{gathered} \tilde{\omega}_{n1}^2 \; \tilde{\omega}_{n2}^2 \; ... \; \tilde{\omega}_{nS}^2 \\ [Y]_1 = [\{Y^{(1)}\}_1 \; \{Y^{(2)}\}_1 \; ... \; \{Y^{(S)}\}_1] \end{gathered} \tag{7}$$

The above results were then brought into Equation (8).

$$\{u\} = [\varphi]\{Y\} \tag{8}$$

After the above linear transformation, the first iteration result of the first S-orders modal vectors at the original system could be acquired, shown in Equation (9).

$$[u]_1 = [\varphi]_1[Y]_1 = [\{u^{(1)}\}_1 \; \{u^{(2)}\}_1 \; ... \; \{u^{(S)}\}_1] \tag{9}$$

Repeated the above iterative calculation until the calculation result and the previous result satisfied with the accuracy requirements.

3. Simulation experiment

As shown in Figure 1, this model can be regarded as a cantilever structure, consisting mainly of two cover plates, two bolts and fixed ends. Plate 1 is above, Plate 2 is below, and the middle is connected by Bolt1 and Bolt 2. The left end face of Plate 1 is approximately fixed at the fixed end.

The sizes of cover plate and bolt are shown in Table 1. And the relevant material parameters are displayed in Table 2.

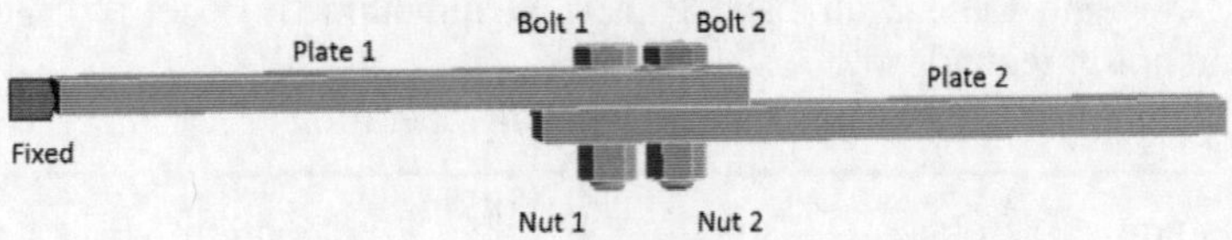

Figure 1. The schematic of the overall bolted connection structure

Table 1. Dimensions of all components

Component	Dimension(mm)
Plate 1	100*20*5
Plate 2	100*20*5
Bolt 1	M6*16
Bolt 2	M6*16
Nut 1	M6*5
Nut 2	M6*5

Table 2. Material properties of all components

Modulus of elasticity(Pa)	Poisson ratio	Density (Kg/m^3)	Volume modulus(Pa)	Shear modulus(Pa)
2e11	0.3	7850	1.67e11	7.7e10

First, according to the dimensions shown in Table 1 above, a three-dimensional parameterized model of the two bolted connection structure is set up and the assembly is formed. Then it is stored into a general three-dimensional model format and imported into the Workbench. Secondly, based the parameters depicted in Table 2, the corresponding parameters of this model is configured: modulus of elasticity denotes 2e11 Pa, Poisson ratio designates 0.3, the volume modulus refer to 1.67e11 Pa, shear modulus indicates 7.7e10 Pa, density connotes 7850 kg/m^3, ultimately the entire finite model is meshed into grids, and the total number of meshes is 3242.

In this paper, there are six components, namely Plate 1, Plate 2, Bolt 1, Bolt 2, Nut 1 and Nut 2. All these components have various contact relationships. The cover Plate 1 and the cover Plate 2 are constantly bonded. Then the complex threaded connection between Bolt 1 and Nut1 is simplified to be bonded. And Bolt 2 and Nut 2 do the same, also set to be bonded. The contact between the lower surface of the Bolt head 1 and the upper surface of the Plate1 is set to bonded, and the upper surface of the Nut 1 and the lower surface of the Plate 2 are set to be bonded. Similarly, the contact between the lower surface of the Bolt head 2 and the upper surface of the Plate 1 is set to bonded, and the upper surface of the Nut 2 and the lower surface of the Plate 2 are set to be bonded.

In addition, Bolt rod 1 and Plate 1 and Plate 2, as well as Bolt rod 2 and Plate1 and Plate 2 can be contacted, or cannot contact. In this article, the contact status between Bolt 1 and Plate1 or Plate 2 can be simply divided into two state: bonded or not bonded, that is, there is clearance between Bolt rod 1 and Plate 1 or Plate 2. Similarly, there is a similar contact relationship between Bolt rod 2 and Plate1 or Plate 2.

The contact between the Bolt 1 and the inner holes of Plate 1 and Plate 2 is independent. Likewise, the contact of the Bolt 2 with the inner holes of the Plate1 and the Plate2 is also unrelated. Therefore, the experimental scheme consists of 16 operating conditions. As shown in Table 3, "0" represents a gap between the bolted rod and the cover plate, whereas "1" embodies no gap between the bolt and the cover plate, i.e., bonded. The 16 working conditions can be represented by four binary digits. For a certain working condition, read data from left to right, then top to bottom, For instance, the working condition 1000 signifies only Bolt1 rod and Plate1 hole is bounded, Bolt 1 rod and Plate 2, Bolt 2 rod and Plate 1, Bolt 2 rod and Plate 2 all have gaps, i.e., unbounded. Other working conditions are analogous in turn and not in redundancy.

Table 3. List of all working operations

Working condition	Position	Bolt 1	Bolt 2	Working condition	Position	Bolt 1	Bolt 2
Case 1	Plate 1	0	0	Case 9	Plate 1	0	1
	Plate 2	0	0		Plate 2	0	0
Case 2	Plate 1	0	0	Case 10	Plate 1	1	0
	Plate 2	0	1		Plate 2	1	0
Case 3	Plate 1	0	0	Case 11	Plate 1	1	1
	Plate 2	1	1		Plate 2	0	0
Case 4	Plate 1	1	0	Case 12	Plate 1	1	1
	Plate 2	1	1		Plate 2	1	0
Case 5	Plate 1	1	0	Case 13	Plate 1	0	0
	Plate 2	0	0		Plate 2	1	0
Case 6	Plate 1	0	1	Case 14	Plate 1	0	1
	Plate 2	1	0		Plate 2	0	1
Case 7	Plate 1	1	0	Case 15	Plate 1	0	1
	Plate 2	0	1		Plate 2	1	1
Case 8	Plate 1	1	1	Case 16	Plate 1	1	1
	Plate 2	0	1		Plate 2	1	1

Annotation: "0" represents a gap between the bolted rod and the cover plate, whereas "1" embodies no gap between them

4. Result analysis

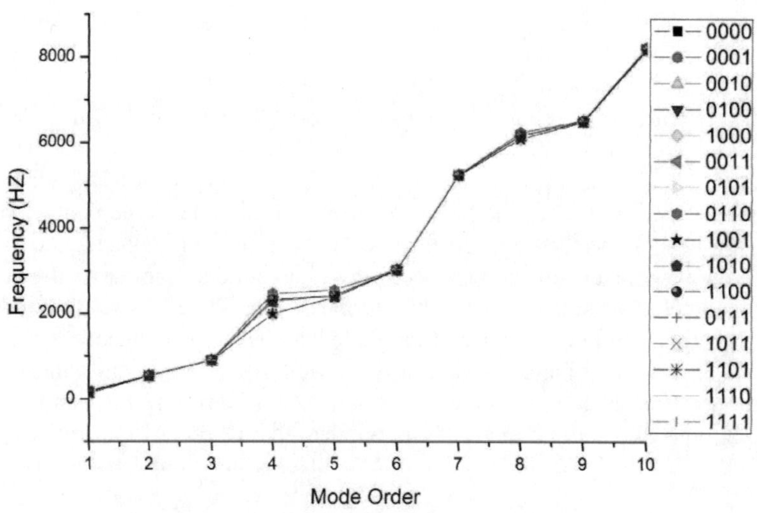

Figure 2. Change trend of natural frequency in 16 working conditions

The first 10-order natural frequencies of 16 operating conditions are shown in Figure 2. As can be seen from Figure 2, although the working conditions are different from each other, their respective first 10-order natural frequencies show certain regularity, both upsurge with the increase of modal order, and their first 10-order natural frequencies range from 150 Hz to 8000 Hz. In the 4th order, 5th order and

8th order modes, the natural frequencies have significant changes, whereas the corresponding natural frequencies basically coincide in other modes.

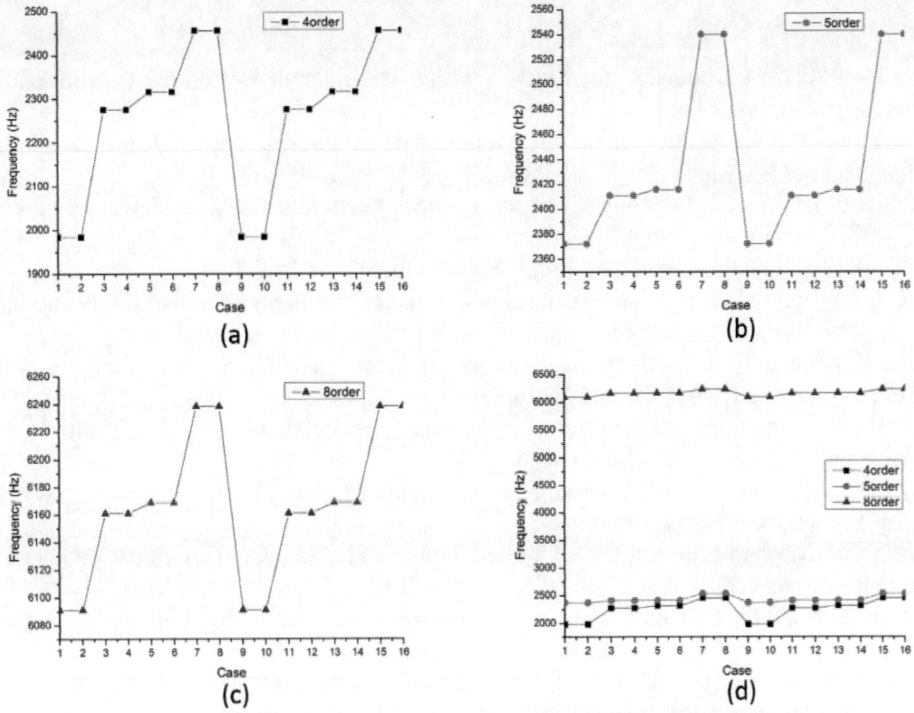

Figure 3. Detailed changes in the natural frequency mutation

The 4th, 5th and 8th order modes of the 16 working conditions are exhibited in Figure 3. From Figure 3 (d), it can be seen that the 5th and 8th order modal frequency values of the 16 conditions are relatively stable, which are about 2300-2500 Hz and 6000-6200 Hz, respectively. The 4th mode frequencies of the 16 working conditions vary fairly sharply, ranging from 2000 to 2500 Hz. It reveals that though the frequency is low, the vibration status of the system will be easily changed.

From Figure 3 (a) - (c), it can be seen that modes of the 16 operating condition vary periodically whatever 4th, 5th and 8th modes. For example, in the 4th order mode, the modal frequency of 4 conditions is 1983.7 Hz, the modal frequency of 4 operating conditions is 2275.1 Hz, the modal frequency of the 4 working conditions is 2315.4 Hz, and the modal frequency of the other 4 operating conditions is 2455.7 Hz. This demonstrations that the vibration is homogeneous and periodic under dissimilar connect status.

5. Conclusion

In this paper, the double-bolted connection structure is taken as the research object. And according to the gap between bolt rod and bolt holes of the cover plate, 16 kinds of connection conditions are designed to execute the finite element experiment. The experimental results indicate that though the frequency is low, it is effortless to trigger the severe vibration of the overall structure. And in a certain mode, modal frequencies at different operating conditions are regular and periodic, which can lay a foundation for the future investigation of vibration characteristics of double-bolted connection structure as well as the location and diagnosis of faults.

Acknowledgement

The authors would like to express thanks to the National Natural Science Foundation of China (Grant No: 11702318) and Beijing Natural Science Foundation (Grant No: 3184053) for their financial support of this research.

Reference

[1] Bickford J H 1990 An introduction to the Design and Behavior of Bolted joints, 2nd Edition (New York: Marcel Dekker Inc.)
[2] Lehnhoff T F, Ko K I and MaKay M L 1994 Member stiffness and contact pressure distribution of bolted joints ASME J. Mech. Des. 116:550-557
[3] Gaul L and Nitsche R 2001 The role of friction in mechanical joints Appl. Mech. Review 54: 93-106
[4] Breards C F 1982 Damping in structural joints Shock vib. Digest 14:9-11
[5] Inrahim R A and Pettit C L 2005 Uncertainties and dynamic problems of bolted joints and other fasteners J. Sound Vib. 279:857-936
[6] Yao X Y and Wang J J 2017 Effects of Load and Structure parameters of aero-engine bolted joints on joint stiffness J.Propul. Tech. 2:424-433
[7] Wang L 2017 Modeling and verification of nonlinear dynamic characteristics of bolted joints (Liao Ning: Shenyang Aerospace University)
[8] Liu J H 2016 Research on the self-loosening mechanism of bolted joints under axial excitation (Chongqing: Southwest Jiaotong University)
[9] Klok A J 2012 Mechanical behavior of bolted joints under impact rates of loading (USA: Michigan State University)
[10] Tan D D 2012 The effect of bolt spacing on the tightness behavior of bolted flange joints (Canada: Supreme Technology University)
[11] He K 2011 Vibration-based structural damage detection with applications to structures with bolted joints (USA: University of Maryland)
[12] Ma H 2013 Nonlinear vibration characteristics of a rotor system with pedestal looseness fault under different loading conditions J. Vibroeng 15(1): 406-418
[13] Zhang M Y 2017 Application of sub-harmonic resonance for the detection of bolted joint looseness Nolinear Dyn. 88(3): 1-11
[14] Barad K H, Sharma D S and Vyas V 2013 Crack detection in cantilever beam by natural frequency based method Procedia Eng. 51:770-775
[15] Tse K T and Song J 2015 Modal analysis of a linked cantilever flexible building system J. Struc. Eng. 141
[16] Huang Z and Zhao W 2015 Theoretical calulation and experimental analyses of natural frequency for high frequency dynamometer China Mech. Eng. 26(1):7-11
[17] Wang H Y, Zhao J Y and Shangguan W B 2013 Calculation and analysis of natural frequency sensitivity for accessory drive system J.South China Uni. Tech. 41(5):125-130
[18] Zeng P 2004 Finite element analysis and application (Beijing: Tsinghua University Press)
[19] Zhang H C 2013 Finite Element Analysis of Ansys13.0 (Beijing: China Machine Press)
[20] Liu Y Z, Chen L Q and et al 2011 Vibration Dynamics (Beijing: Higher Education Press)

Flexible Assembly Tooling Structure and Positioning Accuracy Analysis of Aircraft Fuselage Framework

Q Bai[1, a], W Tian[1], Z H Shi[2], Y J Sun[1], J X Shen[1], W H Liao[1] and H B Shi[1]

[1]College of Mechanical and Electronical Engineering, Nanjing University of Aeronautics and Astronautics, Nanjing 210016, P.R. China
[2]AVIC Chengdu Aircraft Industrial (Group) Co., Ltd., Chengdu 610092, P.R. China

[a] baiquseit@163.com

Abstract. Aiming at the problems of low efficiency, poor positioning accuracy and unreconstructible of the aircraft assembly process at present, according to the structure of the fuselage section of the aircraft, a flexible tooling for the fuselage frame is developed based on the modular idea. Positioning of different models of the fuselage is achieved by servo motor, manual adjustment and positioner. According to the way of the fuselage section, the structure of the flexible tooling is determined. At the same time, the layout of the automatic drilling and riveting system of the flexible tooling and the composite processing machine tool and the assembly process of the flexible tooling are introduced. Finally, the positioning accuracy of the flexible tooling test platform was analyzed, and the experimental verification was carried out through the experimental platform.

1. Introduction

In view of the characteristics of large number of aircraft parts, low stiffness, poor openness and high machining accuracy, a large number of tooling is needed to ensure its accurate geometric shape in the process of machining. The special fixture used in the traditional aircraft manufacturing process can not achieve the goal of rapid conversion and shortening the development cycle of new products [1]. Traditional rigid tooling usually adopts rigid structure and has poor reconfigurability: an aircraft product requires a large number of special tooling to be designed and manufactured. It has great disadvantages in the cost, cycle and storage space occupancy of aircraft manufacturing. With the increase of many new aircraft manufacturing tasks, tooling management and storage costs will become higher [2-4] . Flexible assembly tooling has high degree of digitalization and reconfigurability, which can realize the positioning of different sizes of different products; flexible tooling uses numerical control technology, so that the positioning accuracy is high, it is conducive to the realization of automatic assembly; product digital analog data source is the basis for tooling reconstruction to ensure the high positioning accuracy of flexible tooling [5-6].

Since the early 1990s, scholars have carried out a large number of research on flexible assembly technology of aircraft, the most representative of which are JAM (Jig less Aerospace Manufacture) and ADFAST (Automation for drilling, fastening, assembly, system integration, and tooling) projects. The automatic drilling technology has been studied and applied in the aircraft assembly, which improves the production capacity of aviation enterprises. The concept of box connection tooling [7] was put forward by Linköping University in Sweden. The box connection and passive parallel mechanism were combined to form a low-cost reconfigurable flexible tooling [8]. Flexible tooling with high positioning accuracy and good openness can meet the use of various automation equipment [9], mainly used in aircraft panel and wing beam assembly [10-12]. M-Torres of Spain has developed a series of array multi-lattice vacuum sucker-type flexible tooling for positioning aircraft panel components, which is mainly used for assembling elastic curved surface components. The tooling consists of arrays of POGO column modules with vacuum suckers, by adjusting POGO column arrays and vacuum sucker-based adsorption clamps. Tight efforts to achieve product retention [13].

Based on the analysis of the fuselage structure characteristics, assembly coordination relationship and manufacturing process, a part-level digital flexible assembly worker is developed to meet the requirements of small fuselage size, complex coordination relationship, dense assembly parts and high requirement of openness in different aircrafts. Finally, the positioning accuracy of flexible tooling is analyzed.

2. Flexible tooling structure for aircraft parts product skeleton

The parts that need to be located are generally the intersection holes on the fuselage frame and the fuselage beam, and the worker's operation space is fully considered when assembling. The mid-fuselage products are interchangeable. The front and rear fuselage are connected with the front fuselage and the rear fuselage respectively. The two sides are connected with the wing of the aircraft. The landing gear is located in the mid-fuselage abdomen and requires it to be retracted smoothly. The main structure after the fuselage is simplified is shown in Figure 1.

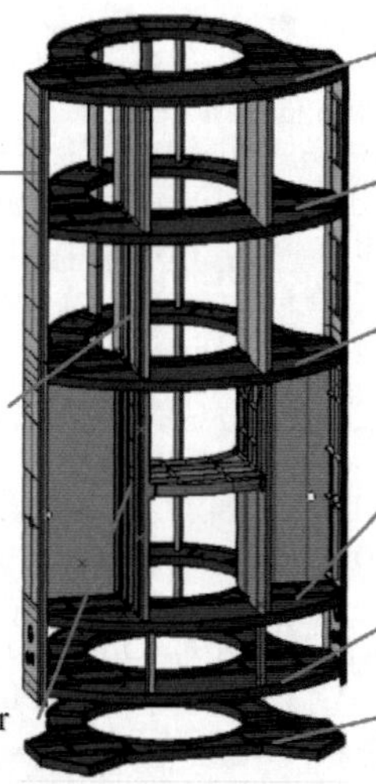

Figure 1 The main structure of the aircraft fuselage after simplification

The flexible tooling developed in this paper adopts gantry structure. Based on the intersection holes in each area of the fuselage reinforcement frame, a gantry structure is formed on both sides. The gantry structure is connected by a snap, the positioning column structure of 900 beams is strengthened, the active rotating shaft is installed at the bottom of the tooling, and the positioning of 10460 frame and 10060 frame is close to the main rotating shaft. The positioner is a rigid positioner, and it is installed on the same supporting column as the main crankshaft positioner. The main guard plate positioner is installed on the 900 beam, and the positioner on the 8700 frame is a rigid positioner. The structure of tooling is shown in Figure 2. The scheme has the advantages of good openness, large worker operating space, strong rigidity, small tremor and high precision in the follow-up drilling process of the fuselage products, but its disadvantages are relatively low flexibility.

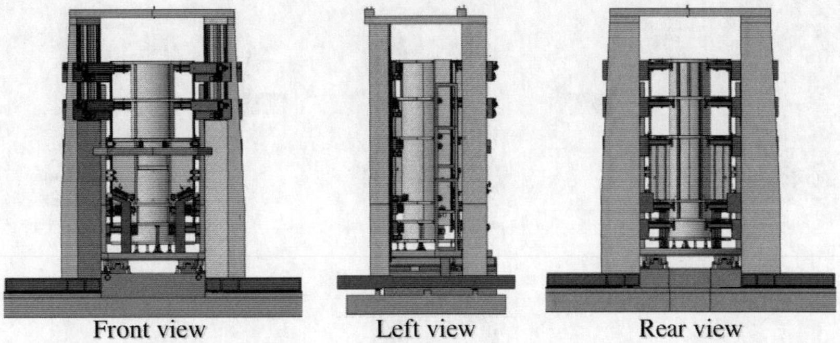

Figure 2 Flexible tooling structure

3. Automatic layout of automatic drilling and riveting system

The automatic drilling and riveting system integrates the equipment of compound machining machine, multi-sensor fusion end-effector, flexible tooling, high-precision rotary positioning table, AGV parts transportation, tool testing station, ground auxiliary equipment, etc. The overall structure is shown in Figure 3.

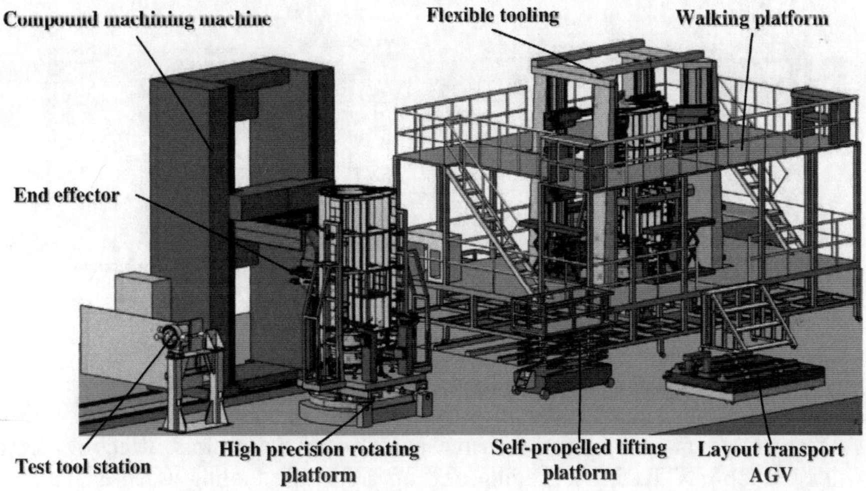

Figure 3 Overall layout of the automatic drilling and riveting system

Firstly, according to the information of the intersection holes on the product digital module and the tool digital module, the displacement of each moving unit along the guide rail and the distance between the left and right directions are planned, and then two gantry columns are moved to move a specified distance along the closure direction of the ground rail. After reaching the position, the guideway brake and the gantry latch are locked; secondly, the moving unit is controlled to move the specified distance along the planned path in front, back, up and down directions, and then locked after reaching the position; thirdly, the fuselage frame and the beam are installed; finally, the positioning of the rest of the components is completed. The assembly process of flexible tooling is shown in Figure 4.

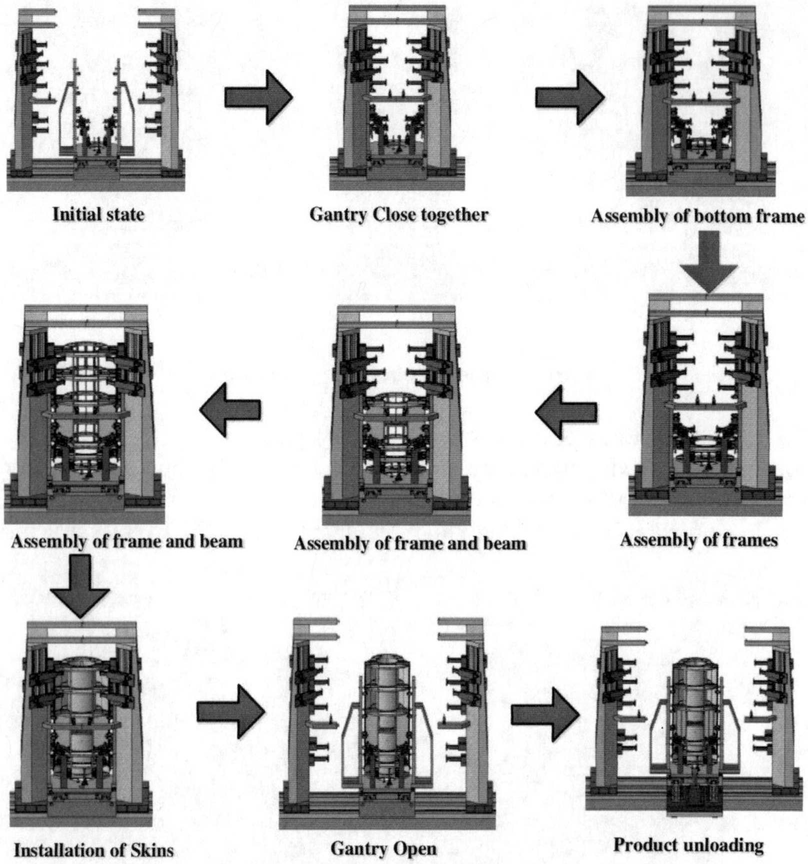

Figure 4 Assembly flow chart

4. Positioning characteristics of flexible tooling test platform

In order to reach the accuracy index and realize the accurate assembly and positioning of tooling and frame, it is necessary to make a comprehensive analysis of the factors affecting the positioning accuracy of flexible tooling. Taking a moving unit on a flexible tooling as an example, in order to improve the positioning accuracy of the center point of the locating pin for the quick-change connector of the moving unit, the laser tracker is used to set up the tool coordinate system, measure the actual position of the point, and establish the positioning error model of the moving unit in two moving directions. Combined with manual adjustment error, the comprehensive positioning error is calculated, and its value is compensated to the initial feed value of the positioning unit.

4.1. Positioning characteristics of flexible tooling test platform

The modular design idea is adopted in the tooling test platform, and the flexible assembly of product samples is realized by using eight moving units. The moving unit is driven in two perpendicular directions (X direction and Y direction) by servo motors. On the X direction, the moving unit is driven by a gear and rack to slide on the guide rail. On the Y direction, the actuator cylinder is driven down by a screw. And on the Z direction, the U-shaped groove hole is used to fine-tune the quick-change connector, as shown in Figure 5. The positioning process is generally as follows: firstly, the fixture coordinate system is established, and then based on the positioning data generated in DELMLA simulation, the X-axis and Y-axis are driven to move to any positioning point in the XOY plane within the range of effective travel from the location of origin. At the same time, with the help of laser tracker, real-time measurement of the actual position of the point P of the quick-change connector is used to compensate the actual spatial position error by comparison. At last, the Z direction fine-tuning is realized by adjusting the positioning joint manually to achieve the positioning accuracy of 0.05 mm in space, and the intersection hole of the fuselage frame can be positioned.

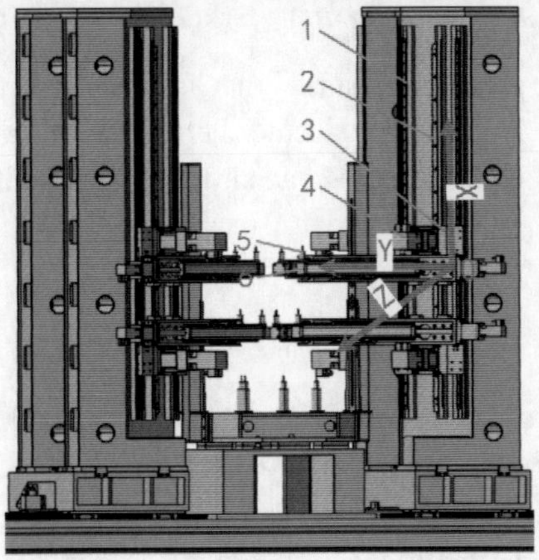

Figure 5 Flexible tooling test platform mobile unit

4.2. Positioning accuracy analysis of moving unit in test platform

4.2.1. Positioning error model in X direction. A rectangular coordinate system $D_1(O_1, X_1, Y_1, Z_1)$ is established at the symmetrical center of the pedestal of the moving unit, and a rectangular coordinate system $D_2(O_2, X_2, Y_2, Z_2)$ is established at the symmetrical center of the sliding table, whose direction of the coordinate axis coincides with the direction of feed. At the symmetrical center of the slide table, a rectangular coordinate system $D_3(O_3, X_3, Y_3, Z_3)$ is established, and its origin coincides with the coordinate system D_2, its coordinate axis is parallel to the coordinate axis D_1, and It does not rotate with the slide table's feed motion, as shown in Figure 6.

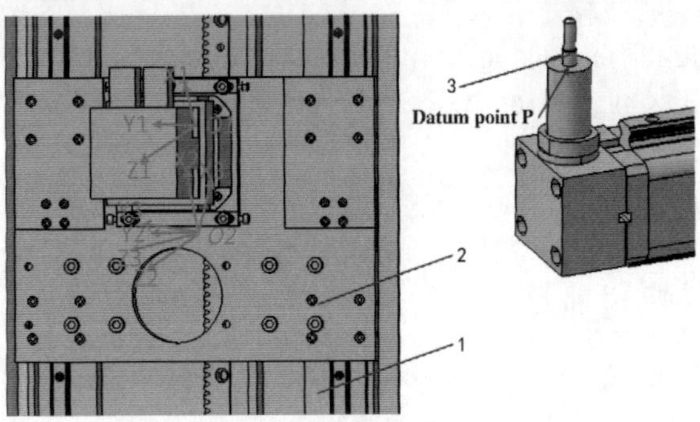

Figure 6 Positioning error model in X direction

When there is no installation orientation error at the lead rail, the coordinate system D_2 coincides with D_3. It is assumed that the coordinate of the datum point's position vector OP in D_2 and D_3 are (x, y, z), and the coordinates of P in D_1 are $(x + x_0, y, z)$. Due to the guiding error of the lead rail in actual motion, the position vector OP at the P point translates Δy and Δz along the Y_1 and Z_1

directions as the coordinate system D_2 and rotates the angles α, β and γ around the axes X_1, Y_1 and Z_1, thus the spatial position of point P is offset and moves to P', resulting in the positioning error, As shown in Formula 1:

$$OP' = R(X_3,\alpha)R(Y_3,\beta)R(Z_3,\gamma)OP = E_1 OP \tag{1}$$

Where: $R(i,j)$ represents the rotation matrix that rotates the angle j around the axis i, $i = X_1, Y_1, Z_1, j = \alpha, \beta, \gamma$. Since the angular displacement errors are all micro, As shows in Formula 2, rotation transformation matrix E_1 can be obtained according to differential rotation transformation:

$$E_1 = \begin{bmatrix} 1 & -\gamma & \beta \\ \gamma & 1 & -\alpha \\ -\beta & \alpha & 1 \end{bmatrix} \tag{2}$$

In the formula, the signs of α, β, γ are determined by the right hand rule. And the results are independent of the order of rotation around the axis X_3, Y_3, Z_3.

Ideally, when the sliding table moves to x_0, the actual coordinate of point P in the coordinate system D_2 transformed to the coordinate system D_1 is $(x'+x_0, y', z')$, as shown in Formula 3.The linear displacement error of point P transformed from angular displacement error is:

$$OP' - OP = (E_1 - I)OP = \begin{bmatrix} \Delta x_r \\ \Delta y_r \\ \Delta z_r \end{bmatrix} = \begin{bmatrix} z\beta - y\gamma \\ x\gamma - z\alpha \\ y\alpha - x\beta \end{bmatrix} \tag{3}$$

Considering the axial feed error Δx_t, the feed value x_0 is input to the motion' direction X of the column positioning unit control system. And considering all the guiding errors and the linear displacement errors of the feed system, the linear displacement error ΔX_P of point $P(x,y,z)$ fixed on the positioning unit is shows as follows in Formula 4:

$$\Delta X_P = \begin{bmatrix} \Delta x \\ \Delta y \\ \Delta z \end{bmatrix} = \begin{bmatrix} \Delta x_r \\ \Delta y_r \\ \Delta z_r \end{bmatrix} + \begin{bmatrix} \Delta x_t \\ 0 \\ 0 \end{bmatrix} \tag{4}$$

Similarly, by inputting the feed value y_0 into the control system of the moving unit Y, as shown in Figure 7, the linear displacement error ΔY_P of the point $P(x,y,z)$ fixed on the positioning unit is as follows in Formula 5:

$$\Delta Y_P = \begin{bmatrix} \Delta x \\ \Delta y \\ \Delta z \end{bmatrix} = \begin{bmatrix} \Delta x_r \\ \Delta y_r \\ \Delta z_r \end{bmatrix} + \begin{bmatrix} 0 \\ \Delta y_t \\ 0 \end{bmatrix} \tag{5}$$

As shown in Figure 8, the linear displacement error ΔZ_P of the point $P(x,y,z)$ caused by manual operationin the direction Z is as follows in Formula 6:

$$\Delta Z_P = \begin{bmatrix} \Delta x \\ \Delta y \\ \Delta z \end{bmatrix} = \begin{bmatrix} \Delta x_r \\ \Delta y_r \\ \Delta z_r \end{bmatrix} \tag{6}$$

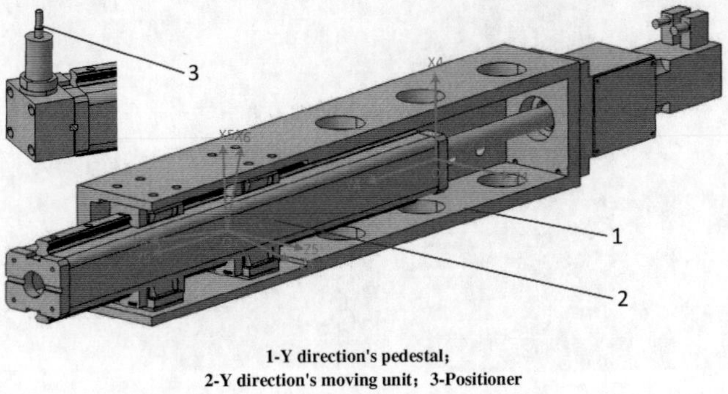

Figure 7 Positioning error model in Y direction

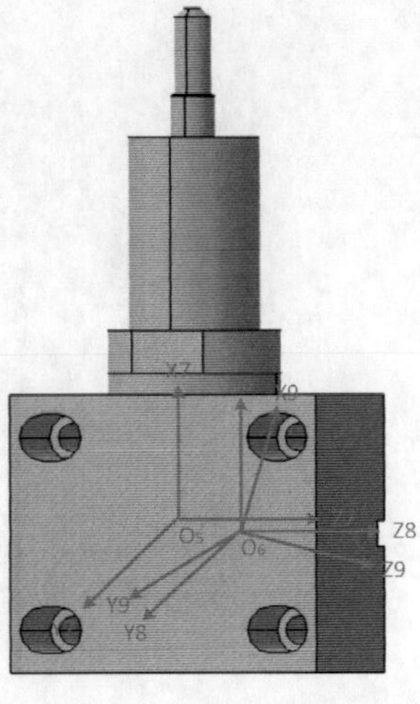

Figure 8 Positioning error model in Z direction

4.2.2. Spatial positioning error model. The total positioning error of datum point P on the quick-change connector is the sum of the linear displacement errors caused by the direction of X, Y, Z to the datum point P. As shows in Formula 7, The error of the positioning datum points in each direction of motion is:

$$[\Delta X, \Delta Y, \Delta Z]^T = \Delta X_P + \Delta Y_P + \Delta Z_P \tag{7}$$

As follows in Formula 8, the difference between the location of the datum hole on the quick-change connector and the position in the digital module is:

$$\Delta P = \sqrt{\Delta X^2 + \Delta Y^2 + \Delta Z^2} \tag{8}$$

4.3. Experimental verification

Four laser tracker target sockets are installed on the pedestal of the test platform, and four target balls, TB_1, TB_2, TB_3, TB_4, are placed on them. In the laser tracking and measuring system, the Cartesian coordinate system of the tooling is automatically generated according to the data of the four target balls. On the basis of the relationship between the tooling pedestal and the column 1, the coordinate system of the column 1 is established. The positioning accuracy of the moving unit before and after the measurement error compensation of the laser tracker is used to verify the correctness of the error model. The laser tracker measurement experiment is shown in Figure 9. Taking one of the datum point P of the moving unit as an example, the position accuracy of it in the column 1 coordinate system is measured to verify the accuracy of the error model. The specific experimental steps are as follows:

- Measure the coordinate value of the datum point P in CATIA digital model
- Measure the coordinate values of multiple sets of the datum point P at the origin of three coordinate axes by using the laser tracker, and then take their average values
- Compare the deviation between the coordinate value of the datum point P in digital model and the coordinate value at the origin of the actual axis, and then input the deviation value into the initial feed value of the control system
- Compensate the spatial error ΔP calculated by the error model to the initial feed value of the control system
- Drive the moving unit to the theoretical spatial position, measure the positioning accuracy after compensation by using the laser tracker, and verify the correctness of the error model

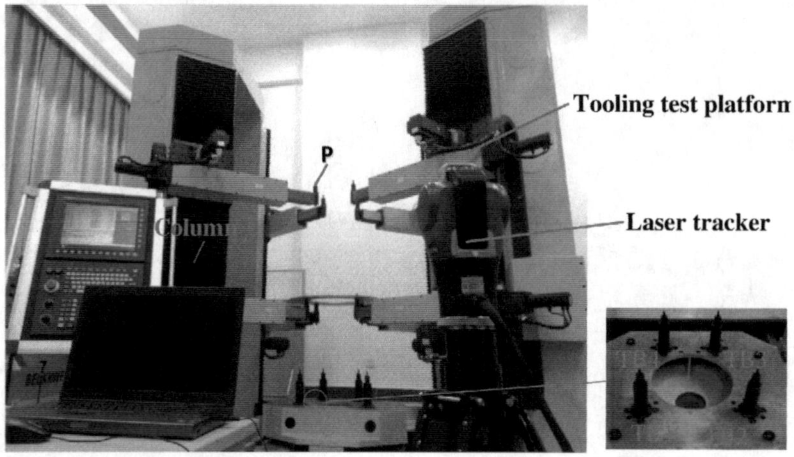

Figure 9 Laser Tracker Measurement Experiment

Through the measurement and calculation of the installation accuracy of the mobile unit, the orientation error of the rail mounting is obtained. The following is an error compensation process for the measurement of the positioning point of the mobile unit, as shown in Table 1.

Table 1 Error compensation process of positioning datum point P

Theoretical coordinates (1334.686,847.886,382.875)	Mean value at origin (1284.371,767.245,382.343)	Initial feed value (50.315,80.641,0.532)
The coordinate of point P in D_3 in the motion of axis X (10.968,647.563,182.287)	The coordinate of point P in D_3 in the motion of axis Y (60.367,343.631,1.256)	The coordinate of point P in D_3 in the motion of axis Z (92.572,23.021,1.236)
The Linear displacement error of point P in the motion of axis X (0.061,0.230,0.131)	The Linear displacement error of point P in the motion of axis Y (0.045,0.085,0.023)	The Linear displacement error of point P in the motion of axis Z (0.012,0.013,0.022)
Spatial integrated error of point P (0.118,0.328,0.176)	Corrected feed value (50.433,94.092,0.708)	Location error value (0.021,−0.032,−0.025)

5. Conclusion

Aiming at the practical problems existing in aircraft fuselage assembly at present, this paper develops a gantry-type flexible tooling structure for machine tools, and designs a flexible tooling for fuselage skeleton assembly. The flexible tooling is suitable for aircraft products with a length of 3-5 meters. The assembly process of this kind of flexible tooling is introduced. Through Electronics Machine servo, manual adjustment and positioner adjustment to achieve the positioning of the new aircraft fuselage. The positioning accuracy of the flexible tooling system is analyzed, and the test bench verifies that the positioning accuracy of the whole flexible tooling system meets the design requirements.

IOP Conf. Series: Materials Science and Engineering **470** (2019) 012025 doi:10.1088/1757-899X/470/1/012025

Acknowledgment
The author would like to thank National Science and technology major project "High-end CNC machine tools and basic manufacturing equipment" (**Grant No.2014ZX04001071**) for sponsoring this project.

References

[1] Naing S, Burley G, Odi R, Williamson A and Corbett J 2000 *Society of Automotive Engineers* **01** 4
[2] E Pezzuti, G Piscopo, A Ubertini, PP Valentini 2001 *Investigation on pin-hole connection in flexible assembly* (Italy : University of Rome) p 1-9
[3] JinHong L, Cheng Z 2001 *LECTURE NOTES IN COMPUTER SCIENCE.* **5315** 680
[4] TAKAYUKI YONEDA, TETSUYA SODA 2005 *Technical Review* **42** 2
[5] Munk C 2002 *Society of Automotive Engineers* **01** 3
[6] Kihlman H, Magnus E 2002 *Society of Automotive Engineers* **01** 3
[7] Millar A, Kihlman H 2009 *Society of Automotive Engineers* **01** 4
[8] Naing S 2001 *Feature Based Design for Jigless Assembly* (UK: Cranfield University) p 45
[9] Kihlman H, Engstrom M 2010 *Society of Automotive Engineers* **01** 2
[10] Branko S 2000 *Society of Automotive Engineers* **01** 2
[11] Hartmann J, Meeker C, Weller M and Izzard N 2004 *Society of Automotive Engineers* **01** 3
[12] Hartmann J, Meeker C, Minshull A and Smith A 2000 *Society of Automotive Engineers* **01** 4
[13] Zieve P, Rudberg T, Vogeli P, Smith A and Moore L 2004 *Society of Automotive Engineers* **01** 3-4

IOP Conf. Series: Materials Science and Engineering **470** (2019) 012026 doi:10.1088/1757-899X/470/1/012026

Influence of Micro-texture Parameters on Bone Drilling Force and Torque: A Finite Element Model

Yahui Hu[1,2,a]**, Xiangjun Li**[1]**, Chuncheng Chen**[1]**, Lili Su**[1] **and QingchunZheng**[1]

[1] Tianjin Key Laboratory for Advanced Mechatronic System Design and Intelligent Control, School of Mechanical Engineering, Tianjin University of Technology, Tianjin 300384, China

[2] Corresponding author: Yahui Hu, Tianjin University of Technology, 391 Binshui West Road, Xiqing District, Tianjin 300384, China.

[a] E-mail: huyahuitj@163.com

Abstract. The micro-texture on the tool surface is an effective way to improve the quality of the drilling. While high-performance micro-texture plays an important role in improving the quality of working surface and prolonging the life of tool. In this study, a three-dimensional finite element model was established to study the relationship between drilling thrust, torque and microstructure parameters (micro-pit depth, micro-pit diameter and micro-pit spacing) during drilling of cortical bone. The model was validated by cortical bone drilling using a micro-structured tool. On basis of this, the thrust and torque prediction models of the micro-texture parameters were obtained. The results indicate that the thrust and torque increased with the increase of micro-pit diameter; while the increase of the micro-pit spacing caused the thrust and torque to decrease. When the depth of the micro-pit increased, the thrust and torque decreased first and then increased, and we found that the effect of micro-pit spacing and micro-pit depth on force and torque is insignificant.

1. Introduction

Bone drilling as an essential step in bone surgery, widely used in various types of orthopedic surgery. Excessive drilling thrust force and torque can raise the drilling temperature, which leads to bone necrosis, as well as poor pore quality that is important to stable screw fixation. Therefore, optimization of thrust force and torque is an important part of clinical problems. Karaca [1] concluded that spindle speed, thrust force and bone density were significant parameters in affecting temperature and borehole quality by an integrated multidisciplinary vitro experiment. Vishal Gupta [2] interpreted how to measure force and torque in the course of surgical drilling into bone. In order to reduce the thrust force and torque during drilling process, a number of researchers have conducted that the effect of different methods on thrust force and torque [3].

In recent years, with the development of micro-texture in the field of tribology, the micro-texture on the tool surface can improve the quality of the surface, extend tool life and improve tool cutting performance. Many researchers have begun to study the application of micro-texture in the field of cutting [4,5]. Shen Dian analyzed the whole process of periodic changes of the surface friction coefficient of the micro-texture, and revealed the friction mechanism of the micro-texture [6]. Andersson's study showed that under the condition of oil lubrication, the micro-texture can significantly reduce the surface friction and wear [7].

Previous studies showed the surface micro-texture which can improve the friction between the interface and reduce the friction coefficient [8]. In this paper, a simulation model of micro-texture drill in drilling the cortical bone was established. And an experimental platform for drilling cortical bone with a micro-texture drill was built to verify the correctness of the model. Then analyzed the relationship between drilling thrust, torque and micro-texture parameters during the process of drilling cortical bone, and a prediction formula of thrust force and torque was given with the response surface method.

2. Experimental procedure

2.1 Sample preparation

Medical twist drills were selected as drilling tool for this study, the parameters of them are shown in Table 1. Surface texture with different parameters on the minor flank which was close to the side-cutting edge of the twist drill was made by laser machining (type: YLP-F20), Figure 1 shows the geometric model of the micro-texture twist drill.

Table 1. The geometric parameters of micro-texture twist drill

diameter (mm)	point angle	helix angle	chisel edge angle
4	118°	25°	125°

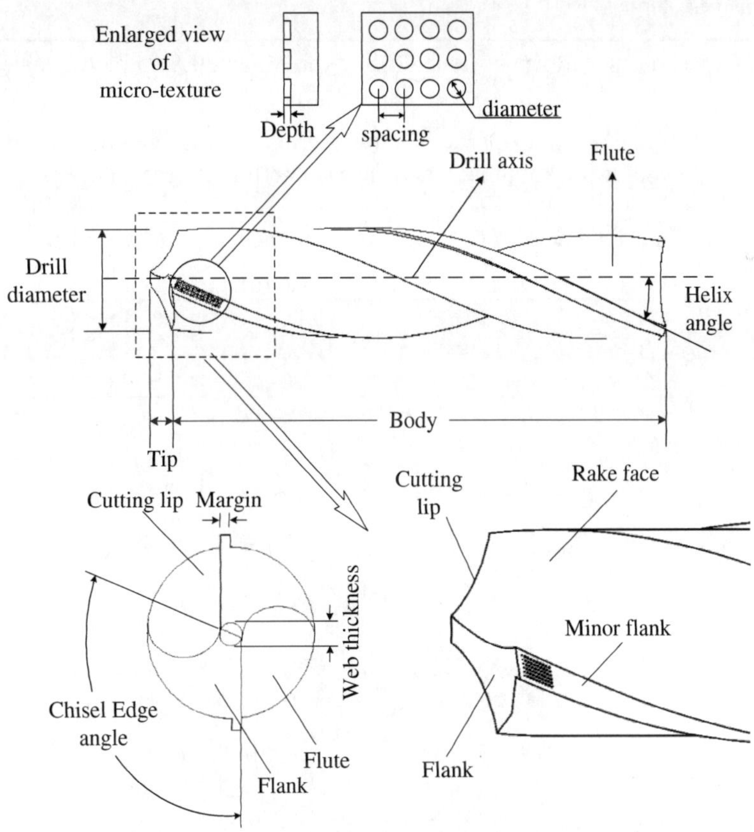

Figure 1. Geometric model of micro-texture twist drill.

The samples of cortical bone were the male pig bones which were obtained from the local butcher, because of its close resemblance to human bones, and the tests were performed within 2 h after the specimens of bone were obtained [9,10].

2.2 Experimental platform

An experimental platform of bone drilling was designed as shown in Figure 2. Hanchuan Machine Tool XK714D was used, the thrust force and torque were measured by a three-channel Kistler dynamometer.

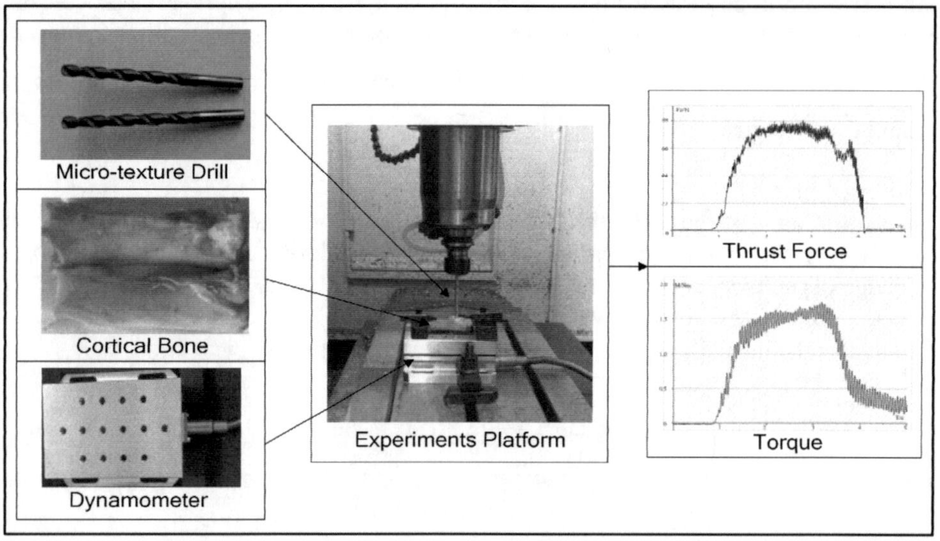

Figure 2. Experiments platform of bone drilling (drill diameter:4mm, drill speed:900rpm, feed rate:60mm/min).

The experiment adopted three kinds of medical twist drills with different micro-texture parameters, which were processed by laser marking machine, and the micro-texture parameters are shown in Table 2.

Table 2. Parameters of micro-texture on the minor flank of drill

Group	Depth(μm)	Diameter(μm)	Spacing(μm)
A	20	80	150
B	20	90	200
C	30	80	200

2.3 FE model of bone drilling

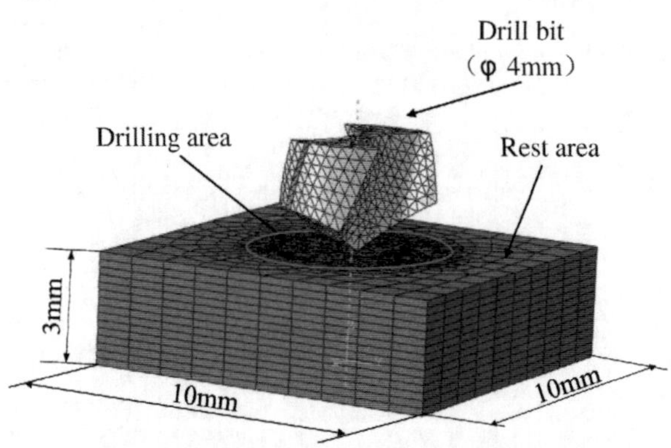

Figure 3. The mesh of the cortical bone and the twist drill.

A geometric model of micro-texture twist drill with complex surface was established by UG is shown in Fig.1, the geometric parameters are shown in Table 1. Since the changes of complex forms of bone on the simulation results are negligible, a simplified geometry of bone is used for the simulation and the results are satisfying [11]. In this model, a solid cuboid is used for the bone, with a thickness of 3 mm and length of 10 mm (Figure 3). The material properties [12] of cortical bone and twist drill used in the model are shown in Table 3.

Table 3. Material properties of the cortical bone

Properties	Values
Density (t/mm^3)	2.0×10^{-9}
Young's modulus (Mpa)	2.0×10^4
Poisson's ratio	0.36
Specific heat capacity (mJ·t^{-1}·K^{-1})	1.64×10^9
Thermal conductivity (W·m^{-1}·K^{-1})	0.56

In this paper, the samples of cortical bone were considered as an elastic-plastic material with nonlinear strain hardening, the Johnson-Cook material model that accounts for the strain rate sensitivity and temperature effect was used for the cortical bone [13]:

$$\sigma_Y = (A + B\varepsilon_p^n)(1 + C\ln \varepsilon^*)(1 - T^{*m}) \quad (1)$$

Where A, B, C and n are constants and $T^* = \frac{(T - 298)}{(T_{melt} - 298)}$; σ_Y is effective stress; ε_p and ε^* are the effective plastic strain and strain rate, respectively; and the T_{melt} is melting temperature. The model parameters [14,15,16] of the cortical bone are shown in Table 4.

Table 4. Johnson-Cook parameters of the cortical bone

A (MPa)	B (MPa)	C	n	m	T_{melt}
50	101	0.03	0.08	1.03	1300

As for the meshing, two kinds of mesh density were used for bone to improve the accuracy of calculation and reduce the processing time. The significant part of the cortical bone involved in the calculation was the area that contacts with the twist drill. Therefore, its mesh was refined. While for the rest of bone model, a low mesh density was enough. The mesh of the cortical bone and the twist drill was shown in Figure 3.

2.4 Simulation verification

Drilling thrust force and torque with different micro-texture parameters are obtained by the experiments of bone drilling, which are compared with the results of simulation to modify the finite element model. The results are shown in Figure 4.

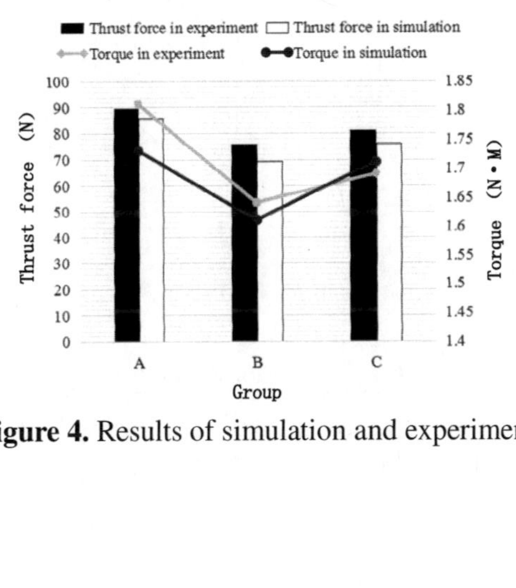

Figure 4. Results of simulation and experiment.

It can be concluded from Fig. 4, the experimental values are similar to the simulated values, so the simulation model is reliable, which provides the basis for the parametric study of micro-texture.

2.5 Experimental design based on response surface methodology

Response surface method (RSM) is an important statistical method to solve the multi-variable problem by using regression equation analysis to find the optimal target parameters [17]. Center composite design (CCD) is the most commonly used RSM. It can provide many information about experimental variables and experimental errors with the least number of experimental cycles. The five grades of all variables in CCD are represented by code (-α, -1,0, + 1, +α). where (0), (-α) and(+α), represent the average, the minimum and the maximum, respectively. The independent variables and their related levels and codes are shown in Table 5. In the study, three factors are associated with micro-texture parameters.

Table 5. Coding and level of design factors

Factor	Levers				
	-1.68	-1	0	+1	1.68
Depth (μm)	13.18	20	30	40	46.82
Diameter(μm)	73.18	80	90	100	106.82
Spacing(μm)	99.55	120	150	180	200.45

Table 6. DOE results for thrust force and torque for bone drilling process

Simulation no.	Design factors			Simulation results	
	Depth (μm)	Diameter (μm)	Spacing (μm)	Thrust force (N)	Torque (N·m)
1	46.82	90	150	75.4	1.53
2	20	80	180	85.8	1.68
3	30	100	150	69.3	1.48
4	30	90	150	69.3	1.48
5	30	90	150	69.3	1.48
6	30	73.18	150	79	1.58
7	40	100	120	76	1.57
8	40	80	120	92.9	1.71
9	30	106.82	150	58	1.39
10	20	100	120	75.4	1.54
11	20	100	180	62.1	1.41
12	30	90	150	69.3	1.48
13	40	80	180	83.2	1.62
14	20	80	120	94.7	1.73
15	40	100	180	63	1.43
16	30	90	150	69.3	1.48
17	30	90	150	69.3	1.48
18	30	90	200.45	69.7	1.49
19	30	90	99.55	81.2	1.61
20	13.18	90	150	73.2	1.52

A total of 20 runs and corresponding response data designed by CCD can be expressed as Table 6.

The empirical model is fitted with the response data. In this paper, a quadratic model is suitable to determine a critical function point using Equation (2):

$$y = \beta_0 + \sum_{i=1}^{k} \beta_i x_i + \sum_{i=1,j=1}^{k} \sum_{i<j}^{k} \beta_{ij} x_i x_j + \sum_{i=1}^{k} \beta_{ii} x_i^2 + \varepsilon \quad (2)$$

Where y is the predicted response; k is the number of variables;x_i, $x_j$is the design variable; $\beta_0$is the constant term; β_i is the coefficients of the linear termsx_i; $\beta_{ii}$is the coefficients of the quadratic termsx_{ii}, and ε is the residual associated with the experiments.

3. Results and discussion

3.1 Development of mathematical models

To obtain the relationship among micro-texture parameters and thrust、torque, we obtain the perform response data obtained from 20 simulation runs and use regression to generate an approximate mathematical model. Equation (3), (4) represent the regression response surface models of the thrust force and torque that can be expressed as follows:

$$F_Z = 229.328 - 1.858H - 1.133D - 0.414L + 13.686 \times 10^{-3}HD - 1.25 \times 10^{-4}HL - 9.625 \times 10^{-4}LD + 0.023H^2 + 3.102 \times 10^{-3}D^2 + 1.465 \times 10^{-3}L^2 \quad (3)$$

$$M = 3.235 - 0.023H - 0.014D - 3.336 \times 10^{-3}L + 8.125 \times 10^{-5}HD - 1.25 \times 10^{-5}HL - 1.625 \times 10^{-5}LD + 2.38 \times 10^{-4}H^2 + 4.182 \times 10^{-5}D^2 + 1.588 \times 10^{-5}L^2 \quad (4)$$

where F_Z is the thrust force; *M* is the torque; and *D, L, H* are the micro-pit diameter, spacing, and depth, respectively.

3.2 Analysis of variance (ANOVA)

In this paper, ANOVA analysis was used to test the quadratic model. The results of the analysis of variance for the quadratic model of thrust are shown in Table 7. The result shows that the value of R^2 is 0.9637, it means that 96.37% of the total variations were explained by the model. The model F-value of 7.4 indicated the significance of the regression model. The P-values of the developed model, which were less than 0.05, indicated that the model terms were statistically significant and the model was credible. The model items with a P-values greater than 0.1 have no significant impact on our research. Similarly, the results of the variance analysis of the quadratic model of torque are shown in Table 8. It shows that the value of R^2 was 0.9721, the P-values less than 0.005.

Table 7. Variance analysis of the thrust force

Source	Sum of Squares	Degree of freedom	Mean square deviation	F-value	P-value
Model	1546.19	9	171.8	7.4	0.0022
H	5.65	1	5.65	0.24	0.6323
D	975.42	1	975.42	42.02	0.0027
L	299.03	1	299.03	12.88	0.0049
HD	4.35	1	4.35	0.19	0.6742
HL	0.031	1	0.031	1.35E-03	0.9715
DL	7.41	1	7.41	0.32	0.5845
H^2	75.18	1	75.18	3.24	0.1021
D^2	22.19	1	22.19	0.96	0.3512
L^2	193.35	1	193.35	8.33	0.0162
Residual	232.11	10	23.21		
Missing item	232.11	5	46.42		
R^2=0.9637					

Table 8. Variance analysis of the torque

Projects	Sum of squares	Degree of freedom	Mean square deviation	F	P
Model	0.15	9	0.017	7.39	0.0022
H	4.74E-004	1	4.74E-004	0.21	0.6593
D	0.090	1	0.090	39.24	<0.0001
L	0.027	1	0.027	11.93	0.0062
HD	2.113E-003	1	2.113E-003	0.92	0.3602
HL	3.125E-004	1	3.125E-004	0.14	0.7200
DL	2.122E-003	1	2.122E-003	0.92	0.3602
H^2	8.163E-003	1	8.163E-003	3.55	0.0887
D^2	4.033E-003	1	4.033E-003	1.76	0.2147
L^2	0.023	1	0.023	9.89	0.0104
Residual	0.023	10	2.298E-003		
Missing item	0.023	5	4.595E-003		
R^2=0.9721					

3.3 Model fitness check

Residual analysis was used to study the accuracy of the model. The normal probability plots for the residuals of thrust force and torque can be seen in Figure 5(a) and (b), respectively. We concluded that the points on the normal probability plots of the residuals should be a straight line when the model was adequate. Consequently, the quadratic model with regard to the micro-texture parameters could be used to determine optimal values.

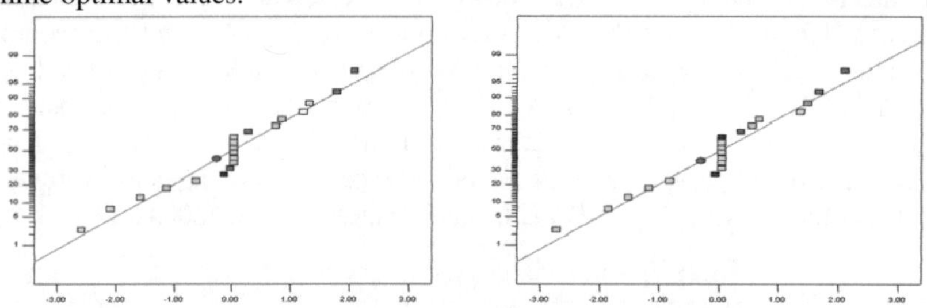

(a) (b)

Figure 5. Normal probability plots for the residuals of thrust force and torque.

3.4 Effect of the parameters on the responses

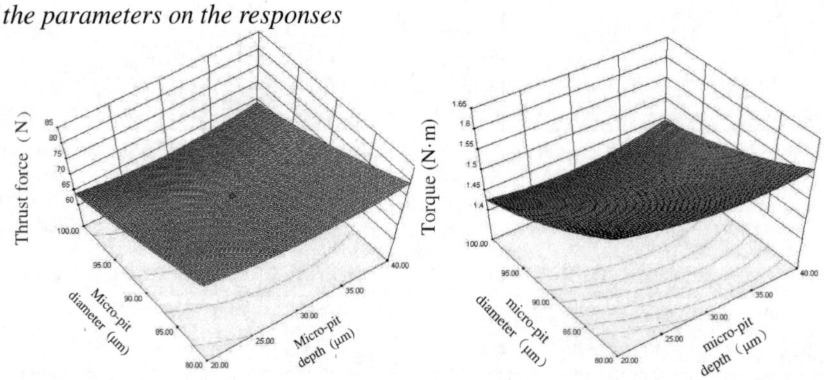

(a)

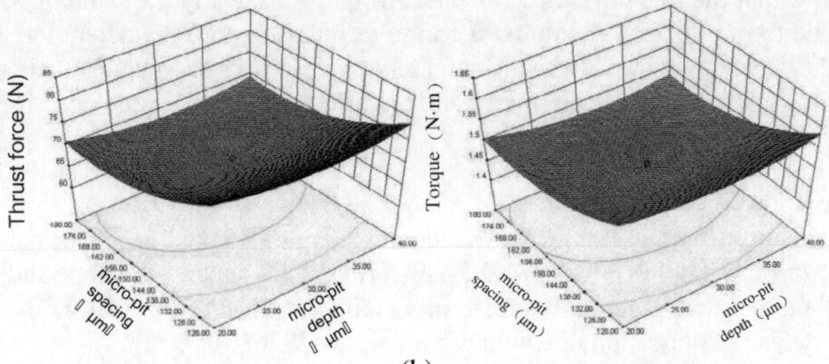

(b)

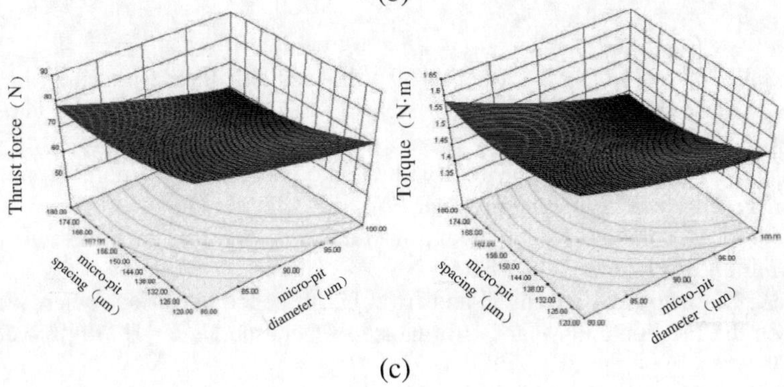

(c)

Figure 6. Interaction effects of micro-texture parameter on thrust force and torque.

Effects of the micro-texture parameters on the thrust force and torque were mathematically expressed through Eq. (3) and Eq. (4). To visualize these effects, the 3D surface profile of the interaction between the parameters by Design Expert is as follows in Fig.6.

Figure 6(a) shows the surface plots of thrust force and torque with variation in the values of micro-pit diameter and depth. It is observed that as the micro-pit diameter increases from 80 to 100 μm, the thrust force and torque reduce further. When the diameter of the micro-pit increases, the friction state between the drill bit and the chip improves, and the contact area between the drill bit and the chip becomes smaller, thus the friction is reduced, the thrust force and torque is reduced.

Similarly, the surface plots are obtained for the thrust force and torque of micro-pit spacing and depth as shown in Figure 6(b). It is observed that as the micro-pit depth increases from 20 to 30 μm, the thrust force and torque reduce; Thereafter, as the micro-pit depth increases the thrust force and torque also raise slightly. Because the micro-pit of tool surface can store micro bubbles in the drilling process, micro bubbles can improve the friction state of contact interface, but too large or too small micro-pit depth is not conducive to the stability of micro bubbles. Further the response plots are obtained for the micro-pit diameter and spacing as shown in Figure 6(c). It is observed that as the micro-pit spacing increases from 120 to 160 μm, the thrust force and torque reduce. Subsequently the changes are not obvious. Because of the smaller micro-pit spacing, too many micro-pits may reduce the surface strength of the tool, and the surface of the tool will change from elastic-plastic deformation to plastic deformation, which reduces the thrust force and torque; When the micro-pit spacing is sufficiently large, it has merely effect on the friction state, so the change of thrust force and torque is not obvious.

4. Conclusions

In this paper, the drilling process was simulated using a general purpose finite element software ABAQUS/explicit. And this model has been proven to effectively replicate the drilling process. The thrust and torque prediction models of the micro-texture parameters were obtained, which can accurately predict the drilling thrust force and torque.

The results show that the trend of thrust force and torque is basically the same, It was observed that the thrust force and torque increased with an increase in micro-texture diameter; The thrust force and torque decreased at the beginning with an increase in micro-texture spacing then the change was negligible; when the micro-texture depth increased, the thrust force and torque decreased first and then increased, but the effect is not significant.

Acknowledgments
This work was supported by the Tianjin Key Laboratory for Advanced Mechatronic System Design and Intelligent Control, School of Mechanical Engineering. This paper was supported partially by the Key Program of the Natural Science Foundation of Tianjin (Grant No. 15JCZDJC32800) and the National Natural Science Foundation of China (Grant No. 11432016).

References

[1] Karaca F, Aksakal B and Kom M. Influence of orthopaedic drilling parameters on temperature and histopathology of bovine tibia: an in vitro study[J]. *Medical Engineering & Physics*, 2011, **33**(10): 1221-7
[2] Vishal G, Pulak M. P, and Vadim V. S. Rotary ultrasonic bone drilling: Improved pullout strength and reduced damage[J]. *Medical Engineering and Physics*, 2017,**41**: 1-8
[3] Khurshid A, A V Mitrofanov, Martin B and Vadim V. S. Stresses in ultrasonically assisted bone cutting[J]. *Journal of Physics*, 2009, **181**(1): 1-8
[4] Zeyu L, Xin W, Xiaozhu X, Xiangang H and Jiwei H. Influence of Surface Micro Texture with Laser Processing on the Friction and Wear Performance of Ceramic Cutter[J]. *Surface Technology*, 2015, **44**(10): 33-39
[5] Chao Y, Xiaojun L, Haidong Y and Kun L. Effect of the textured surface on the cutting performance of the tool and the friction property for the rake face[J]. *Tribology*, 2015, **35**(2): 228-235
[6] Dian S, Wengzhong W and Lingjia K. Effect of surface texturing on lubrication performance of Ball-on-Disc point contacts[J]. *Tribology*, 2012, **32**(6): 570-576
[7] Andersson P, Koskinen J and Varjus S. Micro-lubrication effect by laser-textured steel surfaces[J]. *Wear*, 2007, **262**(3-4): 369-379
[8] Bichong C. The design of micro texture tool and experimental research of its cutting performance[D]. *Beijing Institute of Technology*, 2005
[9] Augustin G, Davila S, Mihoci K and Antabak A. Thermal osteonecrosis and bone drilling parameters revisited[J]. *Arch Orthop Trauma Surgery*, 2008, **128**(1): 71-77
[10] Sun-Jong K, Jaeyoung Y, Young-Soo K and Shin W.S. Temperature change in pig rib bone during implant site preparation by low-speed drilling[J]. *Applied Oral Science*, 2010, **18**(5): 522-527
[11] Waqas L, Kaddour B.M and Ian A. Drilling in cortical bone:a finite element model and experimental investigations[J]. *Journal of the mechanical behavior of biomedical materials*, 2015, **42**: 32-42
[12] Yuan-Kun T, Li-Wen C, Ji-Sih C and Chih-Kun H. Finite Element Simulations of Bone Temperature Rise During Bone Drilling Based on a Bone Analog[J]. *Journal of Medical & Biological Engineering*, 2013, **33**(3): 269-274
[13] A.V. Mitrofanov, Vladimir B and VadimV. S. Finite element analysis of ultrasonically assisted turning of Inconel 718[J]. *Journal of Materials Processing Technology*, 2004, 135-154: 233-239
[14] Alam K, Mitrofanov A V and Silberschmidt V V. Thermal analysis of orthogonal cutting of cortical bone using finite element simulations[J]. *International Journal of Experimental & Computational Biomechanics*, 2010, 1: 236-251
[15] Santiuste C, Rodríguez-Millán M and Giner E. The influence of anisotropy in numerical modeling of orthogonal cutting of cortical bone[J]. *Composite Structures*, 2014, **116**(9): 423-431
[16] Alam K, Mitrofanov A V, Silberschmidt V V. Finite element analysis of forces of plane cutting of cortical bone[J]. *Computational Materials Science*, 2009, **46**(3): 738-743
[17] Pittalà G M and Monno M. A new approach to the prediction of temperature of the workpiece of face milling operations of Ti-6Al-4V[J]. *Applied Thermal Engineering*, 2011, **31**(2–3):173-180

IOP Conf. Series: Materials Science and Engineering **470** (2019) 012027 doi:10.1088/1757-899X/470/1/012027

Estimation of Grinding Force with Consideration of Rupture Factor

Jianliang Guo*, Jun Chi and Lianqing Chen

School of Mechanical Engineering, Ningbo University of Technology, Ningbo, 315336, China

*guojianliang76@gmail.com, +86-13857879737

Abstract. Grinding force has been treated as an indication of the accuracy and surface quality of ground workpiece. Building a model which can be used to estimate or predict accurately the grinding force will allow us to monitor indirectly and on-line the accuracy and surface quality of grinding, which lays the foundation for implementation of intelligent grinding. Based on analysis of stress concentration in this work, a model of grinding force is proposed, in which the rupture factor of chip at the interface between the grain and workpiece is considered quantitatively and independently. The effect of stress concentration on the grinding force is included in the rupture factor. Least square method and experimental tests under orthogonal test design are used to determine the coefficients in the model. The model is validated through additional experimental tests and the results show that the average prediction accuracy reaches 81.42% for specific normal grinding force and 78.58% for specific tangential grinding force, and the trend in the measured results was well predicted by the proposed model.

1. Introduction

Because of the multiplicity of cutting points and their irregular geometry, the high cutting speeds, and the small depths of cut which vary from grain to grain, any attempt to analyze the mechanisms of grinding might appear to be a hopeless task [1]. Based on some simplification or assumption, a model can be constructed and then used as an entry to describe the complicated physical laws in grinding process. In fact, in international research the modeling and simulation of processes is established as an excellent tool for assessing and optimizing grinding processes [2]. Industrial practice shows that grinding force is closely related to the final accuracy and surface quality of the products. Generally, high normal component of grinding force causes erratic part tolerance and self-induced vibration of the grinding system while high tangential component leads to high grinding heat generation and consequently thermal damage of the workpiece. Building a realistic model and hence estimating the force before the actual grinding process would allow the grinding performance to be evaluated and optimized in advance, which would also be quite helpful to the implementation of intelligent grinding in the future.

A number of models have been developed in the literature to describe analytically or empirically the grinding force, in which the stochastic nature of grinding process was generally taken into account. Badger and Torrance developed a 2-D model and a 3-D model to predict grinding force respectively [3], both of which consider the grit-workpiece interface as a rigid plastic contact whose behavior depends on the asperity slope distribution and the interfacial coefficient of friction obtained from wheel profiles. Karkalos et al. employed regression analysis and artificial neural networks to model and predict grinding forces respectively [4]. Chang and Wang formulated the dynamic grinding force as the convolution of a single-grit force and the grit density function [5]. In the frequency domain, the power spectrum density of the grinding force was deduced to be the product of the energy spectrum density of the single-grit force and the power spectrum density of the grit density function. The single-grit force was modeled to be a deterministic impulse response of the grinding process. Mishra and Salonitis compared several existing models in the relevant literature [6], and then made some slight modifications to one of them to obtain the best description of grinding force in advanced grinding processes such as creep-feed grinding and high-efficiency deep grinding. Based on an undeformed chip thickness model, Agarwal and Rao built a stochastic model of grinding force which considers the real contact length in terms of the elastic properties of the wheel and workpiece material and other grinding parameters [7]. Karkalosa et al. conducted a molecular dynamics simulation of nanometric peripheral grinding to determine the grinding force [8]. They found that grinding forces increased about two times at larger depth of cut whereas a much smaller variation was reported for grinding speed variation. Zhang et al. presented a theoretical force model by considering the material-removal mechanism, the plastic-stacking mechanism, and also the influence of lubricating conditions [9], in which the aggregate force was derived through the synthesis of each single-grain force. Durgumahanti et al. presented a model of grinding force by incorporating the effects of variable coefficient of friction and the plowing force [10], based on the fact that the coefficient of friction varies with the process parameters and the plowing force would become more predominant at very low depth of cuts.

Due to the stochastic nature of grinding process, which gives rise to a lot of complicated influencing factors on grinding force, several coefficients are usually employed in the existing models to represent the effects of the undiscovered influencing factors. It would be helpful to make clear all these undiscovered factors in grinding force model step by step for predicting grinding force. If the quantitative contribution of the unknown factors can be identified, then the research of grinding mechanism will be pushed forward considerably. Also, understanding these factors will allow us to make ideal design for the grinder and to optimize the grinding process parameters, which avoids the material and energy waste resulted from the commonly used conservative design nowadays. As one of the unknown factors, stress concentration was seldom discussed in the literature due to its complicated development mechanism. Generally, stress concentration dictates partly how easily the material can be cut off by the grains during grinding operation. Consequently it would improve the comprehensiveness of the existing grinding force models to characterize the contribution of stress concentration to grinding force. Only Lin et al. was found to establish a model of grinding force based on stress concentration theory [11]. However, Lin et al. did not consider the stress concentration phenomenon within the plastic region [11], which actually occurs during the chip formation. Also, Lin et al. did not relate systematically the grain geometry to the shape and size of the notch at the interface between the grain and workpiece [11], which is the foundation of the effect of stress concentration on the grinding force. Considering this, through the analysis of the rupture factor in which the effect of stress concentration within the plastic region is included, we improved the grinding force model proposed by Durgumahanti et al. [10].

2. Grinding force model

Durgumahanti et al. proposed a grinding force model shown in Eq. (1) and (2) [10], in which both the specific normal force Fn' and specific tangential force Ft' are separated into the chip formation force, friction force, and plowing force.

$$F_n^{'} = K_1 \frac{v_w}{v_s} a_p + K_2 \frac{v_w}{v_s} \left(\frac{a_p}{d_e} \right)^{1/2} + K_3 \left(\frac{v_w}{v_s} \right)^{a_0} (d)^{b_0} (a_p)^{c_0} C_s (a_p d_e)^{1/2} \tag{1}$$

$$F_t^{'} = K_4 \frac{v_w}{v_s} a_p + \left(K_5 + K_6 \frac{v_w}{d_e v_s} \right) (a_p d_e)^{1/2} + K_7 \left(\frac{v_w}{v_s} \right)^{a_0} (d)^{b_0} (a_p)^{c_0} C_s (a_p d_e)^{1/2} \tag{2}$$

The chip formation process can essentially be described using the model of metal cutting as shown in Fig. 1. Within the shear slip plane OM, the material stress progressively increases until the chip forms and breaks from the workpiece. Fig. 2 presents the development of the correlation of stress and strain.

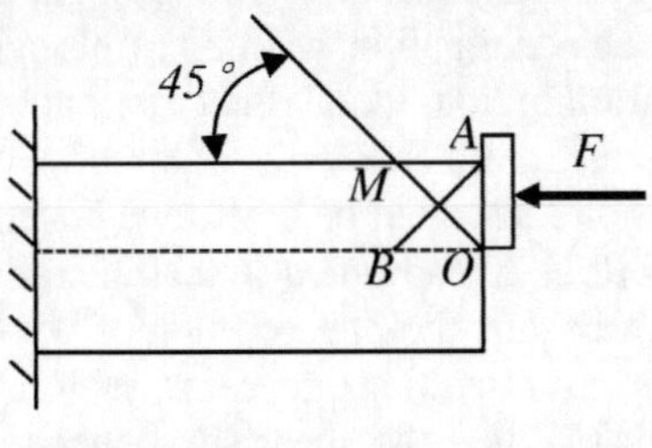

Figure 1. Model of metal cutting

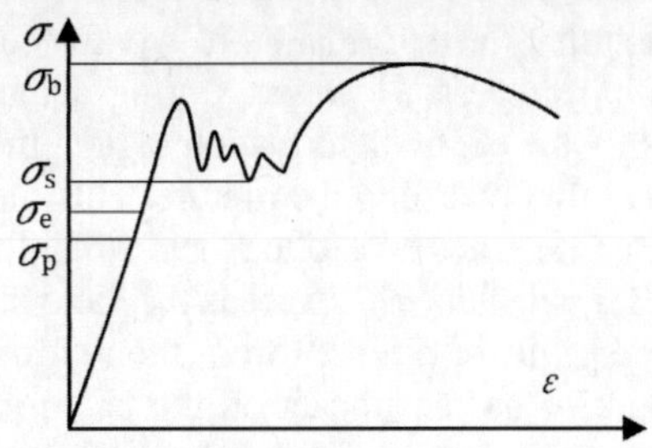

Figure 2. Deformation curve of metal cutting

During the chip formation process, phenomenon of stress concentration occurs in the contact zone of an active grain and the workpiece. Basically the interaction between an individual grain and the workpiece is considered to be quite similar to geometrically defined cutting process like turning or milling process except that the grain has a negative rake angle. As shown in the equivalent approximation of the interaction in Fig. 3, the abrasive grit acts as an indenter and squeezes gradually the material to form a shallow notch, which leads to stress concentration with the largest stress occurring at the bottom of the notch. Based on this feature of stress concentration, the small portion of material surrounding the notch can be equivalently characterized by the tensile model of a plate with a shallow notch shown in Fig. 4. In this model, the ρ denotes the curvature radius at the bottom of the notch. The t represents the depth of the notch. The σ_0 is the reference stress, which occurs far away from the notch.

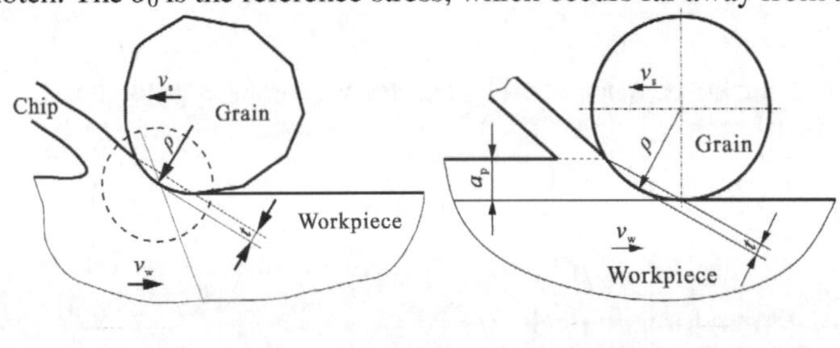

(a) Actual interaction (b) Equivalent approximation of the interaction

Figure 3. Approximation of the interaction of abrasive grain and workpiece.

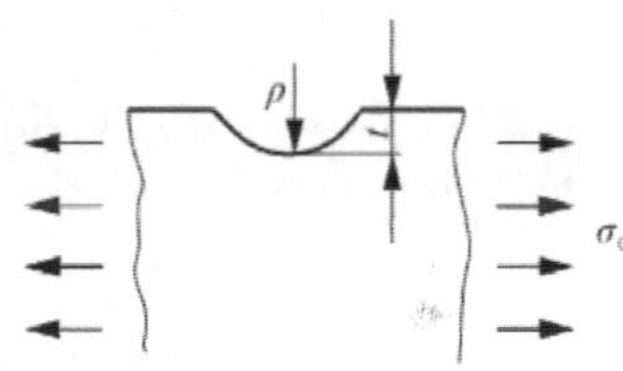

Figure 4. Tensile model of the Interaction

The radius ρ of curvature at the notch bottom can be approximated to be the equivalent radius of the grain. The depth t of notch can be determined through geometrical relationship in Fig. 1.

Li et al. presents the stress concentration factor α corresponding to the model in Fig. 4 [12].

$$\alpha = \frac{\sigma_{max}}{\sigma_0} = 1 + 2\sqrt{\frac{t}{\rho}} \tag{3}$$

in which the σ_{max} represents the maximum stress which occurs at the bottom of the notch. The t can be determined from Fig. 1.

$$t = \rho - \sqrt{\rho^2 - \frac{\rho a_p}{2}} \tag{4}$$

Due to the fact that the notch generates from the squeezing action of the abrasive grit, it is reasonable to consider ρ to be equal to the radius of the abrasive grit. In fact, each nominal grit number M corresponds to a range of grit size, but the grit dimension for a specific grinding wheel might be characterized by the average value d of the grit sizes. Malkin and Guo give the average value in Eq. (5) [1], in which the d is defined to be 60% of the average spacing between adjacent wires in a sieve whose mesh number equals the grit number M.

$$d = 15.2M^{-1} \tag{5}$$

Consequently, the curvature radius ρ can be assigned as

$$\rho = 7.6M^{-1} \tag{6}$$

It should be noted that the expression of stress concentration factor in Eq. (3) is based on the assumption that the deformation of workpiece material does not go beyond the elastic region throughout the grinding process. However, during actual grinding process the chip formation would go through successively the elastic deformation region, the elastic-plastic deformation region, the plastic deformation region, and finally lead to the rupture. This means that only when the maximum stress σ_{max} at the bottom of the notch does not exceed the yield strength σ_s, the stress concentration factor can be determined using Eq. (3). But when the σ_{max} exceeds the yield strength, the stress field will be redistributed so that the ratio of σ_{max} to σ_0 will be different from the ratio in elastic region. Apart from this, the phenomenon of work hardening also has an impact on the σ_{max} to some extent. Like other metal-cutting processes, material removal by grinding involves a shearing process of chip formation [1]. When the shear slip generates within the workpiece material, the work hardening increases the shear slip resistance and hence the failure strength will also be increased. This means that the description of the actual stress state that leads to chip formation entails considering comprehensively all the above-mentioned influencing factors. The rupture factor β acts as a parameter involving these influencing factors, and hence is taken into account to determine the specific chip formation force in this work. As such, the expressions of the specific normal chip formation force F_{nc}' and the specific tangential chip formation force F_{tc}' in Eq. (1) and (2) would be improved to become Eq. (7) and (8) respectively.

$$F_{nc}' = \frac{1}{\beta}\left(\frac{K_1' v_w}{v_s} a_p\right) \tag{7}$$

$$F_{tc}' = \frac{1}{\beta}\left(\frac{K_4' v_w}{v_s} a_p\right) \tag{8}$$

in which the K_1' and K_4' are the undetermined coefficients. Neuber formula relates β to the stress concentration factor α as shown in Eq. (9) [12].

$$\beta = 1 + \frac{\alpha - 1}{1 + \sqrt{\frac{\varepsilon_0}{\rho}}} \tag{9}$$

in which ε_0 is a constant which is equal to 0.48 mm for mild steel.

Another mechanism associated with grinding process is plowing. Plowing usually refers to the side flow of material from the cutting path into ridges shown in Fig. 5. From Fig. 5, it can be observed that material breakage and stress concentration also occur in the plowing process. Thus the normal and tangential plowing force components in Eq. (1) and (2) can be revised to be Eq. (10) and (11).

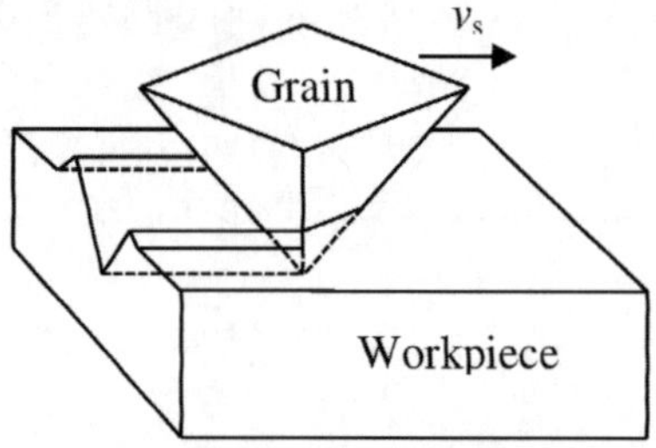

Figure 5. Illustration of plowing

$$F_{np}' = \frac{1}{\beta} K_3' \left(\frac{v_w}{v_s}\right)^{a_0} (d)^{b_0} (a_p)^{c_0} C_s (a_p d_e)^{1/2} \tag{10}$$

$$F_{tp}^{'} = \frac{1}{\beta} K_7^{'} \left(\frac{v_w}{v_s} \right)^{a_0} (d)^{b_0} (a_p)^{c_0} C_s (a_p d_e)^{1/2} \tag{11}$$

So far, the equations in [10] for the specific normal and tangential grinding force can be rewritten to be

$$F_n^{'} = \frac{\sqrt{\rho}+\sqrt{\varepsilon_0}}{\alpha\sqrt{\rho}+\sqrt{\varepsilon_0}} \left(\frac{K_1^{'} v_w}{v_s} a_p \right) + \frac{K_2^{'} v_w}{v_s} \left(\frac{a_p}{d_e} \right)^{1/2} + K_3^{'} \frac{\sqrt{\rho}+\sqrt{\varepsilon_0}}{\alpha\sqrt{\rho}+\sqrt{\varepsilon_0}} \left(\frac{v_w}{v_s} \right)^{a_0} (d)^{b_0} (a_p)^{c_0} C_s (a_p d_e)^{1/2} \tag{12}$$

$$F_t^{'} = \frac{\sqrt{\rho}+\sqrt{\varepsilon_0}}{\alpha\sqrt{\rho}+\sqrt{\varepsilon_0}} \left(\frac{K_4^{'} v_w}{v_s} a_p \right) + \left(K_5^{'} + \frac{K_6^{'} v_w}{d_e v_s} \right) (a_p d_e)^{1/2} + K_7^{'} \frac{\sqrt{\rho}+\sqrt{\varepsilon_0}}{\alpha\sqrt{\rho}+\sqrt{\varepsilon_0}} \left(\frac{v_w}{v_s} \right)^{a_0} (d)^{b_0} (a_p)^{c_0} C_s (a_p d_e)^{1/2} \tag{13}$$

in which the values of plowing force coefficients a_0, b_0, c_0, and C_s are taken identical to those in [10], which were determined through the single-grit test and the imprint test. This is because that these coefficients are dependent on the workpiece material and the grinding wheel composition and the same workpiece material and the grinding wheel composition are used in this work as in [10]. The remaining coefficients K_1', K_2', K_3', K_4', K_5', K_6' and K_7' will be determined through grinding experiments.

3. Coefficient determination and model validation

3.1. Experimental scheme and setup

For determining the model coefficients and then validating the model, a total number of 26 experimental tests in surface grinding configuration were performed. Firstly, the coefficients were determined by using 16 experimental tests given in Table 1, which consists of three variables and four levels arranged under orthogonal test design. Afterwards, the prediction accuracy of the model was estimated through another 10 experimental tests with randomly selected values of the parameters shown in Table 3.

Table 1. Experiments for identifying model coefficients.

Exp. no.	a_p (mm)	v_w (m/min)	v_s (m/s)	F_n (N/mm)	F_t (N/mm)
1	0.01	5	21	7.53	3.59
2	0.01	10	24	11.61	5.61
3	0.01	15	27	18.15	5.58
4	0.01	20	30	22.88	5.61
5	0.02	5	24	11.26	6.27
6	0.02	10	21	25.19	9.92
7	0.02	15	30	22.32	9.26
8	0.02	20	27	34.60	12.25
9	0.03	5	27	14.65	8.88
10	0.03	10	30	21.19	10.94
11	0.03	15	21	46.84	14.13
12	0.03	20	24	45.67	17.45
13	0.04	5	30	18.02	9.39
14	0.04	10	27	30.87	12.54
15	0.04	15	24	52.04	16.88
16	0.04	20	21	71.38	25.71

The grinding experiments have been conducted on a self-developed surface grinding machine using the aluminum oxide grinding wheel A60M6V with an outer diameter of 350 mm and a width of 40 mm. Before each grinding experiment, the dressing operation was carried out by using a single point diamond tool with a dressing depth of 0.02 mm. The axial feed rate of the dressing tool per wheel revolution was 0.04 mm/r. Two spark-out passes were taken after each dressing operation. The workpieces were made of mild steel 20# (Chinese model) with a dimension of 40×20×20 mm^3. Normal and tangential grinding forces were measured throughout each test. The data were not recorded until the forces reached a steady-state value. Usually the dynamometer for grinding operation can be divided into the strain type and the piezoelectric type. The strain dynamometer is especially good for measuring the static force or the slowly varying force, whereas the piezoelectric dynamometer cannot be used for measuring the static force because it is of alternating current coupling. In this work the strain dynamometer was adopted to measure the grinding force under the scheme shown in Fig. 6. The workpiece is held on the strain dynamometer, in which a strain foil is bonded to an elastic body. As the grinding force acts on the elastic body, a deformation occurs on the elastic body and an identical elastic deformation is transferred to the strain foil. The deformation leads to a change in resistance value of the strain foil. The resistance strain gauge is used to transform this change in resistance value into a small electric signal and then to amplify it. Finally, the amplified electric signal is obtained by the industrial personal computer using the data acquisition card for the following storage and processing.

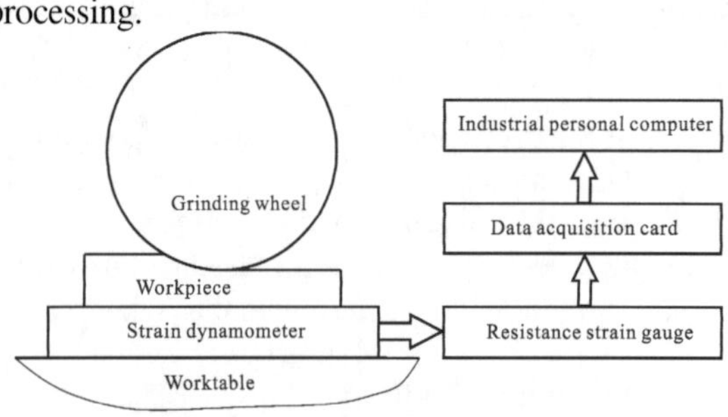

Figure 6. Scheme of measuring the grinding force

3.2. Identification of the model coefficients

As shown in Table 1, the 16 experiments are intended to obtain the values of the model coefficients K_1', K_2', K_3', K_4', K_5', K_6' and K_7'. The considered process variables are the depth of cut a_p, the workpiece velocity v_w, and the grinding wheel velocity v_s. Eq. (14) and (13) use the least square method to identify the coefficients K_1', K_2', K_3', K_4', K_5', K_6' and K_7' in Eq. (12) and (13).

$$\begin{cases} \dfrac{\partial}{\partial K_1'} \sum_{i=1}^{16} \left(F_{n,ip}' - F_{n,im}'\right)^2 = 0 \\ \dfrac{\partial}{\partial K_2'} \sum_{i=1}^{16} \left(F_{n,ip}' - F_{n,im}'\right)^2 = 0 \\ \dfrac{\partial}{\partial K_3'} \sum_{i=1}^{16} \left(F_{n,ip}' - F_{n,im}'\right)^2 = 0 \end{cases} \tag{14}$$

in which $F_{n,ip}'$ and $F_{n,im}'$ refer to the ith predicted and the ith measured specific normal grinding force respectively.

$$\begin{cases} \dfrac{\partial}{\partial K_4'} \sum_{i=1}^{16} \left(F_{t,ip}' - F_{t,im}'\right)^2 = 0 \\ \dfrac{\partial}{\partial K_5'} \sum_{i=1}^{16} \left(F_{t,ip}' - F_{t,im}'\right)^2 = 0 \\ \dfrac{\partial}{\partial K_6'} \sum_{i=1}^{16} \left(F_{t,ip}' - F_{t,im}'\right)^2 = 0 \\ \dfrac{\partial}{\partial K_7'} \sum_{i=1}^{16} \left(F_{t,ip}' - F_{t,im}'\right)^2 = 0 \end{cases} \tag{15}$$

in which $F_{t,ip}'$ and $F_{t,im}'$ refer to the ith predicted and the ith measured specific tangential grinding force respectively. Solving Eq. (14) and (15) gives the values of the model coefficients shown in Table 2.

Table 2. Values of model coefficients.

K_1'	K_2'	K_3'	K_4'	K_5'	K_6'	K_7'
97961	166950	14.30	39564	5780	0.08	15.73

3.3 Model validation and result discussion

Table 3 gives the parameter values of the ten experimental tests for validating the model. The predicted grinding force was compared with the corresponding measured result in Fig. 7. It can be observed that the correlation between the predicted and measured grinding forces is adequate with an average prediction accuracy of 81.42% for the specific normal grinding force and 78.58% for the specific tangential grinding force, and that the trend in the measured results was well predicted by the proposed model.

Table 3. Experiments for model performance test.

Exp. no.	a_p (mm)	v_w (m/min)	v_s (m/s)
1	0.03	15	27
2	0.04	6	21
3	0.01	18	23
4	0.04	19	31
5	0.03	15	22
6	0.01	16	28
7	0.02	16	27
8	0.03	11	24
9	0.04	15	30
10	0.04	8	21

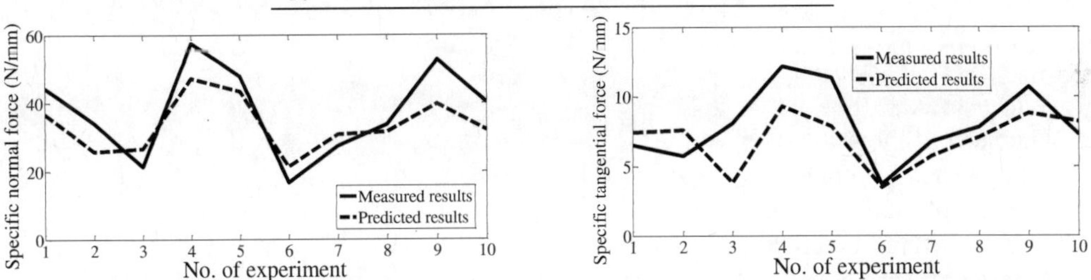

(a) Specific normal force (b) Specific tangential force

Figure 7. Comparison between the measured and predicted grinding force.

There are several possible explanations for the deviation between the predicted and measured values. One reason probably results from the limitation of the available theory in the area of stress concentration. The model shown in Fig. 4 was employed because it can be used as an entry point to explore the regime of the stress concentration in grinding. However, it should be noted that it can not perfectly explain some of the actual phenomena within the grinding zone. For example, as shown in Fig. 4, the stress concentration within the interface of the grain and the workpiece was assumed to be a two-dimension problem. This means that the stress and strain distribution along the third dimension is uniform. In reality, due to the irregular geometry of the grains along with the tiny size of them, the distribution is not probably uniform along the third dimension. Therefore, the effectiveness of the above assumption would be deteriorated by the boundary effects to some extent.

Secondly, throughout the grinding process, self-sharpening of the wheel continues with the generation of wear flats and stochastic grit pullout, which means that the interface conditions between the wheel and workpiece is continuously changing. The effective contact area between the grain and workpiece and the attack angle of the grain are changing all the time accordingly. The magnitude of the grinding force would be inevitably affected by the changes in the contact area and the attack angle. Unfortunately, these influencing factors are not yet able to be considered independently in the proposed model due to the limited conditions.

Another reason of the prediction deviation is due to the temperature rise in grinding zone, which leads to a variation in hardness of the workpiece surface and a variation in friction coefficient between the workpiece and the wheel. The variation of workpiece hardness and friction coefficient will generate a change in grinding force.

4. Conclusion

Based on the model from [10], an improved model for estimating the grinding force is proposed with quantitative and independent consideration of the rupture factor at the interface between the grain and workpiece, in which the contribution of stress concentration is included. The experimental results show that the average prediction accuracy of the model reaches 81.42% for the specific normal grinding force and 78.58% for the specific tangential grinding force, and the trend in the measured results was well predicted by the proposed model. For further improving the prediction accuracy, the remaining important influencing factors need to be considered independently, which include the boundary condition of the stress concentration model, the effects of self-sharpening, and the temperature rise in the grinding zone.

Acknowledgments
This work was supported by the Ningbo Natural Science Foundation of P. R. China under Grant number 2015A610149 and the Zhejiang Provincial Natural Science Foundation of P. R. China under Grant number LY14E050003.

Nomenclature

F_n': Specific normal grinding force
F_t': Specific tangential grinding force
K_1, K_2, K_3, K_4, K_5, K_6, K_7, a_0, b_0, c_0, K_1', K_2', K_3', K_4', K_5', K_6' K_7': Undetermined coefficient
a_p: Depth of cut
v_w: Workpiece feed speed
v_s: Wheel speed
d: Diameter of the grit
d_e: Equivalent wheel diameter
C_s: Number of grains per unit area
α: Stress concentration factor
σ_{max}: Maximum stress
σ_0: Reference stress
σ_p: Proportional limit
σ_e: Elastic limit
σ_s: Yield limit
σ_b: Ultimate strengh
t: Depth of the notch
ρ: Curvature radius of the notch
M: Grit number
σ_s: Yield strength
β: Rupture factor
F_{nc}': Specific normal chip formation force
F_{tc}': Specific tangential chip formation force
ε_0: A constant
$F_{n,ip}'$: The ith predicted specific normal grinding force
$F_{n,im}'$: The ith measured specific normal grinding force
$F_{t,ip}'$: The ith predicted specific tangential grinding force
$F_{t,im}'$: The ith measured specific tangential grinding force

References

[1] Malkin, S. and Guo, C., *Grinding technology: theory and application of machining with abrasives*, Second Ed. Industrial Press: New York, USA, (2008).
[2] Brinksmeier, E., Aurich, J.C., Govekar, E., Heinzel, C., Hoffmeister, H.W., Klocke, F., Peters, J., Rentsch, R., Stephenson, D.J., Uhlmann, E., Weinert, K., and Wittmann, M., Advances in modeling and simulation of grinding processes, *Annals of the CIRP*, 55 (2) (2006) 667-696.
[3] Badger, J.A. and Torrance, A.A., A comparison of two models to predict grinding forces from

wheel surface topography, *International Journal of Machine Tools and Manufacture*, 40 (2000) 1099-1120.

[4] Karkalos, N.E., Markopoulos, A.P., and Dossis, M.F., Application of statistical and soft computing techniques for the prediction of grinding performance, *Journal of Robotics and Mechanical Engineering Research*, 1(2) (2015) 6-16.

[5] Chang, H.C. and Wang, J.J.J., A stochastic grinding force model considering random grit distribution, *International Journal of Machine Tools and Manufacture*, 48 (2008) 1335-1344.

[6] Mishra, V.K. and Salonitis, K., Empirical estimation of grinding specific forces and energy based on a modified Werner grinding model, in: *14th CIRP Conference on Modeling of Machining Operations* (2013).

[7] Agarwal, S. and Rao, P.V., Predictive modeling of force and power based on a new analytical undeformed chip thickness model in ceramic grinding, *International Journal of Machine Tools and Manufacture*, 65 (2013) 68-78.

[8] Karkalos, N.E., Markopoulosa, A.P., and Kundrák, J., Molecular dynamics model of namo-metric peripheral grinding, *Procedia CIRP*, 58 (2017) 281-286.

[9] Zhang, Y., Li, C., Ji, H., Yang, X., Yang, M., Jia, D., Zhang, X., Li, R., and Wang, J., Analysis of grinding mechanics and improved predictive force model based on material-removal and plastic-stacking mechanisms, *International Journal of Machine Tools and Manufacture*, 122 (2017) 81-97.

[10] Durgumahanti, U.S.P., Singh, V., and Rao, P.V., A new model for grinding force prediction and analysis, *International Journal of Machine Tools and Manufacture*, 50 (2010) 231-240.

[11] Lin, K., Xu, X., Li, Y., and Fang, C., Model of grinding force based on stress concentration theory, *Transactions of the Chinese Society for Agricultural Machinery*, 43 (11) (2012) 261-266.

[12] Li, A., Guo, T., Zhang, C., Chen, W., Ye, K., and Lv, Z., *Stress concentration*, China Machine Press: Beijing, China, (1986).

Effect of the Aging Treatment on the Tensile Strength and Impact Toughness of Friction Welded Super304H Joints

Lu Yuan[1, a], Zhang Jianlong[1, 2], Li Debiao[1], Bi Cheng[1] and Ning Tao[1]

[1] Xi'an Special Equipment Inspection Institute, Xi'an710065, China
[2] School of Mechanical Engineering, Xi'an University of Science and Technology, Xi'an 710054, China

[a] luyuan19801104@163.com

Abstract: In this paper, the tensile strength and impact toughness of the friction welded Super304H joints were studied after aging at 625°C for different time. As the aging time increased, the welding zone, heat affected zone and base metal were still austenite, and the number of precipitates gradually increased. The impact toughness of the joints decreased, the tensile strength of the joints increased firstly, than decreased after aged for 500 h, this is attributed to the aggregation and growth of the $Cr_{23}C_6$ carbides. The rupture mode of tensile strength and impact toughness samples changed from ductile fracture to brittle fracture.

1. Introduction

Energy shortage and environmental pollution around the world urges people to develop new techniques of clear and high efficiency thermal power generation. Nowadays, the ultra super critical power generation technique with much higher parameters including vapor temperature and pressure, has been generally accepted as the most feasible and reliable thermal power generation technology [1-3]. In order to fulfill the tough parameters of the ultra super critical power generation, traditional heat-resistant steels used in the subcritical and supercritical units have been replaced by those new heat-resistant steels, such as Super304H. To meet the critical demands for the ultra super critical power generation, a new austenitic stainless steel Super304H (0.1C-18Cr-9Ni-3Cu-Nb-N) has been developed. The Super304H has higher strength at elevated temperatures are required for super heater tubes in fossil fired boilers. However, in this situation, a questions is how to join these new austenitic stainless together. The conventional fusion welding joints of austenitic stainless steel exhibit inferior mechanical properties due to the formation of intermetallic compounds at the joint interface, and the excessive residual stresses. The friction welding can be more suitable than fusion ones since many problems associated with melting are eliminated or reduced [4-6]. In order to use friction welded Super304H joints widely in higher temperature environment, the effect of aging time on the properties of the joints is investigated. This study aims to investigate the tensile strength and impact toughness of the friction welded Super304H joints after aged at 625°C. The analysis of mechanical properties at different aging time provides the more fundamentals for wide application of the friction welded Super304H joints at higher temperatures.

2. Experimental procedures

The pipes of Super304H austenitic stainless steel that in the sizes of Φ44.5mm x 9mm were used as base metals. Friction welding was carried out using a continuous drive friction welding machine. During the friction welding operations, the friction welding parameters were set to the following combinations: a friction speed of 1500 rpm, a friction pressure of 150 MPa, an upset pressure of 200 MPa and a friction time of 5s. These specimens were heated respectively in a tube furnace at 625 °C for 0h, 500h, 1000h, 1500h, 2000h, 2500h, and 3000 h, respectively. The microstructures of the aged samples were observed and analyzed by using scanning electron microscopy. The tensile samples were tested using a testing instrument. The impact samples were tested using a impacting instrument.

3. Results and Discussion

Fig 1-3 shows morphology aged for 1000h, 2000h and 3000 h, respectively. During the friction welding process, the welded zone produced agglutinate and shear tear behavior, which led to the deformation of austenite grains, the dynamic recrystallization driving force increased. The thermoplastic deformation temperature and free energy of recrystallized grains decreased, which produce a large number of recrystallized nucleations that leading to the fine grained dynamic recrystallization structure of the weld zone, the effect of dynamic recrystallization was obvious. The welded zone of austenitic stainless steel welded by fusion welding was typical dendritic morphology with large grains. Compared with the conventional fusion welding joints, the friction welded austenitic stainless steel joints were mainly composed of fine equiaxed austenite crystals. The friction welding was solid state welding techniques, the melting process did not occur in the welded zone and heat affected zone. So the heat input during the friction welding process was much less than the heat input during the fusion welding process. So the friction welded joints with the small grains showed excellent microstructure. After aged at 625°C, the grain size change of the joints was not detectable. However, the precipitation and growth of the $Cr_{23}C_6$ carbide particles along the austenitic grain boundaries is evident. Meanwhile, the precipitation of the $Cr_{23}C_6$ carbides inside the austenitic grains occured with a rather smaller size[7]. As the aging time increased, the amount of $Cr_{23}C_6$ carbides increased obviously.

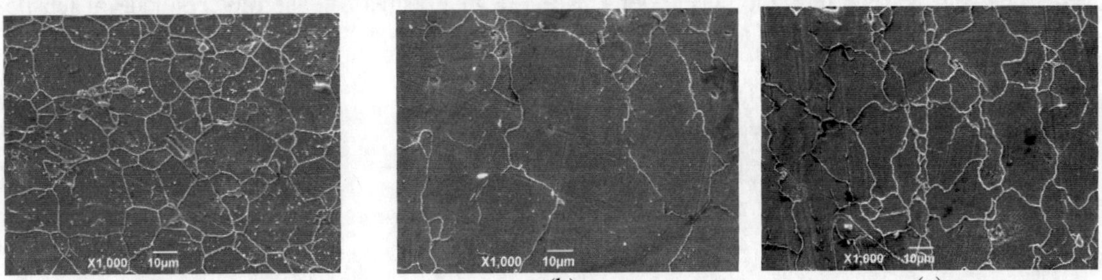

(a) (b) (c)

Figure.1 The metallograph aged for 1000h: (a) the welding zone, (b) the heat affected zone, (c) the base metal

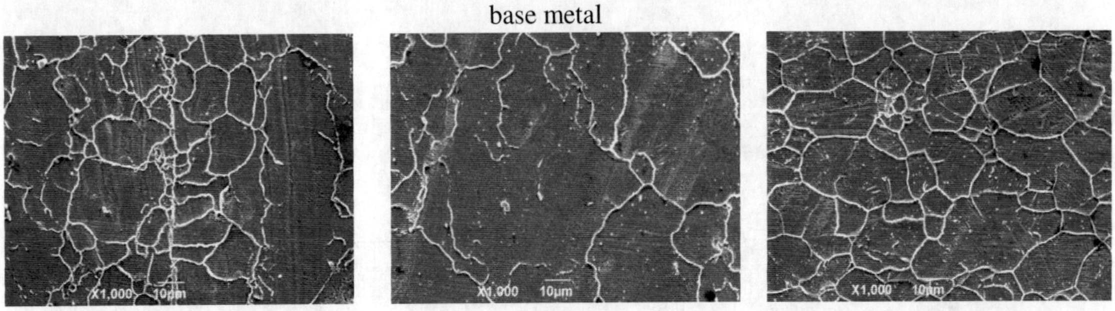

(a) (b) (c)

Figure.2 The metallograph aged for 2000h: (a) the welding zone, (b) the heat affected zone , (c) the base metal

(a) (b) (c)

Figure.3 The metallograph for 3000h: (a) the welding zone, (b) the heat affected zone , (c) the base metal

Fig. 4 depicts the variation of yield strength and tensile strength as a function of the aging time. It was detected that the fracture position of the joints was in the heat affected zone after aged for different time. As the aging time increased, the tensile strength of the joints increases sharply (<500 h), and then decreases (500-3000 h). In the aging process, the mechanical property change of the joints is certainly corresponding to the microstructure evolution. Precipitation, aggregation and growth of the $Cr_{23}C_6$ carbides phases are main reasons of the microstructure evolution of the joints [8]. In the early aging stage, the dispersion strengthening effect of the $Cr_{23}C_6$ carbides may be overwhelming, the yield strength and tensile strength of the joints increased. As the aging time increased continuously, aggregation and growth of the $Cr_{23}C_6$ carbides takes place, the dispersion strengthening effect of the $Cr_{23}C_6$ carbides decreases simultaneously, resulting in the apparent drop of the yield strength and tensile strength. As shown in Fig. 5, the fractographs of the joints change with the ageing time. After aged for 1000h, fine and shallow dimples distribute on the facture surface. As the aging time increased, the density of the dimples on the facture surface reduces. The shear deformation on the facture surface turns to be severe. Therefore, the plastic deformation degree decreases. The rupture mode of tensile strength samples changed from ductile fracture to brittle fracture.

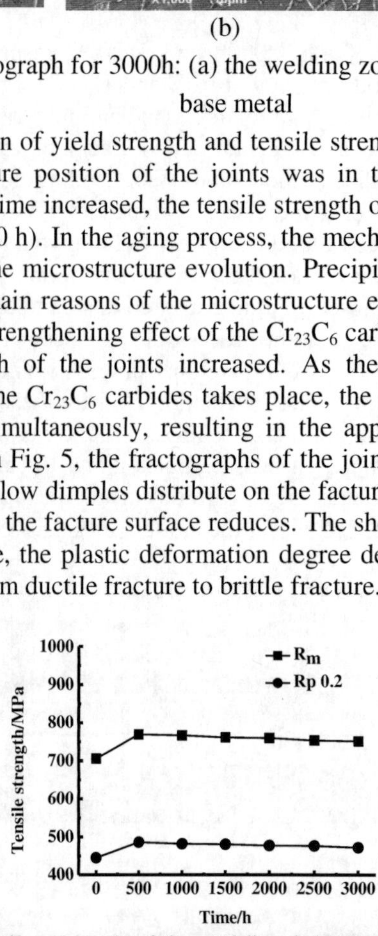

Figure.4 The tensile strength of the welding joints after aged for different time

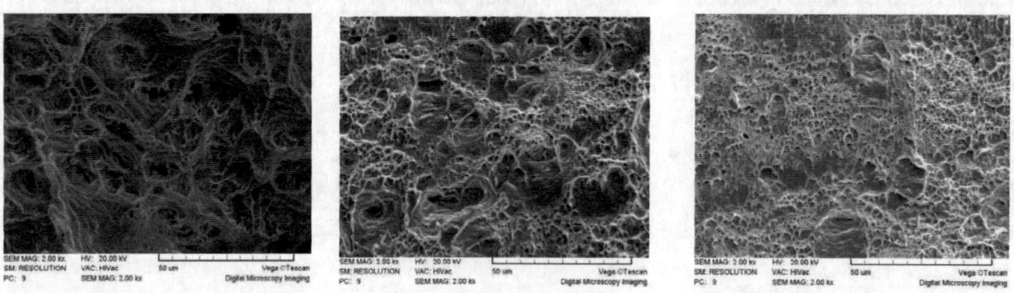

(a) (b) (c)

Figure.5 The rupture appearance of the the welding joints: (a) 1000h, (b) 2000h, (c) 3000h

Fig. 6 depicts the impact toughness of the joints drop monotonously with increasing the aging time, indicating the impact property of the welding joints is much sensitive to the higher temperature aging process. The precipitation, aggregation and growth of the $Cr_{23}C_6$ carbides are the main reason that the impact toughness of the joints decreased with the aging time increased. The grain size of austenite grains has a great influence on the impact toughness of each part of the welded joints. When the grains size was smaller, the impact toughness was higher. The grains size of the heat affected zone was the largest, the grains size of the base metal was the smallest. So the impact toughness of the base metal was the highest, and the impact toughness of the heat affected zone was the lowest. The heat affected zone with coarse austenite grains was the weakest part of the welding joints. As shown in Fig 7- Fig 9, the fractographs of the welding joints change with the aging time. The joints aged for 1000h, a dimple aggregation fracture mechanism was observed. The dimples on the fracture surface were shallower with more $Cr_{23}C_6$ carbides on the bottom of them. As the aging time increased, more and larger $Cr_{23}C_6$ carbides precipitated. Correspondingly, the rupture mode of impact toughness samples changed from ductile fracture to brittle fracture.

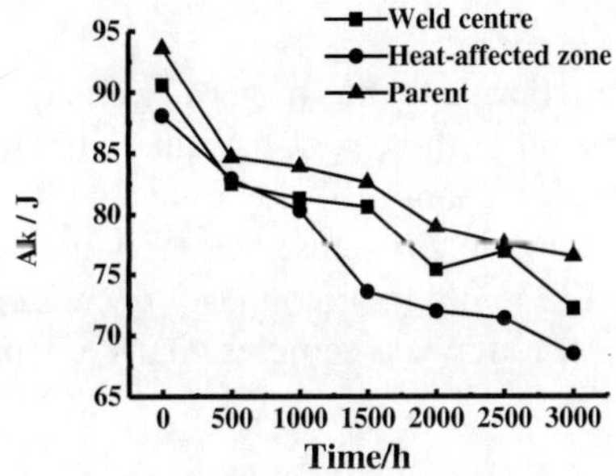

Figure.6 The impact toughness of the welding joint after aging for different time

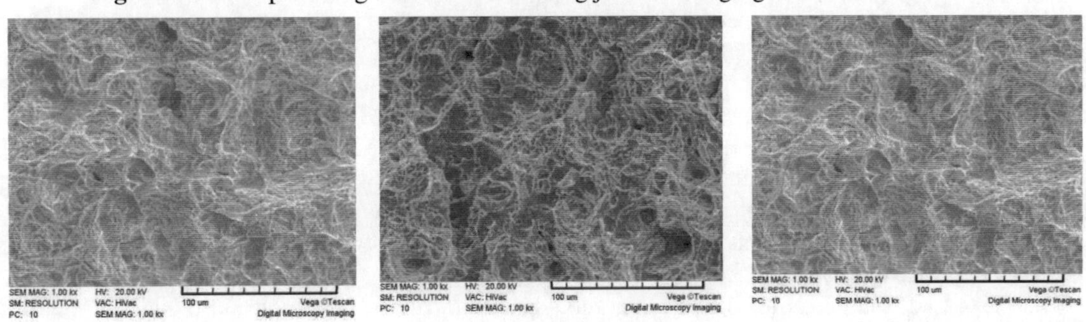

(a) (b) (c)

Figure.7 The rupture appearance of aged for 1000h: (a) the welding zone, (b) the heat affected zone , (c) the base metal

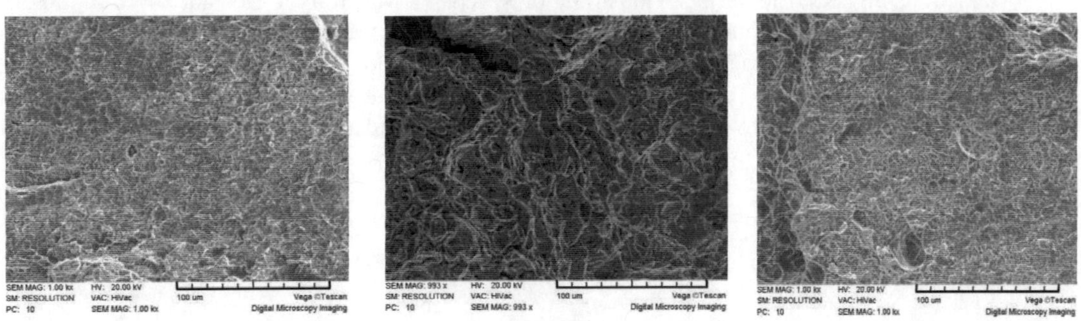

(a) (b) (c)

Figure.8 The rupture appearance of aged for2000h: (a) the welding zone, (b) the heat affected zone, (c) the base metal

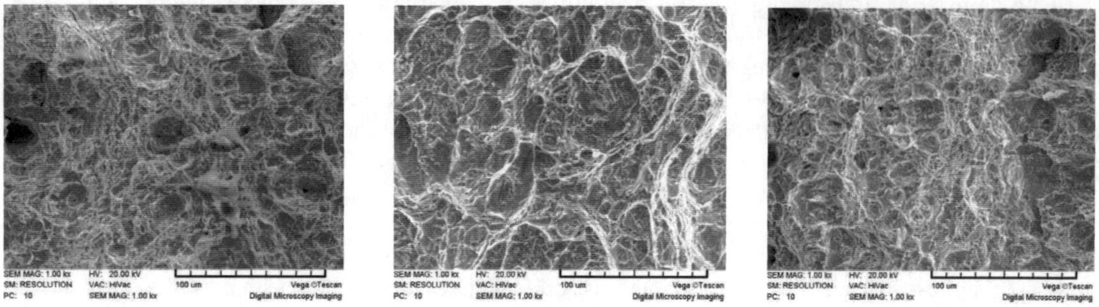

(a) (b) (c)

Figure.9 The rupture appearance of aged for3000h: (a) the welding zone, (b) the heat affected zone , (c) the base metal

4. Conclusions

In the aging process, microstructure of the friction welded Super304H joints was exhibited. Precipitation, growth and aggregation of the $Cr_{23}C_6$ carbides were the main microstructure mechanism, resulting in the mechanical property change of the joints with the aging time. As the aging time increased, the tensile strength of the joints increased in the initial stage, than decreased after aged for 500 h. The impact toughness of the joints monotonously decreased with the aging time. The rupture mode of tensile strength and impact toughness samples changed from ductile fracture to brittle fracture.

References

[1] Zielinski A. Structure and properties of Super 304H steel for pressure elements of boilers with ultra-supercritical parameters[J]. Journal of Achievements in Materials and Manufacturing Engineering,2012,55(2):403-409.

[2] Xinmei Li, Yong Zou, Zhongwen Zhang, et al. Microstructure Evolution of a Novel Super304H Steel Aged at High Temperatures[J]. Mater Trans, 2014, 608: 164-173.

[3] Indrani S, Amankwaha E, Kumara S, et al. Microstructure and mechanical properties of annealed SUS 304H austenitic stainless steel with copper[J]. Mater Sci Eng A, 2011, 528: 4491-4499.

[4] Li P, Li J, Li X, et al. A study of the mechanisms involved in initial friction process of continuous drive friction welding[J]. Journal of Adhesion Science and Technology, 2015, 29(12): 1246-1257.

[5] Li X, Li J, Liao Z, et al. Effect of rotation speed on friction behavior and radially non-uniform local mechanical properties of AA6061-T6 rotary friction welded joint[J]. Journal of Adhesion Science and Technology, 2018, 32(18): 1987-2006.

[6] Li X, Li J, Liao Z, et al. Microstructure evolution and mechanical properties of rotary friction welded TC4/SUS321 joints at various rotation speeds[J]. Materials & Design, 2016, 99: 26-36.

[7] Xinmei Li, Yong Zou, Zhongwen Zhang, et al. Microstructure Evolution of a Novel Super304H Steel Aged at High Temperatures[J]. Mater Trans, 2014, 608: 164-173.

[8] MyungYeon K, Sukchul K, InSuk C, et al. High-temperature tensile and creep deformation of cross-weld specimens of weld joint between T92 martensitic and Super304H austenitic steels[J]. Mater Charact, 2014, 97: 161-168.

IOP Conf. Series: Materials Science and Engineering **470** (2019) 012029 doi:10.1088/1757-899X/470/1/012029

Effect of the Aging Treatment on the Precipitates and Vickers Hardness of Friction Welded Super304H Joints

Lu Yuan[1, a], Li Debiao[1], Zhang Jianlong[1, 2], Shang Kang[1], Yang Yang[1] and Cao Bo[1]

[1] Xi'an Special Equipment Inspection Institute, Xi'an710065, China
[2] School of Mechanical Engineering, Xi'an University of Science and Technology, Xi'an 710054, China

[a] luyuan19801104@163.com

Abstract: In this paper, the precipitates and vickers hardness of the friction welded Super304H joints were studied after aging at 625°C for different time. $Cr_{23}C_6$ and NbC is denoted precipitated phases in the X-ray diffraction profiles. Precipitation, growth and aggregation of the $Cr_{23}C_6$ carbides were the main precipitated phase evolution mechanism. The Vickers hardness of the joints increased firstly, than decreased when after aged for 500 h, this is attributed to the aggregation and growth of the $M_{23}C_6$.

1. Introduction

The increase in operating steam pressure and temperature of the conventional boiler/steam turbine power plant will result in increase in efficiency of the power plant. The increase in efficiency leads to less fuel consumption and thereby reduces CO_2 emission. Critical research is ongoing in the area of development of newer creep resistant, oxidation resistant and corrosion resistant materials which are capable of operating under high stresses. For these reasons, the austenitic stainless steels are the primary candidates in the superheater/reheater tubing, where oxidation resistance and fireside corrosion resistance become important in addition to the creep strength [1]. The recently developed Super304H (0.1C-18Cr-9Ni-3Cu-Nb-N) austenitic steel has shown considerable promise due to its high oxidation resistance, corrosion resistance and creep resistance. The Super304H has higher strength at elevated temperatures are required for super heater tubes in fossil fired boilers. This excellent creep rupture strength is based on finely precipitated particles such as Nb(C, N) and NbC [2,3]. Friction welding is a solid state welding process where the welding takes place well below the melting temperature of the alloy, combined with the autogenous nature of this welding process nullifies adverse effects of low temperature eutectics segregation and the uncertainties in the filler metal selection. The process is a automatic process which can produce quality welds independent of the operator's skill and can offer advantages on environmental issues. This illustrates the benefits associated with establishing a welding procedure for joining of pipes/tubes by friction welding. The friction welding can be successfully applied for welding of austenitic stainless steel joints for higher temperature application [4-6]. In order to use friction welded Super304H joints widely in higher temperature environment, the effect of aging time on the properties of the friction welded Super304H joints is investigated. This study aims to investigate the precipitates and Vickers hardness of the friction welded Super304H joints after aged at 625°C. The analysis of precipitates and Vickers hardness at different aging time provides the more fundamentals for wide application of the friction welded Super304H joints at higher temperatures.

2. Experimental procedures

The pipes of Super304H austenitic stainless steel that in the sizes of Φ44.5mm x 9mm were used as base metals. Friction welding was carried out using a continuous drive friction welding machine. During the friction welding operations, the friction welding parameters were set to the following combinations: a friction speed of 1500 rpm, a friction pressure of 150 MPa, an upset pressure of 200 MPa and a friction time of 5s. These specimens were heated respectively in a tube furnace at 625 °C for 0h, 500h, 1000h, 1500h, 2000h, 2500h, and 3000 h, respectively. The microstructures of the aged samples were observed and analyzed by using a scanning electron microscopy (SEM) equipped with an energy dispersive spectroscopy (EDS) system. Vickers hardness values of the joints were tested using a Vickers hardness testing instrument with a loading of 5 kg and an enduring time of 15 s.

3. Results and Discussion

Fig 1- Fig 4 shows precipitates morphology of the welding joints aged for 0h, 1000h, 2000h and 3000 h, respectively. Compared with the conventional fusion welding joints, the friction welded austenitic stainless steel joints were mainly composed of fine equiaxed austenite crystals, and the welded zone has a narrow width. The friction welding was solid state welding techniques, the melting process did not occur in the welded zone and heat affected zone. So the heat input during the friction welding process was much less than the heat input during the fusion welding process. So the friction welded joints with the small grains showed excellent microstructure [7]. It is known that Super304H is famous for its excellent thermal stability. After aged at 625°C, the austenitic grain size change of the welding joints was not detectable. When the sample was not aged, the precipitated phases that composed of fine spherical particles were formed occasionally, and were mainly distributed along the grains boundary. There were very few precipitated phases in the interior of grains. When the sample was aged for 1000h, it can be observed more precipitated phases distributed along the grain boundary, and the shape of precipitated phases include the lath-like and fine spherical. The precipitated phases was still less in the interior of grains. When the sample was aged for 2000h, the amount of the precipitated phases increased, the size change of the precipitated phases was not detectable obviously. The precipitated phases began distributed along the grains boundary discontinuously, the number of the precipitated phases in the interior of grains increased. When the sample was aged for 3000h, the precipitated phases distributed along the grains boundary continuously, the number of the precipitated phases in the interior of grains increased obviously and distributed in chains. The aggregation and growth of the precipitated phases were main styles of the microstructure evolution. Some precipitated phases that distributed isolated were coarse and lath-like particles [8].

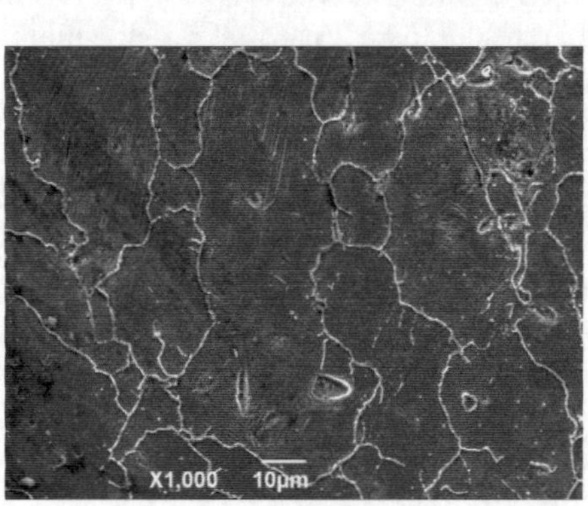

(a)

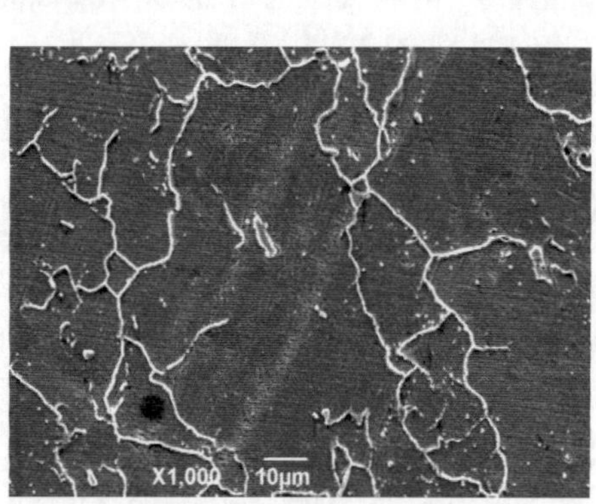

(b)

IOP Conf. Series: Materials Science and Engineering **470** (2019) 012029 doi:10.1088/1757-899X/470/1/012029

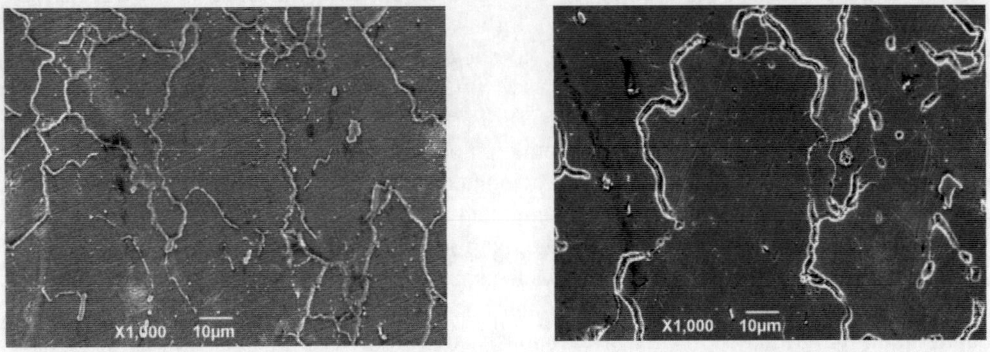

(c) (d)

Figure.1 The precipitated phases of the welding joints: (a) 0h, (b) 1000h, (c) 2000h, (d) 3000h

Fig 2 shows the X-ray diffraction profiles of the welding joints aged for 1000h, 2000h and 3000 h, respectively. It can be sure that the peaks with the highest intensity correspond to the γ-Fe phase and the precipitated phases for all the samples. Therefore, the welded joints after the aging treatment, the main phase is the γ-Fe phase and the precipitated phases are $Cr_{23}C_6$ and NbC. With the increasing of aging time, the lattice parameter and the amount of the precipitated phases $Cr_{23}C_6$ and NbC is increasing. The Cr element has a tendency to form carbides, some C element and Cr element precipitated from matrix and formed the precipitated phases $Cr_{23}C_6$ at the grain boundary preferentially. The X-ray result indicated that the precipitated phases included the Cr and Fe elements, and the Cr element is the major metal element. Therefore, the $Cr_{23}C_6$ is denoted $M_{23}C_6$ precipitated phases. Some N element replaces the element C, and a part of the NbC is converted into Nb(C, N). It is difficult to index the Nb(C, N) due to their lower intensities peaks. Therefore, $Cr_{23}C_6$ and NbC is denoted precipitated phases in the X-ray diffraction profiles. The Super304H welded joints show the excellent creep rupture strength, and the austenite matrix still keep the fine microstructure. This could be attributed to the dispersively distributed of the strengthening phase Nb(C, N) and NbC in the interior of grains, which can effectively pin dislocations and grains boundaries, than impede the movement of dislocations and grains boundaries. The Nb(C, N) and NbC became the core of the subsequent precipitation of $Cr_{23}C_6$, which avoids the precipitation of $Cr_{23}C_6$ and the depletion of Cr at the grain boundaries. So the precipitation of Nb(C, N) and NbC improves the resistance to intergranular corrosion of the welded joints.

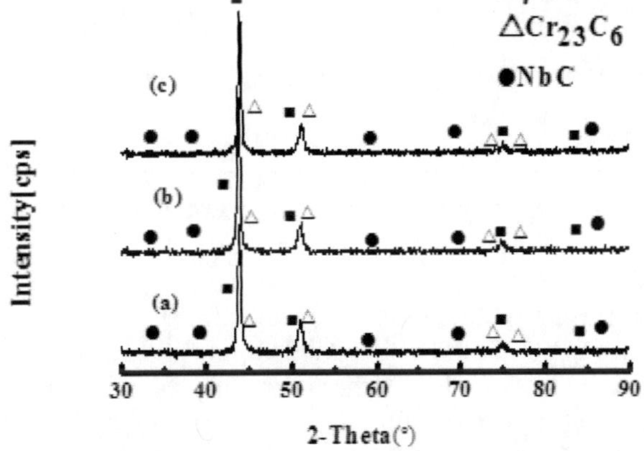

Figure.2 XRD patterns of the welding joint after aging for different time(a)1000h; (b)2000h; (c)3000h;

Fig 3 shows the precipitated phase of the welding joints aged for different times with EDS to show the elements of the precipitated phases. Table 1 shows the EDS analysis results of the precipitated phases at positions shown in Figure 3. Fig 3 (a) shows the shape of precipitated phases of welded joints aged for 1000h, the precipitated phase A is Nb(C, N) and the precipitated phase B is $Cr_{23}C_6$. It can be observed some fine $Cr_{23}C_6$ and Nb(C, N) distributed along the grains boundaries and in the interior of grains. With the increasing of the aging temperature, the size and amount of $Cr_{23}C_6$ increase continuously. Fig 3 (b) shows the shape of precipitated phases of welded joints aged for 2000h, the precipitated phase A is Nb(C, N) and the precipitated phase B is $Cr_{23}C_6$. The amount of $Cr_{23}C_6$ and Nb(C, N) along the grains boundaries and in the interior of grains increased, the size change of Nb(C, N) was not detectable obviously. Fig.3(c) shows the shape of precipitated phases of welded joints aged for 3000h, the precipitated phase A is $Cr_{23}C_6$ and the precipitated phase B is Nb(C, N). The $Cr_{23}C_6$ distributed along the grains boundaries continuously, and the Nb(C, N) distributed in the interior of grains in chains. The aggregation and growth of $Cr_{23}C_6$ were obviously. Some $Cr_{23}C_6$ that distributed isolated were coarse and lath-like.

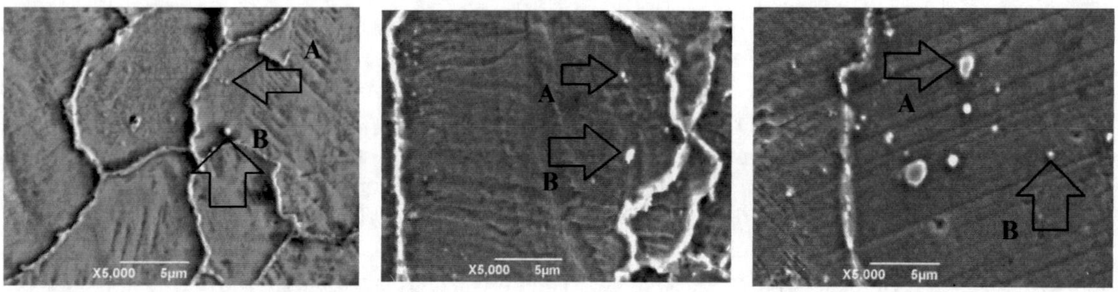

(a) (b) (c)

Figure.3 The shape of precipitated phases of welded joints: (a)1000h; (b)2000h; (c)3000h;

Table 1 The energy spectrum analysis of the precipitated phase

	C	Cr	Fe	Ni	Cu	Nb
The precipitated phase A (aged for 1000h)	10.08/45.84	1.23/1.29	1.47/1.44	0.34/0.32	0.06/0.06	86.81/51.05
The precipitated phase B (aged for 1000h)	13.55/44.55	16.26/12.35	42.97/30.37	4.19/2.82	0.63/0.39	22.40/9.52
The precipitated phase A (aged for 2000h)	14.82/49.3	12.05/9.56	26.48/19.56	2.44/1.72	1.19/0.76	43.01/19.10
The precipitated phase B (aged for 2000h)	15.13/45.26	18.19/13.00	51.65/34.35	5.66/3.58	0.40/0.23	8.97/3.59
The precipitated phase A (aged for 3000h)	25.04/65.83	11.62/7.05	22.65/12.81	2.06/1.11	0.49/0.24	38.14/12.96
The precipitated phase B (aged for 3000h)	11.83/49.94	1.72/1.68	2.21/2.01	0.29/0.25	0.73/0.69	83.21/45.42

Fig 4 shows the Vckers hardness variations in the direction perpendicular to the weld interface. Fig 5 shows the austenite grains size of the welding zone, the heat affected zone and the base metal. Because the size of austenite grains has a great influence on the Vickers hardness, the Vickers hardness decrease with the increasing of the austenite grains size. Compared with the conventional fusion welding joints, the welded zone of the friction welded joints with a narrow width were mainly composed of fine equiaxed austenite crystals. Furthermore, the strain hardening effect during the friction process at the welded zone is another reason. So the maximum Vickers hardness value is obtained at the welded zone. The friction welding was solid state welding techniques, the melting process did not occur in the welded zone and the heat affected zone. But, the grains size of the heat affected zone was the larger than the remaining parts of the welded joints, because that the heat input during the friction welding process. So, the minimum Vickers hardness value is obtained in the heat affected zone. Fig 4 depicts the Vickers hardness as a function of the aging time. As the aging time increased, the Vickers hardness of the joints increases sharply (<500 h), and then decreases (500-3000 h). In the aging process, the Vickers hardness change of the welding joints is certainly corresponding to the microstructure evolution. Precipitation, aggregation and growth of the precipitated phases are main reasons of the microstructure evolution of the welding joints. In the early aging stage, the strain hardening effect of the $Cr_{23}C_6$ and Nb(C,N) may be overwhelming, the Vickers hardness of the joints increased. As the aging time increased continuously, aggregation and growth of the $Cr_{23}C_6$ takes place, the hardening effect of the precipitated phases decreases simultaneously, resulting in the apparent drop of the Vickers hardness.

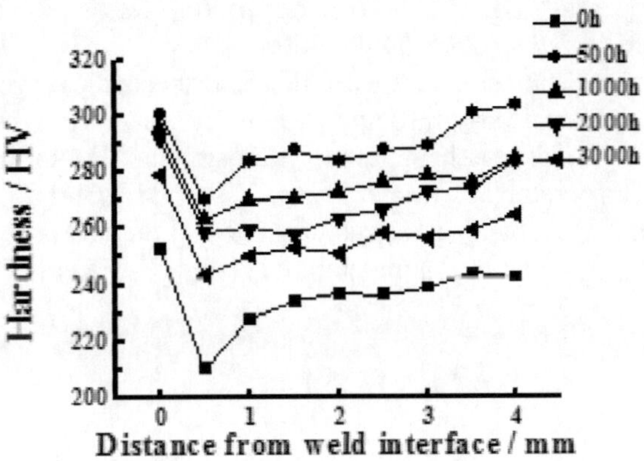

Figure.4 The Vickers hardness of the welding joints after aging for different time

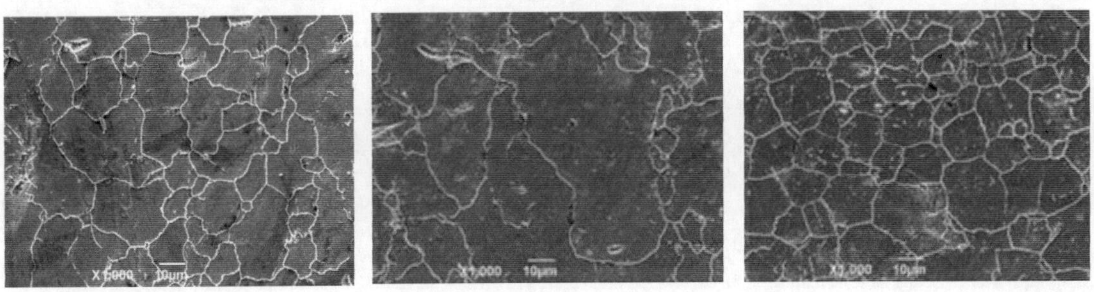

(a) (b) (c)

Figure.5 The austenite grains size of the welding joints: (a) the welding zone, (b) the heat affected zone, (c) the base metal

4. Conclusions

In the aging process, the precipitated phase evolution of the friction welded joints was exhibited. $Cr_{23}C_6$ and NbC is denoted precipitated phases in the X-ray diffraction profiles. Precipitation, growth and aggregation of the $Cr_{23}C_6$ were the main precipitated phase evolution mechanism. The welded joints shows the excellent creep rupture strength, it could be attributed to the dispersively distributed of the strengthening phase Nb(C, N) and NbC in the interior of grains, which can effectively pin dislocations and grains boundaries. The maximum Vickers hardness value is obtained in the welding zone. As the aging time increased, the Vickers hardness of the joints increased in the initial stage, than decreased after aged for 500 h.

Reference

[1] Zielinski A. Structure and properties of Super 304H steel for pressure elements of boilers with ultra-supercritical parameters[J]. Journal of Achievements in Materials and Manufacturing Engineering,2012,55(2):403-409.
[2] Xinmei Li, Yong Zou, Zhongwen Zhang, et al. Microstructure Evolution of a Novel Super304H Steel Aged at High Temperatures[J]. Mater Trans, 2014, 608: 164-173.
[3] Indrani S, Amankwaha E, Kumara S, et al. Microstructure and mechanical properties of annealed SUS 304H austenitic stainless steel with copper[J]. Mater Sci Eng A, 2011, 528: 4491-4499.
[4] Li P, Li J, Li X, et al. A study of the mechanisms involved in initial friction process of continuous drive friction welding[J]. Journal of Adhesion Science and Technology, 2015, 29(12): 1246-1257.
[5] Li X, Li J, Liao Z, et al. Effect of rotation speed on friction behavior and radially non-uniform local mechanical properties of AA6061-T6 rotary friction welded joint[J]. Journal of Adhesion Science and Technology, 2018, 32(18): 1987-2006.
[6] Li X, Li J, Liao Z, et al. Microstructure evolution and mechanical properties of rotary friction welded TC4/SUS321 joints at various rotation speeds[J]. Materials & Design, 2016, 99: 26-36.
[7] Xinmei Li, Yong Zou, Zhongwen Zhang, et al. Microstructure Evolution of a Novel Super304H Steel Aged at High Temperatures[J]. Mater Trans, 2014, 608: 164-173.
[8] Guohong C, Qi Z, Junjian L, et al. Microstructures and mechanical properties of T92/Super304H dissimilar steel weld joints after high-temperature ageing[J]. Mater design, 2013,44: 469–475.

Optimized Control Based on Nonlinear Modelling and Parameter Estimation of the Electro-Hydraulic Servo System

Meng Li[1] **and Yong Zhang**[1]

[1] National Engineering Laboratory for Automotive Electronic Control Technology, Shanghai Jiao Tong University, No.800, Rd. Dongchuan, Minhang, Shanghai, China

lemon_lee@sjtu.edu.cn

Abstract. In this paper, a nonlinear physical model of the electro-hydraulic system is built and described in Bond Graph in order to show its physical structure and flow of energy. The nonlinear model is simplified based on response bandwidth of different parts of system and nonlinear degree. A series of typical parameters of the model are chosen based on the effect on nonlinear response of system and estimated with the experiment response. A simple and useful model based nonlinear feedforward PID control methods is proposed based on the steady state response of system and dynamic stribeck friction force. The simulation and experiment results of parameter estimation and feedforward control are presented, which shows that the model response with estimated parameters can fit the experimental data very well, and the simulation and experiment response of the optimized control can eliminate the nonlinear response very well. Therefore, the nonlinear model and control method used in this paper have a guiding significance and application value, and have theory value for the further research of hydraulic servo system with the similar structure.

Key words: electro-hydraulic, Bond Graph, nonlinear model, stribeck friction, parameter estimation, feedforward PID.

1. Introduction

Because of its operation characteristics, the displacement servo system for the automotive vibration test equipment should have the capability of large load, large movement, high accuracy, fast response and long time working stability[1-3]. Electro-hydraulic system fits the requirements and has advantages on cost and reliability; therefore, it is widely used in vibration test equipment[4, 5]. The dual-nozzle force feedback two-stage electro-hydraulic servo valve and the symmetric hydraulic actuator are commonly used in electro-hydraulic servo system[6, 7].

The physical principles of hydraulic system and hydraulic components have been well researched[8, 9]. For the servo valve, from the first beginning many kinds of linear model have been developed, like the 2-order model which only contains the main valve dynamic characteristic[10-14], the 3-order model which adds the 1 order torque motor model[8, 14]. In order to get more accurate results, higher order model including linear and nonlinear model is necessary, like 5-order servo valve model which adds 1 order torque motor and 2-order pilot valve[15, 16], 5-order servo valve which adds 2-order torque motor and 1 order pilot valve[14], 7-order model which adds 2-order torque motor, 2-order pilot valve and 1-order nozzle output pressure feedback[17]. For hydraulic actuator, it also has many kinds of model, like 2-order model which mostly represents actuator's piston moving characteristics[18, 19], 4-order actuator which contains 2-order compression and 2-order piston kinematic[13, 20], the multi-body model with stribeck friction model[21], the thermal-hydraulic model based on the electro-hydrostatic theory[22].

Generally speaking, when the model has higher order, it will get higher simulation accuracy, better dynamic characteristics and be closer to the real equipment, while at the same time, it will get less stability and need higher calculation capacity. Considering the needs of the vibration test and combined with the actual characteristics of equipment and predecessors' experience, the order and structure of model in this paper are determined.

This paper can be divided into 4 parts. Firstly, the nonlinear model is structural analysed with Bond Graph and described as simultaneous differential equations, especially the nonlinear elements like hydraulic flow, valve opening and stribeck friction[23]; And the nonlinear model is simplified base on the frequency bandwidth of different parts of the system, so the main valve flow and whole actuator which have smallest bandwidth and effect the system output mostly are remain. Secondly, based on the simplified model, a series of parameter including characteristics of the main valve flow and actuator motion that hard to measure and important to system response are chosen and estimated with experiment response. Thirdly, based on the system steady state transfer characteristics, the control method of the damping and stribeck friction feedforward is proposed. Finally, the results of sine wave response with initial and estimated parameters as well as original PID[24, 25] and model based feedforward PID control are presented.

2. Nonlinear model

2.1. *Structure*

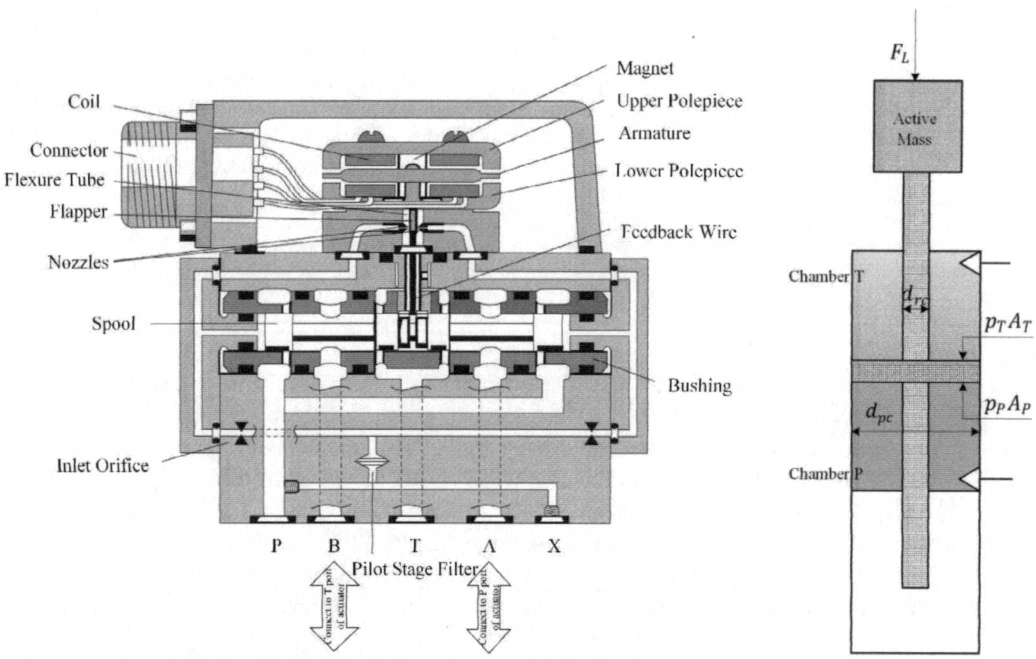

Figure 1 Electro-hydraulic servo valve and actuator profile[26]

In the figure 1, it's the profile of the MOOG G761 series servo valve, and is force feedback 2 stage electro-hydraulic servo valve. Its pilot valve is dual-nozzle flapper valve, which driven by permanent magnet moving-iron torque motor. Its main valve is 4-way spool valve, and valve core's displacement is connected to armature of the torque motor by feedback wire, to constitute force feedback loop of spool valve.

As you can see, the actuator is a symmetric actuator, so we set chamber on the bottom side as the Chamber P, the other side as Chamber T. The pressure force difference between two active chambers will make the piston move.

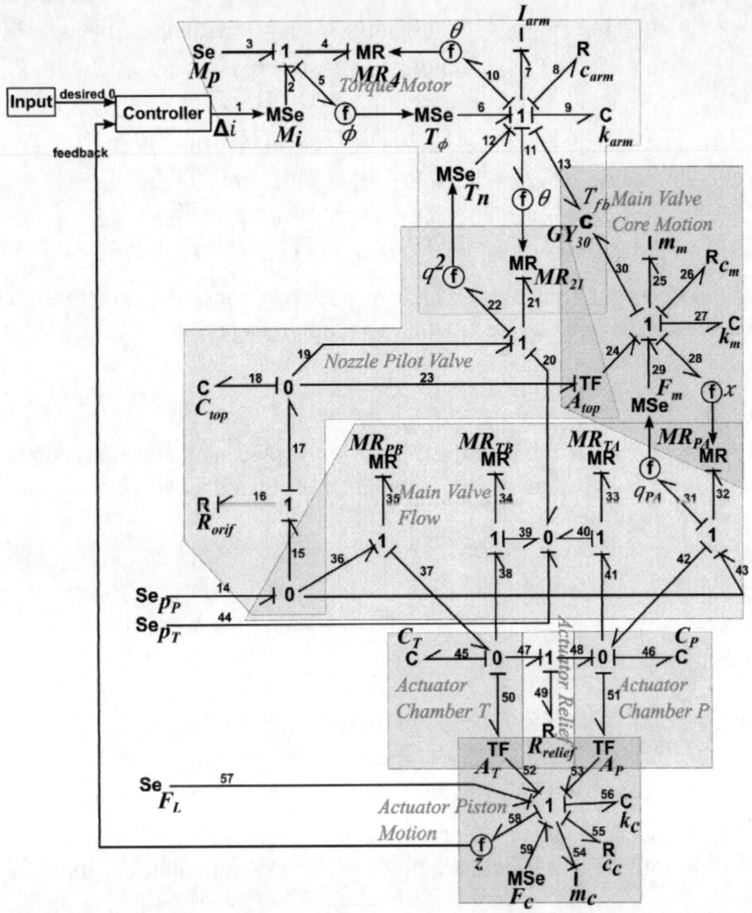

Figure 2 System bond graph, for the sake of observation, the magnetic air gap of the torque motor in the diagram has only one quadrant, the pilot valve only has the left half, and the fluid forces of the main valve only contain PA valve port.

As shown in the figure 2, the inputs of controller are the desired displacement (node 0) and feedback displacement (node 58) of the actuator, and the output of controller is the servo current (node 1) that inputted into the servo valve;

The servo valve contains torque motor, pilot valve and main valve;

The actuator contains two compression chambers and the piston;

The 0 junction connected with the nodes 14,15,36 and 43 represents the chamber of oil supply, while the 0 junction connected with the nodes 20,39,40 and 44 represents the chamber of oil return flow;

2.2. *Full model*

The full model is the summary of the nonlinear model, which written as:

$$\dot{\boldsymbol{x}} = f(\boldsymbol{x},\boldsymbol{u}), \boldsymbol{y} = H\boldsymbol{x} \tag{1}$$

The vector of independent variables $\boldsymbol{x}$, inputs $\boldsymbol{u}$ and outputs $\boldsymbol{y}$ are:

$$\boldsymbol{x} = [\dot{\theta} \quad \theta \quad p_{t,L} \quad p_{t,R} \quad \dot{x} \quad x \quad p_{CP} \quad p_{CT} \quad \dot{z} \quad z]^{\mathrm{T}}, \boldsymbol{u} = [\Delta i \quad F_L]^{\mathrm{T}}, \boldsymbol{y} = [z] \tag{2}$$

The variables $\boldsymbol{x}$ contains the rotational speed $\dot{\theta}$ and angle θ of torque motor, the pressure of left $p_{t,L}$ and right $p_{t,R}$ chambers of pilot valve, the speed $\dot{x}$ and displacement x of the main valve, the pressure of P p_{CP} and T p_{CT} chambers of the actuator, as well as the speed $\dot{z}$ and displacement z of the piston of the actuator.

The differential equations are

$$f = \begin{cases} \mathrm{I}_7\ddot{\theta} + \mathrm{R}_8\dot{\theta} + \mathrm{K}_\theta\theta = \mathrm{K}_{\Delta i}\Delta i + \mathrm{MSe}_{12}[q^2_{n,L|(\theta,p_{t,L})} - q^2_{n,R|(\theta,p_{t,R})}] + \mathrm{GY}_{13}x \\ \dot{p}_{t,L} = \mathrm{C}_{18}q_{L|(\theta,p_{t,L})} - \mathrm{C}_{18}\mathrm{TF}_{23}\dot{x} \\ \dot{p}_{t,R} = \mathrm{C}_{18}q_{R|(\theta,p_{t,R})} + \mathrm{C}_{18}\mathrm{TF}_{23}\dot{x} \\ \mathrm{I}_{25}\ddot{x} + \mathrm{R}_{26}\dot{x} + (\mathrm{C}_{27} + \mathrm{GY}_{30})x = \mathrm{TF}_{24}p_{t,L} - \mathrm{TF}_{24}p_{t,R} + \mathrm{GY}_{30}l_{fb}\theta + F_{m|(\dot{x},x,p_{CP},p_{CT})} \\ \dot{p}_{CP} + \mathrm{K}_\mathrm{C}p_{CP} = \mathrm{C}_{46}q_{P|(x,p_{CP})} + \mathrm{K}_\mathrm{C}p_{CT} - \mathrm{TF}_{50}\dot{z} \\ \dot{p}_{CT} + \mathrm{K}_\mathrm{C}p_{CT} = \mathrm{C}_{46}q_{T|(x,p_{CT})} + \mathrm{K}_\mathrm{C}p_{CP} + \mathrm{TF}_{50}\dot{z} \\ \mathrm{I}_{54}\ddot{z} + \mathrm{R}_{55}\dot{z} + \mathrm{C}_{56}z = \mathrm{TF}_{52}p_{CP} - \mathrm{TF}_{52}p_{CT} + F_L + F_{c|(\dot{z})} \end{cases} \quad (3)$$

The f contains 7 differential equations, which respectively represent armature swing, pressure of the left and right chambers of pilot valve, main valve moving, pressure of the P and T chambers of actuator, and actuator piston moving.

2.3. *Simplified Model*

Because the relationship between input current and the displacement response of value core of the servo valve is nearly linear, and its bandwidth is much larger than the cylinder response.

So the system can be simplified as:

$$\dot{\boldsymbol{x}}_s = \tilde{f}(\boldsymbol{x}_s, \boldsymbol{u}), \boldsymbol{y} = \tilde{H}\boldsymbol{x}_s \quad (4)$$

The vector of simplified independent variables $\boldsymbol{x}_s$, inputs $\boldsymbol{u}$ and outputs $\boldsymbol{y}$ are:

$$\boldsymbol{x}_s = [p_{CP} \quad p_{CT} \quad \dot{z} \quad z]^\mathrm{T}, \boldsymbol{u} = [\Delta i \quad F_L]^\mathrm{T}, \boldsymbol{y} = [z] \quad (5)$$

The expressions of simplified model are:

$$\tilde{f} = \begin{cases} x \cong \mathrm{K}_x\Delta i \\ \dot{p}_{CP} \cong \mathrm{C}_{46}\mathrm{K}_q x - \mathrm{TF}_{50}\dot{z} \\ \dot{p}_{CT} \cong -\mathrm{C}_{46}\mathrm{K}_q x + \mathrm{TF}_{50}\dot{z} \\ \mathrm{I}_{54}\ddot{z} + \mathrm{R}_{55}\dot{z} + \mathrm{C}_{56}z = \mathrm{TF}_{52}(p_{CP} - p_{CT}) + F_L + F_{c|(\dot{z})} \end{cases} \quad (6)$$

The K_C which represents leakage of the actuator is ignored, and the flow of main value to the cylinder chambers is linearized at the balance point of the system when the input is zero, as shown below:

$$q_P \cong \mathrm{K}_q x \, , q_T \cong -\mathrm{K}_q x \quad (7)$$

where the K_q the linearized flow coefficient of the valve opening, at the actuator's pressure balance point, which the leak is ignored and pressure is stable:

$$\mathrm{K}_q = C_d\pi(d_s + d_c/2)\sqrt{\rho_{oil}(p_P - p_T)} \quad (8)$$

3. Parameter Estimation

In the parameter analysis, the sine signal is chosen because the response is highly nonlinear and the features are easy to observe. The input signal is sine signal of 0 mean, 0.04m amplitude and 1Hz frequency.

3.1. *Parameter choosing*

The main valve opening parameters discussed in this paper include: valve core-seat gap d_c, valve core edge round radius r_c, and valve opening bias x_{bias}, which affect main valve flow. These parameters in the main valve is hard to measure, by the same time, they have great influence on the nonlinear characteristics of the response, so these parameters need to be discussed.

The actuator parameters discussed in this paper include: actuator coulomb force F_{cc}, actuator strict friction force F_{sc}, actuator inactive length l_{iac}, actuator damping c_c.

In practical use, the friction and damping coefficients will vary greatly with environmental parameters, oil deterioration and equipment wear. In addition, although stroke of the hydraulic cylinder can be measured, but connecting path between the servo valve and the hydraulic cylinder is very complex, at the same time, the hydraulic cylinder seal stiffness is much smaller than the piston and cylinder block itself, and it is difficult to measure.

3.2. *Estimation*

The cost function chosen in this paper for the parameter estimation is sum absolute error (SAE). The optimization method used to estimate the parameter is nonlinear least squares[27], and the algorithm is trust region reflective[28].

The iteration process is shown in the figure 3 below:

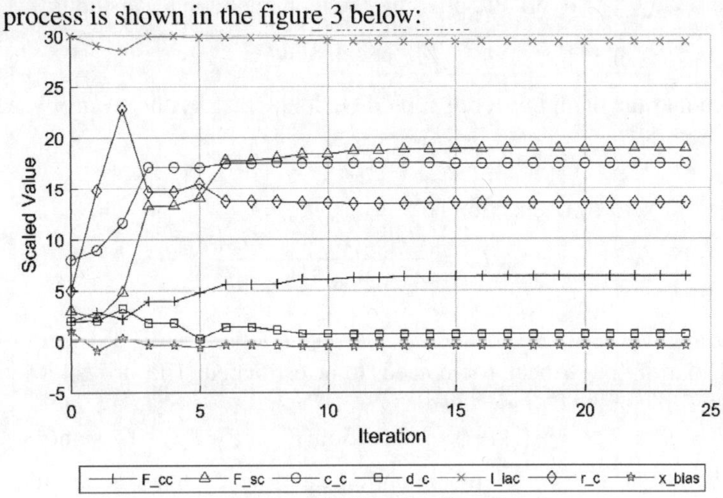

Figure 3 Estimation iteration

After 24 times of iteration for the parameter estimation, the parameters tend to be stable. The values in the **Figure 3** are scaled by the 'Scale gain' in the **Table 1** to distinguish values more clearly.

The initial, boundary and estimated value of the parameters are shown in the table 1 below:

Table 1 Estimated parameters

Parameter	Initial value	Boundary value	Estimated value	Scale gain	Dimension
$\boldsymbol{d_c}$	1×10-6	[1×10-6, +∞)	8.4918×10-6	5×10-7	m
$\boldsymbol{r_c}$	1×10-6	[1×10-7, +∞)	1.5238×10-6	2×10-7	m
$\boldsymbol{x_{bias}}$	1×10-6	[-6×10-6, 6×10-6]	-3.6659×10-8	1×10-6	m
$\boldsymbol{F_{cc}}$	200	[50, +∞)	1.0469×103	100	N
$\boldsymbol{F_{sc}}$	300	[100, +∞)	2.2019×103	100	N
$\boldsymbol{l_{iac}}$	0.15	[0.105, +∞)	1.4417×10-1	0.005	m
$\boldsymbol{c_c}$	8000	[2000, +∞)	1.610×104	1000	Ns/m

The time responses with initial and estimated parameters are shown in figure 5 and figure 6.

4. Control method

The sketch of the control is shown in the figure 4 below:

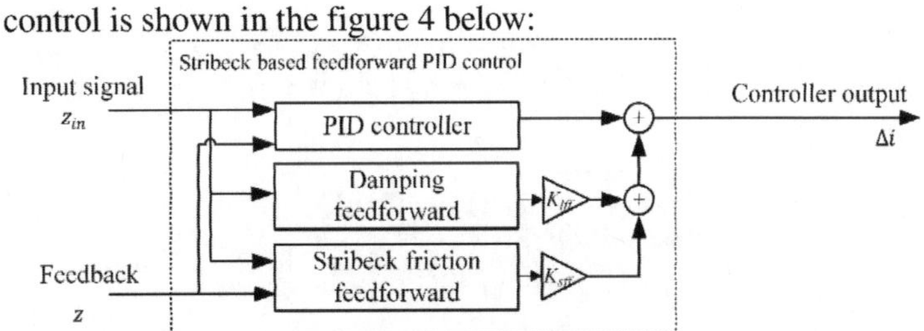

Figure 4 Controller sketch

The controller contains PID controller, damping feed forward and stribeck friction feed forward.

4.1. *Feedforward*

When the time of force is short, then the movement of cylinder piston could be ignored, so the pressure function of the chambers of the cylinder can be simplified as:

$$\dot{p}_{CP} - \dot{p}_{CT} = 2\mathrm{C}_{46}\mathrm{K}_q x \tag{9}$$

Therefore, the steady state gain between the integration of the input current and the hydraulic pressure force on the cylinder piston is:

$$\mathrm{K}_p = 2\mathrm{TF}_{52}\mathrm{C}_{46}\mathrm{K}_q\mathrm{K}_x \tag{10}$$

Which means the steady state hydraulic pressure force on the cylinder piston is

$$F_p = \mathrm{K}_p \int \Delta i \, \mathrm{d}t \tag{11}$$

The stribeck force and the damping force should be counteract by the feedforward pressure force, as shown below:

$$F_{psc} = -F_c + c_{visc}\dot{z} \tag{12}$$

Therefore, the feedforward compensation is:

$$\Delta i_{sc} = \frac{1}{2\mathrm{TF}_{52}\mathrm{C}_{46}\mathrm{K}_q\mathrm{K}_x}\frac{\mathrm{d}F_{psc}}{\mathrm{d}t} \tag{13}$$

4.2. *Stribeck estimation*

For the feedforward control, the stribeck force needs to be estimated. The stribeck friction is a function of speed and resultant force of the cylinder piston,

$$F_c = f_c(\dot{z}, F_{oc}) \approx -[(1 - \xi_{fc})\mathrm{saturation}(F_{oc}, -F_{sc}, F_{sc}) + \xi_{fc}\mathrm{sign}(\dot{z})\, F_{cc}] \tag{14}$$

where the ξ_{fc} is stribeck transition between column force and steady state friction force,

$$\xi_{fc} = [\tanh(\mathrm{abs}(\dot{z})/v_c - 1) + 1]/2 \tag{15}$$

and the F_{oc} is the resultant force of the actuator except the friction force. If do not consider the load force, the resultant force is pressure force as expressed in (12).

So the stribeck friction force is:

$$F_c = f_c(\dot{z}, F_{oc}) \cong f_c(\dot{z}, F_p) \tag{16}$$

4.3. *Delay between signal and response*

Generally, the delay between signal and response is constant, and can be expressed as:

$$z = e^{-\tau s} z_{in} \tag{17}$$

where the τ is delay time, z_{in} is input signal.

For feedforward control, the response is unknown, and can be replaced by delayed input signal if the delay can be measured.

4.4. *Control method summery*

The control output of the controller is the summery of PID output and feedforward:

$$\Delta i = \mathrm{PID}(z - z_{in}) + \Delta i_{sc} \tag{18}$$

where the PID() is:

$$\mathrm{PID}(\Delta z) = K_p \Delta z + K_i \int \Delta z \mathrm{d}t + K_d \frac{\mathrm{d}\Delta z}{\mathrm{d}t} \tag{19}$$

and the feedforward output is:

$$\Delta i_{sc} = \frac{1}{2\mathrm{TF}_{52}\mathrm{C}_{46}\mathrm{K}_q\mathrm{K}_x}\frac{\mathrm{d}}{\mathrm{d}t}\left[-f_c\left(e^{-\tau s}\dot{z}, \mathrm{K}_p \int \Delta i \, \mathrm{d}t\right) + c_{visc} e^{-\tau s}\dot{z}\right] \tag{20}$$

5. Results

5.1. *Test Settings*

For vibration and fatigue test equipment, sine signal is often used, so it is chosen as input signal.

The input is sine signal of 0 mean, 0.04m amplitude and 1Hz frequency.

In order to observe the contrast between the responses and the input signal, the simulation and experiment curve are translated by -0.016s. The PID parameters for the simulation and experiment are [0.75 0.051 0.00049] which separately are the gain of proportion, integration and derivation.

5.2. *Response with initial and estimated parameters*

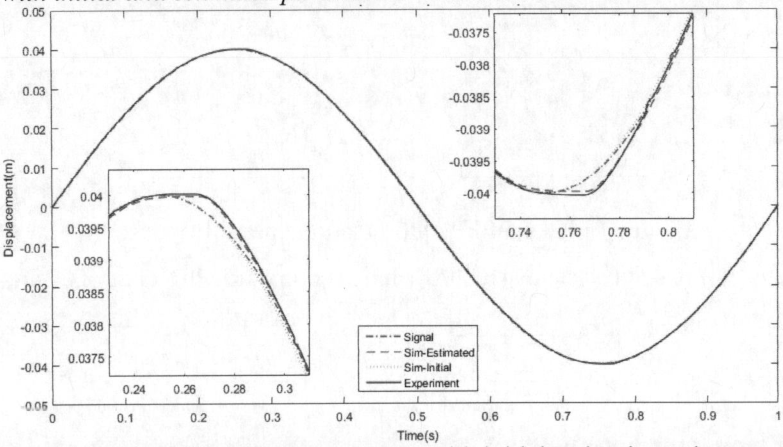

Figure 5 Actuator displacement response with initial and estimated parameters

As shown in figure 5, the displacement response of the experiment has ceiling phenomenon; the simulation response with initial parameter is very close to signal but different from the experiment response, while the simulation response with estimated parameter is close to the experiment results.

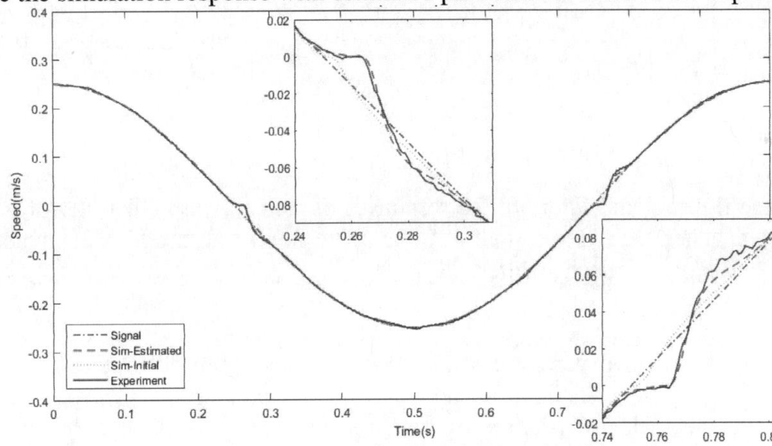

Figure 6 Actuator speed response with initial and estimated parameters

As shown in figure 6, the speed response of the experiment has dead-zone phenomenon; the dead-zone phenomenon of the simulation response with initial parameters is very small, while the simulation response with estimated parameter is close to the experiment.

5.3. *Response with PID and feed forward PID controller*

The simulation responses with PID and feedforward PID controller are shown in the figure 7 below:

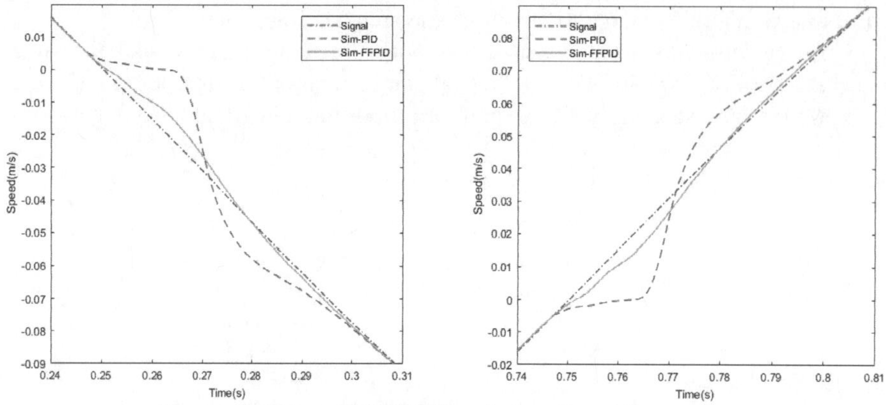

Figure 7 PID and FFPID Simulation Speed Response

The experiment response of system with PID and feedforward PID controller are shown in the figure 8 below:

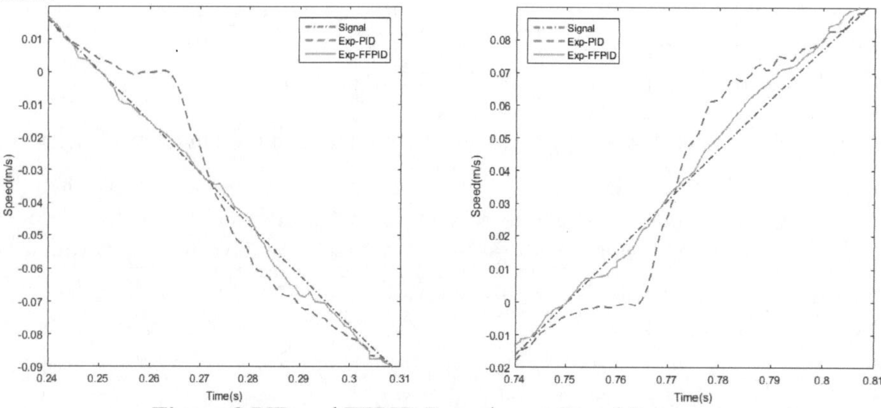

Figure 8 PID and FFPID Experiment Speed Response

As shown in the figure 7 and figure 8, the responses of feed forward (FF) PID of both simulation and experiment are better than the normal PID, and the nonlinear dead-zone phenomenon of the response near the zero point is much smaller.

5.4. *Discussion*
The results of simulation and experiment shows that the model can accurately describe the actual equipment, and after parameter estimation, the response of simulation is close to the experiment results. Because of the stribeck friction of actuator piston, displacement tracking has ceiling phenomenon and speed tracking has dead-zone phenomenon. When the input is small, the hydraulic drive force of actuator is very small. After actuator stops moving, it needs to overcome static friction force that is bigger than sliding friction to start moving again.

As shown in the results, the nonlinear response with initial parameters is very different from the experiment, and the response with estimated parameters is close to the experiment, which means the parameter estimation is usable.

With the estimated parameters and the stribeck friction feed forward, the nonlinear dead-zone phenomenon of the response is much smaller than the original PID control. Though the feedforward PID control cannot fully eliminate the nonlinear characteristics, because the steady state assumption of the feedforward controller. However, considering computation requirement and availability on the actual test bench, the feedforward controller in this paper is valuable and useful.

6. Conclusion

In this paper, the physical structure and energy flow of the typical hydraulic servo system are described in the Bond Graph, and with the graph, a 10 order nonlinear physical model are built. The nonlinear elements mainly include the flow nonlinearity of the valve port, the nonlinear opening of the valve, and the stribeck nonlinear friction of the hydraulic actuator.

The nonlinear model is simplified based on the difference of the frequency response bandwidth of the servo valve and actuator. Some parameters of the simplified model that are important and hard to measure are estimated based on the speed response of the experiment. The simulation and experiment results of estimation and feedforward control is presented.

To sum up, in this paper, and a high order nonlinear hydraulic servo model is set up, and is consistent with actual situation. The model response with estimated parameters can fit the experimental data very well, and the simulation and experiment response of the optimized control can eliminate the nonlinear response very well. Therefore, the nonlinear model and control method used in this paper have a guiding significance and application value, and have theory value for the further research of hydraulic servo system with the similar structure.

7. Declaration on conflict of interest

The authors declare that there is no conflict of interests regarding the publication of this article.

References

[1] P. Xu, D. Wong, P. LeBlanc, and G. Peticca, "Road Test Simulation Technology in Light Vehicle Development and Durability Evaluation," in *2005 SAE World Congress* Detroit, Michigan: SAE International, 2005, p. 12.

[2] C. J. Dodds and A. R. Plummer, "Laboratory Road Simulation for Full Vehicle Testing: A Review," in *Symposium on International Automotive Technology 2001*: SAE International, 2001, pp. 487-494.

[3] B. Cryer, P. Nawrocki and R. Lund, "A Road Simulation System for Heavy Duty Vehicles," in *SAE Technical Paper* 1976 Automotive Engineering Congress and Exposition: SAE International, 1976.

[4] A. Akers, *Hydraulic power system analysis*. Boca Raton, Fla.: Boca Raton, Fla. : CRC/Taylor & Francis, 2006.

[5] D. Van Den Bossche, "The A380 flight control electrohydrostatic actuators, achievements and lessons learnt," in *25th international congress of the aeronautical sciences*, 2006, pp. 1-8.

[6] M. Inc., "Electrohydraulic Valves... A Technical Look,". vol. 2018, 2018.

[7] F. Yeaple, "Hydraulic and Electrohydraulic Cylinders and Actuators," in *Fluid Power Design Handbook* New York: Marcel Dekker, Inc., 1995, pp. 111-115.

[8] H. E. Merritt, *Hydraulic control systems*. New York: John Wiley and Sons, Inc., 1967.

[9] J. Wang, H. Zhang and Y. Huang, *Hydraulic and Pneumatic Transmission*, 2 ed. Beijing: China Machine Press, 2005.

[10] S. Zhou, X. Cao and Y. Cen, "Optimal Control of Speed Conversion of a Valve Controlled Cylinder System," *Journal of Dynamic Systems, Measurement, and Control,* vol. 113, pp. 691-695, 1991-12-01 1991.

[11] J. T. William, *Transfer functions for moog servovalves*. New York: Moog Controls Division, 1965.

[12] K. Dasgupta, J. Watton and S. Pan, "Open-loop dynamic performance of a servo-valve controlled motor transmission system with pump loading using steady-state characteristics," *Mechanism and Machine Theory,* vol. 41, pp. 262-282, 2006-01-01 2006.

[13] M. Kalyoncu and M. Haydim, "Mathematical modelling and fuzzy logic based position control of an electrohydraulic servosystem with internal leakage," *Mechatronics,* vol. 19, pp. 847-858, 2009-01-01 2009.

[14] G. Jacazio and L. Borello, "Mathematical Modelling and Analysis for the Air Force," *Mathematical Computational Modelling,* vol. 11, pp. 563-569, 1988.

[15] D. H. Kim and T. Tsao, "A Linearized Electrohydraulic Servovalve Model for Valve Dynamics Sensitivity Analysis and Control System Design," *Journal of Dynamic Systems, Measurement, and Control,* vol. 122, pp. 179-187, 1998-02-02 1998.

[16] D. H. Kim and T. Tsao, "An improved linearized model for electrohydraulic servovalves and its usage for robust performance control system design," in *Proceedings of the 1997 American Control Conference*. vol. 6 Albuquerque, New Mexico, 1997, pp. 3807-3808.

[17] C. Liu and H. Jiang, "A seventh-order model for dynamic response of an electro-hydraulic servo valve," *Chinese Journal of Aeronautics,* vol. 27, pp. 1605-1611, 2014-01-01 2014.

[18] N. Hori, A. S. Pannala, P. R. Ukrainetz, and P. N. Nikiforuk, "Design of an Electrohydraulic Positioning System Using a Novel Model Reference Control Scheme," *Journal of Dynamic Systems, Measurement, and Control,* vol. 111, pp. 292-298, 1989-06-01 1989.

[19] S. R. Lee and K. Srinivasan, "Self-Tuning Control Application to Closed-Loop Servohydraulic Material Testing," *Journal of Dynamic Systems, Measurement, and Control,* vol. 112, pp. 680-689, 1990-12-01 1990.

[20] D. Sha, V. B. Bajic and H. Yang, "New model and sliding mode control of hydraulic elevator velocity tracking system," *Simulation Practice and Theory,* vol. 9, pp. 365-385, 2002-01-01 2002.

[21] A. Ylinen, H. Marjamäki and J. Mäkinen, "A hydraulic cylinder model for multibody simulations," *Computers and Structures,* vol. 138, pp. 62-72, 2014-01-01 2014.

[22] K. Li, Z. Lv, K. Lu, and P. Yu, "Thermal-hydraulic Modeling and Simulation of the Hydraulic System based on the Electro-hydrostatic Actuator," *Procedia Engineering,* vol. 80, pp. 272-281, 2014-01-01 2014.

[23] M. Woydt and R. Wäsche, "The history of the Stribeck curve and ball bearing steels: The role of Adolf Martens," *Wear,* vol. 268, pp. 1542-1546, 2010-01-01 2010.

[24] K. M. Elbayomy, J. Zongxia and Z. Huaqing, "PID Controller Optimization by GA and Its Performances on the Electro-hydraulic Servo Control System," *Chinese Journal of Aeronautics,* vol. 21, pp. 378-384, 2008-01-01 2008.

[25] K. J. Åström and T. Hägglund, "The future of PID control," *Control Engineering Practice,* vol. 9, pp. 1163-1175, 2001-01-01 2001.

[26] M. Inc., "Servo Valves Pilot Operated Flow Control Valve with Analog Interface G761/761 Series, Size 04,", 2004.

[27] Jr. J. E. Dennis, *"Nonlinear Least-Squares." State of the Art in Numerical Analysis*: Academic Press.

[28] T. F. Coleman and Y. Li, "An Interior Trust Region Approach for Nonlinear Minimization Subject to Bounds," *SIAM Journal on Optimization,* vol. 6, pp. 418-445, 1996-05-01 1996.

Adaptive Neuro-Fuzzy Inference System Based Orientation Control of an Intra-operative Ultrasound Robot

Shuangyi Wang[1], James Housden[1], Davinder Singh[2] and Kawal Rhode[1]

[1] School of Biomedical Engineering & Imaging Sciences, King's College London
[2] Xtronics Ltd., Gravesend, Kent, UK

shuangyi.wang@kcl.ac.uk

Abstract. Trans-esophageal echocardiography (TEE) is a miniatured intra-operative ultrasound system, widely used in routine diagnosis and interventional procedure monitoring, to assess cardiac structures and functions. As a way to assist the operation of TEE remotely, we have developed an add-on robotic system to actuate a commercial TEE probe. For the proposed robot, understanding the inverse kinematics (IK) which relates the probe pose to the joint parameters is the fundamental step towards automatic control of the system. Rather than using conventional numerical-based techniques which may have problems with speed, convergence, and stability when applying to the TEE robot, this paper explores a soft computing approach by constructing an Adaptive Neuro-Fuzzy Inference System (ANFIS) to learn from training data generated by the forward kinematics (FK) and then computing the inverse kinematics in order to control the orientation of the TEE probe. With 1900 training data over 40 epochs, the minimum training error for each joint parameter was found to be less than 0.1 degree. Validation using a separate data set has indicated that the maximum error was less than 0.3 degree for each joint parameter. It is therefore concluded that the ANFIS-based approach is an effective way, with acceptable accuracy, to compute the inverse kinematics of the TEE robot.

1. Introduction

Trans-esophageal echocardiography (TEE) is an important type of intra-operative ultrasound system that has been widely utilized for cardiac disease diagnosis and interventional procedure guidance [1, 2]. TEE uses a cable actuated steerable probe system, in which the bi-directional bending of the distal rigid end of the probe in a bending plane is controlled by two independently rotatable control knobs on the handle of the system. The ultrasound transducer is embedded into the distal rigid end of the probe, providing unique features of this probe for imaging and assessing heart structure and function via esophagus. To facilitate the use of TEE system, we have developed the first add-on robot [3] working with a commercial TEE probe to enable remote control of the system, and therefore keeping the operator (known as echocardiographer) away from radiation exposure when TEE is used in conjunction with X-ray system during interventional procedure guidance. The robotic system has four degrees of freedom (DOFs) mechanically, including the translation of the probe along the esophagus, rotation of the probe about the long axis of the handle, and bi-directional bending of the probe tip at the distal end of the robot-probe system, similar to the manual control of the probe.

Working towards automation, understanding the kinematics of the TEE robot is a fundamental step. In our previous work, a model for forward kinematics (FK), which relates the joint parameters to the location of the probe tip, i.e., its position and orientation, was descripted and validated experimentally.

While the forward kinematic model is established, the inverse kinematics (IK) problem, which aims to calculate the joint parameters when giving a target probe tip location, is found to be a challenging task, since the mapping between Cartesian space and joint space is non-linear in nature and involves redundancy problem with multiple solutions caused by the combination use of handle rotation axis and bi-directional bending axes of the TEE robot in adjusting the orientation of the probe tip.

While the exact IK solution of the TEE robot is seldom covered in the literature, several existing works have been presented for the similar types of robot where the actuated device is similar to the TEE probe mechanically, such as the intra-cardiac ultrasound catheter robot [4], the endoscopic robot for natural orifice transluminal surgery [5], and the cardiac ablation catheter robot [6]. Among these works, the most common approach for solving the IK problem is to use Jacobian-based inverse technique [7], which uses the iterative optimization based on the differential changes between the joint parameters and the end-effector's location in the Cartesian space. In addition to these works, we also explored the gradient-descent search scheme in solving the IK problem for the TEE robot in our previous work described in [8]. Although these methods have demonstrated successful solution to the IK problem, numerical-based iteration methods are time consuming and do not guarantee convergence to an accurate solution.

In this paper, we aim to explore a soft computing approach by constructing an Adaptive Neuro-Fuzzy Inference System (ANFIS) [9-11] to learn from training data generated by the FK of the TEE robot and then compute the IK using the trained system. With the recent works using ANFIS to test the arm-based robotic systems, such as described in [12-16], this approach has been proved to be an efficient way in solving IK problem and would provide a faster solution without the need of numerical iterations in real-time application after the network is trained. Since the use of ANFIS to solve the IK problem for the endoscopic robot has not been clearly reported, the objective of this paper is to test the effectiveness and accuracy of the method in solving IK problem for the TEE robot, which also provides the feasibility of using such method in solving other similar types of endoscopic robots.

2. FK of the TEE robot

The developed add-on TEE robot (Figure 1) holds and manipulates the probe handle with a commercial TEE probe (x7-2t, Philips, Amsterdam, The Netherlands) plugged into the robot. The system comprises three structures [3]: the handle control structure to rotate the two knobs in an actuating chamber; a probe rotation mechanism for rotating the shaft about the long axis of the probe; and a linear belt mechanism mounted on a rail to provide translation movement.

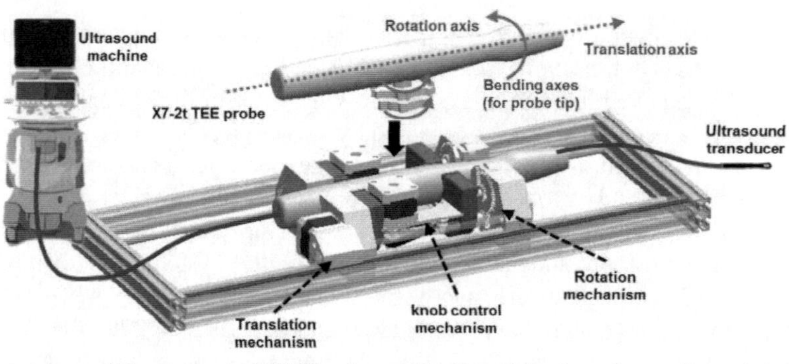

Figure 1. Overview of the proposed TEE robot with the original probe and moving axes shown.

With the translation of the probe modelled by an interpolation of the esophagus's centre line, the probe base coordinates could be easily decided mathematically. The FK then models rotational movements of the TOE probe handle with the rotation angle v, which can be described by a standard rigid transformation. The bi-directional bending of the probe tip has been previously described in [3]. The inputs are the rotation angle of the first bending knob, φ_x, which controls the bending tip pitch in the posterior-anterior plane; and the rotation angle of the second bending knob, φ_y, which controls the bending tip yaw in the left-right plane. The geometric illustration of the FK of the TEE robot is shown in Figure 2. The transformation matrix between the probe tip coordinates and the probe base coordinates, after translation of the probe, is expressed as:

$$\mathbf{T}=\begin{bmatrix} C_t(C_\beta C_\alpha^2+S_\alpha^2)-C_\alpha S_\alpha S_t(C_\beta-1) & C_\alpha S_\alpha C_t(C_\beta-1)-S_t(C_\alpha^2+C_\beta S_\alpha^2) & C_\alpha S_\beta C_t-S_\alpha S_\beta S_t & C_t(L_uC_\alpha S_\beta-(L_fC_\alpha(C_\beta-1)/\beta)-S_t(L_uS_\alpha S_\beta-(L_fS_\alpha(C_\beta-1)/\beta) \\ S_t(C_\beta C_\alpha^2+S_\alpha^2)+C_\alpha S_\alpha C_t(C_\beta-1) & C_\alpha S_\alpha S_t(C_\beta-1)+C_t(C_\alpha^2+C_\beta S_\alpha^2) & C_\alpha S_\beta S_t+S_\alpha S_\beta C_t & S_t(L_uC_\alpha S_\beta-(L_fC_\alpha(C_\beta-1)/\beta)+C_t(L_uS_\alpha S_\beta-(L_fS_\alpha(C_\beta-1)/\beta) \\ -C_\alpha S_\beta & -S_\alpha S_\beta & C_\beta & L_uC_\beta+(L_fS_\beta/\beta) \\ 0 & 0 & 0 & 1 \end{bmatrix} \quad (1)$$

where α is the angle between the bending plane and the X-Z plane, β is the bending angle in the bending plane, L_f is the length of the bending section of the probe tip, and L_u is the length of the rigid section of the probe tip. In the matrix, *Sx* and *Cx*, respectively, denote *sin*(*x*) and *cos*(*x*).

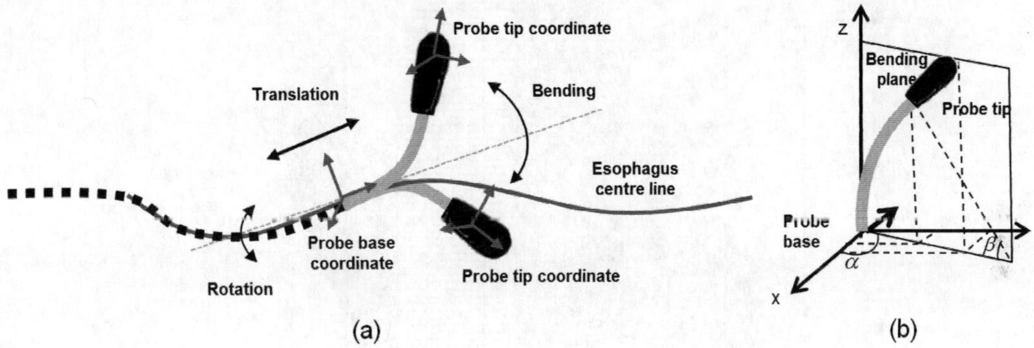

Figure 2. (a) Overall geometric illustration of the movements of the TEE robot modelled in the FK. (b) The projection view in the bending plane for the bi-directional bending axes.

3. Proposed ANFIS architecture

ANFS is a combination of two soft-computing methods of artificial neural network (ANN) and fuzzy inference system (FIS), which constructs an adaptive network that uses supervised learning on learning algorithm. By using ANN to automatically adjust parameters of membership functions (MFs) of fuzzy inference systems, the network brings learning capabilities of neural networks to fuzzy inference systems. Using ANFIS, the inputs are mapped through input MFs and associated parameters, and then through output MPs and associated parameters to outputs. The parameters associated with the MFs changes through the learning process. The computation of these parameters is facilitated by back propagation algorithm alone or in combination with a least squares type of method. The structures of the ANFIS network can be divided into fuzzification, inference, implication, and aggregation:

- *Fuzzification*: Every node in this layer adapts to a function parameter. The output from each node is a degree of membership value that is given by the input of the membership functions.
- *Inference*: Every node in this layer is a fixed or non-adaptive node. In each node, the result of multiplying of signal coming into the node is firstly calculated, which represents the firing strength for each rule. Then the ratio between the i-th rules firing strength and the sum of all rules' firing strengths, known as the normalized firing strength, is the output from each node.
- *Implication*: Every node in this layer is an adaptive node to an output. The output from each node is decided by the output membership function.
- *Aggregation*: The single node in this layer is a fixed or non-adaptive node that computes the overall output as the summation of all incoming signals from the previous node.

In the specific application of TEE, where the probe is constrained by the esophagus (less than 2 cm in diameter) and orientated towards the heart for ultrasound imaging, the translation movement along the esophagus can be easily decoupled and compensated from the kinematics model mathematically. As for the remaining Cartesian parameters of the probe tip location: Trans-X, Trans-Y, Rot-X, Rot-Y, and Rot-Z, the control of the orientation of the probe tip is important for visualizing the heart, while the translation offsets due to the orientation change are small and have less impacts to the ultrasound visualization with the constraint from the esophagus. Therefore, the inputs of the considered ANFIS network are set to be the three rotation angles representing the orientation of the probe tip and the outputs are the joint parameters, the handle rotation angle v and the rotation angle of the bending knobs φ_x and φ_y. In total, the ANFIS architecture consists of three ANFIS networks, with one for each joint parameter. In the network, four input membership functions are utilized based on our experimental trials. Accordingly, 64 numbers of fuzzy rules and output membership functions are constructed. The structure of the considered ANFIS network is shown in Figure 3.

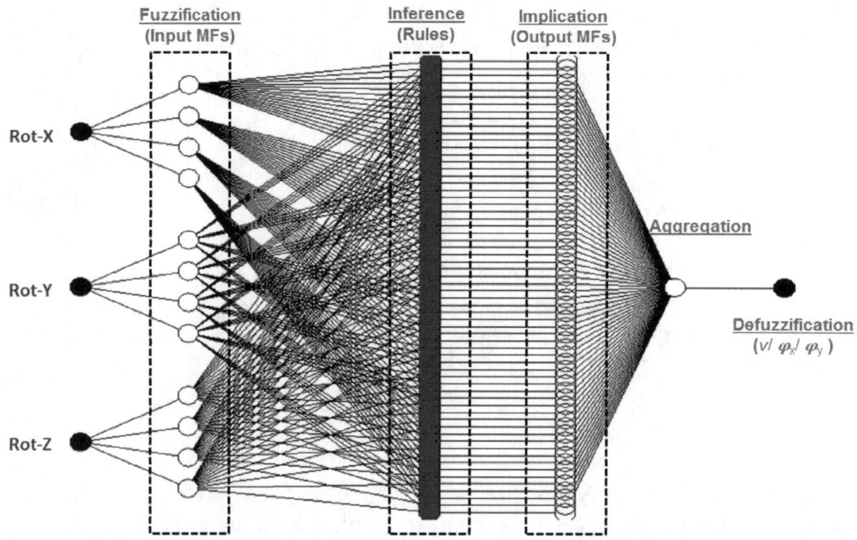

Figure 3. Proposed ANFIS architecture for computing the IK problem of the TEE robot.

4. Training and validation

Using the FK solution described by formula (1), training data (1900 sets) are generated for different combinations of the robot joint parameters (axial handle rotation angle and bi-directional bending angles) at certain intervals within the working space and the resulted probe tip orientation (three decomposed rotation angles). After our experimental trials, 40 training epochs were selected for training of each ANFIS network. The change of the training error after each epoch is shown in Figure 4. The step size for the training process of each network has been tuned to an optimal profile (Figure 4), which increases initially, reach a maximum, and then decrease for the rest of the training.

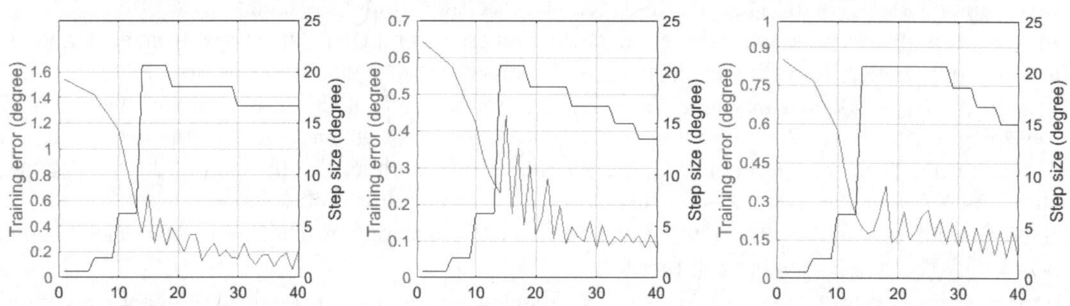

Figure 4. Change of training error (in red) and step size (in blue) over training epochs for each ANFIS network. From left to right: probe axial rotation, probe left-right bending, and probe up-down bending.

To validate the performance of the ANFIS-based IK after each network is trained, a separate validation data set is created using the FK model with 100 randomly generated probe tip locations within the workspace. The according joint parameters were recorded and treated as the desired outputs of the IK solution. The three decomposed rotation angles were fed into each ANFIS network and the combined outputs will result in a set of output joint parameter. The differences between the desired joint parameters and the resulted output joint parameter from the ANFIS networks are treated as the errors of the proposed method, which are summarized in Figure 5.

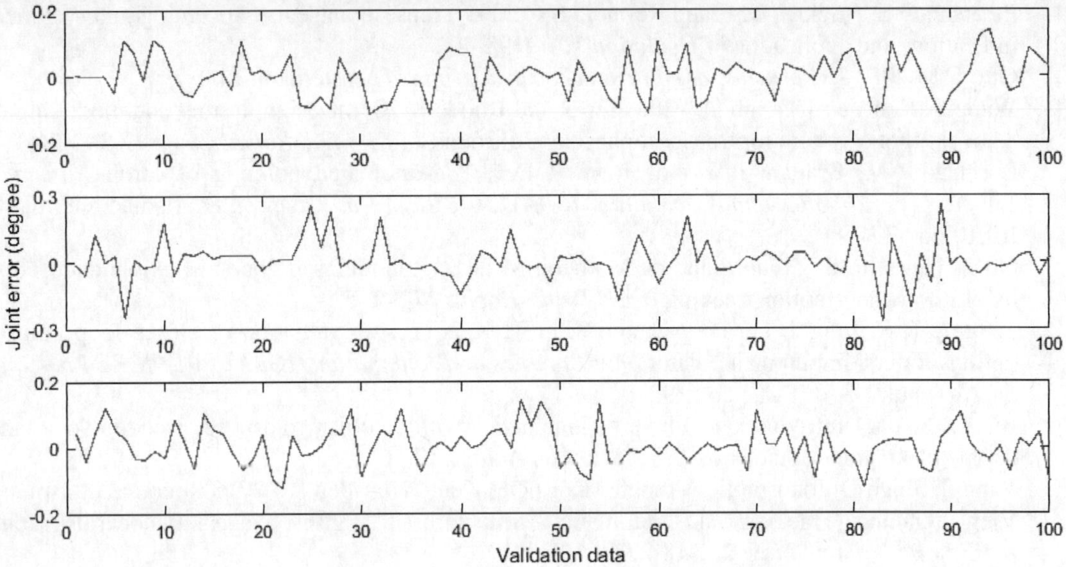

Figure 5. Joint errors for each ANFIS network validated using a separate data set. From top to bottom: probe axial rotation, probe left-right bending, and probe up-down bending.

5. Discussion and Conclusion

This paper introduces an ANFIS-based IK solution of a newly developed TEE robot, aiming at control the orientation of the TEE probe to visualize the heart. Three ANFIS networks were established with orientation of the probe tip as the input and the three joint parameters as the outputs. The networks were successfully trained using the data set generated by the FK solution with errors converged to the values less than 0.1 degree. A separate randomly generated data set was used for validation with maximum errors found to be less than 0.3 degree for all the joint parameters. The ANFIS-based IK solution has shown a reliable performance with an acceptable error range. Compared with the conventional numerical-based IK solutions, the ANFIS-based soft computing approach do not require heavy interactive computing for calculation of the joint parameters once the networks are trained. Therefore, such solution would be ideal for real-time applications, such as the feedback control to automatically adjust the TEE probe in order to keep track of a moving structures or interventional devices. Compared with geometry-based analytical IK solutions which are usually difficult to derive, the ANFIS-based solution is easy to implement with the training data generated using the FK, thus providing a more generic solution. This paper has verified the effeteness and accuracy for the ANFIS-based IK method used for a TEE robot. Considering the similar mechanical configuration, it is believed that several other types of endoscopic robots could use the same solution for calculating IK.

Based on the results from the training and the validation experiments, it is concluded that the ANFIS-based IK solution is feasible and accurate for control the orientation of the TEE robot. Future works will focus on performing more task-driven experiments to further validate the method and also incorporating the additional DOF of ultrasound image steering controlled by the ultrasound system into the ANFIS networks to provide an extended IK solution to the image space.

Acknowledgments
This work was supported by the Wellcome/EPSRC Centre for Medical Engineering [WT 203148/Z/16/Z]. This research was also supported by the KCL NIHR Healthcare Technology Centre,

KCL-China Scholarship Scheme, the National Institute for Health Research (NIHR) Biomedical Research Centre at Guy's and St. Thomas' NHS Foundation Trust and King's College London.

References

[1] Peterson G E, Brickner M E and Reimold S C 2003 Transesophageal echocardiography: clinical indications and applications *Circulation* **107** 2398-402
[2] Otto C M 2012 *The practice of clinical echocardiography* (Amsterdam: Elsevier)
[3] Wang S, Housden J, Singh D, Althoefer K and Rhode K 2016 Design, testing and modelling of a novel robotic system for transoesophageal ultrasound *Int. J. Med. Robot. Comp* **12** 342-54
[4] Loschak P M, Brattain L J and Howe R D 2013 Automated pointing of cardiac imaging catheters. In: *Robotics and Automation (ICRA), 2013 IEEE Int. Conf. on* (Karlsruhe, Germany: IEEE) pp 5794-9
[5] Ott L, Nageotte F, Zanne P and de Mathelin M 2011 Robotic assistance to flexible endoscopy by physiological-motion tracking *IEEE Trans. Robot.* **27** 346-59
[6] Penning R S, Jung J, Ferrier N J and Zinn M R 2012 An evaluation of closed-loop control options for continuum manipulators. In: *Robotics and Automation (ICRA), 2012 IEEE Int. Conf. on,* (St. Paul, USA: IEEE) pp 5392-7
[7] Buss S R 2004 Introduction to inverse kinematics with jacobian transpose, pseudoinverse and damped least squares methods *IEEE J. Robot. Autom.* **17** 16
[8] Wang S, Singh D, Johnson D, Althoefer K, Rhode K and Housden R J 2016 Robotic Ultrasound: View Planning, Tracking, and Automatic Acquisition of Transesophageal Echocardiography *IEEE Robot. Autom. Mag.* **23** 118-27
[9] Jang J-S 1993 ANFIS: adaptive-network-based fuzzy inference system *IEEE Trans. Syst., Man, Cybern.* **23** 665-85
[10] Takagi T and Sugeno M 1993 *Readings in Fuzzy Sets for Intelligent Systems,* (Amsterdam: Elsevier) pp 387-403
[11] Wang L-X 1993 Stable adaptive fuzzy control of nonlinear systems *IEEE Trans. Fuzzy Syst.* **1** 146-55
[12] Howard D W and Zilouchian A 1998 Application of fuzzy logic for the solution of inverse kinematics and hierarchical controls of robotic manipulators *J. Intell. Robot. Syst.* **23** 217-47
[13] Wei L-X, Wang H-R and Li Y 2003 A new solution for inverse kinematics of manipulator based on neural network. In: *Machine Learning and Cybernetics, 2003 Int. Conf. on,* (Xi'an, China: IEEE) pp 1201-3
[14] Alavandar S and Nigam M 2008 Inverse kinematics solution of 3DOF planar robot using ANFIS *Int. J. Comput. Commun.* **3** 150-5
[15] Aghajarian M and Kiani K 2011 Inverse kinematics solution of PUMA 560 robot arm using ANFIS. In: *Ubiquitous Robots and Ambient Intelligence (URAI), 2011 8th Int. Conf. on,* (IFEZ Songdo, Korea: IEEE) pp 574-8
[16] Duka A-V 2015 ANFIS based Solution to the Inverse Kinematics of a 3DOF planar Manipulator *Procedia Technology* **19** 526-33

Research of High Precision Humidity Fuzzy Control Algorithm of One Kind Humidity Verification Device

Guangda Lu, Qiuyue Zhang, Tongtong Qie and Qihui Feng

Key Laboratory of information sensing and intelligent control, Tianjin University of Technology and Education. Tianjin 300222

E-mail: Lugd1229@163.com

Abstract. Aiming at different accuracy requirements of humidity in a variety of environment, a fuzzy control algorithm is applied to design a humidity fuzzy control system. Based on the quantification of humidity, a membership function, fuzzy rules and a rule table are established. The experimental results show that this method has high control accuracy, small overshoot and good stability, and achieves high-precision intelligent control of humidity.
Key words. high precision, humidity, fuzzy control, intelligence.

1. Introduction

At present, humidity control has been widely used in various fields such as agricultural production, industrial manufacturing, and daily life in China. The temperature and humidity test is an experiment that evaluates the performance of a product placed in an artificially set specific environment[1]. Chen Shucheng proposed that vegetables in the agricultural greenhouse growth process, its output and quality directly affected by temperature, humidity and other parameters, the design completed a temperature and humidity fuzzy control system, and make the actual observation[2] of the temperature and humidity control effect. Li Shuangxiu proposed that although the living standards of the people are significantly improved, the understanding of air quality and the degree of understanding of pollutants purging have lagged far behind of western countries. Through the use of fuzzy mathematics and grey theory to evaluate air quality, a focused study of air purification methods has been conducted[3]. He Wenbo et al. proposed that some warehouses are located in a complex environment, with high temperature and high humidity, large variations in temperature difference between day and night, and difficulty in control of temperature and humidity. An adaptive fuzzy PID controller that can adjust parameters online is designed to have higher control accuracy and stronger ability to adapt to the changes in the environment[4]. Ruilan Wang proposed to control the hatching system through fuzzy control algorithm. This algorithm can improve the robustness of control algorithm and self-adaptive ability[5].J.Y. Zheng obtained experimentally that DBHCBM can control the humidity performance better, and the porosity is smaller, providing the experimental basis for the preparation of high-performance humidity control materials[6]. Xiao Min Shan applied fuzzy inference to the greenhouse environmental control system and established a fuzzy PID system with temperature and humidity as control variables. Through simulation, the PID system has a fast response speed and strong anti-interference ability[7]. Liu Jinhua et al proposed to adopt the fuzzy control algorithm to solve the requirements of temperature and humidity control of the equipment warehouse. Effectively solves the adverse effects of strong coupling and large time delay on the control process and improves the control accuracy of the system[8].

Therefore, research and development of highly intelligent humidity control systems with superior performance, reliable operation, and low cost are hot topics of our research[9]. Because the humidity control system is a non-linear, strong coupling, multi-interference, time-delay system, the relationship during the production process is complex, and it is difficult to control the humidity. It is difficult to establish an accurate data model and control it. In recent years, the fuzzy control algorithm has the characteristics of higher control precision because it does not need to know the exact mathematical model of the controlled object. Therefore, the fuzzy control algorithm has been widely researched and used in dealing with the problems of nonlinearity and model uncertainty in complex systems. Based on this, this article has carried on the thorough research to the humidity control algorithm, has designed a set of humidity fuzzy control system, has completed the application of the fuzzy control algorithm, has carried on the actual observation to the humidity control effect. The feasibility of high precision humidity fuzzy control algorithm is verified. The algorithm has fast response speed, rapid and stable control process, small overshoot, high control precision, and satisfies control error of ±1% RH, which has a good application prospect.

2. Fuzzy control

Fuzzy control is a kind of intelligent control method that uses the knowledge of fuzzy mathematics to imitate the way of thinking of the human brain, recognizes and judges the fuzzy phenomenon, gives precise control quantity, and controls the controlled object [10].

2.1 Determining the Fuzzy Control System

The temperature and humidity sensors are mainly used to measure the air humidity. The fuzzy controller determines the working position within a certain period of time based on the data of the air humidity. The fuzzy control flow is shown in Figure 1

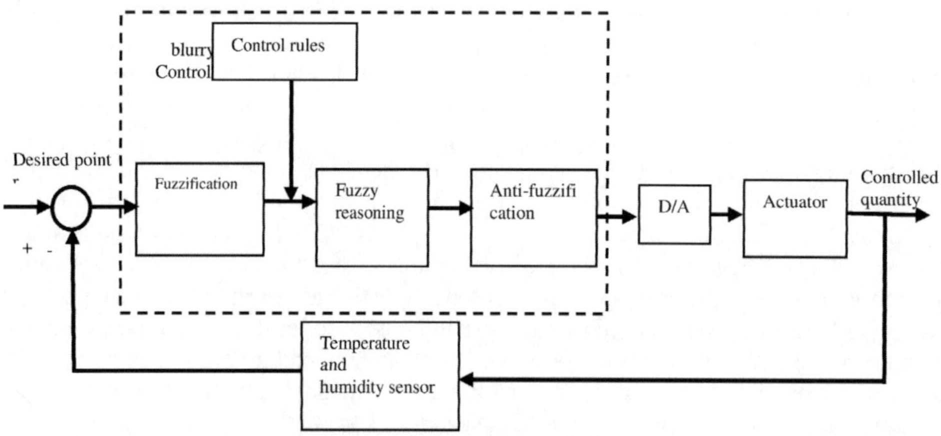

Figure 1 Fuzzy control chart

2.2 Input variables and output variables of the fuzzy controller

Under the condition of changing humidity t0, the stable humidity is measured as t(K), then the error e(K)=t0-t(K) is taken as the input variable of the fuzzy control.

The output variable is the control voltage K, which can be achieved by changing the thyristor conduction angle or PWM ratio.

2.3 Fuzzy Language Description of Input Variables and Output Variables

Let the fuzzy subset of the language value describing the input and output variables be: {negatively large, negatively small, 0, positivesmall, positive large} or denoted as {NB,NS,0,PS,PB}

Let the domain of error e be X, and divide the error size into seven levels: -3, -2, -1, 0, 1, 2, 3, then there are: X = {-3, -2, -1,0,1,2,3}

The controversy of the control variable λ is Y, and it is also divided into seven levels: -3, -2, -1, 0, 1, 2, 3, then: Y = {-3, -2 , -1,0,1,2,3}

Define its membership function as shown in Figure 2.

This gives the fuzzy variable assignment Table 1:

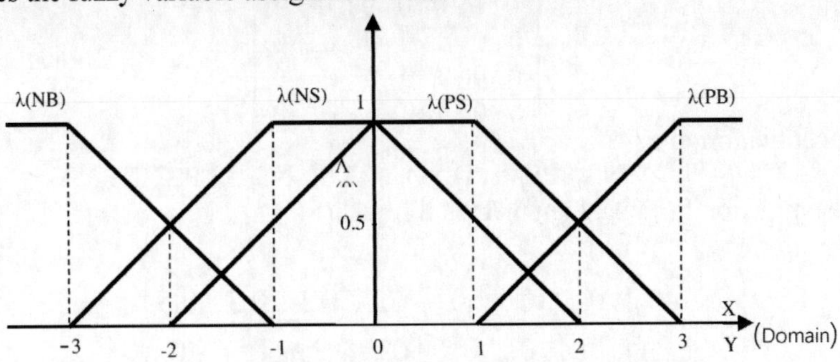

Figure 2 Membership function graph

Table 1 Fuzzy variable assignment tab

Membership	-3	-2	-1	0	1	2	3
PB	0	0	0	0	0	0.5	1
PS	0	0	0	0	1	0.5	0
O	0	0	0.5	1	0.5	0	0
NS	0	0.5	1	0	0	0	0
NB	1	0.5	0	0	0	0	0

2.4 Language Description of Fuzzy Control Rules

According to the strategy rules:

(1) If e is negatively large, then λ is positive large;
(2) If e is negatively small, then λ is positive small;
(3) If e is zero, then λ is zero;
(4) If e is positive small, then λ is negatively small;
(5) If e is positive large, λ is negative large;

This results in the control rule table 2:

Table 2 The table of fuzzy rules

e	λ
NB	PB
NS	PS
0	0
PS	NS
PB	NB

2.5 Matrix Forms of Fuzzy Control Rules

The fuzzy control rule is actually based on the fuzzy relation R between the error domain X and the control quantity domain Y, which yields formula (1):

$$R=(NB_e\times PB_\lambda)+(NS_e\times PS_\lambda)+(O_e\times O_\lambda)+(PS_e\times NS_\lambda)+(PB_e\times NB_\lambda) \quad (1)$$

From the fuzzy assignment table, equations (2) and (3) can be obtained.

$$NB_e \times PB_\lambda = (1,0.5,0,0,0,0,0) \times (0,0,0,0,0,0.5,1) \tag{2}$$

$$= \begin{bmatrix} 0 & 0 & 0 & 0 & 0 & 0.5 & 1 \\ 0 & 0 & 0 & 0 & 0 & 0.5 & 0.5 \\ 0 & 0 & 0 & 0 & 0 & 0 & 0 \\ 0 & 0 & 0 & 0 & 0 & 0 & 0 \\ 0 & 0 & 0 & 0 & 0 & 0 & 0 \\ 0 & 0 & 0 & 0 & 0 & 0 & 0 \\ 0 & 0 & 0 & 0 & 0 & 0 & 0 \end{bmatrix} \tag{3}$$

Similarly, we can get the formula (4)

$$NS_e \times PS_\lambda \quad O_e \times O_\lambda \quad PS_e \times NS_\lambda \quad PB_e \times NB_\lambda \tag{4}$$

Put these into the summation formula, then there is formula (5):

$$\begin{bmatrix} 0 & 0 & 0 & 0 & 0 & 0.5 & 1 \\ 0 & 0 & 0 & 0 & 0.5 & 0.5 & 0.5 \\ 0 & 0 & 0.5 & 0.5 & 1 & 0.5 & 0 \\ 0 & 0 & 0.5 & 1 & 0.5 & 0 & 0 \\ 0 & 0.5 & 1 & 0.5 & 0.5 & 0 & 0 \\ 0.5 & 0.5 & 0.5 & 0 & 0 & 0 & 0 \\ 1 & 0.5 & 0 & 0 & 0 & 0 & 0 \end{bmatrix} \tag{5}$$

2.6 *Fuzzy Decision*

When the fuzzy input is e(K), corresponding to its output, we get the formula (6)

$$\lambda(K) = e(K) o R \tag{6}$$

For example, when e(k)=PS, λ(K)=e(K)oR=(0.5 0.5 1 0.5 0.5 0 0)

2.7 *Anti-fuzzification*

Anti-fuzzification [11] is to transform the fuzzy control quantity obtained by fuzzy inference into a certain control quantity, so that the output can control actual physical things. The specific process of the anti-fuzzification of this system: First, the quantification level is obtained by quantifying the humidity error or its error change rate, and the corresponding control rules in the control rule table are queried according to the quantification level. Then execute the mechanism by calculating the output value of each actuator. Based on this idea, a fuzzy control query table for each implementing agency can be established, and a corresponding program can be written to realize the control of the fuzzy controller to the executing agency.

Table 3 Fuzzy Control Table

e	λ
-3	3
-2	2
-1	1
0	0
1	-1
2	-2
3	-3

2.8 *Fuzzy Output Becomes Accurate Value*

This algorithm uses the average method to obtain formula (7):

$$\lambda = \frac{1}{7}[Rn1*(-3) + Rn2*(-2) + Rn3*(-1) + Rn4*(0) + Rn5*(1) + Rn6*(2) + Rn7*(3)] \tag{7}$$

IOP Conf. Series: Materials Science and Engineering **470** (2019) 012032 doi:10.1088/1757-899X/470/1/012032

2.9 *Fuzzy Control Table*
The above result can be used for actual control, but in order to increase the control speed, various situations are generally calculated in advance and stored in a table to form a fuzzy control table 3. The input and output values can be defined by themselves.

Definition e: -3=-5%RH; -2=-2%RH; -1=-1%RH; 0=0%RH; 1=1%RH; 2=2%RH; 3=5%RH.

Definition λ: 3=100%W; 2=50%W; 1=10%W; 0=0%W; -1=10%W; -2=50%W; -3=100%W.

If the negative deviation of the relative humidity is set larger, the power output of the humidifier is larger; if the negative deviation of the relative humidity is smaller, the output of the humidification power is smaller; if the relative humidity deviation is set to zero, the power output of the device is zero; If the relative humidity has a small positive deviation, the power output of the drying device is small; if the relative humidity has a large positive deviation, the drying device has a large power output.

3. Experimental

3.1 *Experimental device*
The experimental device uses a humidity test box. The test box is used to test the hair hygrometer (meter) and other types of humidity sensors. As shown in Figure 3. It is mainly composed of a box body and a measurement and control system. The verification box used humidity standard which adopts the JBB1 type digital standard dry-wet meter. The humidity within the humidity verification box can be automatically adjusted, controlled, data processing of humidity verification and results printing through the communication interface and the measurement and control system. The hardware of the automatic control part is 32-chip STM32F103VET6. The control algorithm is the fuzzy control algorithm designed in this paper.

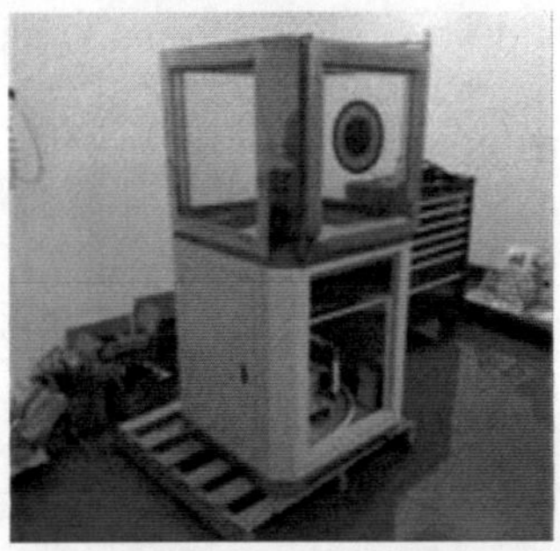

Figure 3 Humidity verification device

3.2 *Experimental Design*
The design and debugging of the high-precision humidity control system is completed and the experiment and verification of the humidity control algorithm are completed on the system. The experimental environment was a 0.5*0.5*0.5 (m^3) closed space, and the ambient air humidity was 15% RH.

Experiments were conducted on the target humidity of high humidity stability test (75% RH), medium humidity stability test (50% RH), and low humidity stability test (30% RH), Designing experimental schemes respectively to verify the effectiveness of the algorithm and collect and record humidity changes as a basis for data analysis.

Experimental software design and debugging, including fuzzy controller design, data acquisition and control system, and communication control, realize high-precision and constant control of environmental humidity.

3.2.1 Full Scale Experiment. Through experiments, the humidity adjustment range of the experimental device was obtained. Humidification: The humidity of the experimental device was reduced to a minimum (13%Rh), and then the humidity was increased to a maximum value (88%Rh). The collected data was as shown in Figure 4. After about 340S, the humidity of the enclosure reaches a steady state.

Dehumidification: The humidity of the experimental device was increased to a maximum (88% Rh), and then the humidity was reduced to a minimum (13%Rh). The collected data was as shown in Figure 5. After approximately 780S, the humidity of the enclosure reaches a steady state.

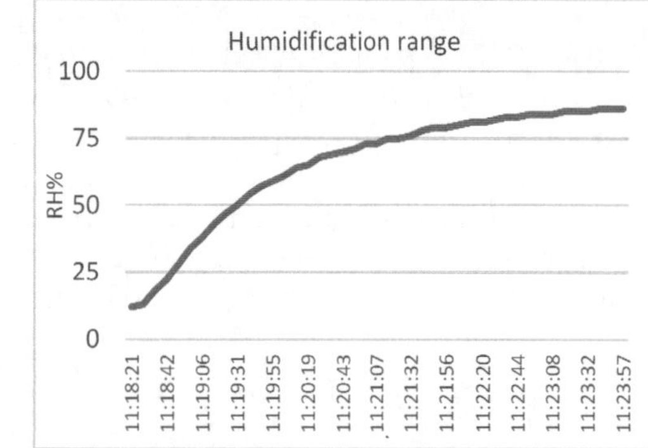

Figure 4 Humidification range under the control of the algorithm

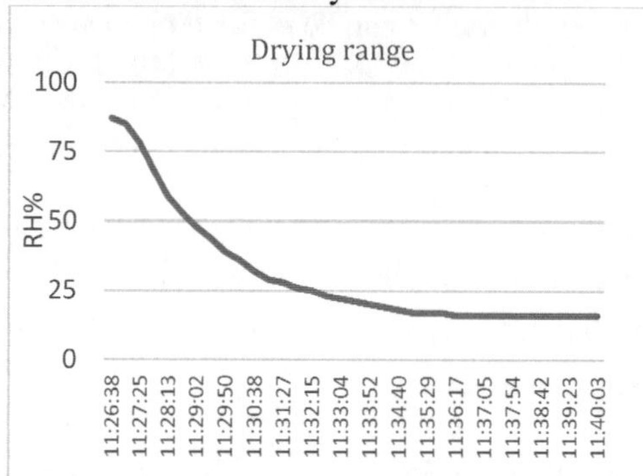

Figure 5 Drying range under the control of the algorithm

3.2.2 The target humidity is a high humidity stability test. Dry environment humidification: The initial humidity is 25% Rh, and the target humidity is 75% Rh. The collected data is as shown in Figure 6. After about 150S, the humidity of the enclosure reaches a steady state, and the maximum humidity is %Rh within 10 minutes of the experiment. The minimum value is 74%Rh, and the error is ±1%Rh, meeting the precision control requirements.

High-humidity environment dehumidification: The initial humidity is 85%Rh, the target humidity is 75%Rh. The collected data is shown in Figure 7. After about 20S, the humidity of the enclosure reaches a steady state. Within 10 minutes of the experiment, the maximum humidity is 76%Rh, the minimum value is 73%Rh, but only once, it is accidental data, the entire process meets the error of ± 1%Rh, to meet the precision control requirements.

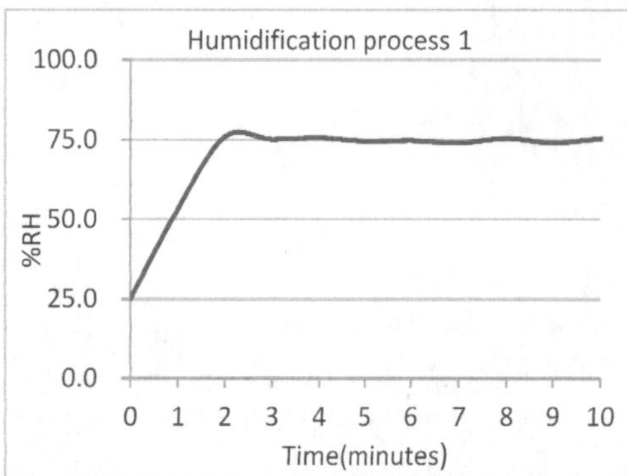

Figure 6 High Humidity Stability Experimental Humidification Process

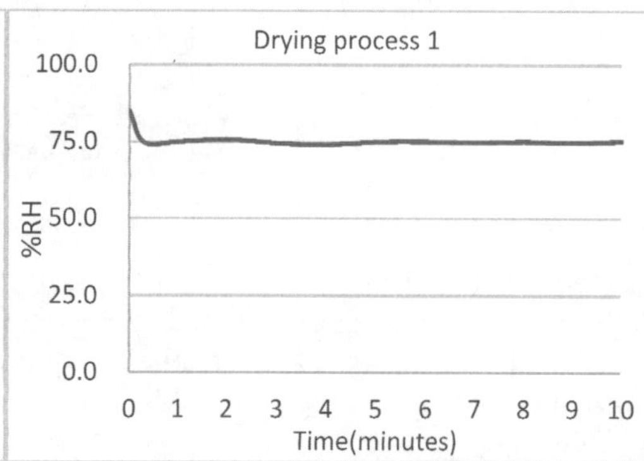

Figure 7 High Humidity Stability Experimental Drying Process

*3.2.3 The target humidity is the medium humidity stability experiment.*Dry environment humidification: the initial humidity is 25% Rh, the target humidity is 50% Rh, and the collected data is shown in Figure 8. After about 100S, the humidity in the cabinet reaches a steady state, and the maximum humidity is 51%Rh in 10 minutes in the experiment. The minimum value is 49%Rh, the error is ±1%Rh, meet the precision control requirements.

High-humidity environment dehumidification: The initial humidity is 85% Rh, and the target humidity is 50% Rh. The collected data is shown in Figure 9. After 90S or so, the humidity of the cabinet reaches a steady state. Within 10 minutes of the experiment, the maximum humidity is 51%Rh, the minimum value is 49%Rh, the maximum error is ±1%Rh, to meet the precision control requirements.

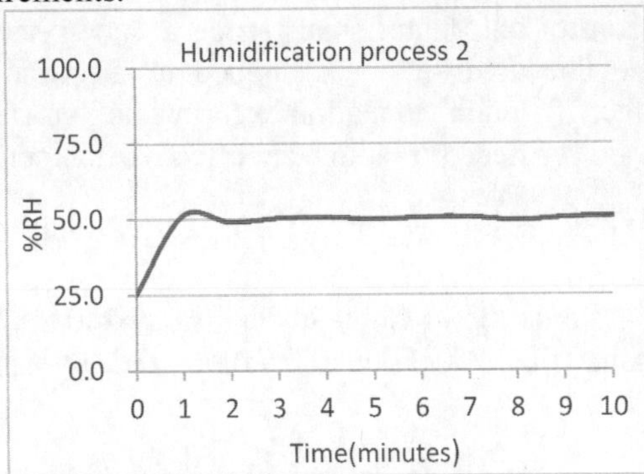

Figure 8 Middle Humidity Stability Experimental Humidification Process

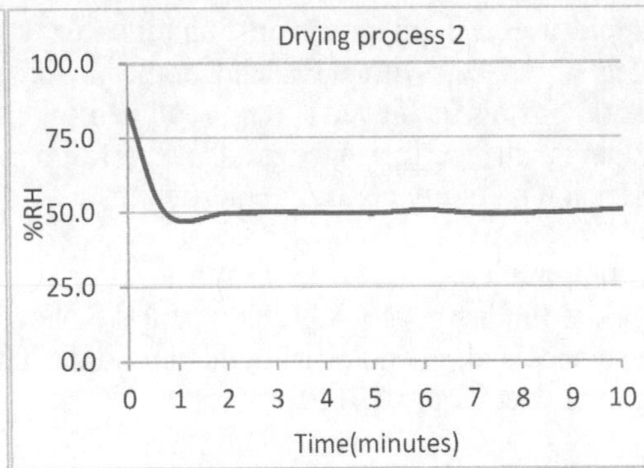

Figure 9 Middle Humidity Stability Experimental Drying Process

3.2.4 The target humidity is a low humidity stability experiment. Dry environment humidification: The initial humidity is 15% Rh, and the target humidity is 30% Rh. The collected data is as shown in Figure 10. After about 60S, the humidity of the enclosure reaches a steady state, and the maximum value of humidity is 31%Rh within 10 minutes in the experiment. The minimum value is 29%Rh, and the error is ±1%Rh, meeting the precision control requirements.

High-humidity environment dehumidification: The initial humidity is 50% Rh, and the target humidity is 30% Rh. The collected data is as shown in Figure 11. After about 50S, the humidity of the cabinet reaches a steady state. Within 10 minutes of the experiment, the maximum humidity is 31%Rh, the minimum value is 29%Rh, the maximum error is ±1%Rh, which meets the precision control requirements.

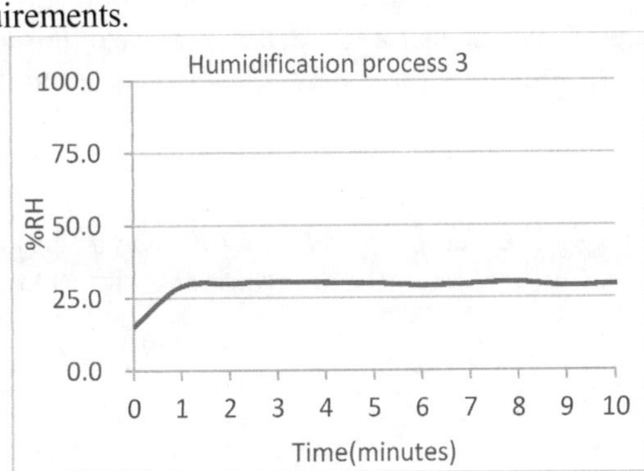

Figure 10 Low Humidity Stability Experimental Humidification Process

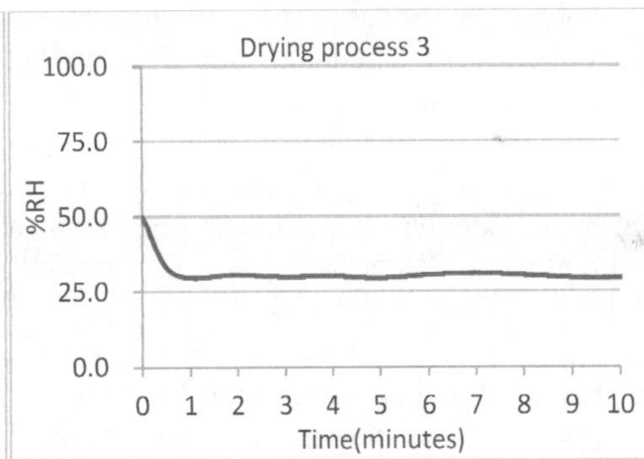

Figure 11 Low Humidity Stability Experimental Drying Process

3.3 Experimental Analysis

The experiment takes humidity as the research object, constructs the fuzzy control algorithm, applies the fuzzy control theory, creates the fuzzy control rule, and carries on the actual observation to the humidity fuzzy control effect. The results show that the humidity adjustment range of the system is between 13Rh and 88Rh. Through the humidity control of the three application environments of high humidity, medium humidity, and low humidity, the algorithm responds quickly, and the steady state time does not exceed 180 seconds. The control process is rapid , smooth, small overshoot, high control accuracy, meet the requirements of ± 1% RH of control error, verify the high-precision humidity fuzzy control algorithm in the three conditions of high humidity, moderate humidity and low humidity performance. The algorithm can be applied in industrial, agricultural, and civil applications, and it can achieve good humidity control and meet high-precision humidity control requirements in various environments.

4. Conclusion

This article uses a high precision humidity fuzzy control algorithm, small space accuracy can be controlled at ± 1%, with advanced domestic level. The algorithm was applied to the humidity verification box and achieved a very good control effect. Through simulation experiments in different environments, the performance of the algorithm was verified, providing an effective method for high-precision humidity control in more fields.

Acknowledgments

This work is funded by the National Natural Science Foundation of China under Grant 61403282, by Tianjin science and technology Foundation under Grants 17ZLZXZF00070, 171006603B, by Tianjin Foundation under Grants 20140714.

References

[1] Shen Ronglin. Temperature and Humidity Control Methods of Constant Temperature and Humidity Box[J].Environmental Technology,1994(1):42-44.

[2] Chen Shucheng, Yang Zhiyong, Wang Jianjia. Design of Agricultural Greenhouse Temperature and Humidity Measuring Instrument of Low Power Consumtion[J]. Agricultural Mechanization Research, 2014(8): 62-66.

[3] Li Shuangxiu. Indoor Air Quality Assessment and Purification Research Postgraduates Thesis, Qingdao University of Science and Technology, 2015(5).

[4] He Wenbo, Song Lizhong. Fuzzy PID Temperature and Humidity Control in Warehouse Management System Design [J].Equipment Manufacturing Technology,2017(07):50-54.

[5] Ruilan Wang. The Design of Temperature and Humidity Control System Based on Fuzzy Control in Multi Incubators[J]. Advances in Electronic Engineering, Communication and Management,2012(139):15-18.

[6] J.Y. Zheng, M.X.Zhang, Z.Q. Chen. Hygrothermal performance of diatomite-based humidity control building material[J]. 6th International Conference on Energy and Environment of Residential Buildings, 2014(8):37.

[7] Xiao Min Shan. Research on Control System of Greenhouse Temperature and Humidity Based on Fuzzy PID. Trans Tech Publications,2014(687):3395-3398,

[8] Jinhua Liu, Guoquan Ren, Guang Tian, Yunguang Qi. The Temperature/Humidity Control System of Equipment Warehouse Based on Fuzzy Control Algorithm. Springer International Publishing,2014(13):127-134.

[9] Liu Fang, Zhen Zhen. Research on the Control Mode of Agricultural Greenhouse Control System in China[J].Journal of Agricultural Mechanization,2008 (10) :223-226.

[10]Li Haofeng, Ma Tingting, Wei Wei, Chen Hang. Flue-cured tobacco room temperature and humidity wireless monitoring system based on fuzzy PID intelligent control algorithm [J]. Software Guide .2011(6):92-94.

[11] Sun Zengqi, et al. Intelligent Control Theory and Technology[M]. Control and Decision, 1996(1).

A Robot Odor Source Localization Strategy Based on Bionic Behavior

Guangda Lu, Qiuyue Zhang, Tongtong Qie and Qihui Feng

Key laboratory of information sensing and intelligent control , Tianjin University of Technology and Education, Tianjin, 300222

E-mail: Lugd1229@163.com

Abstract. In this paper, a robot bionic odor source localization algorithm based on kinesis response and tropotaxis behavior of flatworm is proposed, the algorithm does not need to detect wind direction information and odor absolute concentration information. Experiments show that the proposed algorithm is not affected by both wind direction and wind speed, only by judging the rate of change of odor concentration, which improves the searching efficiency, has good stability and environmental adaptability. it is a reliable initiative olfactory strategy with simple, search speed, high efficiency advantages.
Key Words. flatworm, bio-tropotaxis, Mobile Robot, Indoor environment, odor navigation

1. Introduction

In recent years, the use of mobile robots and gas sensors to achieve the odor source localization has become a research focus. Have olfactory function of the robot can engage in odor-related areas of work. Including the detection of land mines, search for explosives, search and rescue victims, poisonous gas detection, fire alarm, detection of various types of hazardous chemical storage containers or pipeline leakage and repair, can also carry out prospecting. In fact, The most critical problem of the robot's olfactory task is to search, locate and identify the odor source.

The robot can find the odor source by tracing the plume, the so-called plume refers to the feather-like trajectory of smell molecules released by the scent source in air [1]. Hayes [2] decomposed the problem into three sub-tasks, that plume found, plume cross, odor source confirmed. As early as 1991, Rozas [3] used the developed artificial nose to conduct research in this area, installing the artificial nose on a mobile robot to find the source of the odor by tracking the concentration of the gas in the test environment. Similarly, it is to install a pair of gas sensors on the mobile robot to compare the output of the two sensors so that the robot moves in the direction of high concentration. Li suxia proposes the algorithm of odor source detection by using the camera small hole imaging model and the component equation model of odor active olfaction [4]. Chen Yicun (et al) get good result in source localization and active olfaction [5-9]. but due to the influence of turbulence, the concentration distribution in the gas diffusion process is not regular and smooth, so robots use only the gas concentration gradient to search for odor sources less efficiently. To address this issue, Ishida [10] and his colleagues are inspired by the moth's upwind tracking of pheromones. Using four gas sensors and four wind speed sensor made the odor direction detection device, which made full use of the odor information and the wind direction information to complete the taste source search. Based on this, they also studied some new algorithms. By searching the smoke plume along the concentration gradient and tracking the smoke plumes along the wind direction. Pyk [11] developed a mobile person equipped with six arrays of metal oxide gas sensors and wind vane wind direction sensors moth and used it to simulate the moth's tracing movement of pheromone traversing the wind direction and the upwind direction in the wind tunnel. It find the source of taste at a distance of 4 m. In recent years, Marques [12] and Loutfi [13] also made great achievements in using mobile robots to study the taste source localization.

In this paper, we propose odor source localization strategy for robots that is based on the kinesis response and bio-tropotaxis behavior of flatworm [14] and have validated the effectiveness, speediness and independence of the wind direction and the odor of the search strategy in simulation experiments concentration characteristics, this paper validates and analyzes its performance through experiments.

2. Robot Odor Source Localization Strategy Based in Flatworm and Bio-tropotaxis Behavior

2.1 Bionic principle

Many animals look for odor source(food or heterology) by tracking plume, but do not measure the average concentration. For example, the moth can detect the surrounding gas with the antenna while exercising, and there is not enough time to obtain the mean value of a certain point. There are some ways to find the odor source by obtaining the instantaneous change in the concentration of smoke in the plume.

Kinesis response [15]: This is a phototactic reaction mechanism of flatworm, which prefers darker areas to light intensity. Increasing its turning rate as light intensity increases and does not directly depend on current the absolute intensity of light, but the rate of change of light intensity, the faster the light intensity changes, the greater the steering rate, the faster the trend toward darker areas.

Tropotaxis [16]: An tropotaxis mechanism is that an animal uses two or more sensors at the same time to produce an instantaneous estimate of the same or opposite direction of the concentration gradient. It use two or more sensors that distribution in space to detect the odor.It is decapoda crustaceans, flying Insects, reptiles, insects commonly used to track odors stimulate the tropotaxis mechanism. This is useful when tracking turbulent flow plumes.

2.2 Odor source localization algorithm

The flatworm's kinesis response is typical of bio-tropotaxis behavior independent of absolute intensity. When the light intensity increases, it will increase its own turn rate, and does not directly depend on the absolute intensity of the current light, but the rate of change of light intensity, the light intensity changes faster, the turn rate is larger, faster to darker area. A tropotaxis mechanism is that the animal produces an instantaneous estimate of the same or opposite direction of the concentration gradient by comparing two or more sensors at the same time. Based on this bionic principle, the difference in concentration between the two positions is detected to make direction estimation. According to the Kinesis response, bionic odor source localization strategy based on the tropotaxis behavior is proposed, which does not rely on the absolute sensor concentration and improves the searching efficiency of the robot, and can be independent of the wind direction.

Definition C is the current concentration, Cmax is the maximum concentration from the initial to the nth position, V is the concentration change rate, Vmin is the set minimum change rate, and Vn is the concentration change rate at the nth step. Vn = (Cn-Cn -1) / Cn-1, where n is the nth sample to start the search. a is the angle with the current direction of the robot, dmin is the minimum step size, d is the current step, ka, kd is the change factor related to V, which determines the change of a and d. According to the dynamic stimulus response, an = kaan-1, dn = kd * dn-1. When the repetition conditions are not satisfied, and Cn changes slightly with respect to Cmax, it is the source of taste, and ε is the absolute value of the difference between the two. Search strategy as shown in Figure: First, the robot random search to a given step, any angle forward, did not detect the gas, the same step size, random left or right turn an angle a until detected gas, then according to the rate of change of the gas concentration in the last two steps of the robot determines whether the robot turns left or right and changes its step. When the sensor acquisition data to meet the robot stop conditions, that place of taste, as shown in Figure 1.

Sampling, α as the angle between robot and the current direction of advance, d min is the minimum step size, d is the current step size, k is the change factor related to V, which determines the change of α and d. According to dynamic bionic response principle, $\alpha_n = k_\alpha * \alpha_{n-1}$, $d_n = k_d * d_{n-1}$,the initial value is set according to the visual situation, the coefficient value rules such as formula (1), (2). Decided to turn anticlockwise a or clockwise turn a According to the trend of the mechanism to determine that the strategy to determine the steering part of the schematic diagram shown in Figure 2.

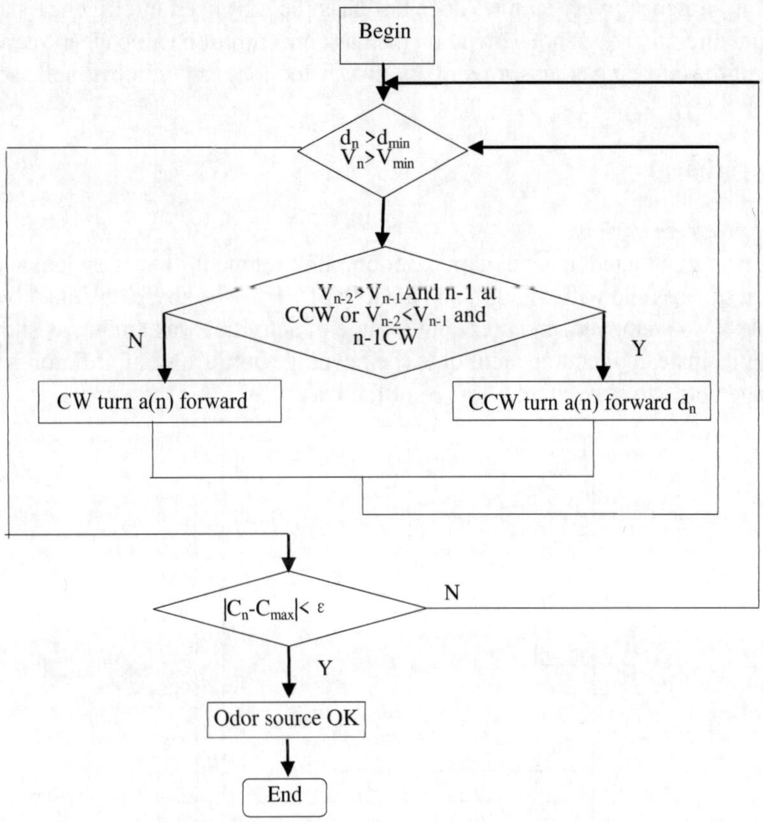

Figure 1 Olfaction searching module based on Bionic behavior

$$\alpha_n = \begin{cases} 1.1 & 0.8 \le |v_n| \\ 1 & 0.4 \le |v_n| < 0.8 \\ 0.9 & 0 \le |v_n| < 0.4 \end{cases} \tag{1}$$

$$k_d = \begin{cases} 1.1 & 0.8 \le |v_n| \\ 1 & 0.4 \le |v_n| < 0.8 \\ 0.9 & 0 \le |v_n| < 0.4 \end{cases} \tag{2}$$

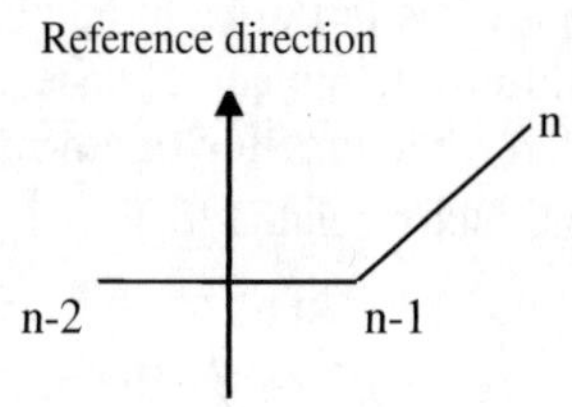

Figure 2 Estimating orientation principle based on tropotaxis

The algorithm's robot steps and corners vary in real time according to the rate of change of concentration. This strategy is very simple, does not pass the advanced intelligence strategy, does not depend on the wind direction, does not depend on the gas concentration value when searching, it avoids the disadvantage of the current gas sensor recovery time is too long, and improves the search efficiency of the robot.

3. Simulation Experiment

3.1 Experimental program designs

The experiment was conducted in a dynamic indoor environment. The gas leakage experimental environment was a square field with a length and a width of 10 m. The space occupied by the items such as the computer desk was removed, and the robot's range of activities was 8m8m, as shown in Figure 3. Smell leakage source instead of adding alcohol to the beverage bottle, a small inflator is used to quickly inflate the beverage bottle to simulate the source of the leak.

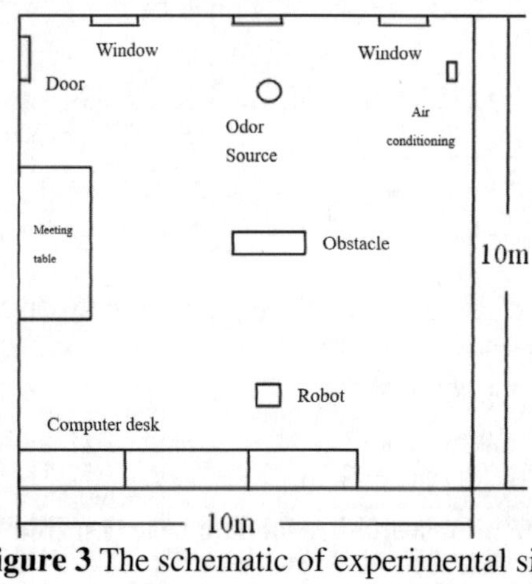

Figure 3 The schematic of experimental site

In the experiment, in order to detect the obstacle avoidance function in the robot's initiative olfactory strategy and whether the obstacle was mistakenly judged as the leak source, a cardboard box was added as an obstacle in the experimental site, the wind field of the site was in a natural state, the experimental gas concentration threshold from random to local search was set to 10 ppm. In order to test the reliability and stability of the initiative olfactory strategy, each experiment was performed 15 times.

3.2 Experiment platform

Experimental platform as shown in Figure 4, the multiple crawler-mounted mobile robot is composed of a crawler-type mobile robot body and a four freedoms anthropomorphic robot head system. The robot head has three-dimensional visual, auditory and olfactory sensory functions. The sensor detects the wind speed and direction in the experiment, providing wind direction and wind speed information for the initiative olfactory task.

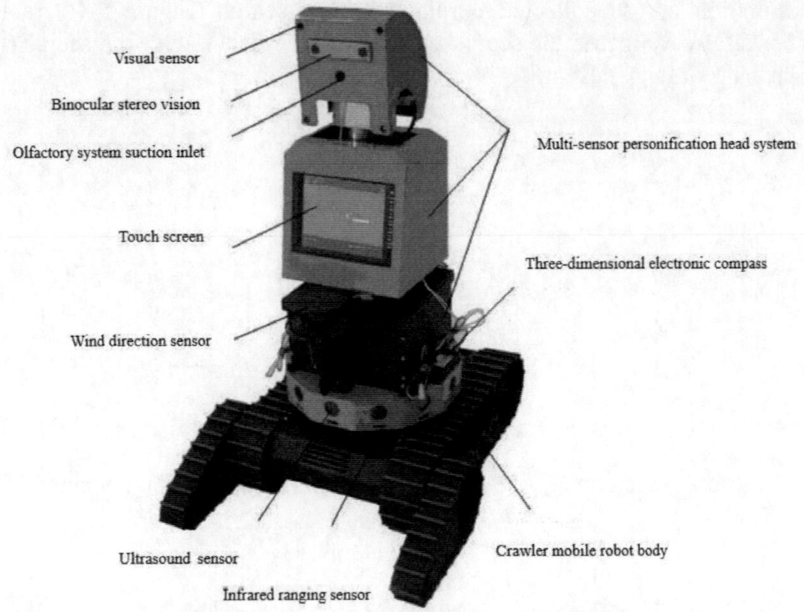

Figure 4 Multi-sense tracked mobile robot

3.3 Experimental process and results

Based on the kinesis response and the phototactic behavior of the robot odor source localization strategy experiments first in a wind-free environment, in order to reduce the randomness of gas concentration changes, the concentration change rate of the weighted value of the six collected values, that is, each step sensor detects the relative concentration changes, the robot forward six steps based on the average rate of change as a basis for steering angle changes and step changes, and then the next step search. The beginning of the middle began, the experimental data is about 30 minutes in the simulation odor diffusion began. Robot search process need to avoid the obstacles set to reach the leakage source that is the odor source, obstacles placed point, the robot starting point, the odor source in a straight line. Figure 5 (a) shows the robot's overall search path. It can be seen from the figure that the robot search confirms the process of the odor source is divided into several stages. At the beginning, when the gas concentration information detected by the robot is lower than the set threshold, the random search of the initial step is executed and the search goes forward by a certain distance. After the robot detected the concentration information, the concentration increased, the absolute value of the rate of concentration change began to increase, the fluctuation was also large, and the robot search step and the angle began to change. When the vicinity of the obstacle is found, the concentration distribution of the leaked gas is complicated due to the presence of the obstacle. When the robot approaches the obstacle attachment, the concentration of the gas fluctuates greatly. When the robot detects the obstacle, as shown in Figure. 5 (b), when the robot is at a certain distance from the obstacle, the detected value of the concentration at this time is also the highest. The expert decision-making system based on the sensor information. The condition determining obstacle is a suspected taste source, the taste source determining module is started to determine whether the object is a taste source. After the robot crosses the obstacle, the detected gas concentration is higher than that before the obstacle, which is different from the condition for determining the odor source identification strategy , The robot determines the target is not the taste source, the robot goes to the implementation of the olfactory search module, the robot continues to smell the search until it reaches the vicinity of the taste, the robot will be the odor source as a suspected source, the implementation of the flavor determination module again, identify the conditions of the strategy, confirm that the suspected target is the source of the leak, stop the execution of the program, and the task ends. The concentration of gas collected during the robot search (Figure 5 (c)) and the rate of concentration change also validated the search process in accordance with the search algorithm and expert decision-making strategy design.

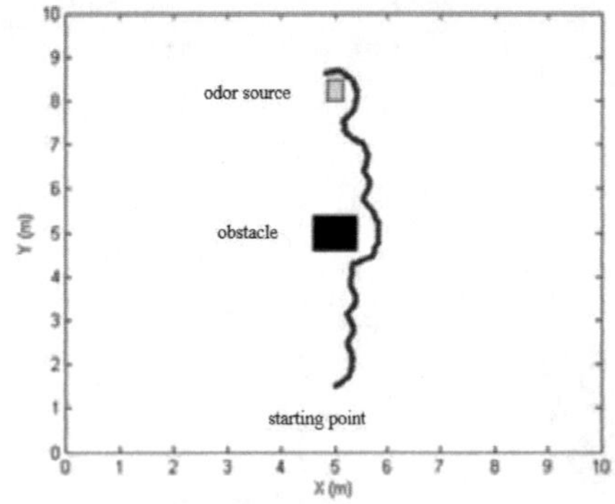

(a) Robot search trace

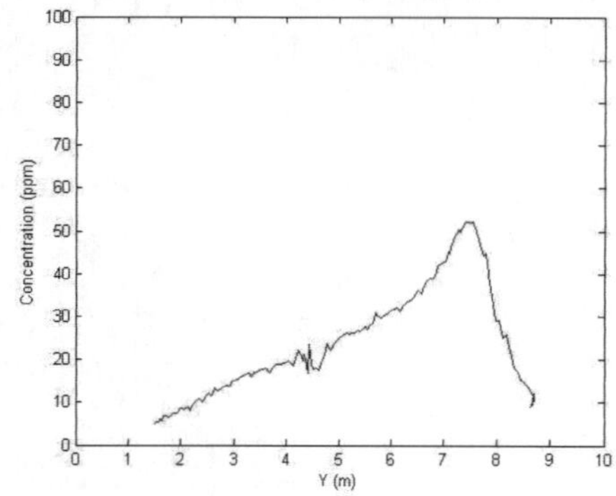

(b) The gas concentration collected by the robot

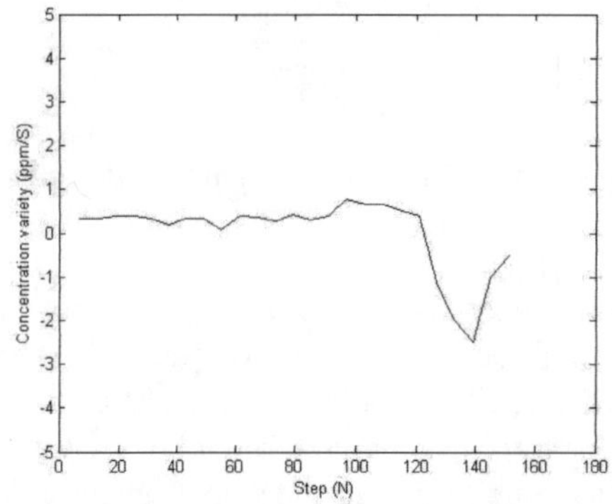

(c) Concentration rate of change

Figure 5 The searching path and collected gas concentration and variety of robot

3.4 Experiment analysis

According to the above experimental process, based on the kinesis response and tropotaxis robot odor source location strategy analysis, as shown in Table 1, this article from the following three aspects of strategy performance evaluation:

- Searching success rate

In the case of unsuccessful search, the robot failed to find a valid olfactory message clues when the odors began to diffuse for a short period of time (the robotic olfactory search experiment started almost simulating diffusion), trapped in the high-concentration area trap, the experimental set-time resulted in two more robots identifying the obstruction as odor source. This is turbulent because of turbulence near the obstacles, with the lowest nearby concentration being less than half the highest concentration, leading to miscarriage of justice.

- Searching efficiency

Robot search efficiency is the time it takes to complete the search task, here is the time the robot search to taste source. Due to inconsistent experimental conditions, such as site size, odor source, obstacle placement and robots search from different starting point, the completion time is not horizontal comparability and cannot be used as a measure of the efficiency of a strategy. Therefore, the ratio of the straight line distance to the leakage hole of the odor source and the total length of the path taken by the robot during the search process is taken as the search efficiency index of the robot's active olfactory strategy. From the search efficiency, it is relatively high, the robot can efficiently accomplish the odor source locating task.

- Average searching speed

The search speed of robots is the distance traveled by robots per unit time, which can reflect the rapid effectiveness of the robot's initiative olfactory strategy. In the data of Table 1, the numerator is the robot's average search distance and the denominator is the average search time, which means the average search speed of the robot in m/s.

Table 1 Performance analysis of odor source location strategy

Average success rate	Average search efficiency	Average search speed
13/15=0.86	6.5/8.46=0.77	8.46/51.8=0.163

4.Conclusion

Based on the kinesis response of flatworm and bio-tropotaxis bionic behavior of the flatfish, this paper designs orienting algorithm the odor source of a robot without depending on the absolute concentration of the odor and the information of the wind direction. Compared with the simulation, the robot's actual path is not particularly accurate, which is due to the accuracy of the robot walking, and the slipping of the crawler on the ground is an important factor. Experiments show that the algorithm is not affected by the wind direction and wind speed, and based on the rate of change of odor concentration to improve the search efficiency, has good stability and environmental adaptability, is a reliable active olfactory strategy, in the dark environment under extreme conditions of very low visibility, such as a fire, this strategy enables robots to work in extreme environments such as darkness, and odor smell tracking can be accomplished by detecting odor alone

Ackonwledgments

This work is funded by the National Natural Science Foundation of China under Grant 61403282, by Tianjin science and technology Foundation under Grants 17ZLZXZF00070,171006603B, by Tianjin Foundation under Grants 20140714.

References

[1] PayneT L, BirchM C and KennedyC E J. Mechanisms in Insect Olfaction. *New York, USA: Oxford University Press*, 1986.

[2] HayesA T, Martinoli A and Goodman RM. Distributed odor source localization. *IEEE Sensors Journal* 2002,2(3): 260-271.

[3] Rozas R, Morales J and Vega D. Artificial smell detection for roboticnavigation. *Fifth International Conference on Advanced Robotics. Piscataway, NJ, USA: IEEE*, 1991. 1730-1733.

[4] LiSuxia. Research on locating gas leakage source By robot detection. *Experimental technology and management*. 2017,34(7):46-59.

[5] Chen Yicun. Time-variant method based on contaminant source localization active olfaction in indoor environment. *Journal of PIA University of Science and Technology (Natural Science Edition)*,2016,(03):257-263.

[6] Mohamed Awadalla,Tien-Fu Lu and Zhao F. Tian,Bassam Dally,Zhenzhang Liu. 3D framework combining CFD and MATLAB techniques for plume source localization research . *Building and Environment*. 2013, 70 (12): 1019.

[7] Tengfei Zhang, Hongzhu Li and Shugang Wang. Inversely tracking indoor airborne particles to locate their release sources . *Atmospheric Environment* . 2012, 55:328-338.

[8] Xiaojun Ma, Xiaoliang Shao, Xianting Li and Yiwei Lin. *An analytical expression for transient distribution of passive contaminant under steady flow field. Building and Environment* . 2011, 52 (6): 98-106.

[9] Zhiqiang (John) Zhai, Xiang Liu, Haidong Wang, Yuguo Li and Junjie Liu. Experimental verification of tracking algorithm for dynamically-releasing single indoor contaminant . *Building Simulation*. 2012, 5 (1): 5-14.

[10] Ishida H, Nakayama G and Nakamoto T, et al. Controlling a gas/Odor Plume-tracking robot based on transient responses of gas sensors J.*IEEE Sensors Journal* 2005, 5(3): 537-545.

[11] Pyk P, Bermúdez I Badia S and Bernardet U, et al. An artificial moth:Chemical source localization using a robot based neuronal model of moth optomotor anemotactic search. *Autonomous Robots*, 2006,20(3): 197-213.

[12] Marques L, Nunes U and DeAlmeida A T. Spread Nose: Distributed agents for environmental monitoring. *Proceedings of the International Workshop on New Developments on Sensors for EnvironmentalControl. Singapore: World Scientific Publishing*, 2003. 234-238.

[13] Loutfi A and Coradeschi S. Smell. Think and act: A cognitive robot discriminating odours. *Autonomous Robots*, 2006, 20(3): 239-249

[14] Lu Guang-da and Zhang Ming-lu. A Biological-inspired Gas/Odor Source Localization Strategy for Mobile Robot. *Journal of Hebei university of technologe*. 2010, 39(05):48-52.

[15] Frankel,G.S.and D.L.Gunn.The Orientation of Aniamals:Kineses,Taxes and Compass Reactions. *New York, Dover Publications Inc.1961.*

[16] FRANK W. GRASSO. Invertebrate-Inspired Sensory-Motor Systems and Autonomous, Olfactory-Guided Exploration.Biol.*Bull*.200(April 2001): 160–168.

IOP Conf. Series: Materials Science and Engineering **470** (2019) 012034 doi:10.1088/1757-899X/470/1/012034

A Modeling Method for Abrupt Change Shafts in Gearboxes and Experiment Verification

Z C Liu, W Song and Q Lu

ABB Corporate Research Center, Kangxin highway 4528, Pudong, Shanghai, China

Zhanchi.liu@cn.abb.com

Abstract. The modeling method of the shaft with abrupt change of section is essential in gearbox design. This study reviewed the proposed piece-wise beam model for abrupt change shaft, which is dedicated to estimate shaft deflection. The proposed model is an equivalent model, which is established by comparing and analysing the distribution of the flexural rigidity of the shaft model using solid element. Based on the proposed model a design tool is developed for creating the piece-wise beam model with high efficiency and convenience. As verification of the piece-wise beam model, a test bench with abrupt change shaft is designed and constructed. The shaft deflection with static load has been measured. Experiment analysis has been made by comparing the results between experiments and simulation, which is obtained using solid model, the traditional beam model and the proposed piece-wise beam model. Result shows that the piece-wise beam model is more accurate than the traditional beam model in shaft deflection estimation. The experiment result reveals the correctness and effectiveness of the piece-wise beam model and the design tool.

1. Introduction

To realize high transmission ratio in limited space, it is quite common to mount a large gear onto a shaft. Shrink fitted gear is normally used to realize a stable matching condition. In design of the gearbox it is important to estimate the deflection of the shaft. For shafts with shrink-fitted disk, an equivalent diameter is well defined using the pitch and bore diameter of the gear according to ISO-6336[1]. The resultant shaft is often a shaft with abrupt change section which means a sudden change of diameter at the equivalent part.

Modeling of the abrupt change shaft is essential in gear design. Solid element is a good choice when performing the simulation using finite element method. However, 1-D beam element is still preferred by engineers due to its simplicity of pre-process and high efficiency. Moreover, with some commercial software focused on gearbox like KissSoft, beam element is the only choice for users. But it is proved that traditional beam element is questionable in shaft bending analysis when it comes to abrupt change shafts. It is because traditional beam element failed to depict the distribution of the rigidity at the abrupt change section. In gearbox design, precise estimation of shaft bending is of great importance because it affects tooth modification which will further influence the wearing and lifetime of the gear. Thus a relative accurate beam model is desired for abrupt change shaft in gearbox design.

Literatures [2-4] investigated constrains of the traditional beam elements which will increase the errors when dealing with shafts or beams with abrupt change sections. It is also pointed out that the reason of the error is the constant rigidity property of the tradition beam element. Actually the distribution of the shaft rigidity is continuous in the neighbour of the discontinuity of the geometry. [3,

5] investigated the distribution of the flexural rigidity for abrupt change shafts. The equivalent second moment of area $I(x)$ was used, which can be obtained using finite element method with solid element. Some dimensionless design curves is supplied, in which the continuous $I(x)$ is supplied.

$$I(x) = \frac{M(x)}{E\frac{d^2y}{dx^2}} \quad (1)$$

Based on the limitation of the previous study work, [6] investigated the application in practical engineering. A generalized shaft model is proposed, which is applicable to the most case of the shaft with spur or helical gears. Using finite elements analysis, a piece-wise beam model for abrupt change shaft is proposed. This model could be used to estimate the deflection of the shaft. In the supplied design case the error between the beam model and solid model could be highly decreased. Figure 1 shows the general idea of development of the piece-wise beam model from an abrupt change shaft solid model.

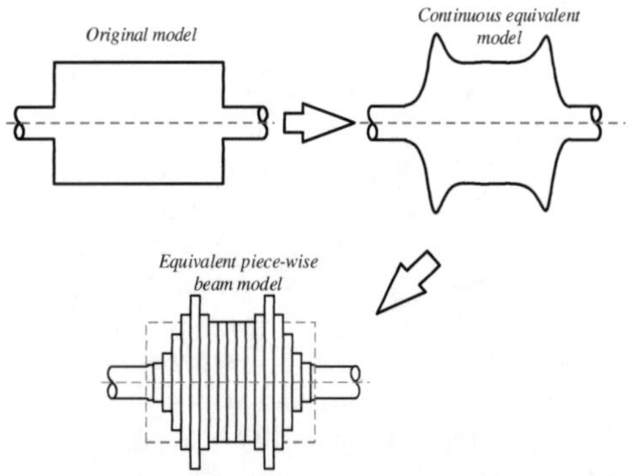

Figure 1. From original model to equivalent piece-wise beam model.

However, there are still limitations of the application in design practice. Obviously, the analysis using solid element will decrease the efficiency of the application of the piece-wise beam model. Furthermore, trying shafts with different geometry parameters might be necessary for designer. The current solution is not convenient to embed the proposed model in the fast design loop. Finally the piece-wise beam model is not verified in real design case so far, which add questionable factors to the propose beam model. Therefore, there is still a gap between the study outcomes and applications. Experiment verification is also necessary for the proposed piece-wise beam model.

The present work is focused on the application method of the piece-wise beam model and experimental verification. Firstly, the establishment of the piece-wise beam model has been reviewed. The application method of the piece-wise beam model is proposed. Then a design tool for abrupt change shaft is developed, in which design data of thousands of shafts is stored. Using this tool the piece-wise beam model for abrupt change shafts in a certain geometry range can be obtained with high efficiency, which make it possible to merge the proposed beam model well into the design loop. A shaft test rig has been designed and constructed. The deflection of a shaft with abrupt change section has been measured under static loading condition. Comparison of test and simulation results has been made, which verifies the effectiveness of the piece-wise beam model in increasing the accuracy of deflection estimation for abrupt change shafts.

2. Equivalent piece-wise beam model

2.1. Shaft model preparation

The generalized shaft model[6] contains three parts: basic shaft 1, stepped shaft and basic shaft 2 as is shown in figure 2. It can be well defined using the following three dimensionless parameters:

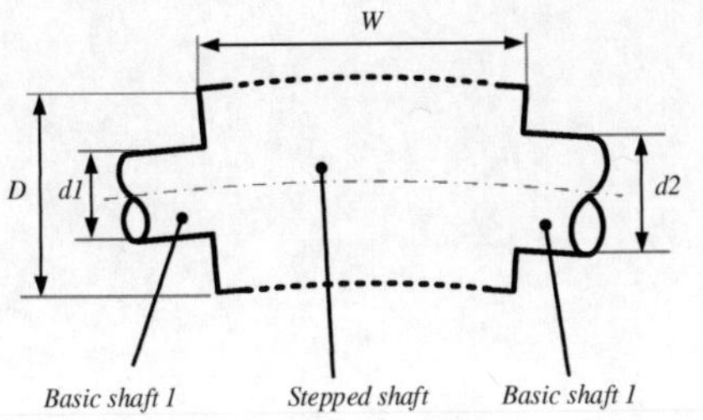

Figure 2. Generalised shaft model with different basic shafts.

$$\begin{cases} W_{dim} = \frac{W}{d_1}, & W_{dim} \in R \\ D_{dim} = \frac{D}{d_1}, & D_{dim} > 1 \\ d_{2dim} = \frac{d_2}{d_1}, & 1 < d_{2dim} < D_{dim} \end{cases} \quad (2)$$

where W and D are the length and diameter of the stepped shaft; d_1 is the diameter of the basic shaft 1; d_2 is the diameter of the basic shaft 2; subscript *dim* means dimensionless. In similarity condition (2), it is assumed that $d_2 \geq d_1$. This condition can be easily met by switching them if $d_2 < d_1$ in design practice.

2.2. Continuous equivalent model

For the abrupt change shaft with specific geometry dimension, a finite element shaft model can be employed with the same geometry parameters at the abrupt change section, which means the W_{dim}, D_{dim} and d_{2dim} are the same with the objective shaft model. Figure 3 shows an example of the shaft model with simple supports and pure bending moments. Then the static analysis can be performed to obtain the shaft deflection. An easy treatment is using the displacement of the shaft centre line as the shaft deflection. Figure 4 shows the deformation of the shaft using solid element. Figure 5 shows the deflection of the shaft which is extracted from the shaft center line of solid model.

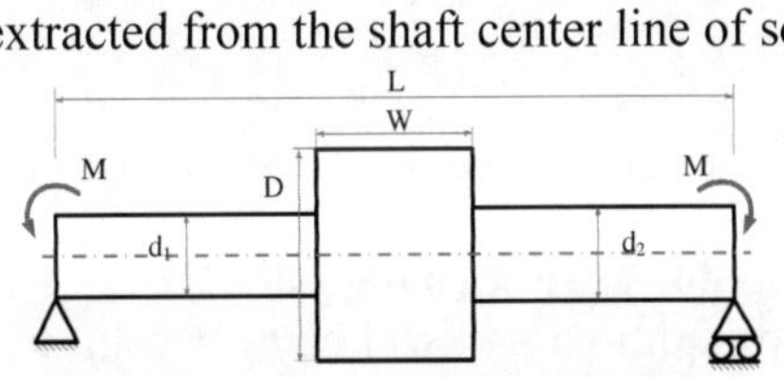

Figure 3. Set-up of the shaft model.

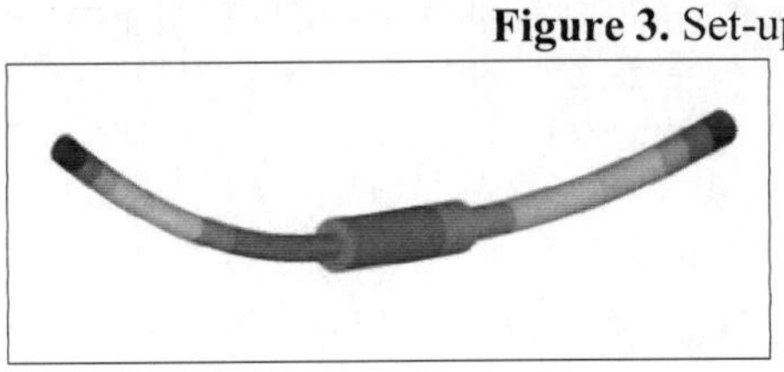

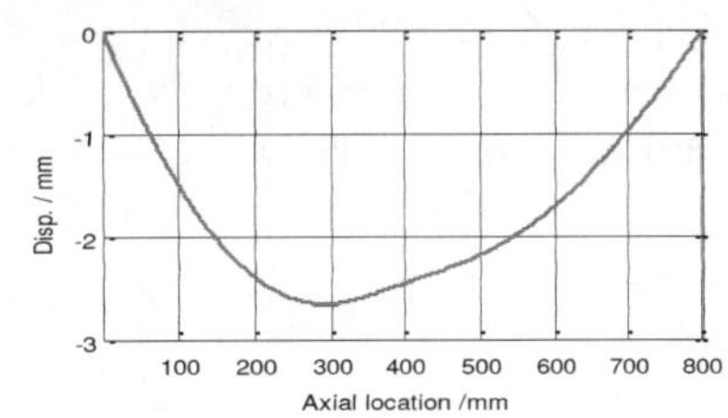

Figure 4. Deformation of the FE shaft model. **Figure 5.** Deflection of shaft center line.

With shaft deflection, the local curvature can be calculated at each node axial place. Then the local second moment of area $I(x)$ can be obtained using formula (1) as shown in figure 6, in which all the parameters has been made dimensionless with r_1. Because $I(x)$ is also a property of the cross section $I = \frac{\pi r^4}{4}$, the equivalent radius of the solid model can be calculated as is shown in figure 7. Using the continuous radius an equivalent solid model is obtained.

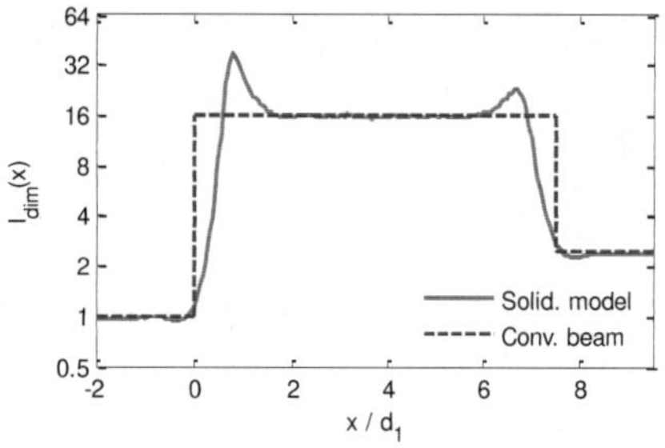

Figure 6. Equivalent second moment of area. Dash line: $I(x)$ of conventional beam, solid line: $I(x)$ of the solid model.

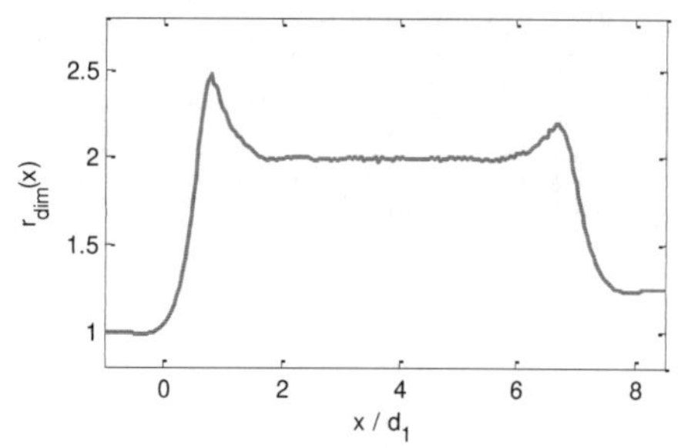

Figure 7. Distribution of dimensionless radius $r_{dim}(x)$.

2.3. Piece-wise beam model

The piece-wise beam model can be obtained by cutting the equivalent continuous model into pieces. Figure 8 demonstrate an example of piece-wise beam model for the continuous shaft, which is shown in figure 7. For further operation, it is easy to make it in to a model with dimension using d_1.

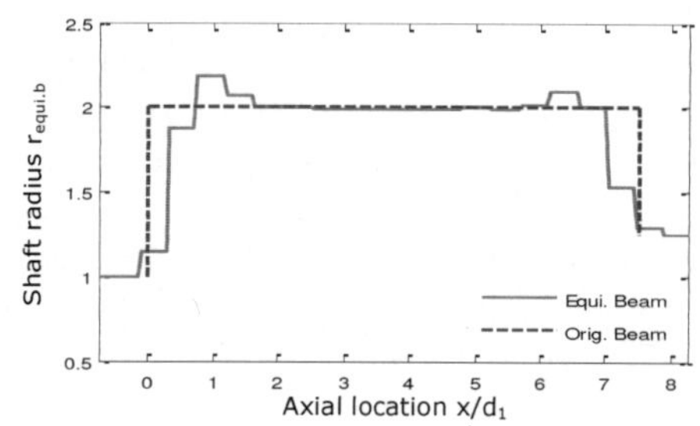

Figure 8. Dimensionless equivalent piece-wise beam model.

3. Development of the design tool for abrupt change shafts

In this section the development of a design tool will be demonstrated, which contains the stored data of simulation for thousands of shafts and a searching algorithm for creating the right piece-wise beam model for later use. Figure 9 shows the basic concept of the design tool. With this tool the piece-wise beam model can be generated instantly based on the in-put shaft dimension, which bridge the gap between the principle of the piece-wise beam model and its application.

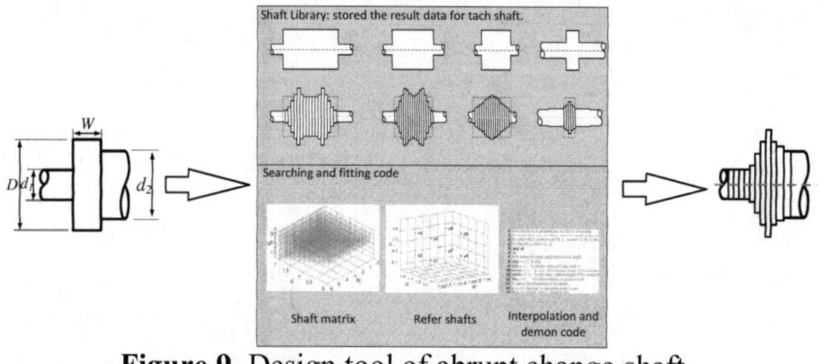

Figure 9. Design tool of abrupt change shaft.

3.1. Model dimension and finite element analysis

The shaft geometry dimension range could be defined with the current product data. Then the range of the abrupt change shaft parameters: W_{dim}, D_{dim}, and d_{2dim} can be defined. With an appropriate step, a shaft matrix containing different combination of shaft parameters can be established. In this demon case the shaft dimension range is defined as:

$$\begin{cases} W_{dim} = \frac{W}{d_1}, & W_{dim} \in [0.5,\ 3] \\ D_{dim} = \frac{D}{d_1}, & D_{dim} \in [1,\ 4] \\ d_{2dim} = \frac{d_2}{d_1}, & d_{2dim} \in [1, 3] \end{cases} \tag{3}$$

With the step 0.2, the shaft matrix can be obtained, which contains 3146 shafts of different geometry. Figure 10 shows the shaft matrix, in which every dot stands for a shaft model.

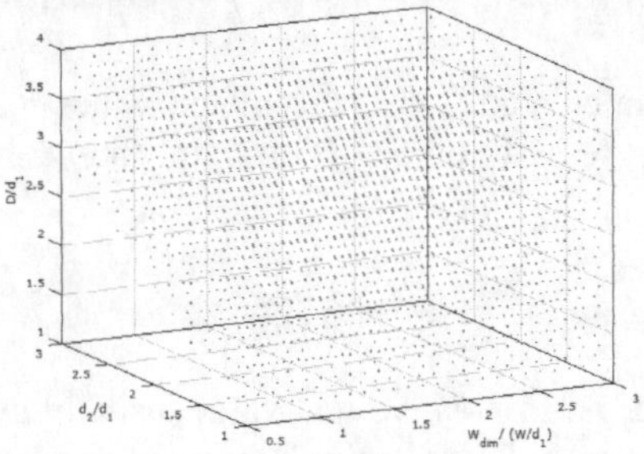

Figure 10. Abrupt change shaft matrix.

For each shaft, static finite element analysis, which contains the same procedure as shown section 2, has been made. In this example static analysis is performed in HyperWorks which supplies the script interface for repetitive work. Figure 11 demonstrates some of the finite element shaft models creating using script. After analysis the deflection of shaft center line is extracted and stored for later usage.

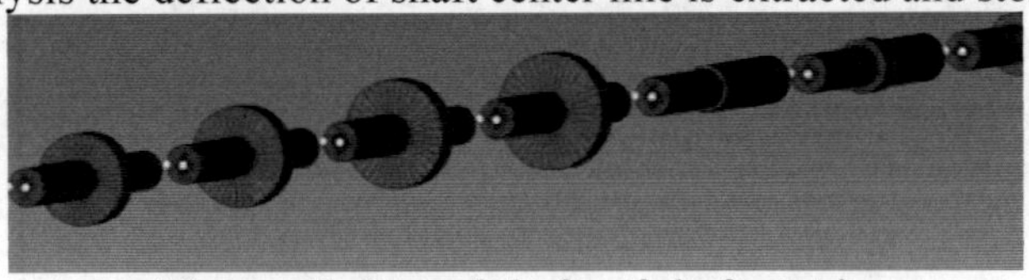

Figure 11. Part of shafts of shaft matrix.

3.2. Design tool development

For any combination of input parameters, which are W_{dim}, D_{dim}, and d_{2dim} of objective shaft, the design tool will supply shaft dimension parameters of the equivalent piece-wise beam model as output. It contains the shaft radius of the each axial sections, which can be defined according to user's requirements.

To achieve this goal, a searching algorithm is developed to obtain the deflection of the objective shaft. When the matrix contains the objective shaft, which means exist a shaft with the same dimensionless parameters as the input shaft, the deflection result can be used directly. When the matrix doesn't contain the shaft with the same dimension with objective shaft. Samples of shaft will be found as envelop shafts. The IDW (Inverse Distance Weighted) methods will be used to generate the deflection of the input shaft. Figure 12 shows the relationship of the input shaft (objective point) and envelop shafts (sample points).

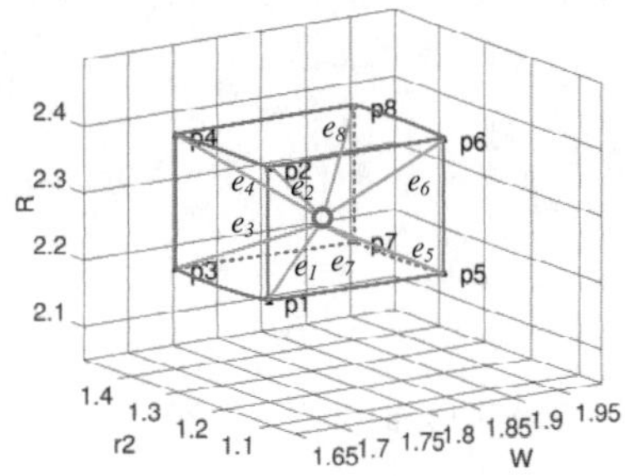

Figure 12. Envelop shafts (sample points) and input shaft (objective point). (○: input shaft; p1-p8: envelop shafts).

The shaft deflection can be calculated with interpolated value w_i. A general form of finding the interpolated values at given distance between objective point (input shaft) and sample points (envelop shafts) as be expressed as:

$$w_i = \frac{e_i^{-p}}{\sum_{j=1}^{n} e_j^{-p}} \tag{4}$$

where i is the index of sample; p is a positive real value, usually use 2; n is the total number of the samples; e_i is the distance of between the objective point and the corresponding sample point. With shaft deflection the piece-wise beam model can be calculated same process demonstrated in section 2. Example of output is shown in the next section.

4. Experiments verification

4.1. Set-up of test rig and experiments

To verify the proposed equivalent piece-wise beam model an abrupt change shaft test rig is designed and constructed. The deflection of shaft can be measured using linear variable deformation transformers (LVDT) under static loading condition. Figure13 and 14 shows 3D model and the test rig, which contains an abrupt change shaft simply supported between two ball bearings. Table 1 lists the shaft parameters of the test rig.

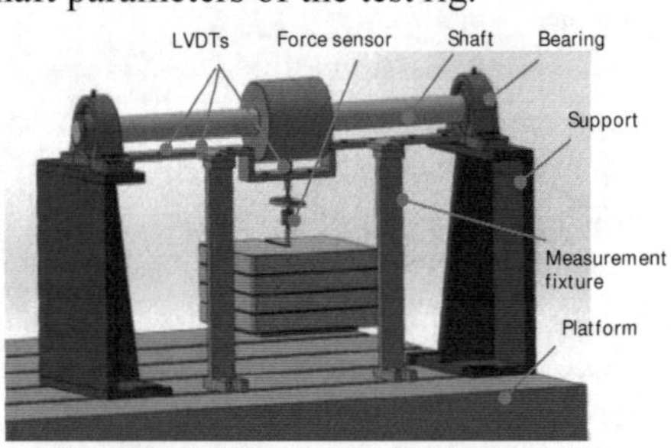

Figure 13. 3D model of the test rig.

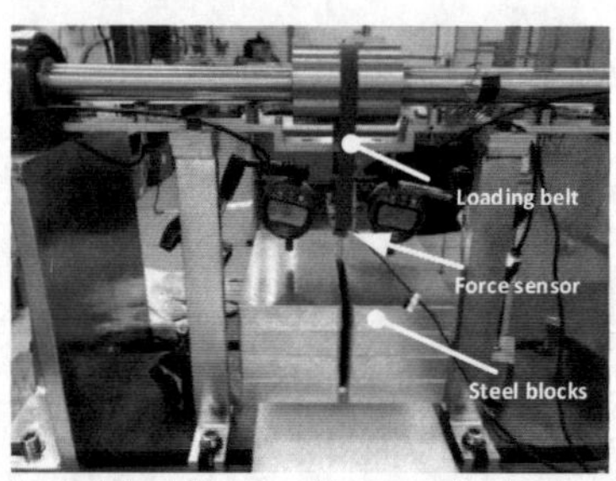

Figure 14. Abrupt change shaft test rig.

Table 1. Shaft parameters of test rig.

Shaft diameter /mm	Diameter of stepped section /mm	Length of the stepped section /mm	Distance of supports L/mm
36	72	100	600

In experiments seven displacement measure points are set along the shaft. The axial position coordinates of the measure points are listed in Table 2. Because the bearing radius stiffness is found to have obvious influence to the test results. Two measuring points near supports are used as the reference points (calibration points). Under this circumstance, the deflection between the reference points can be used to verify the shaft model, without influence of the bearing stiffness. In the experiment result analysis the displacement value relative to the reference points is calculated.

Table 2. Axial location of measuring points

Measure point	1	2	3	4	5	6	7
	l_1/mm	l_2/mm	l_3/mm	l_4/mm	l_5/mm	l_6/mm	l_7/mm
Distance	45	100	210	300	390	500	555

Static loading condition is achieved by hanging weights in the middle of the shaft with force sensor on the loading mechanism as is shown in figure 14. The maximum static loading force is 313.6 N (32kg force). The results of the experiment will be analyzed in 4.3 results analysis.

4.2. Simulation with different shaft models

To verify the effectiveness of the proposed piece-wise beam model, the deflection is estimated using the solid model, traditional and piece-wise beam model. The simulation result is compared with the experiment result.

Firstly the piece-wise beam model is achieved using the design tool. The stepped part of the shaft is cut into 8 pieces. Figure 15 shows the geometry of the piece-wise beam model. Table 3 lists the value of the original and piece-wise beam model in detail.

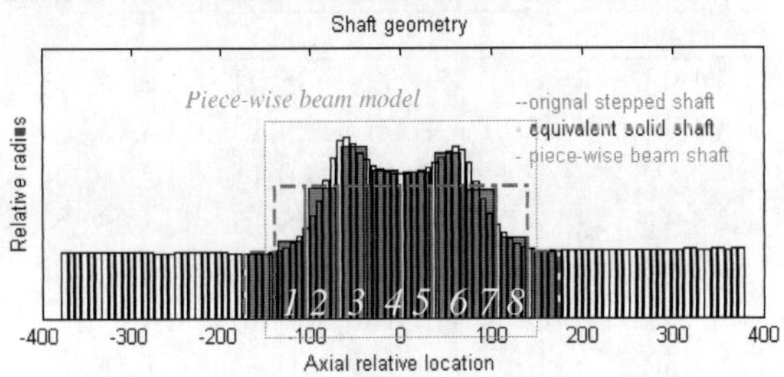

Figure 15. The obtained piece-wise beam model. (Red dash line: original stepped shaft; black: equivalent solid model; pink: piece-wise beam model.)

Table 3. Radius value of the piece-wise beam model

Section	1	2	3	4	5	6	7	8
Length/mm	12.5	12.5	12.5	12.5	12.5	12.5	12.5	12.5
Radius (traditional)/mm	18	18	18	18	18	18	18	18
Radius (piece-wise)/mm	21.33	35.58	46.97	40.40	40.16	45.12	35.40	22.08

Figure 16 and 17 show the comparison of the shafts using traditional beam model and piece-wise beam model. These are also the model using for shaft deflection estimation. A 1-D beam based finite elements code is developed to finish the shaft deflection estimation [7]. Then the displacement of the testing points (5 points) is extracted, which would be compared with the experiment result in the following subsection. For the simulation using solid element, there is no special treatment. Thus the detailed information will not be supplied.

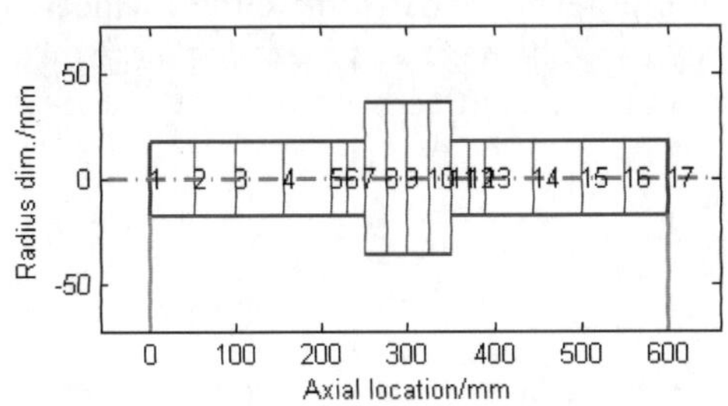

Figure 16. Shaft with traditional beam model.

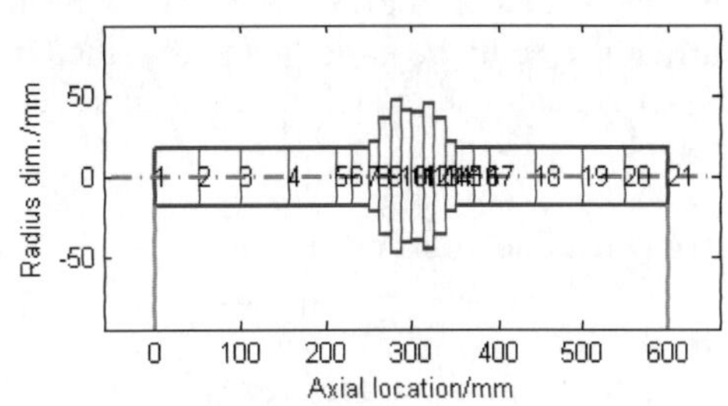

Figure 17. Shaft with piece-wise beam model.

4.3. Results analysis

Figure 18 is the deflection estimation of the shaft between the two calibration points with solid shaft model, traditional beam model, piece-wise beam model and the test result. The maximum displacement, which would be the mesh point for gear in this case, is used to evaluate the accuracy of the shaft with different models. Test result is used as reference for simulation results. The difference of the results is listed in table 4.

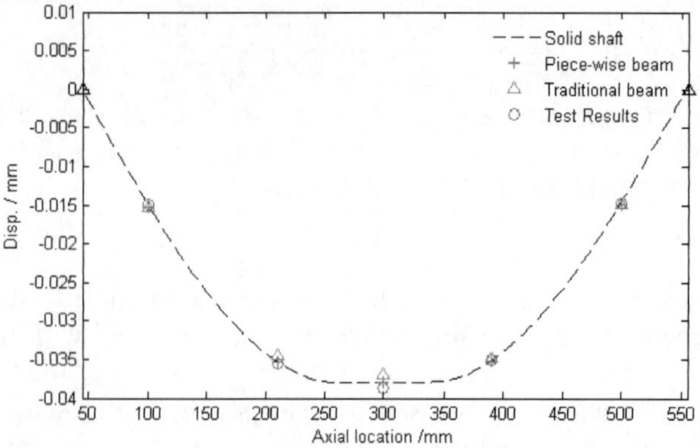

Figure 18. Comparison of shaft deflection using solid model, traditional beam model, piece-wise beam model and the test result.

Table 4. Difference of shaft deflection estimation using different model and experiment result

Testing Point		1	2	3	4	5
Experiments /um		-14.93	-35.61	-38.62	-35.01	-14.78
Solid shaft model	disp/um	-14.90	-35.35	-38.05	-35.15	-14.45
	Diff/%	0.19	0.74	1.46	0.15	2.19
Piece-wise beam model	P.W./um	-15.31	-35.41	-37.93	-35.15	-14.95
	Diff/%	2.5	0.56	1.79	0.16	1.16
Traditional beam model	Tradi./um	-15.31	-34.56	-36.89	-34.80	-14.95
	Diff/%	2.50	2.96	4.51	0.83	1.16

It can be seen that compared with test results, simulation with solid shaft model is with the highest accuracy. The traditional beam model is with the lowest accuracy. The result of the piece-wise beam model is in the between of the two other models, and much closer to the test results than the traditional beam model. The different at the middle point, which could be meshing point for gear, between experiment and estimation using traditional beam and proposed beam model is decreased from 4.51% to 1.79%. It can be seen that with traditional beam model, the stiffness of the shaft can be overestimated. It will result in the estimation of the shaft deflection to be smaller than the real value, which is influenced the tooth modification for gear design.

It can be seen that with the proposed piece-wise beam model, the shaft deflection can be estimated with higher accuracy compared with the traditional beam model. In this case the initial difference between the simulation and the experiment results is in a low level. Under some circumstances the initial difference could be larger than 10%, then the accuracy could be increased much higher [6].

The experiment results and the analysis reveal the correctness and effectiveness of the piece-wise beam model.

5. Future work and conclusion

5.1. Future work

The correctness and the effectiveness of the piece-wise beam in static analysis has been verified. However the dynamic characteristic of the shaft is critical important in the design for rotating machinery. Thus the work of next step will be focused on the dynamic characteristic of the piece-wise beam element. Obviously if the geometry shape of the shaft is varied, the mass and moment of inertia, which are key factors for dynamic characteristic analysis, would all be changed. In this condition some special treatment would be needed to ensure a higher accuracy in simulation. This would be the focus of the work in the future.

5.2. Conclusion

The main focus of present work is on the establishment and verification of the piece-wise beam model for shaft with abrupt change sections. The establishments of the piece-wise beam model is reviewed and constrains of its application is analysis. Based on the proposed beam model a design tool for abrupt change shaft is developed, which can be used to creating the piece-wise with high efficiency and convenience for engineers.

To verify the proposed beam model, a test-bench with abrupt change shaft has been constructed. Shaft deflection has been measured under static loading conditions. Experiment result shows that using the proposed beam model, the deflection of the abrupt change shaft can be estimated with higher accuracy compared with the traditional beam model. At the same time it reveals the correctness and effectiveness of the piece-wise beam model and the design tool.

Acknowledgement

The authors would like to thank Honghua Huang, MinYuan, and Ming Yan, ABB Power Transmission China. Their inputs, comments and discussions on this work are greatly appreciated.

References

[1] Calculation of load capacity of spur and helical gears –Part 1: Basic principles, introduction and general influence factors [S]. Geneva: ISO 6336-1:2006-09

[2] Sato H. Free vibration of beams with abrupt changes of cross-section[J]. Journal of Sound & Vibration, 1983, 89(1), pp. 59-64.

[3] Kitching R, Sanderson N, Hinduja S. Flexibility of rectangular beams with abrupt changes of section[J]. International Journal of Mechanical Sciences, 1975, 17(6), pp. 403-410.

[4] Sanderson N, Reid S R. Photoelastic approach to the determination of the flexural rigidity of rectangular beams with abrupt changes of section[J]. International Journal of Mechanical Sciences, 1976, 18(7), pp. 341-346.

[5] Sanderson N, Kitching R. Flexibility of shafts with abrupt changes of section[J]. International Journal of Mechanical Sciences, 1978, 20(3), pp. 189-199.

[6] Liu ZC, Song W and Lu Q, A Piece-wise Beam Model for Shafts with Abrupt Changes of Section in Gearbox Products[C]. ICAMechS 2017, Xiamen

[7] Yamamoto T, Ishida Y. Linear and Nonlinear Rotordynamics[M]. John Wiley & Sons, Inc., 2013

Fault Diagnosis Method of Autonomous Underwater Vehicle Based on Deep Learning

Yushan Sun, Zikai Wang and Guocheng Zhang

145-11, Nantong Road, Harbin, P.R.CHINA

ricardo_wzk@outlook.com

Abstract. In order to solve the questions that autonomous underwater vehicle(AUV) can't accurately predict thrusters' fault only depend on sensor's information, which is caused by the effect of closed-loop control system, and shallow neural network can't well fit the complex nonlinear system, this paper proposes a new method for the fault diagnosis of AUV's thruster, based on Deep Neural Network(DNN) and Denoising Autoencoder(DAE).In this proposed method the difference between AUV's theoretical state value and measured state value are used as input signal. Considering the disturbance of AUV's working environment, when using DAE to pre-training the DNN, Gaussian noise was added in the input signal to simulate environment. The trained DNN network was finally used to detect the AUV's fault propeller components. The results show that the proposed method is more effective, accurately and robust than other traditional methods.

1. Introduction

Nowadays, when developing offshore and deep-sea resources, the Autonomous Underwater Vehicle (AUV) is becoming an indispensable part of the exploration and development of marine resources, because it can complete the extreme tasks that divers cannot. AUV generally works in an unknown complex marine environment, working offline for a long time, frequent changes in the payload could change the shape and weight balance, which will cause changes in the motion characteristics of the robot and the dynamic performance of the system, and the ability to respond to system commands [1].

In the field of machine learning in recent years, deep learning with a deep hidden layer structure extracts the hidden features of data by mimicking the cognitive processes of the human brain, and has achieved remarkable results in the fields of image and speech recognition. The results show that when the human brain receives an external signal, the brain does not immediately process the signals directly but extracts the characteristics of the signal through a multi-layered neural network[2]. In this paper, deep learning is introduced into the field of underwater robot fault diagnosis, and a deep learning based underwater robot thruster fault diagnosis method is proposed.

2. Auv platform and fault modelling

2.1. AUV platform introduction

The tunnel AUV is equipped with a variety of acoustic detection sensors that can perform a variety of underwater detection tasks. Its actuator uses a propeller and is equipped with five propellers. There is a set of vertical pushes and side pushes on each of the rafts, and a horizontal main thruster is additionally arranged on the raft[3].

2.2. Fault modelling

As mentioned above, the underwater robot is a motion with a multi-degree of freedom for a strong nonlinear system. The existing AUV thruster fault diagnosis methods mostly judge the robot propulsion by comparing whether the difference between the theoretical state value of the robot and the measured state value is exceeded the threshold. However, this method ignores the influence of the compensation effect of the closed-loop control system on the control quantity on the fault detection system[4].

$$\begin{cases} x(k+1) = F(k,u(k),x(k),f(k)) + w(k) \\ y(k+1) = G(k+1,x(k+1)) + n^s(k) \\ w(k) = d(k) + Bn^a(k) \\ v(k+1) = | y(k+1) - x(k+1) | \end{cases} \quad (1)$$

where x(k) and y(k) are the theoretical state output of the AUV at time k and the actual measured output of the sensor, respectively, and the nonlinear function of the AUV system, u(k) is the propeller control amount, and f(k) is The fault vector of the thruster, and the self-noise of the thruster and the sensor, d(k) is the external disturbance of the system, B is the uncertainty coefficient, and v(k) is the difference between the theoretical state output of the AUV and the actual measured output of the sensor at time k. It can be seen from Equation (1) that when the AUV is operating normally and the propeller is not faulty, when the propeller is faulty, by judging whether f(k) is zero or not, we can know if the AUV's thruster has failed[5]. However, in the actual operation and state detection process of AUV, we can't directly measure f(k) directly through the sensor. For the tunnel AUV, the variables that can be directly obtained through software and hardware mainly include AUV speed, target yaw angle, and propeller control. The voltage, the actual measured speed of the sensor and the actual measured turn angle. We will study the measurable data to determine whether f(k) is not equal to zero. [6]

3. Pre-training and tune

3.1. Denoising autoencoder

As shown in Figure 1, the automatic encoder (AE) is an unsupervised three-layer neural network model, which consisting of an input layer, a hidden layer and an output layer. The core idea of the model is to extract the signal characteristics in the input data. Reconstruct the original input signal, and then train the network weights by maximizing the restored input data through the network output[7].

However, when the input layer data is affected by noise, the obtained input data itself may not obey the original distribution. In this case, the results obtained by the encoder will also be incorrect. In order to solve the data deviation caused by noise, we adopt the network structure of Denoising Autoencoder (DAE) [1], as shown in Figure 2. Compared with the automatic encoder, the DAE adds a noise processing process between the input layer and the hidden layer, and the data of the original distribution is processed to obtain the noise data, and then the maximum output of the DAE is restored. Noise data to train network weights to improve network robustness.

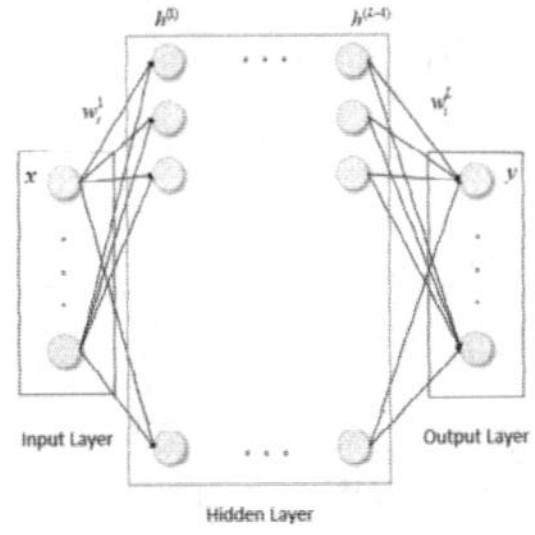

Figure 1 - Autoencoder

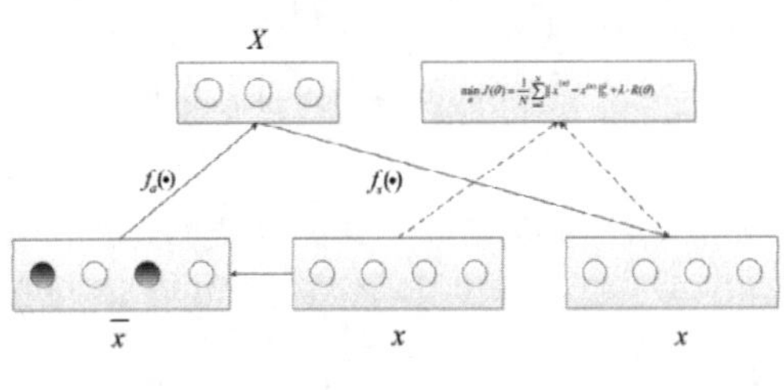

Figure 2 - DAE structure

Where x and z are both m-dimensional input column vectors and X is an n-dimensional output column vector.

$$\begin{cases} \{x^{(n)}\}_{n=1}^{N} \xrightarrow{Noising} \{\overline{x}^{(n)}\}_{n=1}^{N} \\ \overline{x}^{(n)} = x^{(n)} + \varepsilon^{(n)} \\ \varepsilon^{(n)} \sim E(\vartheta) \end{cases} \tag{2}$$

Where $E(\vartheta)$ is the distribution type of noise, considering the characteristics of the working environment of the underwater robot, we use Gaussian noise in the method introduced: $\overline{x}^{(n)} = x^{(n)} + Normal(0,1)$

Then pre-training the weight by minimizing the energy loss function, constructing the optimization objective function based on the energy loss [2] which is shown as follows:

$$\min_{\theta} J(\theta) = \frac{1}{N}\sum_{n=1}^{N} \| \overline{x}^{(n)} - x^{(n)} \|_2^2 + \lambda \cdot R(\theta) \tag{3}$$

In the above formula, the parameter is defined as a regular term as:

$$\begin{cases} \theta = [W_a, b_a; W_s, b_s] \\ R(\theta) = \| W_a \|_F^2 + \| W_s \|_F^2 \end{cases} \tag{4}$$

3.2. Fine-tuning using BP algorithm

After the DNN network pre-training is over, we use the BP algorithm to fine tune the network weights. We reversely update the weight parameters based on the output of the pre-trained DNN network. For a vectors of input x^m, $y^m = f_{\theta_{N+1}}(h_N^m)$ is the corresponding DNN network output, the optimization function is:

$$\varnothing_{DNN}(\Theta) = \frac{1}{M}\sum_{m} L(y^m, d^m) \tag{5}$$

Where Θ is the parameter set of weight, $\Theta = \{\theta_1, \theta_2, \cdots, \theta_{N+1}\}$, the parameter of weight is iterated as follows:

$$\Theta = \Theta - \eta \frac{\partial \varnothing_{DNN}(\Theta)}{\partial \Theta} \tag{6}$$

Where η is the learning rate of Bp algorithm[8].

3.3. Dropout method

Since the fault samples in the AUV state samples are less than the normal samples, it is easy to cause over-fitting of the network structure during network training. In order to prevent the over-fitting phenomenon from affecting the accuracy of the fault diagnosis of the diagnostic system, we introduce the dropout algorithm. Dropout is a method to prevent overfitting during training with a small sample network. In essence, Dropout can be thought of as setting the output of some hidden layer neurons at a certain ratio to 0, so that these selected neurons will not participate in the forward propagation of the training[9].

3.4. Auv's thruster fault diagnosis method based on deep learning

In this section, we apply deep learning to the propeller fault diagnosis of underwater robots. Firstly, the theoretical state value of the AUV yaw angle is obtained by the motion simulation model of the tunnel AUV, and then the residual of the measured state value is used as a fast Fourier transform to obtain 4000 sets of Fourier coefficients, due to the Fourier coefficient. Symmetry We take the first 2000 coefficients as input vectors for the DNN network. Secondly, a deep neural network consisting of multi-layer DAEs is built, and the noise-processed input vectors are used for pre-training, and the weights between the layers of the DNN are trained layer by layer. After pre-training with DAE, the weight parameters of the DNN network are initialized, and then the bp algorithm is used to fine tune the network weights. The technique of dropout is used in the training process to prevent over-fitting caused by the network due to fewer fault samples [10].

4. Experiment test

In order to verify the effectiveness of the deep DAE-based DNN method in the AUV thruster fault diagnosis system, and taking into account that the AUV is vulnerable to personnel and property losses when testing in the external field with fault conditions, the AUV fault motion is adopted. Simulation platform failure data. The data is collected by the semi-physical simulation platform of the laboratory tunnel robot project. The simulation program runs under the VC6.0 environment of 64-bit win10, and the motion control program runs in the PC104 board. The signal acquisition frequency is 2Hz.

The task of this test is to determine the fault degree of the AUV thruster by the residual signal of the theoretical steering angle data predicted by the sensor's measured heading angle and motion simulation during the AUV operation. In this study, we collected 600 sets of samples for each fault level. In each set of samples, the signal was subjected to fast Fourier transform to obtain 2000 Fourier coefficients (sample feature points). 4 Fourier map under fault level.

4.1. Influence of DNN hyperparameter

In the noise reduction automatic encoder, the noise pollution coefficient has a great influence on the final diagnosis result. Therefore, it is listed as one of the network hyperparameters in this paper. The influence of this coefficient on the diagnostic accuracy of DNN network is studied. It has great meaning. As shown in Figure 3, the pollution coefficient changes from 0 to 0.5, and the change of each step is 0.1. It can be seen from the figure that the diagnostic accuracy of DNN increases from the coefficient of 0 - 0.4 with the increase of the pollution coefficient. Out: Before the coefficient reaches 0.4, the pollution resistance of the input signal can increase the anti-interference ability and robustness of the DNN structure. At the same time, when the pollution coefficient is greater than 0.4, the diagnostic accuracy of the DNN network is also obviously degraded. It can be explained that when the pollution coefficient is too large, the quality of the signal is affected, resulting in a decrease in diagnostic accuracy. This confirms that the SAE trained with appropriate noise data can extract more powerful features than the traditional ones.

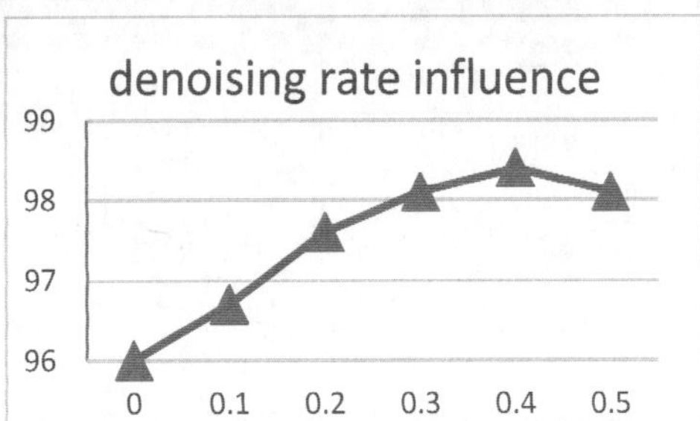

Figure 3 – denoising rate influence

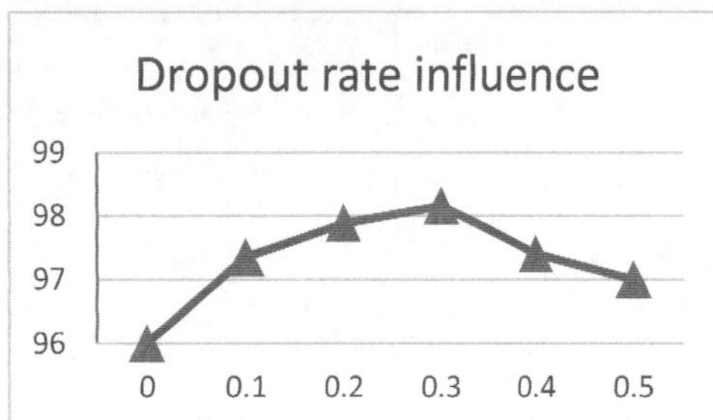

Figure 4 – dropout rate influence

In addition, over-fitting problems due to insufficient samples for deep networks have long been a major problem faced by researchers. This paper also studies the impact of the dropout rate on the diagnostic accuracy of the network structure. Here, the same interval and step size as when studying the noise rate are used. It can be seen from Figure 4 that the DNN network achieves the best performance when the dropout rate is close to 0.3. When the dropout rate exceeds 0.3, the diagnostic accuracy decreases. It can be concluded that the dropout rate of about 0.3 can better compensate the network.

4.2. Analysis of results

In order to prove the processed method can effectively learn the characteristics of each fault level signal of AUV propeller through DAE pre-training and fine-tuning, we show the classification result of DNN network through an effective data visualization technology "t-SNE"[11]. As shown in Figure 5, the same fault types are effectively gathered together, and different fault levels are effectively distinguished.

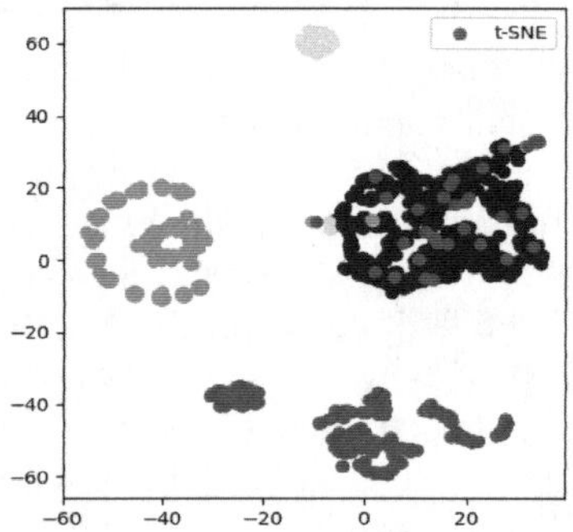

Figure 5 – feature map of four typical faults

Because the motion state of the AUV has great delay, it is difficult to accurately separate the normal state to the fault when extracting the fault data sample. The samples at the time of transition between states, so that in Fig. 5, there will be a case where a small number of data points of each fault state are gathered with normal state data points.

Finally, in order to prove that the selected DNN network structure and the traditional fault diagnosis method SVM, bp network have obvious advantages in diagnostic accuracy, we randomly conducted 50 sets of comparative experiments. The experimental results are shown in Table 1. From Table 1, we can see that the proposed method not only can identify different fault levels of the propeller, but also has obvious advantages in diagnostic accuracy compared with the traditional fault diagnosis method.

Table 1. The experimental results

Method	Classification accuracy
Proposed method	98.21
SVM based	93.45
BP based	73.58

5. Conclusion

In this paper, a fault diagnosis method for intelligent underwater robot propeller based on DNN deep neural network is proposed. The method can intelligently extract the characteristics of the thrust level of the thruster from the residual signal of the theoretical state and the measured state of the AUV, and perform a higher precision distinction. It has been found through experiments that the diagnostic accuracy is higher when the noising rate is around 0.4. Then, according to the characteristics of fewer fault samples in the normal condition of AUV operation, the dropout strategy is introduced to prevent over-fitting during network training. It is found through experiments that when the dropout rate is around 0.3, the DNN network is for AUV.

Acknowledgment

This research was part of the project (Grant No. 51779057,51709061) supported by National Natural

Science Foundation. It has also been partially supported by Equipment pre-research project(Grant 41412030201). Great thanks are addressed to them by the research team.

REFERENCES

[1] Zhang, Ming Jun, et al. "Thruster fault diagnosis in autonomous underwater vehicle based on grey qualitative simulation." *Ocean Engineering* **105(2015):247-255.2**

[2] Chen, Zhouliang, and Z. Li. "Research on fault diagnosis method of rotating machinery based on deep learning." *Prognostics and System Health Management Conference 2017*:**1-4**.

[3] Yu-shan Sun, Xiang-rui Ran, et al." Thruster fault diagnosis method based on Gaussian particle filter for autonomous underwater vehicles". *Naval Architecture and Ocean Engineering* **8(2016):243-251**

[4] Juan Wu. *Research on Multi-fault Diagnosis Technology of Autonomous Underwater Vehicle Propeller and Sensor*. Diss. Harbin Engineering University, 2011.

[5] Zhixian Jin. *Underwater robot thruster fault diagnosis technology research* Diss. Harbin Engineering University, 2006.

[6] Jia, Feng, et al. "Deep neural networks: A promising tool for fault characteristic mining and intelligent diagnosis of rotating machinery with massive data." *Mechanical Systems & Signal Processing* **72-73(2016):303-315.**

[7] Pan, Jun, et al. "LiftingNet: A Novel Deep Learning Network with Layerwise Feature Learning from Noisy Mechanical Data for Fault Classification." *IEEE Transactions on Industrial ElectronicsPP*.**99(2017):1-1**.

[8] Tan, Junbo, et al. "Fault diagnosis method study in roller bearing based on wavelet transform and stacked auto-encoder." *Control and Decision Conference IEEE,* **2015:4608-4613.**

[9] Li, Xun, et al. "Sensor fault diagnosis of autonomous underwater vehicle based on extreme learning machine." *Underwater Technology IEEE,* **2017:1-5.**

[10]K. Yu, L. Jia, Y. Chen, and W. Xu,"deep learning: Yesterday, Today, and Tomorrow", *Journal of Computer Research and Development*, vol**. 50(9), pp. 1799-1804, 2013.**

[11]Maaten, L, and G. Hinton. "Visualizing data using t-SNE." *Journal of Machine Learning Research*. **9.2605(2008):2579-2605.**

IOP Conf. Series: Materials Science and Engineering **470** (2019) 012036 doi:10.1088/1757-899X/470/1/012036

Piezoelectric Energy Harvesting Based on Vertical Vibration of Car Body Induced by Road Surface Roughness

Hongyan Wang[1,a], Fengjuan Miao[2], Zhilong Sun[1] and Liqiu Jing[1]

[1] College of Mechatronic Engineering, Qiqihar University, Qiqihar 161006, China
[2] College of Communications and Electronics Engineering, Qiqihar University, Qiqihar 161006, China

[a] wanghongyan1993@163.com

Abstract. This paper presents the analysis of a cantilevered piezoelectric energy harvester (PEH) under the random vibration excitation of the car body. A two degree-of-freedom (2-DOF) quarter car model is established. The Matlab/Simulink simulation software is used to analyze the dynamic response of the car body under the random road excitation. The transient analysis of the PEH is carried out using ANSYS software to obtain the time-dependent voltage, current and power outputs of the PEH. The root mean square (RMS) power is used to evaluate the electricity generation of the PEH. The effect of the load resistance, the car speed, and the road surface roughness on RMS power are studied. The results show that it exists the optimal load resistance to achieve the maximum RMS power output for the vibration-based PEH. When the road surface roughness is constant, the increase of the car speed can induce the increase of the RMS power of the PEH. When the car speed is constant, higher RMS power outputs of the PEH can be achieved by increasing road surface roughness coefficient.

1. Introduction

With the rapid development of the low power electronic devices, the conventional batteries used to power these devices show some disadvantages, such as large volume, limited lifespan and high cost. Energy harvesting from ambient vibrations provides an exciting solution. Vibration energy can be converted into electricity via electrostatic mechanism [1], electromagnetic induction [2], or direct piezoelectric effect [3]. Among them, the PEHs have attracted more attention from researchers due to small volume, high power density, and easy to integrate with other electronics.

Among the literatures on piezoelectric energy harvesting, the input excitation is usually assumed to be a deterministic harmonic wave [4], while in practical situations, the mechanical excitation is a random signal. In this paper, the piezoelectric energy harvesting based on the vertical random vibration excitation of the car body is studied. Firstly, the dynamic responses of a 2-DOF quarter car under the road random excitation are obtained by using Matlab/Simulink simulation software. Subsequently, with the vertical displacement excitation from the car body, the time-dependent voltage, current and power outputs of the PEH are obtained by using ANSYS software. Furthermore, the effect of the load resistance, the car speed, and the road surface roughness on RMS power of PEH are investigated.

2. Modeling and simulation of road-car system

In this section, we establish the state-space equation for a 2-DOF quarter car model and use Matlab/Simulink simulation software to analyze the dynamic response of the car body under the random road excitation. The acceleration spectrum analysis of the car body provides a reference for the design of the PEH. The vertical displacement response of the car body is used as the excitation source of the PEH.

2.1. 2-DOF quarter-car model

Figure 1 shows a 2-DOF quarter car model. It comprises two moving masses (m_1 and m_2) connected in series with their respective springs (k_1 and k_2) and a damper (c) between m_1 and m_2. m_1 and m_2 are the mass of the car wheel and the car body, respectively. k_1 and k_2 are the spring stiffness of the tyre and the suspension, respectively. c is the damping coefficient of the suspension. q is the excitation displacement of the road surface. u_1 and u_2 are the absolute displacement of m_1 and m_2, respectively.

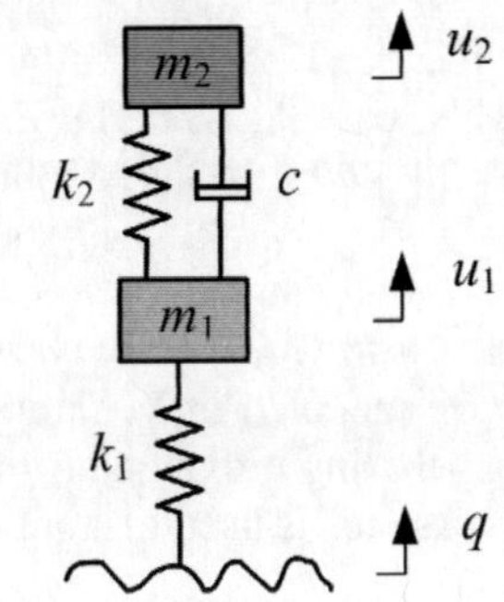

Figure 1. 2-DOF quarter car model.

The governing equations of the 2-DOF quarter car model can be written as

$$\begin{cases} m_2\ddot{u}_2 + c(\dot{u}_2 - \dot{u}_1) + k_2(u_2 - u_1) = 0 \\ m_1\ddot{u}_1 + c(\dot{u}_1 - \dot{u}_2) + k_2(u_1 - u_2) + k_1(u_1 - q) = 0 \end{cases} \tag{1}$$

where $\dot{u}_1$ and $\dot{u}_2$ are the velocities of the car wheel and car body in the vertical direction, respectively. $\ddot{u}_1$ and $\ddot{u}_2$ are the accelerations of the car wheel and car body in the vertical direction, respectively.

Defining the state space vector as

$$\mathrm{x} = [x_1\ x_2\ x_3\ x_4]^T = [\dot{u}_2\ u_2\ \dot{u}_1\ u_1]^T \tag{2}$$

$$\dot{\mathrm{x}} = [\dot{x}_1\ \dot{x}_2\ \dot{x}_3\ \dot{x}_4]^T = [\ddot{u}_2\ \dot{u}_2\ \ddot{u}_1\ \dot{u}_1]^T \tag{3}$$

$$\mathrm{y} = [y_1\ y_2\ y_3]^T = [\ddot{u}_2\ \dot{u}_2\ u_2]^T \tag{4}$$

$$\mathrm{u} = q \tag{5}$$

We can write the governing equations of the 2-DOF quarter car model in the state space form as

$$\begin{aligned} \dot{\mathrm{x}} &= A\mathrm{x} + B\mathrm{u} \\ \mathrm{y} &= C\mathrm{x} + D\mathrm{u} \end{aligned} \tag{6}$$

where

$$A=\begin{bmatrix} -\frac{c}{m_2} & -\frac{k_2}{m_2} & \frac{c}{m_2} & \frac{k_2}{m_2} \\ 1 & 0 & 0 & 0 \\ \frac{c}{m_1} & \frac{k_2}{m_1} & -\frac{c}{m_1} & -\frac{k_1+k_2}{m_1} \\ 0 & 0 & 1 & 0 \end{bmatrix},\quad B=\begin{bmatrix} 0 \\ 0 \\ \frac{k_1}{m_1} \\ 0 \end{bmatrix},\quad C=\begin{bmatrix} -\frac{c}{m_2} & -\frac{k_2}{m_2} & \frac{c}{m_2} & \frac{k_2}{m_2} \\ 1 & 0 & 0 & 0 \\ 0 & 1 & 0 & 0 \end{bmatrix},\quad D=\begin{bmatrix} 0 \\ 0 \\ 0 \end{bmatrix}$$

A time-domain expression of the road surface roughness [5] is employed in this paper, which is written as

$$\dot{q}=0.111\left[-vq+40\sqrt{G_q v}w_0\right] \tag{7}$$

where $\dot{q}$ is the excitation velocity of the road surface. v is the car velocity. w_0 is the white noise with the covariance value of $1\text{m}^2/\text{s}$. G_q is the road surface roughness coefficient. The road surface coefficient G_q is classified into eight levels, e.g., A, B, C, D, E, F, G, H, which have corresponding geometrical mean value as 16, 64, 256, 1024, 4096, 16384, 65536, and 262144, respectively. The unit of G_q is 10^{-6}m^3.

2.2. Vibration simulation of road-car model using Matlab/Simulink software

The theoretical equations in section 2.1 are modeled in Matlab/Simulink, as shown in figure 2. This model includes mainly two modules, e.g., the input excitation module (equation (7)) and the 2-DOF quarter car module (equation (6)). The workspace is used to store the analytical results.

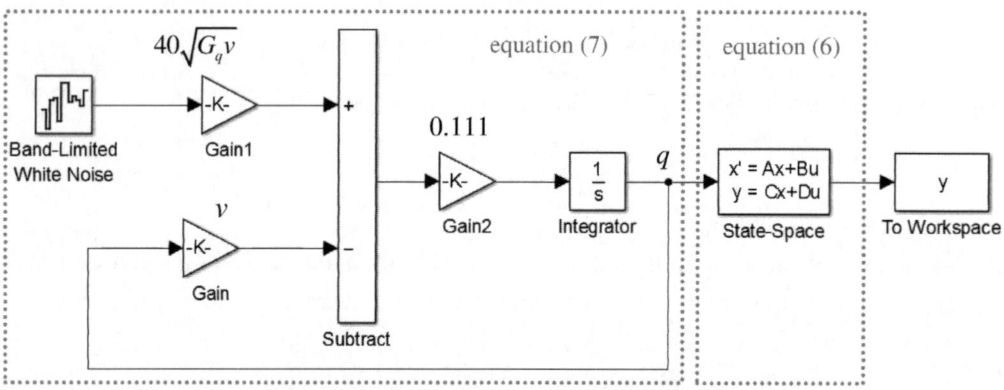

Figure 2. Simulink model for road-car system.

The parameters used in the dynamic analysis of the road-car system are m_1=45.4 kg, k_1=192000 N/m, m_2=317.5 kg, k_2=22000 N/m, c=1520 Ns/m, v=10 m/s, and $G_q=16\times10^{-6}\text{ m}^3$ (Level A road surface). Figures 3(a), 3(b), and 3(c) show the vertical displacement, velocity and acceleration responses of the car body, respectively. The response time is 10 s. It can be seen from figures 3(a), 3(b), and 3(c) that the time-dependent displacement, velocity and acceleration responses are random. Figure 3(d) plots the acceleration power spectral density (PSD) of the car body. It can be seen from figure 3(d) that the acceleration PSD of the car body shows the highest peak at 1.074 Hz and the second highest peak at 11.23 Hz. The two peaks are induced by the resonance of the 2-DOF quarter car system. It is well known that the PEH is efficient near resonance. In this paper, we consider to design the PEH with the natural frequency near to 11.23 Hz to harvest vibration energy from the car body. We do not design the harvester with the natural frequency near to 1.074 Hz. The reason is that the fatigue cracking problem is easier to arise in the harvester with lower natural frequency (1.074 Hz) due to the larger amplitude of oscillation.

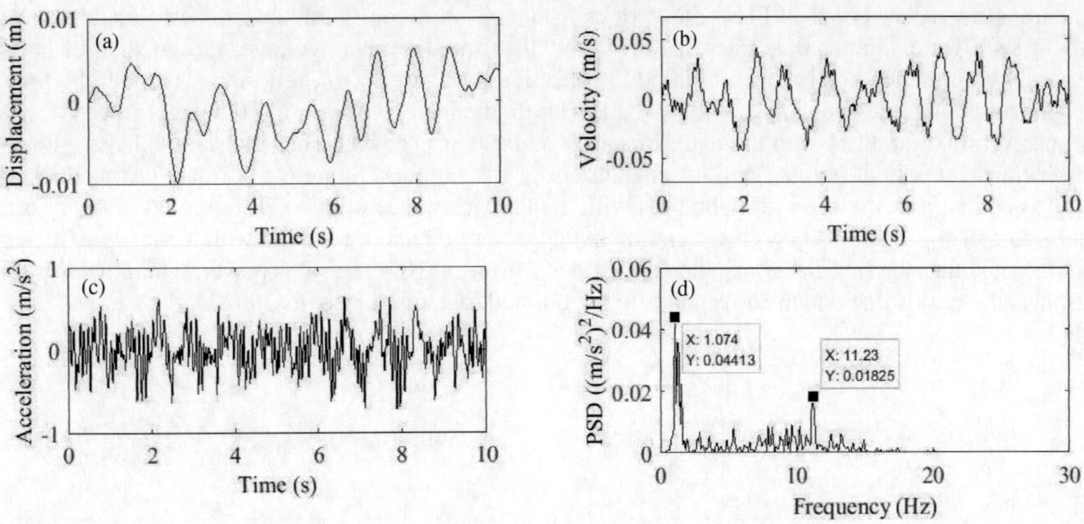

Figure 3. Dynamic responses of car body. (a) Time-dependent displacement; (b) Time-dependent velocity; (c) Time-dependent acceleration; (d) Acceleration power spectral density.

3. Vibration energy harvesting using cantilevered PEH

3.1. Model of cantilevered PEH

Figure 4 shows the structure of the cantilevered PEH. The beam is a piezoelectric sandwich structure with a central brass plate and two piezoelectric plates. The two piezoelectric plates were boned to the top and bottom of the brass plate, respectively. The piezoelectric bimorphs are oppositely polarized in the *z* direction and are connected in series. The output terminal is directly connected to a resistor (*R*). *L*, *w*, h_p, and h_m are the beam length, the beam width, the thickness of piezoelectric plate, and the thickness of brass plate, respectively. M_t is the tip mass. The structural parameters used in the modeling of the PEH are L=66 mm, w=20 mm, h_m=0.3 mm, h_p=0.2 mm, and M_t=0.09 kg. The fundamental natural frequency of the PEH is 11.116 Hz, which is near to 11.23 Hz. Figure 5 shows the finite element (FE) model of the cantilevered PEH in ANSYS software. The 8-node hexahedral coupled-field element SOLID5 is used for the piezoelectric plates. The 8-node linear structural element SOLID45 is used for the brass plate. The circuit element CIRCU94 is used to model the resistor. The point element MASS21 is used for the lumped tip mass. The piezoelectric coefficients of PZT-5H used in ANSYS are available in literature [6]. The polarization direction of the piezoelectric material is represented by the sign of the piezoelectric constants. For the series-type piezoelectric bimorphs, the signs of the piezoelectric constants for the top and bottom piezoelectric plates are opposite. The electrode layer is simulated by coupling the nodal voltage DOF to ensure a uniform electrical potential. The mass-weighted damping and the stiffness-weighted damping are set as 1.3529 and 9.023e-6, respectively. The excitation source is from the vertical displacement of the car body.

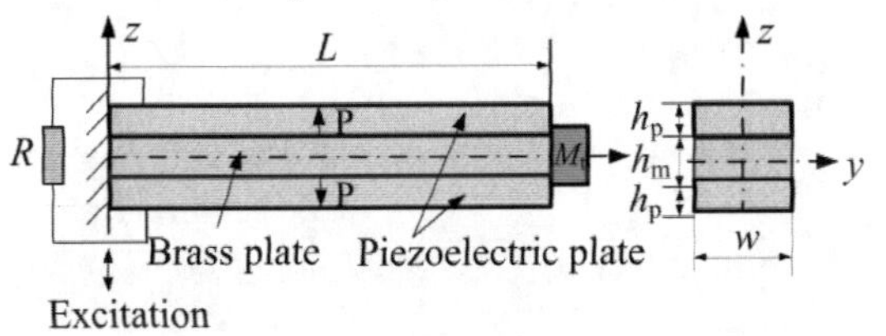

Figure 4. Structure of cantilevered PEH.

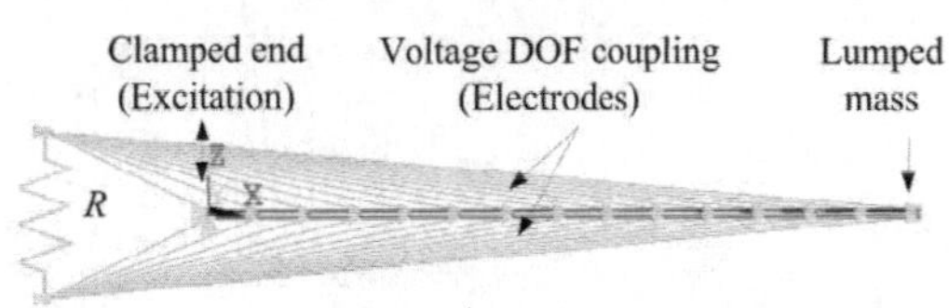

Figure 5. FE model of cantilevered PEH.

3.2. Electrical performance analysis of PEH

The transient analysis of the PEHs with various load resistance configurations are performed by using ANSYS software. Figures 6(a) , 6(c) and 6(e) show the time-dependent voltage, current and power for the PEH with the load resistance of 200 kΩ, respectively. Figures 6(b) , 6(d) and 6(f) show the time-dependent voltage, current and power for the PEH with the load resistance of 1000 kΩ, respectively. It is observed that the PEH with the load resistance of 1000 kΩ shows higher transient voltage outputs, lower transient current outputs, and lower transient power outputs (figure 6(b) , figure 6(d) and figure 6(f)) as compared with those from the PEH with the load resistance of 200 kΩ (figure 6(a) , figure 6(c) and figure 6(e)). Figure 7 shows the variation in the power outputs of the PEH with the load resistance. In figure 7, the power outputs are the RMS values from the10 s transient power. The power graph displays the peak value which corresponds to the optimal load resistance, R=200 kΩ.

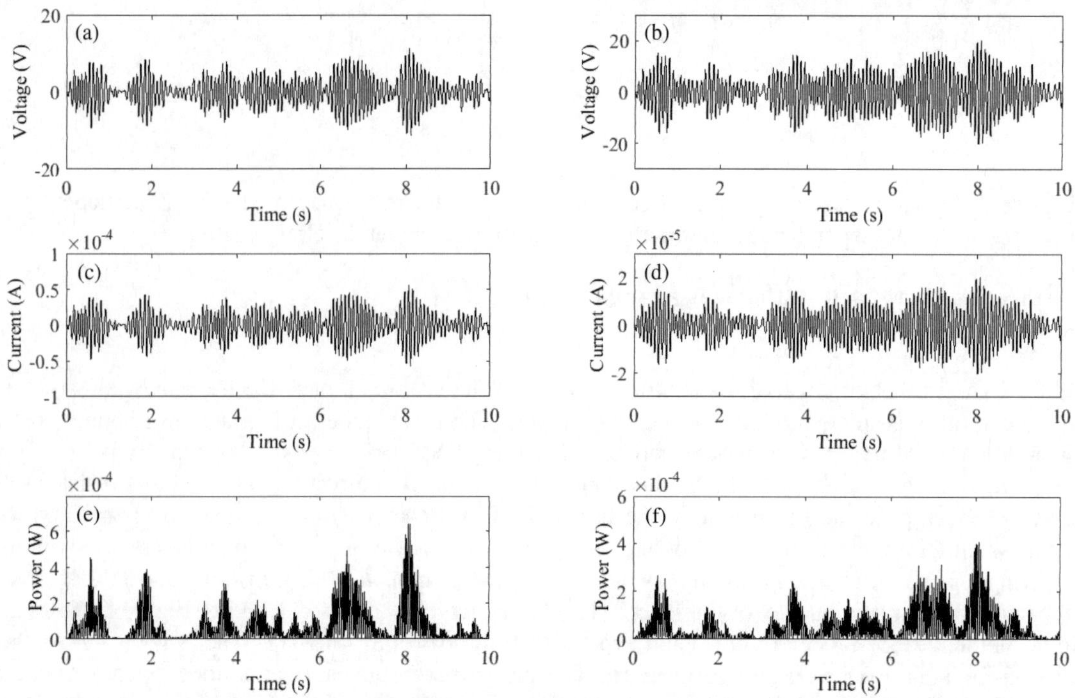

Figure 6. Voltage, current and power outputs of PEH. (a), (c) and (e) R=200kΩ; (b), (d) and (f) R=1000kΩ. Level A road surface. v=10 m/s.

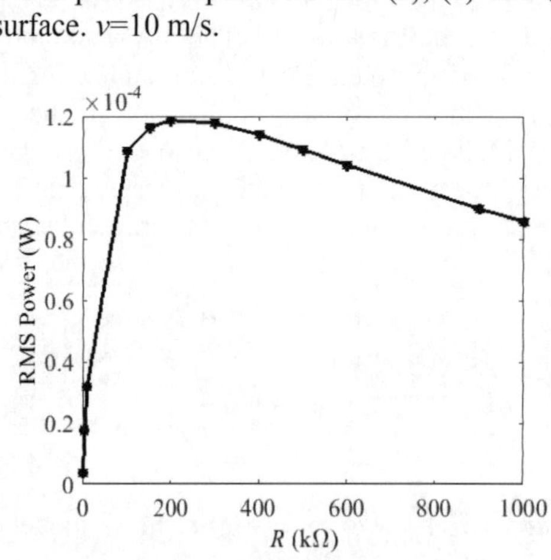

Figure 7. RMS power of PEH versus load resistance. Level A road surface. v=10 m/s.

3.3. Effect of vehicle speed and road roughness on power

With the optimal load resistance, we analyze the effect of car speed and road surface roughness on RMS power of the PEH. Figures 8(a) and 8(b) show the RMS power of the PEH for different car speed configurations and the road surface roughness configurations, respectively. It can be observed from figure 8(a) that with the same road surface roughness level, the increase of the car speed can induce the increase of the RMS power of the PEH. It can be observed from figrue 8(b) that with the same car speed condition, the PEH can harvest higher RMS power when the road surface roughness becomes larger.

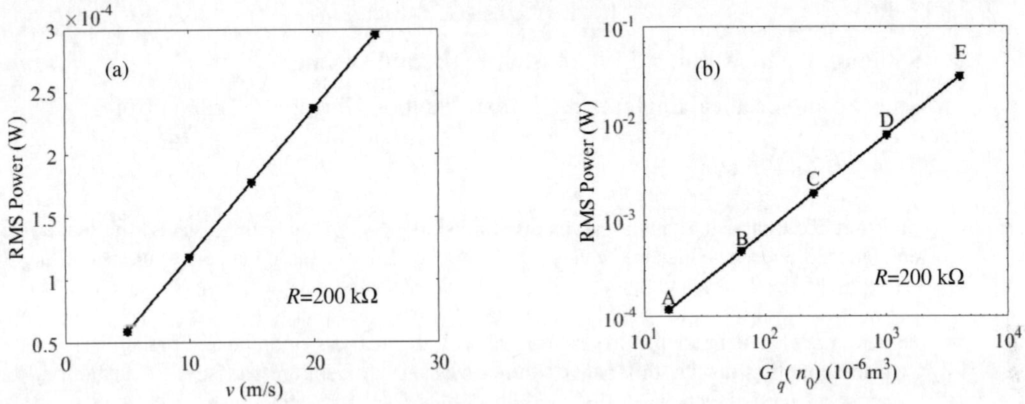

Figure 8. Effect of car speed and road surface roughness on RMS power of PEH. (a) Level A road surface; (b) *v*=10m/s.

4. Conclusions

In this paper, we analyze the dynamic response of the car body under the random road excitation. The vertical displacement of the car body is used as the excitation source for the electrical generation of the vibration-based PEH. The time-dependent voltage, current and power of the PEH are obtained by using the transient analysis in ANSYS software. The RMS power is used to evaluate the electricity generation of the PEH. We obtain the optimal load resistance of the PEH for the maximal power output of the PEH. With the optimal load resistance, the effect of the car speed and the road surface roughness on RMS power are studied. The results show that with the same road surface roughness, the increase of the car speed can induce the increase of the RMS power of the PEH. With the same car speed, larger road surface roughness coefficient can achieve higher RMS power of the PEH.

Acknowledgments

This work is supported by Heilongjiang Provincial Natural Science Foundation of China (No. LC2017028) and Fundamental Research Funds in Heilongjiang Provincial Universities of China (No.135209229).

References

[1] Tao K, Wu J, Tang L, Hu L, Lye S W and Miao J 2017 *J. Micromech. Microeng.* **27** 044002
[2] Samad F A, Karim M F, Paulose V and Ong L C 2016 *IEEE Sens. J.* **16** 1969-74
[3] Fan K, Liu Z, Liu H, Wang L, Zhu Y and Yu B 2017 *Appl. Phys. Lett.* **110** 143902
[4] Wang H and Tang L 2017 *Mech. Syst. Signal Pr.* **86** 29-39
[5] Wu Z, Chen S, Yang L and Zhang B 2009 *Transactions of Beijing Institute of Technology* **29** 795-98
[6] Wang H, Shan X and Xie T 2012 *J. Zhejiang Univ.-Sci A* **13** 526-37

IOP Conf. Series: Materials Science and Engineering **470** (2019) 012037 doi:10.1088/1757-899X/470/1/012037

Friction-thermal-vibration Coupling Analysis of Large-megawatt Wind Turbine Brake

S Zhang, P Xin, J Yin, Y Liu*, Z Sha, F Ma and D Yang

School of Mechanical Engineering, Dalian Jiaotong University, Dalian, China

liuyu_ly12@126.com

Abstract. To establish a mutual contact relationship between the brake disc and the brake pad, simulation of the brake disc in the actual working conditions braking deceleration, according to the wind of the brake and brake piece of actual size and related parameters of brake disc material in ABAQUS to establish 3D transient friction-thermal-vibration coupling finite element model. The brake process is verified with the analysis of brake and brake disc contact, from the coupling results, this paper summarizes the temperature and stress distribution, the research on the temperature in time and space distribution. In this paper, the stress of brake stress is analyzed, and the distribution of stress is summarized, and the importance of the coupling study on the stress analysis of brake is analyzed.

1. Introduction

The braking process of the large- megawatt wind turbine brake under the actual working condition is very complicated. Friction-thermal-vibration coupling will have a great influence on the braking performance. Study on friction-thermal-vibration coupling is of great significance on the design and optimization of the wind turbine brake. In recent years, many studies were carried out on the field of thermal-structural coupling analysis and vibration analysis respectively. The heat conduction equation and thermal stress equation are used to describe the thermal structure coupling process by mathematical analytic method. Zagrodzki[1] et al established the 2D symmetric model of the brake and analyzed the thermo-elastic problem. The finite element method was used to describe the temperature and the thermal stress. Lu[2] et al took the train disc brake as the research object, established a 3D transient thermal structural coupling finite element analysis model and simulate the actual braking conditions. The results showed that the uneven distribution of temperature and equivalent stress on the contact surface leads to the reducing of actual contact area. Ghosh[3] through the analysis of the vibration of the brake disc with variable thickness, it is found that the temperature caused by the friction in the braking process has a certain influence on the vibration. The finite element model of automobile disc brake is set up by Dai[4] showed the influence of the coupling state of temperature field and stress field on the vibration of brake. In this paper, considering friction-thermal-vibration coupling, the vibration coupling dynamic force is put forward, and the friction-thermal-vibration coupling model of the braking process is established with ABAQUS software.

2. Vibration analysis in large-megawatt wind turbine brake

2.1. Vibration analysis of brake pressure during the braking process

Vibration can be regarded as uniform force when the disc brake is braking urgently. Assuming that the braking pressure of the unit area on the upper surface of the brake pad is *p*, the compression force *F* generated by each element is:

$$\mathrm{d}F = kpr\mathrm{d}\theta\mathrm{d}r \tag{1}$$

where *k* is the effective contact area coefficient in the brake disc/pad, *r* is the radius of the brake disc, θ is the angle of the radial boundary of the brake pad.

According to the knowledge of thermodynamics, the friction heat generated during braking will cause thermal expansion of brake pad, and its expansion thickness in axial direction $\Delta\delta_s(r,t)$ is:

$$\Delta\delta_s(r,t) = \frac{\frac{\pi}{6} a_s \delta_s \mu kp\left(r^3 - r_0^3\right)\left[\theta(t) - \theta(t_0)\right]}{C_a \gamma_a\left(\delta_s + \delta_f\right)} \tag{2}$$

where a_s and a_f represent the thermal expansion coefficient in the brake disc/pad respectively. The deformation of the brake pad in the axial direction is a linear function relationship with the three square of the radius of the brake. With the increase of the radius, the deformation of the material will also increase.

2.2. Effect of deformation on brake pressure

The initial state of the brake disc is uniform in shape, and in the working process it will become thin in the outer thickness. The distribution of the pressure on the brake sluice is also small, and the relationship between the braking force and the radius on the brake sluice is obtained by converting the type (2) to the brake disc.

$$p\left(r^3\right)r^3 = \lambda\Delta\delta(r,t) + p\left(r_0^3\right)r_0^3 \tag{3}$$

Considering the integer variable is not very large, the expansion thickness is assumed to be a fixed value. Therefore, a new expression of pressure distribution on the brake sluice is proposed and defined as a vibration coupling dynamic force.

$$p\left(r^3\right)r^3 = C \tag{4}$$

That is to say, the three power product of the brake pressure and the brake disc radius is the fixed value *C*, so that the pressure distribution on the brake pad has certain regularity.

The vibration and noise are caused by the heat deformation of the material, which leads to the change of the braking force during the whole braking process. The expression of the vibration coupling force can better express the distribution of brake pressure on the brake Sluice at every moment. Combined with the temperature field and stress field, the actual braking condition can be simulated more effectively, and the analysis results are closer to the actual working conditions.

3. Establishment of friction thermal vibration coupling model

3.1. Assumptions

During the braking process, a large amount of heat is generated by the friction in brake disc/pad. The heat is distributed in a gradient on the brake disc. The heat is transmitted to the interior of the disc and in the surrounding air. The main heat transfer mode is heat conduction and heat convection. As the temperature of the braking process is very high, there will be a small amount of heat radiation. When models of the brake disc/pad are established, the following assumptions are made, which are material of the brake disc/pad are isotropic, and parameters of materials do not change with temperature variation. The friction coefficient μ in the brake disc/pad is constant during the braking process. Wear of the materials during the braking process is ignored, there is only elastic deformation. During the braking process, all of the friction work is converted to heat, and the heat dissipation of the brake disc is non-uniform.

3.2. Determination of related parameters

According to a certain type of large-megawatt wind turbine brake, material of the brake pad is copper-based powder metallurgy, and the brake disc's is high thermal conductivity cast iron. The relevant parameters of the brake disc/pad under different temperature are shown in Table 1 and Table 2:

Table 1. Material parameters of copper-based powder metallurgy.[5]

Temperature °C	Thermal conductivity w/(kg·°C)	Specific heat capacity J/(kg·°C)	Modulus of elasticity MPa	Linear expansion coefficient /°C
100	380	436	1.72×10^5	1.11×10^{-5}
200	364	436	1.57×10^5	1.12×10^{-5}
300	352	436	1.41×10^5	1.12×10^{-5}

Table 2. Material parameters of highly conductive cast iron.[6]

Temperature °C	Thermal conductivity w/(kg·°C)	Specific heat capacity J/(kg·°C)	Modulus of elasticity MPa	Linear expansion coefficient /°C
100	57	497	0.98×10^5	1.12×10^{-5}
200	74	525	0.88×10^5	1.20×10^{-5}
300	74	574	0.84×10^5	1.23×10^{-5}
400	74	634	0.71×10^5	1.27×10^{-5}
500	74	688	0.67×10^5	1.30×10^{-5}

3.3. Determination of vibration boundary conditions

When the friction-thermal-vibration coupling model is established, the influence of the spindle vibration on brake pressure should be fully considered. The braking pressure on the brake pad is applied dynamically when the model is established. In this paper, the vibration boundary of the brake pad is based on the data of the whole acceleration a of the brake pad through dynamic simulation with ADAMS software, which is transformed by the Newton's second law. The initial brake pressure F is 17000N, the initial spindle speed ω is 1000rmp and the friction coefficient is 0.3, and the acceleration-time variation curve of the brake pad is shown in Figure 1, in which the positive and negative scale of the ordinate represent the direction of acceleration (up is positive and down is negative). Combined with the proposed brake vibration coupling dynamic equation, the braking pressure at each moment is the force at the center radius of the brake pad, and the brake pressure on the brake pad is rearranged at every moment. Same method is used with different working conditions. The friction-thermal-vibration FE analysis model is presented in Figure 2.

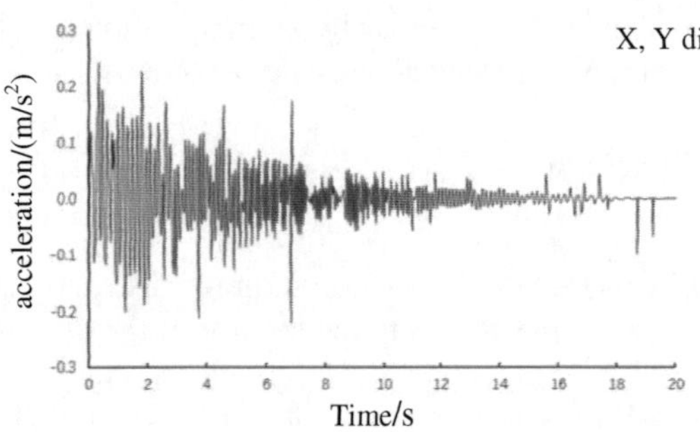

Figure 1. Acceleration-time variation curve of the brake pad.

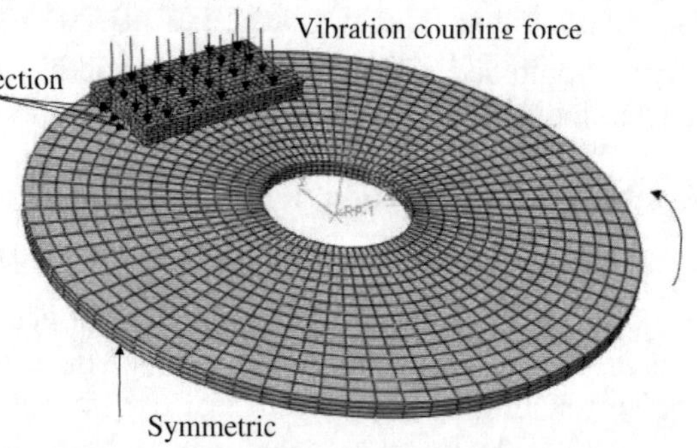

Figure 2. Friction-thermal-vibration coupling finite element model of the disc brake.

4. Simulation results and analysis of brake disc temperature and stress

The distribution contour of the transient temperature of the brake disc when the initial brake pressure is 17000N, the initial brake disc speed is 1000rpm, and the friction coefficient is 0.3 is shown in Figure 3. As can be seen, the high temperature is concentrated in the friction area, and the temperature rise in the non-friction area is not obvious. The maximum temperature is related to the position of the contact area. The initial contact temperature was the highest and then concentrated in the central annular area where the radius was 300mm. The temperature rise near the friction area spreads to both sides.This is because at the initial stage of braking, frictional heat flux input dominates the distribution of the temperature field. Because the brake is a three-dimensional model, as the braking process progresses, due to the thermal deformation of the friction material, the heat flow input to the brake disc is not axis symmetric. As the brake disc speed decreases, the heat input intensity gradually decreases, and the effect on the distribution of the temperature field gradually decreases.

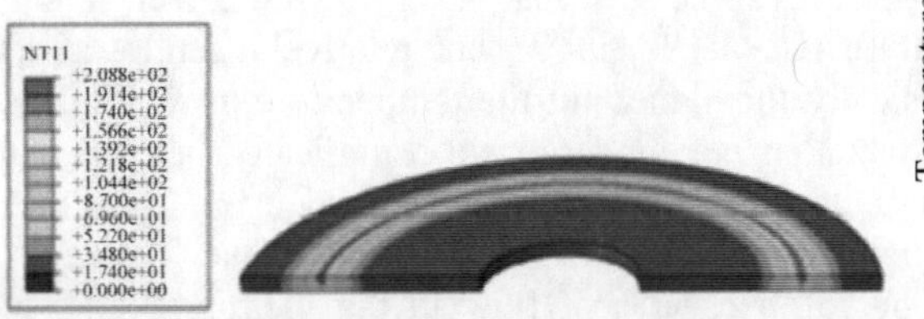

Figure 3. Brake disc temperature distribution contour.

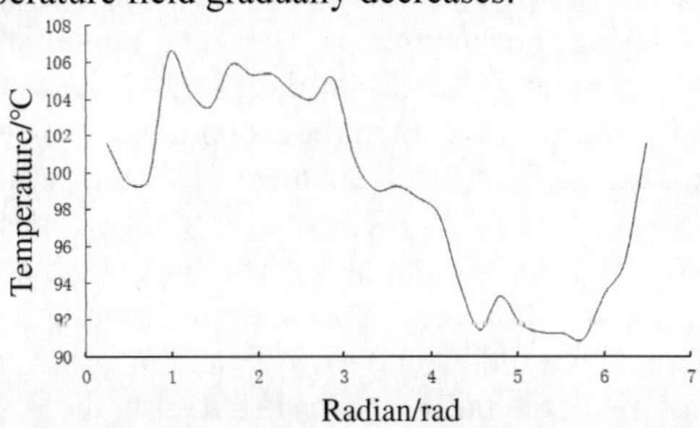

Figure 4. Temperature variation curve of the direction of rotation of the brake disc (0.50s).

To quantitatively describe the circumferential distribution of the temperature shown in Figure 3, 26 equally spaced nodes at the disc surface contact radius R_1=300mm in the counterclockwise direction are chosen to draw the temperature-arc curve in Figure 4. The starting point on the curve is the node at the corresponding radius in the positive direction of the Y axis. The ending point on the curve is the node closest to the Y axis on the right side of the Y axis. The temperature difference at the time of 0.50s is not obvious, and the temperature fluctuates counterclockwise from the left end of the contact with the brake pad until it contacts the right end of the brake pad again.

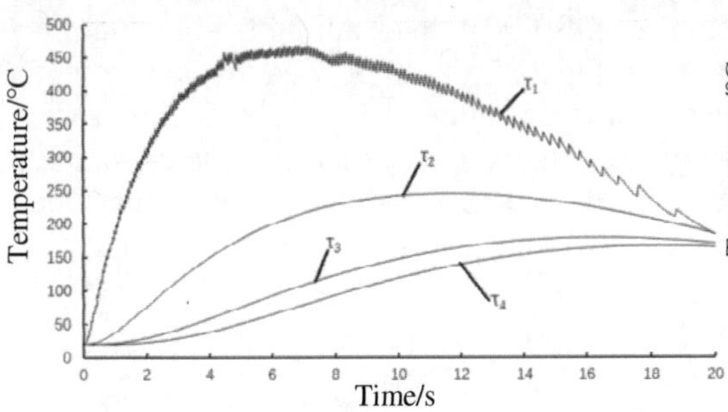

Figure 5. Nodes temperature of the brake disc in the axis-direction.

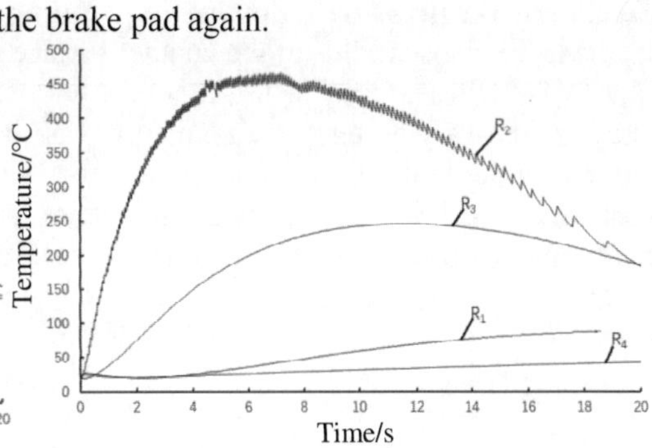

Figure 6. Nodes temperature of the brake disc in the radial-direction.

Figure 5 is the curve of the temperature of nodes along the Z axis selected at the radius of the disc R_1=300mm along the Z axis at τ_1=0, τ_2=6mm, τ_3=12mm, τ_4=20mm and plot the temperature of the node over time. The temperature rise of each node is different in Figure 5. The temperature rise at the contact surface of the brake disc is relatively rapid and much higher than that of the internal node of the brake disc. It can be found that only the temperature rise of the contact region node fluctuates. In the axial direction, the nearer the contact surface is, the more obvious the temperature rise during the braking process and the gradual transfer of frictional heat flow to the inside are. The temperature rise of the inner node of the brake disc far from the contact surface is obviously delayed compared with that of the node near the brake disc contact area. This is because the internal node heat flow input along the Z-axis direction except the nodes in contact area are mainly derived from the thermal conduction of the upper nodes. However, the heat transfer distance is proportional to the time, and the upper layer needs a certain amount of time to accumulate heat flow to the internal materials. Node τ_2 will output to the lower layer in addition to the input of the upper layer heat flow during the braking process. In the first 12 s, the heat flow input of Node τ_2 is larger than the heat flow output. The node temperature rises and reaches the maximum temperature in the 12th second. After that, the heat flow output is larger than the heat flow input, and the node temperature begins to decrease.

Select nodes with the radius R_1=275mm, R_2=300mm, R_3=325mm, and R_4=350mm in the positive direction of the Y-axis of the brake disk contact surface. Extract the temperature value of each node at each moment and draw the curve of the node temperature as shown in Figure 6. It can be seen that there is a temperature gradient with a large temperature difference in the radial direction of the brake disc during the braking process. The temperature changes at R_2 and R_3 in a parabolic shape. The temperature at R_2 varies in the form of fluctuations during the braking process, and the decreasing trend is more slowly than the rising trend. However, it can be seen that the temperature at R_2 is higher than that at R_3 at each moment. The temperature rise at R_1 and R_4 are relatively gentle, and the temperature of R_1 and R_4 in the first 4s are basically the same, but the temperature at R_1 after 4s is always higher than that at R_4. The brake disc contact surface undergoes a complicated heat exchange process, and the three heat exchange strengths occurring at every moment and every position are different. The temperature rise R_2 and R_3 is not only due to the frictional heat flow contact between the brake disc and the brake pad, but also the input of friction heat flow in the vicinity of the high temperature conduction form, which causes the obviously temperature rise. A temperature gradient with a large temperature difference is formed on the contact surface of the brake disc. After the above-mentioned nodes separating from the brake pad, the heat exchange form becomes heat dissipation into the air. Due to the deceleration, the input of heat flow generated by friction is reduced in each circle, but the output gradually increases, the input of heat flow in the early stage is larger than the output, the temperature of the node rises and reaches the maximum value, and the output of heat flow in the later stage is greater than Input, the node temperature decreases.

5. Effect of brake pressure on transient coupling of the brake disc

Select the node whose radius of the contact surface of the disc is R=300mm in the positive direction of the Y-axis as the research object. Extract the temperature at different times during the braking process and draw a time-varying temperature diagram as shown in Figure 7. It can be seen that under the three conditions, as the brake pressure increases, the brake disk temperature change rate also increases. In the early stage of braking, the stress fluctuation amplitude is small, and then the change amplitude gradually increases. The larger the initial brake pressure, the greater the stress change. This situation will aggravate the thermal instability of the brake disc and reduce the service life of the brake disc.

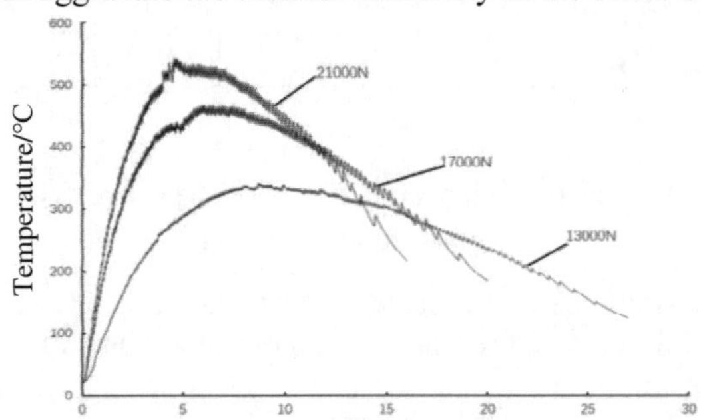

Figure 7. Nodes temperature of the brake disc at different initial pressure(R=300mm).

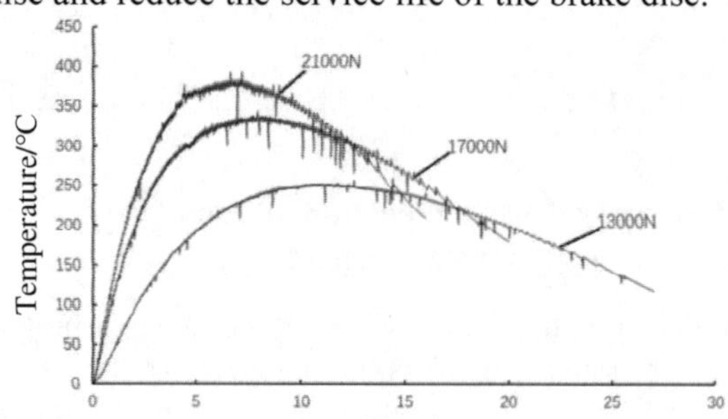

Figure 8. Nodes von mises stress of the brake disc at different initial pressure(R=300mm).

This is mainly because the relative speed of the brake is relatively high and the change frequency of stress is high, so the initial equivalent stress increases rapidly, but with the braking process, the speed of the brake disc gradually decreases and the equivalent stress change frequency of the contact area of the brake disc becomes smaller, so the stress variation characteristics of the large value fluctuation shown in the diagram show. In addition, the greater the stress of the larger braking force will increase the amplitude of the equivalent stress fluctuation. The equivalent stress value of the node in the positive direction of the Y axis at the radius of R=300mm under the three working conditions is drawn, and the equivalent stress time variation of the node under the action of different initial brake pressure shown in Figure 8 is drawn. It can be seen from the image that the equivalent stress time variant curve is lagging behind the corresponding temperature time curve.

6. Conclusion

Based on the heat and stress analysis of the megawatt wind electric brake, the coupling problem of the sliding temperature field, the stress field and the vibration between the brake disc and the brake disc is fully considered. The finite element model of the friction thermal vibration coupling analysis is established, and the braking deceleration process is simulated, and the disc brake is obtained by using the non linear multi physical field method. The transient temperature field and the stress field. Through the analysis of the results of the analysis, the following conclusions are drawn. The temperature value in the circumferential direction of the brake disc is decreasing from the contact position of the brake disc and brake disc. The temperature field of the friction region of the brake disc is circular distribution and produces a clear temperature gradient in the radial and axial direction. The temperature rise near the contact surface of the brake disc is far away from the temperature at the contact surface. The deeper the depth is, the more obvious the hysteresis characteristic of temperature is rise. Increasing the initial brake pressure will aggravate the uneven distribution of the temperature and stress field of the brake disc, and with the increase of the initial brake pressure, the maximum of the temperature and stress will increase and the change rate increases obviously in the process of braking, so the use of brake pressure can be reduced to increase the use of the brake. In addition, by comparing the equivalent stress time-varying curve and temperature curve, we can see that the equivalent stress change has obvious hysteresis compared with the temperature.

Acknowledgement

The authors gratefully acknowledge the financial support of National Natural Science Foundation of China (No. 51475066 and No. 51675075), Natural Science Foundation of Liaoning Province (No. 2015020114).

References

[1] ZAGRODZKI P.Analysis of thermomechanical phenomena in multidisc clutches and brakes[J].*Wear*,1990,140:291-308.

[2] Lu Chao, Gao Chenghui, Huang Jianmeng. Study on contact pressure distribution of train brake pad contact interface [J]. *Mechanical design and manufacturing engineering*, 2013, 7:47-51. (In Chinese)

[3] Ghosh N C.Thermal effect on the transverse vibration of spinning disk of variable thickness[J].*Journal of Applied Mechanics*,1975,42(2):358.

[4] Dai changju. Analysis of vibration and noise generation mechanism and transfer path of automotive disc brake [D]. *Shenyang: Northeastern University*, 2014. (In Chinese)

[5] Liu Jianxiu. Research and development of copper based powder metallurgy friction materials and their high temperature fatigue and impact properties [D]. *Henan: China Institute of engineering and physics*, 2003-12-10. (In Chinese)

[6] Liu Li. High thermal conductivity cast iron for brake discs [J]. *World automobile*, 1992 (5):62-67. (In Chinese)

The Longitudinal Dynamic Characteristics of a V/STOL Aircraft Based on Deflected Slipstream during the Hovering Phase

Shi Chen and Weijun Wang

School of Aeronautical Science and Engineering, Beihang University, Beijing 100191, China

chenshi@buaa.edu.cn

Abstract. Compared with general V/STOL (vertical/short taking-off and landing) aircrafts, the V/STOL aircraft based on deflected slipstream shows extremely different dynamic characteristics during the hovering phase. In this paper, a certain new type of V/STOL aircraft based on slipstream deflection is designed as research object. According to a simplified longitudinal dynamic model for the V/STOL aircraft based on deflected slipstream, it is proved that trim flap angle and centre of gravity position are two important factors during the hovering phase. Different flap angle means different trim state while the variation of centre of gravity position could change the range of trimming. Besides, modes of aircraft during the hovering phase are drawn and the law of eigenvalues changing as the variation of trim flap angle and center of gravity position is discovered. Regarding the findings as reference, the design of V/STOL based on deflected slipstream and control law design for V/STOL based on deflected slipstream during the hovering phase can be greatly improved.

1. Introduction

V/STOL (vertical/short taking-off and landing) aircraft is a kind of special vehicle which combines the advantages of helicopter and fixed-wing Aircraft [1, 2]. In recent years, plenty of developing design schemes for V/STOL aircraft are under researching. One of those schemes is the V/STOL aircraft based on deflected slipstream. Deflected slipstream via deflecting the flap angle forces propeller slipstream changing its flow direction, then changing direction of the aerodynamic resultant force on the wing eventually[2-5], as showed in figure 1. It is believed that NACA (U.S. National Advisory Committee on Aeronautics) conduct amount of early research works for STOL aircraft and VTOL aircraft based on deflected slipstream in 1950s[3]. However, limited to technical level that time, NACA's early VTOL testing aircraft based on deflected slipstream couldn't accomplish hovering and transition between hovering and horizontal flying stably. In recent years, a new type of V/STOL aircraft based on deflected slipstream appears[5, 6], which has advantages of simple structure, no variable-pitch/tilt mechanism and high manoeuvrability. This paper aims at deeply analysing the longitudinal dynamic characteristics of the new type of V/STOL aircraft based on deflected slipstream. In this paper, a certain new type aircraft based on deflected slipstream is designed as research object. According to a reasonable simplified mathematical model, its trim features and longitudinal dynamic characteristics around trim point during hovering phase are mainly and deeply discussed.

IOP Conf. Series: Materials Science and Engineering **470** (2019) 012038 doi:10.1088/1757-899X/470/1/012038

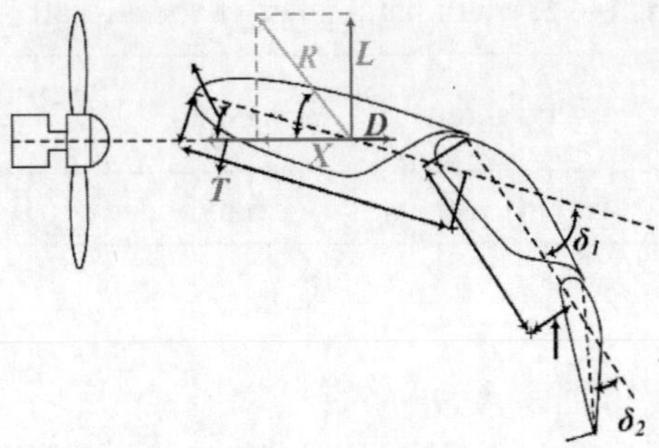

Figure 1. Schematic diagram of deflected slipstream principle.

2. Research object

The research object of this paper is a certain type V/STOL aircraft based on deflected slipstream, as showed in figure 2, and its aerodynamic configuration is canard configuration with strake-wing. The two electromotive propellers installed at outboard of canard, called head propellers, could provide lift force and nose-up pitching moment, besides, the two head propellers could assist aircraft in realizing lateral motion at differential rotational speed. There are another four electromotive propellers, called main propellers, installed at the wing leading edge, overcoming most gravity in hovering phase or drag in horizontal flying phase. Double slotted flaps are symmetrically installed at the inboard of wings, cooperating with main propellers realizing deflected slipstream. During the hovering phase, double slotted flaps deflect at a certain angle, causing the result force vector of propeller-wing-flap (hereinafter referred to as PWF) combination directing upward, keeping balance with the pull force of head propellers and gravity, as in figure 3. Elevons are installed at the outboard of wings. During the hovering phase, elevons angle could continuously change, improving its lateral stability and controllability. The main parameters of the aircraft are showed in table 1.

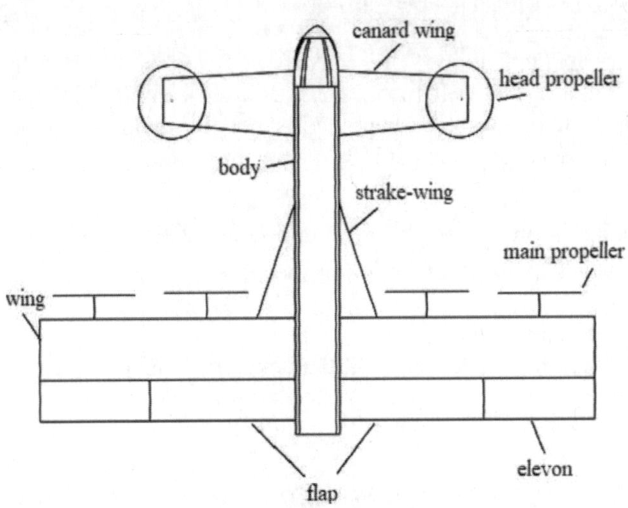

Figure 2. The schematic diagram of a certain type aircraft based on deflected slipstream.

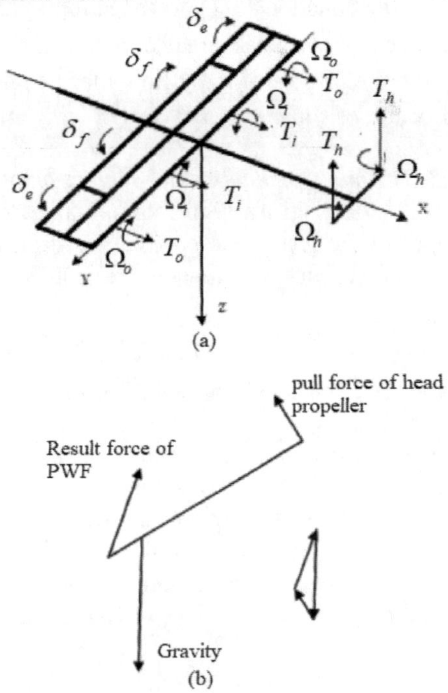

Figure 3. Force diagram.

Table 1. Main parameters of the aircraft

Takeoff Weight	Longitudinal Inertia	Wing Area	Wingspan	Aerodynamic Chord Length	Canard span
5.2 kg	0.469 kg·m^2	0.480 m^2	1.6 m	0.3 m	0.7 m

3. Mathematical model

3.1. The mathematical model of PWF combination

This paper only discusses the dynamic characteristics during the hovering phase. It is believable that airspeed of the aircraft is approximately zero during the hovering phase, and the influence of incoming flow on aerodynamic force is negligible. Thus, only the effect of deflected slipstream is considered for setting up mathematical model of PWF combination.

Due to the effect of deflected slipstream, the result force on the PWF combination is nonlinear and strongly coupled[4]. For simplifying complex degree of setting up mathematical model, the result force on the PWF combination is resolved into pull force of four main propellers, lift and drag which generated by propeller slipstream flowing through the wing. According to momentum theory of propeller, the relational expression between lift and pull force and the relational expression between drag and pull force could be derived, as in equation(s) (1) and equation(s) (2). The equation(s) (1) and equation(s) (2) is validated by elementary experiment on the ground. These equations could ensure enough accuracy when the flap angle is large enough.

$$L = L_0 + T\frac{\kappa C_{L\alpha'} S}{A}\alpha' + T\frac{\kappa C_{L\delta} S}{A}\delta \tag{1}$$

$$D = D_0 + T\frac{\kappa C_{D\alpha'} S}{A}\alpha' + T\frac{\kappa C_{D\delta} S}{A}\delta \tag{2}$$

In the equation(s) (1) and equation(s) (2), T represents the pull force of electromotive propeller. A represents rotor disk area. S represents wing area. κ represents loss coefficient,which is used to weigh the loss causing by flow tube of slipstream constricting. α' represents the angle between flow direction of slipstream and cord of wing and δ represents the angle of flaps.

3.2. The mathematical model of electromotive propeller

The relationship between rotation speed of propeller and pull force is exceedingly complex. In this paper, the effect of its complex relationship on dynamics characteristics is not the main object of discussion. Thus, relational expression is simplified as equation(s)[7] (3).

$$T = T_{\max}\bar{n}^2 \tag{3}$$

In equation (3), $T_{\max}$ represents the maximal pull force of propeller, $\bar{n}$ represents relative rotation speed,which is a ratio of real-time rotation speed and the maximal rotation speed.

3.3. The longitudinal dynamic model of the V/STOL aircraft

Refer to the same way of setting up dynamic model for general aircraft, the new type V/STOL aircraft based on deflected slipstream is considered as rigid body, and coordinate definition in this paper is same as in normal way[8, 9]. According to 3.1 and 3.2, the 3-DOF longitudinal dynamic equations could be derived. As in equation(s) (4)

$$\begin{cases} m(\dot{u} + qw) = 2(T_{\max,i} + T_{\max,o})\bar{n}^2 - 2D - mg\sin\theta \\ m(\dot{w} - qu) = -2L - 2T_{\max h}\cdot\bar{n}_h^2 + mg\cos\theta \\ I_y\dot{q} = 2T_{\max,h}\cdot\bar{n}_h^2\cdot l_h - 2L\cdot l_{pwf} \end{cases} \tag{4}$$

The expressions of L and D are showed in equation(s) (5)

$$\begin{cases} L = L_0 + \kappa C_{L\alpha} \alpha' (S_o T_{max,o} + S_o T_{max,o}) \bar{n}^2 A^{-1} + \kappa C_{L\delta} (S_i T_{max,i} + S_i T_{max,i}) \bar{n}^2 \delta_f A^{-1} \\ D = D_0 + \kappa C_{D\alpha} \alpha' (S_i T_{max,i} + S_o T_{max,o}) \bar{n}^2 A^{-1} + \kappa C_{D\delta} (S_i T_{max,i} + S_o T_{max,o}) \bar{n}^2 \delta_f A^{-1} \end{cases} \tag{5}$$

4. Results and analysis

4.1. Trim feature at different flap angles

According to longitudinal dynamic equations (4), a significant difference contrast with general aircraft is that there are four irrelevant variables θ , $\bar{n}_h$, δ_f and $\bar{n}$, but only three mutually independent equations are available. It means that the new type V/STOL aircraft could trim at different situation. In order to analyze its trim feature, the flap angle is selected as a parameter variable which could artificially change from 0° to 90° , and the trim equations of this aircraft could have the only determination solution at a certain flap angle. For different flap angles, the trim equations could get different solution. The relationship between flap angle and other variables is showed in figure 4.

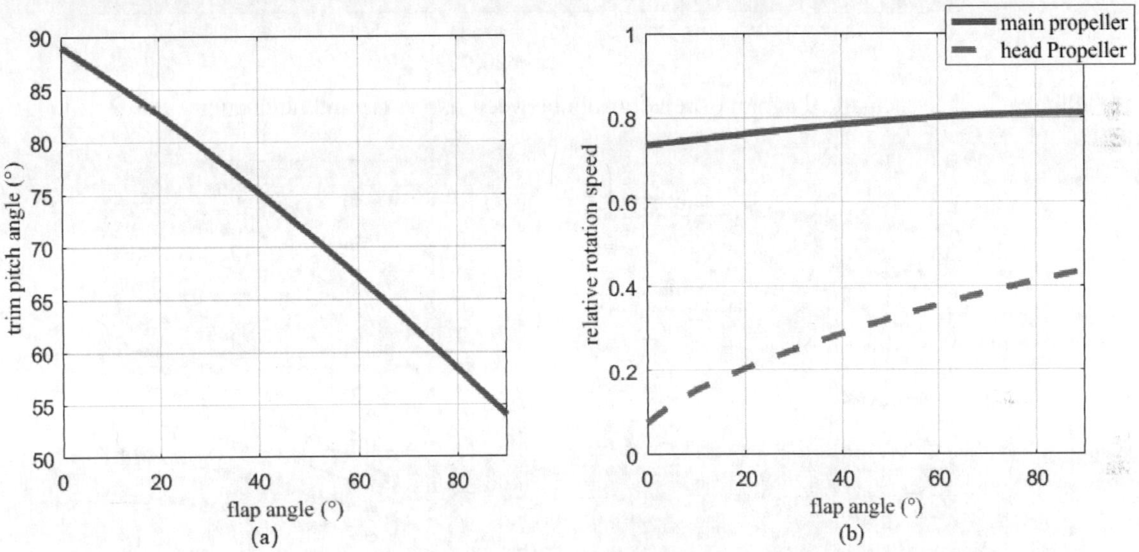

Figure 4. The schematic diagram of relationship between flap angle and other variables.

The relationship between flap angle and trim pitch angle is showed in figure 4. (a). From this figure, the trim pitch angle almost linearly descends with the increase of flap angle. The maximal trim pitch angle is near 90° and the aircraft can hovering like a tail sitter VTOL aircraft at the maximal trim pitch angle. The minimal trim pitch angle is near 55°. The relationship between flap angle and relative rotation speed and flap angle is showed in figure 4. (b). According to this picture, as flap angle increasing, the relative rotation speed of main propeller increases slightly, from 0.78 to 0.81, while the rotation speed of head propeller increases more violently, from 0.08 to 0.42. it seems that as trim pitch angle descending, the head propeller takes more important role in overcoming gravity during hovering phase. And the reason of relative rotation speed of main propeller increasing might owe to the drag of flap increasing as its angle increasing.

4.2. Trim feature at different positions of center of gravity

The relationship between the position of center of gravity and the position of aerodynamic focus is crucial for fixed-wing airplane and V/STOL aircraft in horizontal flying phase. It is worth discussing that the effect of center of gravity position changing on the new type V/STOL based on deflected slipstream. The result is showed in figure 5 and figure 6.

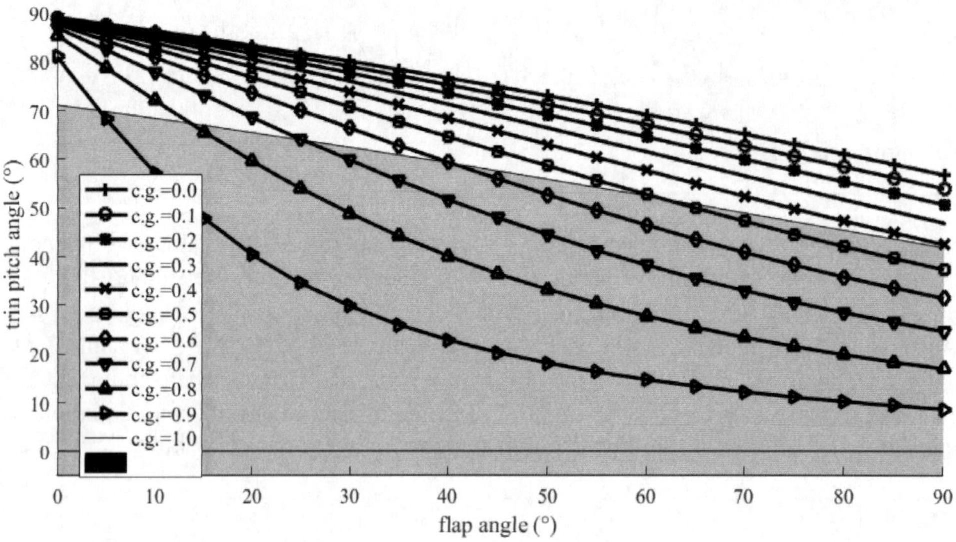

Figure 5. The schematic diagram of relationship between flap angle and pitch angle at different c.g..

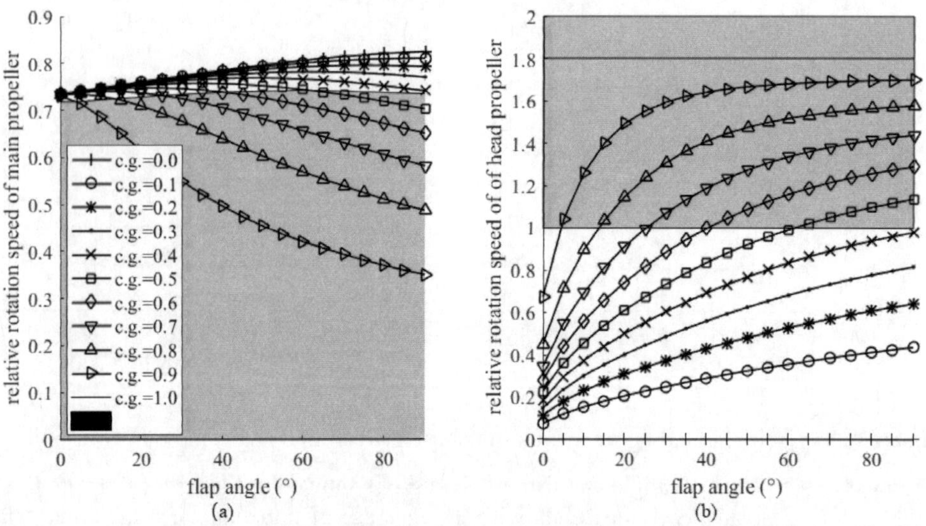

Figure 6. The schematic diagram of relationship between flap angle and relative rotation speed of main propeller and head propeller at different c.g..

In the figure 5 and figure 6, c.g. represents the ratio between the length of position of center of gravity to action point of result force of PWF combination and the length of aircraft nose to action point of result force of PWF combination. For being limited by the maximal pull force of head propeller, the shadow area in the figure 5 and figure 6 represent areas that cannot be trimmed. It is obvious that increasing maximal pull force of head propeller could reduce shaded area in the figure 5 and figure 6. The smaller shaded area means the wider range for trimming in hovering phase.

From figure 5, it indicates that as c.g. increasing, position of center of gravity moving ahead, the range of trim pitch angle is wider. At c.g.=0.4, the trim pitch angle is range from approximate 42° to 90°. However, according to figure 6 (b) as c.g. increasing, the load on the head propeller seems heavier.

4.3. Dynamic characteristics at different flap angles

In order to analyze the dynamic characteristics of the V/STOL aircraft during the hovering phase, it is necessary to linearizing the longitudinal dynamic equations around trim point. The result of linearization of equation(s) (4) is equation(s) (6).

$$\begin{cases} m\dot{u} = 4(T_{\text{max,o}} + \text{T}_{\text{max,i}})\bar{n}\Delta\bar{n} - 2\dfrac{\partial D}{\partial \delta_{\text{f}}}\Delta\delta_{\text{f}} - \dfrac{4D}{v_2}u - 2\dfrac{\partial D}{\partial \alpha'}\dfrac{w}{v_2} - 2\dfrac{\partial D}{\partial \bar{n}}\Delta\bar{n} - 2\dfrac{\partial D}{\partial q} - mg\cos(\theta)\Delta\theta \\ m\dot{w} = -2\dfrac{\partial L}{\partial \delta_{\text{f}}}\Delta\delta_{\text{f}} - \dfrac{4L}{v_2}u - 2\dfrac{\partial L}{\partial \alpha'}\dfrac{w}{v_2} - 2\dfrac{\partial L}{\partial \bar{n}}\Delta\bar{n} - 2\dfrac{\partial L}{\partial q}q + 4T_{\text{max,h}}\bar{n}_{\text{h}}\Delta\bar{n}_{\text{h}} - mg\sin(\theta)\Delta\theta \\ I_{\text{y}}\dot{q} = 4T_{\text{max,h}}\bar{n}_{\text{h}}l_{\text{h}}\Delta\bar{n}_{\text{h}} - l_{\text{pwf}}(2\dfrac{\partial L}{\partial \delta_{\text{f}}}\Delta\delta_{\text{f}} + \dfrac{4L}{v_2}u + 2\dfrac{\partial L}{\partial \alpha'}\dfrac{w}{v_2} + 2\dfrac{\partial L}{\partial \bar{n}}\Delta\bar{n} + 2\dfrac{\partial L}{\partial q}q) \end{cases} \quad (6)$$

In equation(s) (6), $\Delta\bar{n}$, $\Delta\bar{n}_{\text{h}}$ and $\Delta\delta_{\text{f}}$ are control variables, and u, w, θ and q are state variables. According to the trim result of 4.1, the eigenvalues could be calculated. The result is showed in table 2.

Table 2. Eigenvalues at different flap angles.

flap angle (°)	Eigenvalue λ_1	Eigenvalue λ_2	Eigenvalue λ_3	Eigenvalue λ_4
0	-1.490 + 0.7845i	-1.490 - 0.7845i	0.0111	0.9514
5	-1.515 + 0.7596i	-1.515 - 0.7596i	0.0072	0.9480
10	-1.541 + 0.7329i	-1.541 - 0.7329i	0.0033	0.9448
15	-1.567 + 0.7042i	-1.567 - 0.7042i	-0.0007	0.9419
20	-1.594 + 0.6732i	-1.594 - 0.6732i	-0.0046	0.9392
25	-1.620 + 0.6397i	-1.620 - 0.6397i	-0.0086	0.9368
30	-1.647 + 0.6033i	-1.647 - 0.6033i	-0.0126	0.9345
35	-1.673 + 0.5635i	-1.673 - 0.5635i	-0.0165	0.9325
40	-1.700 + 0.5196i	-1.700 - 0.5196i	-0.0204	0.9305
45	-1.726 + 0.4707i	-1.726 - 0.4707i	-0.0241	0.9287
50	-1.752 + 0.4152i	-1.752 - 0.4152i	-0.0277	0.9269
55	-1.778 + 0.3501i	-1.778 - 0.3501i	-0.0312	0.9252
60	-1.803 + 0.2691i	-1.803 - 0.2691i	-0.0345	0.9235
65	-1.827 + 0.1486i	-1.827 - 0.1486i	-0.0375	0.9217
70	-2.019	-1.683	-0.0403	0.9199
75	-2.154	-1.595	-0.0429	0.9180
80	-2.254	-1.539	-0.0452	0.9160
85	-2.338	-1.498	-0.0472	0.9138
90	-2.412	-1.465	-0.0489	0.9116

Table 2 shows that at least a positive real number eigenvalue exists. Thus, it is not a stable dynamics system. Among the four Eigenvalues, there are a pair of conjugate negative eigenvalues. The absolute values of real part of conjugate negative eigenvalues is largest among the four eigenvalues. As the increasing of flap angle, the absolute values of imaginary part of conjugate negative eigenvalues descend and degenerate into two negative real numbers eventually. One of the four eigenvalues is a small positive real number. As the increasing of flap angle, it changes from positive eigenvalue to negative eigenvalue. The last eigenvalue is a positive root.

The trim state at 60° flap angle is selected as an example to research characteristics of mode of the V/STOL aircraft based on deflected slipstream during the hovering phase. The system has a pair of conjugate negative eigenvalues, a small negative real number eigenvalue and a positive number eigenvalue. Thus, it is undoubtable that the dynamics system has three different longitudinal motion modes. The corresponding eigenvectors are showed in table 3. The motion modal excitation response curve is showed in figure 7

Table 3. Eigenvectors at 60° flap angle.

eigenvalue	eigenvector
λ_1, λ_2	$[-0.6565 \pm 0.0609i \quad -0.7305 \quad -0.1527 \pm 0.0264i \quad 0.0812 \mp 0.0268i]^T$
λ_3	$[0.9537 \quad -0.3007 \quad 0 \quad -0.0011]^T$
λ_4	$[0.1923 \quad -0.8992 \quad 0.2667 \quad 0.2888]^T$

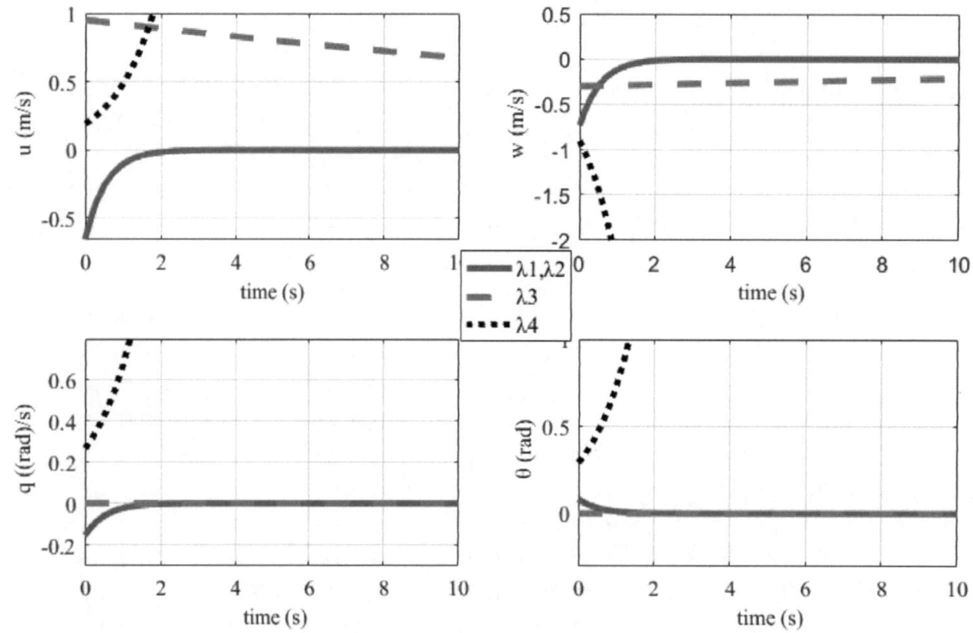

Figure 7. Motion modal excitation response curve at 60°flap angle.

According to table 3 and figure 7, the first motion modal matched with the pair of conjugate negative eigenvalues is all state variables converging to zero rapidly. The second motion modal matched with the small negative eigenvalue is state variables u and w converging to zero slowly and state variables q and θ almost unchanged. Thus, the major motion variables are u and w. The third motion modal matched with the positive eigenvalue is all state variables diverging to infinity.

4.4. Eigenvalues changing at different positions of center of gravity

The law of eigenvalues changing as positions of center of gravity moving is worth analyzing. Eigenvalues changing curves as position of center of gravity moving at 60° flap angle are showed in figure 8.

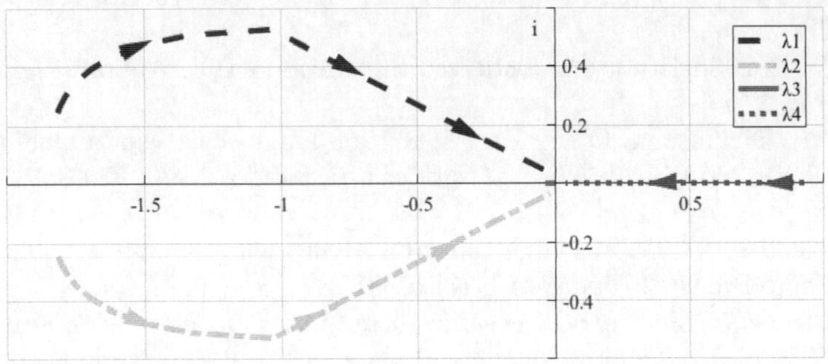

Figure 8. Eigenvalues changing curves as position of center of gravity moving.

In figure 8, the arrow direction presents the changing direction as position of center of gravity moving ahead. It indicates that all eigenvalues are moving close to imaginary axis. Because of positive eigenvalue decreasing, the dynamics system is close to a stable system. The absolute values of imaginary part of conjugate negative eigenvalues tend to raise from 0.24 to 0.52 at first, and then decreasing to near zero rapidly.

5. Conclusion

In this paper, the longitudinal dynamic characteristics of a certain type of V/STOL aircraft based on deflected slipstream are mainly and deeply discussed. According to a simplified longitudinal dynamic model, following five conclusions are drawn:

1). Compared with general V/STOL aircraft, the new type V/STOL aircraft based on deflected slipstream could trim at innumerable state. Flap angle is selected as a parameter variable to analyse the trim range of the new type V/STOL. As flap angle increasing from 0° to 90°, the trim pitch angle descends from near 90° to near 55°.

2). Position of centre of gravity is an important factor of the range of trimming. Appropriately moving position of centre of gravity ahead could extend the range of trim. For example, at c.g.=0.4, the minimal trim pitch angle decreases to 42°. However, limited by the maximal pull force of head propeller, if c.g. exceed 0.4, the minimal trim pitch angle increases and the range of trim shrinks.

3). The dynamics system has four different eigenvalues. Two of them are a pair of conjugate negative eigenvalues when flap angle is less than 70°. when flap angle exceeds 70°, conjugate negative eigenvalues degenerate into two negative real numbers. One of the remaining two eigenvalues is a small real number changing from positive to negative as flap angle increasing. The last one is a positive real number. It indicates the dynamics system is instable.

4). Motion modal excitation response curve and eigenvectors show its motion mode characteristic during hovering phase. In the motion modal matched with conjugate negative eigenvalues, all state variables converge to zero rapidly. In the motion modal matched with small negative eigenvalue, u and w are the main motion variables and converge to zero extremely slowly while state variables q and θ almost unchanged. In the last motion modal matched with positive eigenvalue, all state variables diverge to infinity rapidly.

5). Position of centre of gravity is also an important factor for dynamic characteristics. As position of centre of gravity moving ahead the positive eigenvalue moves close to imaginary axis. Thus, the dynamics system is close to be a stable system.

References

[1] Ransone R 2002 An overview of experimental VSTOL aircraft and their contributions *Biennial Int. Powered Lift Conf. and Exhibit (Williamsburg)* AIAA-2002-5979

[2] Weiberg J A, Giulianetti D, Gambucci B and Innis R C 1973 Takeoff and landing performanceand noise measurements of a deflected slipstream STOL airplane with

interconnected propellers and rotating cylinder flaps *NASA Technical Memorandum* NASA-TM-X62320
[3] Bertelsen W D 2005 History of Deflected Slipstream VTOL Aircraft *Helicopter Society 61st Annual Forum* **1** 2-3
[4] Deckert W H, Dickinson S O and Page V R 1968 Large-scale wind-tunnel tests of a deflected slipstream STOL model with wings of various aspect ratios *NASA TECHNICAL NOTE* NASA-TN-D-4448
[5] Wang W J and Liu J T 2013 Light airscrew flying wing aircraft capable of taking off and landing in short distance *CN patent* CN103158856B
[6] Pan L Y 2016 *Beijing Society of Theoretical and Applied Mechanics (Beijing)* vol 7 pp 60-61
[7] Гессоу A 1959 *Helicopter aerodynamics* (Beijing: National Defence Industry Press)
[8] Etkin B and Reid LD 1996 *Dynamics of flight: stability and control* Vol 3 (New York:Wiley)
[9] Fang Z P, Chen W C and Zhang S G 2005 *Flight dynamics of aircraft* (Beijing: Beihang University Press)

Adaptive Control Algorithm for Steady Flight of Mars UAV with Four Rotors

Yangyang Liu[1,*], Lei Zhang[1] and Jingyi Shao[1]

[1] Beijing Institute of Spacecraft Environment Engineering,100094,China

[*]lyycasc@163.com

Abstract: Mars exploration is a hot issue at present, and Mars aircraft will become a new tool in Mars exploring. Four-rotor aircraft will be a powerful method in Mars aircraft investigation, because of its simple structure, high reliability, besides, it can also hover in the air, take off and land repeatedly. This paper combines Mars atmospheric environment with influence of UAV flight control, and improve PID to make controller parameters adjust themselves to temperature, pressure and atmospheric density on Mars .The effectiveness of adaptive control algorithm has been checked by using MATLAB method, which make Mars UAV with four rotors fly stably.

1. Introduction

As the closest planet to earth in the solar system, Mars is the hotspot of international deep space exploration [1]. The atmospheric of Mars, which is only equivalent to the high atmospheric pressure of the earth's 30~35km, has a density of about 0.0155kg/m^3, but this still makes it possible to use an aircraft on Mars. The United States, Russia and Europe have been studying Mars probe for many years. There are main types of Mars UAV [2], such as American ARES aircraft, American MATADOR aircraft, Swiss solar Martian aircraft, GTMARS aircraft developed by Georgia Technology University and American Prandtl-m aircraft.

The fixed wing aircraft [3-5] needs very high flight speed to meet the lift requirements, and it is difficult to repeat falling off and down, and is not conducive to scientific detection task; the helicopter structure is very complex, difficult to achieve, and low reliability.

Therefore, the paper selects four-rotor UAV as the research object of Mars aircraft which has simple structure, high reliability, hovering, multiple take-off and landing characteristics. Compared to the traditional Martian vehicle, the Mars four-rotor probe extends the Mars exploration dimension from the plane to the three-dimensional space, which is a powerful complement to the Mars ground detector.

In this paper, according to the characteristics of the Mars environment[6], the improved PID control algorithm is designed for the first time so that the aircraft adapts the environment to automatically adjust the parameters of the controller, and carries out the control algorithm simulation test on the Matlab/Simulink platform, so that the aircraft can quickly respond, adjust quickly and fly stability, after the aircraft is disturbed.

2. Control structure analysis and theoretical basis

When the four-rotor aircraft is hovering or slow flying, the change rate of attitude angle is small, which makes it possible to design PID controller for the system. The line motion depends on the angular motion but the angular motion is not dependent on the line motion. The control system can divide into the inner loop attitude controller and the outer loop position controller, according to the semi coupling relationship. There is no coupling relationship between the height control and the yaw angle, so it can be a control channel separately. And the position controller is divided into the horizontal controller and the height controller, as shown in Figure 1.

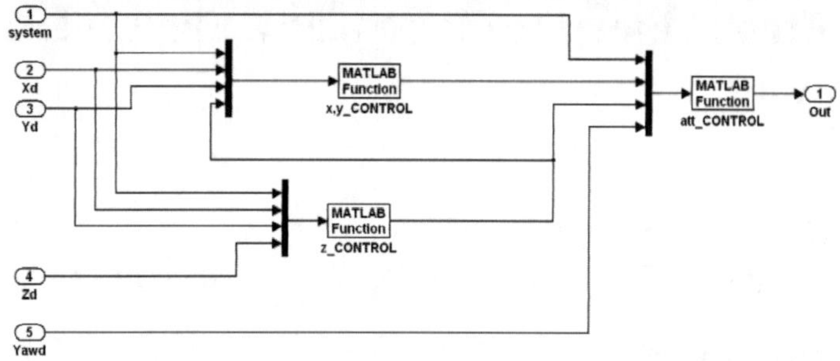

Figure 1. Controller.

In the continuous time domain, the four-rotor aircraft controller follows the PID control theory .According to the PID control algorithm(1), the control algorithm of four-rotor aircraft can be obtained.

$$u(t) = k_p(t)e(t) + k_i(t)\int e(\tau)d\tau + k_d(t)\frac{de(t)}{dt} \quad (1)$$

U_1 and $U_{2\text{-}4}$ are respectively output signal of the attitude control and position control .They are input into the motor module, then on the model of the aircraft. The position and attitude signal of the four-rotor are used as feedback, and the closed-loop control system is formed. By constantly adjusting the PID parameters, the system, which is shown in Figure 2, can reach the target position and keep stable.

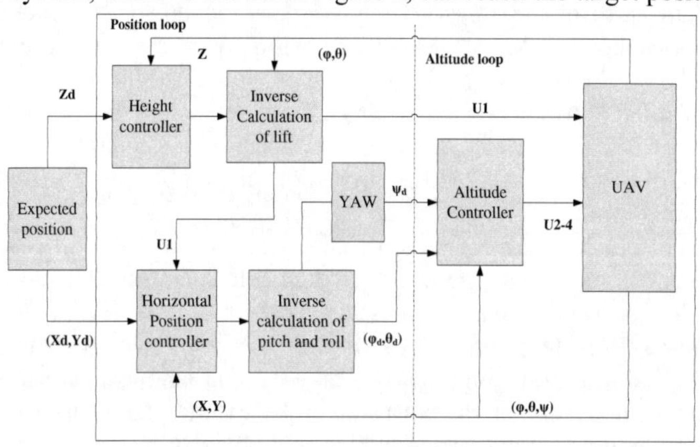

Figure 2. System structure diagram.

2.1. Analysis of the influence of Mars environmental changes on Control

The research of Mars UAV is difficult. Mars' environmental gas composition is different from that of the earth, and the density is far below the earth's surface atmosphere. Due to the thin atmosphere of Mars, it is difficult to transfer the surface heat through the atmosphere, the surface temperature varies greatly within one day or over a long period of one year. Assuming that the surface pressure is basically unchanged, the temperature change will lead to the change of air density according to the ideal gas state equation $pM = \rho RT$. Because the atmospheric density is directly related to the derivation of the lift and torque of the four-rotor [7], the aerodynamic performance of the propeller is changed. It caused changes in the aerodynamic and dynamic properties of the four-rotor UAV. In addition, atmospheric parameters may vary in the altitude of 0~50 meters, which will also affect flight stability. Therefore, it is critical to the impact of the controller that the relationship between atmospheric density, pressure, temperature and altitude of the aircraft in the Martian.

In order to analyse the influence of environmental changes on UAV flight stability, the controller is tested for environmental factors by PID simulation under varying atmospheric conditions.

Fig. 3 shows attitude angle response at different temperatures. At 280k (the standard temperature based on the controller design), the controller has a better attitude stabilization effect and fast response. But at 250K, the attitude angle overshoot is serious, and the time to reach equilibrium is seriously delayed. As the 250K temperature is the daytime temperature often appearing on the surface of Mars, it can be seen that the fixed parameters of the PID controller cannot meet the control requirements of this common.

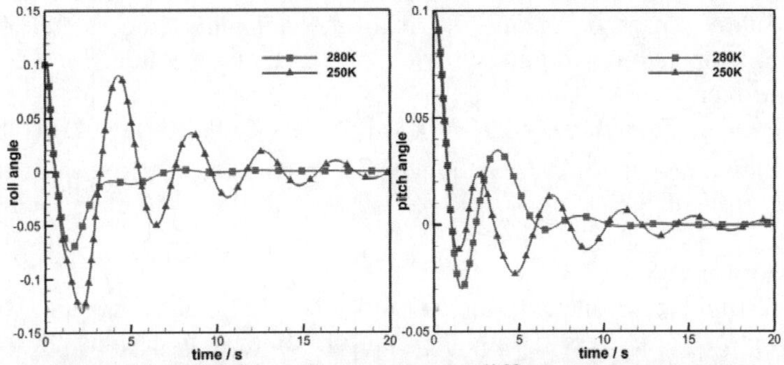

Figure 3. Attitude angle response at different temperatures.

The flight environment will change with the change of altitude. When the researchers set up a three-dimensional model about Mars atmosphere density, the height of the Mars reference radius 0~147km is evenly divided according to the interval of the Mars reference radius according to the interval of 1km, and the regularity of the density is obtained in each high layer with height variation. In this paper, the planned flight height of Mars UAV is 0~50m, which is in the surface flight range. According to the formula (2) (3), when h=0m, T=241K, P=700Pa, ρ=0.0154kg/m^3; when h=50m, T=240.95K, P=696.86Pa, ρ=0.0153kg/m^3. According to the calculation results, it is too small to worry about temperature, pressure and atmospheric density of Mars in this space range.

3. Adaptive PID controller design

Nihon University provides a simplified model [8] of the Martian atmosphere through data fitting,. using Martian atmosphere database in 2010, including atmospheric density, temperature, pressure and many other atmospheric parameters.which can reflect the atmospheric characteristics of Mars in a more comprehensive way, as shown in formula (2) (3).

$$\text{At that time } h < 7000m, \quad \begin{cases} T = 241.0 - 0.999(h/1000) \\ P = 700\exp\left[-0.09(h/1000)\right] \\ \rho = P/188.95110711075T \end{cases} \tag{2}$$

$$\text{At that time } h \geq 7000m, \quad \begin{cases} T = 249.5 - 2.22(h/1000) \\ P = 700\exp\left[-0.09(h/1000)\right] \\ \rho = P/188.95110711075T \end{cases} \tag{3}$$

Among (2) (3), T, P and ρ are respectively the air temperature, pressure and air density of the aircraft, besides H is the flying height of the aircraft.

According to the model, the atmospheric density is changing as the flight position changes, then the dynamic model makes the PID controller unable to meet stable flight of the UAV. On the basis of the designed PID controller, it is necessary to improve the controller so that it can adapt to the environmental changes and automatically adjust the controller parameters.

Reference to the landing-point temperature of the Martian vehicle "Spirit" [9], it is obtained conclusion through studying the change of short-term temperature and seasonal temperature using data from the first 240 days of Spirit.

(1) Martian surface temperature is different between the day and night ,in 10 to 12 PM warming faster,15 to 19 PM cooling faster.

(2) Measurable to the highest temperature is about 304K between 13 and 14 during the day in Mars.

(3) The lowest average temperature is 240K during the day in190th days.

(4) Martian surface average maximum temperature is 280K appearing in the first few days.

The Mars UAV plans to use energy sources like solar cells, only perform a flight mission during the day within the temperature range of 240K-280K ,to design a controller. The four-rotor body installs the temperature sensor to obtain the real-time ambient temperature. According to the relationship between temperature and atmospheric density, the temperature can be used as the input condition of the controller.

Under P=700Pa, when T=240K, ρ=0.0154 kg/m^3; when T=280K, ρ=0.0132 kg/m^3. Therefore, the atmospheric density range of Mars UAV flight environment is 0.0129 kg/m^3 (earth density is about 1%) to 0.0154 kg/m^3.

4. Simulation results analysis

Due to the special flight environment, the core of the PID controller for Mars UAV is the determination of control parameters. Using the control strategy, it designs the attitude controller parameters first according to the semi coupling relationship of the internal and external loops, and then designs the PID parameters of the height and level controllers. At last, observe the change of the position and attitude angle by making UAV control simulation test under 7 groups of atmospheric density evenly.

The initial state of the UAV is set to: position (0,0,0) m, line speed (0,0,0) m/s, attitude angle (0.1,0.1,0.1) rad, angular rate (0,0,0) rad/s. The control target is from location point (0,0,0) to target location point (1,1,1), and hover is maintained. Through the simulation of the operation, observe the response curve of the system, constantly optimize and adjust, and finally determine the parameters of the PID controller.

Table 1. Variable value of controller parameters.

ρ	roll			pitch			z			X, y		
	k_p	k_i	k_d	k_p	k_i	k_d	k_p	k_i	k_d	k_p	k_i	k_d
0.0129	50	5	50	50	5	50	8	10	2	1	0	3
0.0133	60	5	50	60	5	50	7.5	10	2	1	0	3
0.0137	70	5	50	70	5	50	7	9.8	2	1	0	3
0.0141	75	5	50	75	5	50	6	9.5	2	1	0	3
0.0145	78	5	50	8	5	50	5	9	2	1	0	3
0.0149	80	5	50	80	5	50	4	8	2	1	0	3
0.0153	80	5	50	80	5	50	4	7	2	1	0	3

The four-rotor yaw motion is produced by means of rotors' torque, rather than the aerodynamic force. Thus, changes in the density of Martiant atmosphere do not affect the yaw motion, and the yaw curve will not change obviously, that is, parameters of yaw controller need not be adjusted. When the air density is 0.0129kg/m^3, 0.0141kg/m^3 and 0.0154kg/m^3, the control effect is basically unchanged when the controller parameters are fixed, as shown in Figure 4. Finally, the yaw control parameter are determined such as kp=3, k_i =1, k_d=4.

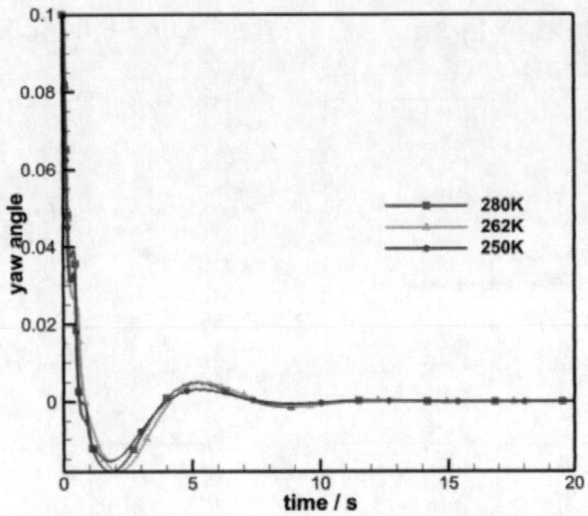

Figure 4. Yaw control curve.

According to the change trend of curve under 7 sets of atmospheric density, we can see the general trend of change. Use MATLAB/Curve Fitting toolbox for curve fitting.
The equation of fitting curve is as follows:

$$k_{p\phi} = k_{p\theta} = -7.319e^6\rho^2 + 2.19e^5\rho - 1557 \quad (4)$$

$$k_{pz} = -1800\rho + 31.24 \quad (5)$$

$$k_{iz} = -7.056e^5\rho^2 + 1.867e^4\rho - 113.5 \quad (6)$$

The correlation coefficient R-square is above 0.96, and has high correlation. Therefore, some parameters of the PID controller satisfy the above curve equation with the variation of atmospheric density, and the controller parameters can be adaptively adjusted according to the atmospheric density.

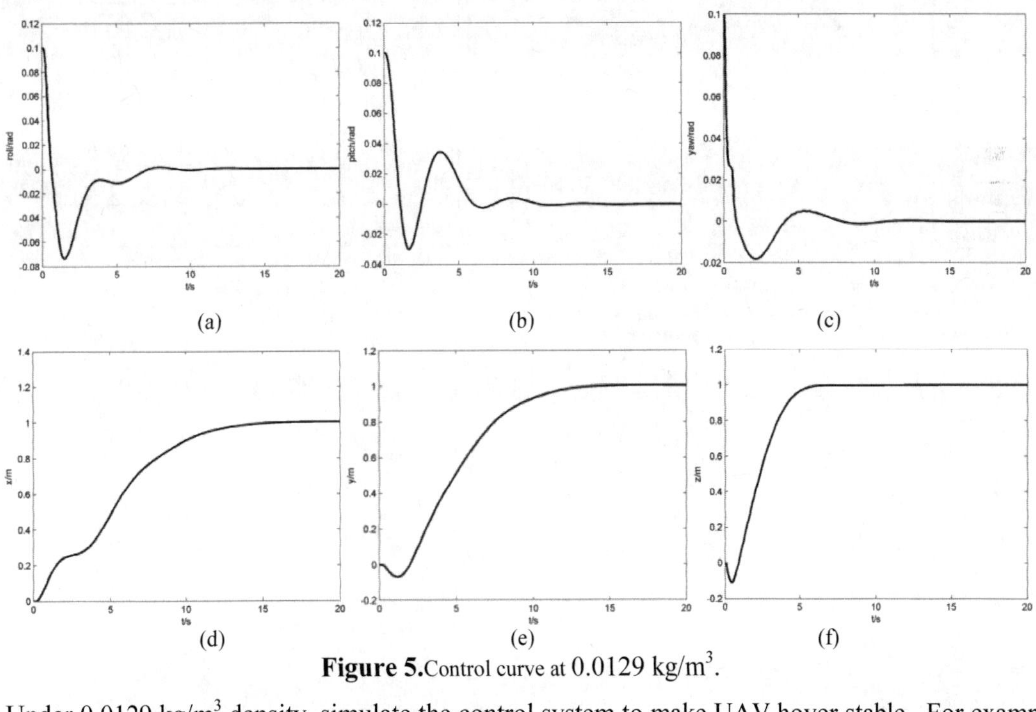

Figure 5. Control curve at 0.0129 kg/m^3.

Under 0.0129 kg/m^3 density, simulate the control system to make UAV hover stable. For example, Figure 5 (a) (b) (c) is the control curve of roll, pitch and yaw respectively; Figure 5 (d) (e) (f) is the position control curve of x, y and z, respectively. Based on the control effect, the simulation of the four-rotor system is carried out with fixed controller parameters and various atmospheric density.

IOP Conf. Series: Materials Science and Engineering **470** (2019) 012039 doi:10.1088/1757-899X/470/1/012039

At 0.0141 kg/m^3 **At 0.0154 kg/m^3**

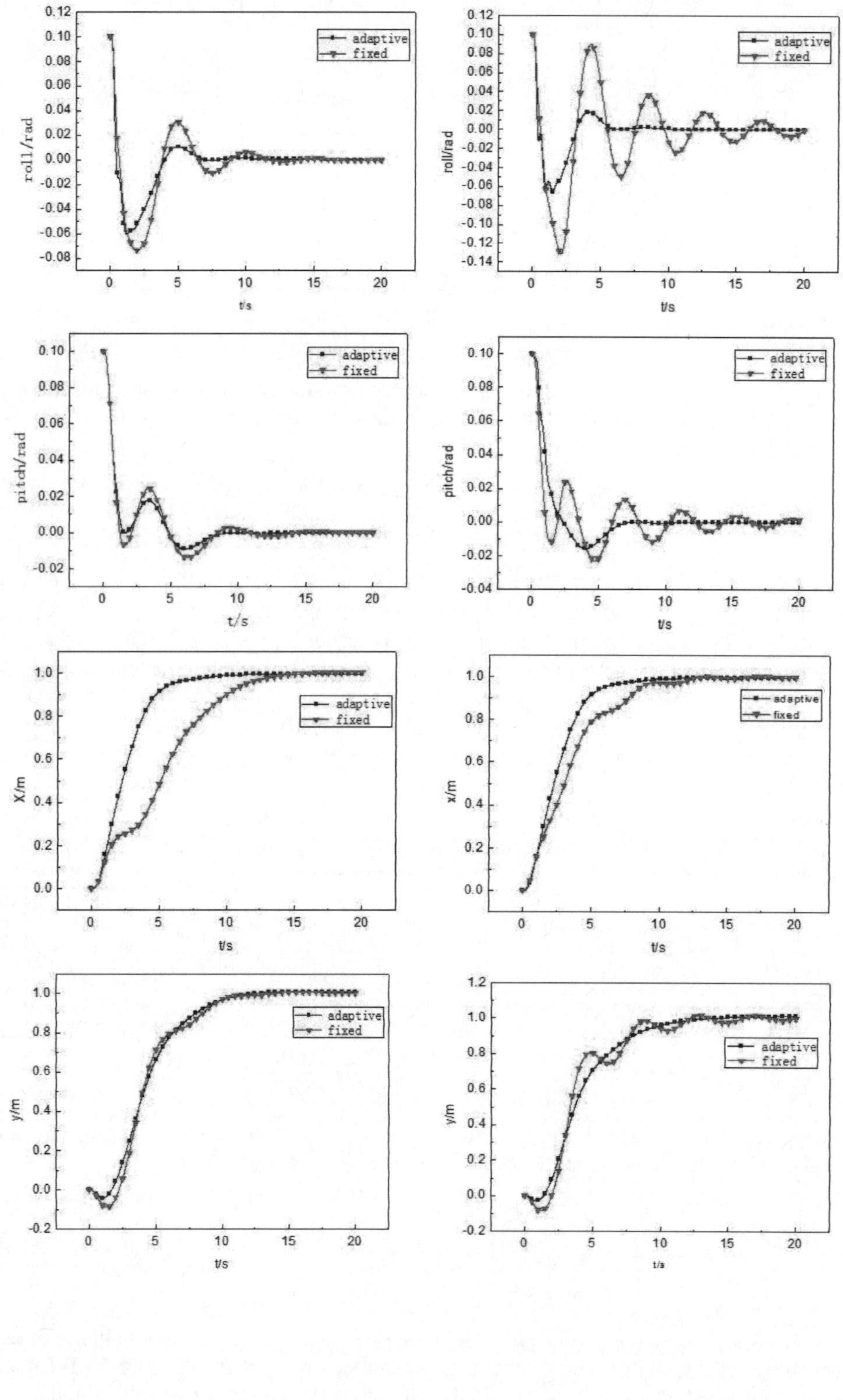

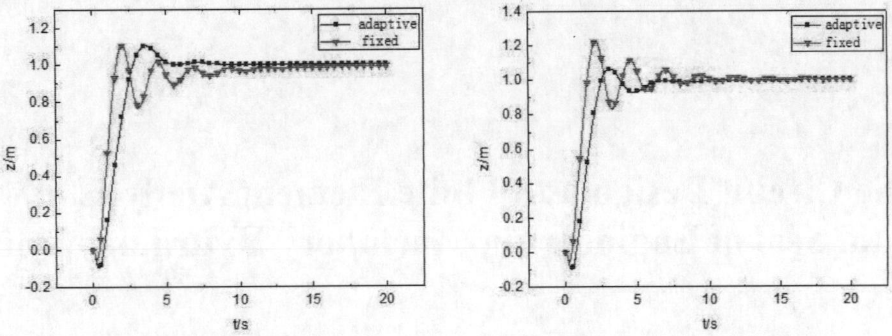

Figure 6. Simulation contrast diagram under two kinds of density.

In two different densities, fixed parameter controller and adaptive parameter controller are used to compare the policy. As shown in Figure 6, the atmosphere density is increased, at 0.0141kg/m^3, the attitude and position control curve fluctuates, the aircraft system is unstable, and the control curve fluctuates more obviously when the atmosphere density value continues to increase at 0.0154kg/m^3. According to the PID setting method, when the curve oscillates frequently, the ratio coefficient should be increased. When the overshoot is increased, the proportional band is reduced and the differential term is reduced properly. Adjust proportional gain properly, and then adjust integral coefficient to the steady state. The controller parameters meet the setting principle in Table 1. Therefore, the improvement of the controller, the PID parameters can be adjusted according to the above fitting equation. From Figure 6, it can be seen that after the parameters are adjusted, the system response is faster and the control effect is better. The designed PID adaptive controller has good control effect.

5. Conclusion

The research of Mars four-rotor UAV has great significance for future deep space exploration. The control algorithm of UAV directly affects the flight stability of the aircraft. In order to satisfy the environment of Mars flight, a PID controller is designed, and a control algorithm is improved that adjusts the parameters of the controller adaptively to the Mars environment. The simulation results show that the adaptive PID control method has stable control effect and has application value.

Reference

[1] Ouyang Ziyuan, Xiao fugen. Major scientific issues involved in Mars exploration[J]. Spacecraft Environment Engineering, 2011, 28(3):205-217.

[2] N. Brown, A, Samuel, R. Colgen. Mars Exploration Airplane: Design, Construction, and Flight Testing of aStability, Control, and Performance Demonstrator. AIAA-2007-2723.

[3] Braun Robert D. Wright Henry S. Croom Mark A. et al. Design of the ARES Mars Airplane and Mission Architecture[R], 2008.

[4] Kuhl Christopher A. Design of a Mars Airplane Propulsion System for the Aerial Regional-Scale Environmental Survey (ARES) Mission Concept [R]. NASA/TM-2009-215700, 2009.

[5] Kuhl Christopher A. Gayle Steven W. Hunter Craig A. et al. Trade Study of Multiple Thruster Options for the Mars Airplane Concept[R]. NASA/TM-2009-215699, 2009.

[6] Ouyang Ziyuan, Xiao fugen. The Mars snd its environment. Spacecraft Environment Engineering, 2012, 29(6): 591-600.

[7] Fay G. Derivation of the Aerodynamic Forces for the Mesicopter Simulation[J]. 2001.

[8] Qin Tong, Wang Shuo, Gao Ai, Zhu Shengying, CuiPingyuan. Three-Dimensional Analytical Model for Mars Atomospheric Density[J].Journal of Deep Space Exploration,2014,1(2):117-122.

[9] XuZhou, ZhaoHuijie, Study on Temperature Characteristic of Mars Surface Where Spirit Landed Based on TES Algorithm[J]. Journal of Astronautics, 2009, 30(2):780-785.

Magnetic Circuit Design and Finite Element Analysis of Ferrofluid Seal of Engineering Machinery Hydraulic Cylinder

Fan Chen[1], Xiaolong Yang[1,*] and Shanghan Gao[1]

[1]School of Mechanical Engineering, Guangxi University of Science and Technology, Liuzhou, China; 545006

*Corresponding author: Xiaolong Yang. Tel:+86 18307721513; E-mail address: 09116324@bjtu.edu.cn

Abstract. In order to solve the problem that the construction machinery hydraulic cylinder seal has a short life and cannot achieve zero leakage, This paper creatively applies ferrofluid sealing technology to engineering machinery hydraulic cylinders. In order to obtain the maximum magnetic energy product in the sealing gap of the ferrofluid sealing structure of the engineering machinery hydraulic cylinder, the pressure resistance performance and life expectancy of the ferrofluid seal of the engineering machinery hydraulic cylinder are improved. Based on the theory of magnetic circuit and the theory of ferrofluid seal, the magnetic circuit design of a ferrofluid seal structure with a uniform pole piece with eight magnetic sources is designed. The finite element analysis method was used to verify the pressure resistance of the ferrofluid seal of engineering machinery hydraulic cylinder, and the calculation results were analyzed and discussed. The results show:The design method of the magnetic circuit structure is reasonable and feasible, and has important reference significance for the design of the ferrofluid seal of the engineering machinery hydraulic cylinders.

Key words: Magnetic circuit; Engineering machinery hydraulic cylinder; ferrofluid; Seal; Finite element

1 Introduction

Ferrofluid sealing technology is widely used in aerospace, machinery, petroleum, chemical and instrumentation because of its advantages of zero leakage, long life and simple structure. However, the application of ferrofluid sealing technology in the field of engineering machinery hydraulic cylinders is also proposed for the first time. Due to the current low life of hydraulic cylinders and the inability to guarantee zero leakage, Therefore, improving the life of hydraulic cylinders is one of the current research focus. ferrofluid sealing technology is an effective way to improve the life of hydraulic cylinder seals. Studying the ferrofluid seal structure of hydraulic cylinders is of great significance for broadening the application field of ferrofluid seal technology[1-5].

Magnetic circuit design is the key to the design of ferrofluid seals for engineering machinery hydraulic cylinders. The use of magnetic circuit design ensures the reliability of sealing performance and reduces production costs. Although researchers have done a lot of research on the magnetic circuit design, magnetic field analysis and pressure resistance of ferrofluid seals, this is the first time to study the magnetic circuit design of the ferrofluid seal structure of engineering machinery hydraulic cylinders. The author designed the ferrofluid seal structure of engineering machinery hydraulic

IOP Conf. Series: Materials Science and Engineering **470** (2019) 012040 doi:10.1088/1757-899X/470/1/012040

cylinder by magnetic circuit design. The size of the magnetic source working at the position of the maximum magnetic energy product is obtained. And carried out finite element theory analysis and verification. Thereby providing a basis for the design of the ferrofluid seal of the engineering machinery hydraulic cylinder[6-7].

2 Ferrofluid seal theory

The total sealing capacity of the ferrofluid seal can be approximately expressed as

$$\Delta P = \int_{H_{i\min}}^{H_{i\max}} \mu_0 M dH \approx \sum_{i=1}^{i=N} M_s (B_{i\max} - B_{i\min}) \tag{1}$$

where μ_0 the permeability of vacuum, $H_{i\min}$ and $H_{i\max}$ are the minimum and maximum magnetic field strengths under the i pole tooth respectively, M the magnetization of ferrofluid, M_s the saturation magnetization of ferrofluid, N the number of pole tooth, $B_{i\max}$ the maximum magnetic flux density under the i pole tooth, and $B_{i\min}$ the minimum magnetic flux density under the i pole tooth. Under the condition of injecting a proper amount of ferrofluid into the sealing gap, The multi-stage ferrofluid seal pressure resistance is the sum of the seal pressure resistance of each stage[8-9].

3 Magnetic circuit design of ferrofluid seal of engineering machinery hydraulic cylinder

On the basis of the original hydraulic cylinder rubber seal, A ferrofluid seal structure of an engineering machinery hydraulic cylinder with eight magnetic sources was preliminarily designed. As shown in Figure 1, its equivalent magnetic circuit is shown in Figure 2. F_m represents the magnetic potential of the permanent magnet; R_m represents the reluctance of the permanent magnet; R_p represents polar resistance; R_t represents pole tooth reluctance; R_g represents the reluctance of the seal gap; R_a represents The reluctance of the shaft.

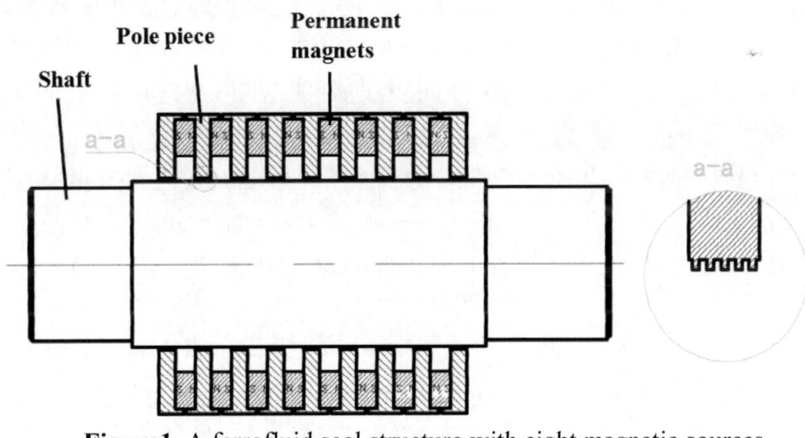

Figure1. A ferrofluid seal structure with eight magnetic sources

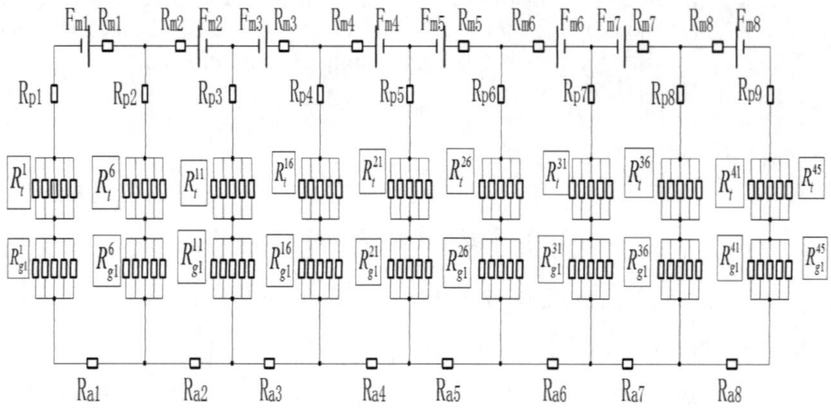

Figure 2. equivalent magnetic circuit of hydraulic cylinder ferrofluid seal

It can be concluded from Fig.2 that the multi-stage sealed magnetic circuit is mainly composed of pole piece, ferrofluid, permanent magnet and shaft. The permanent magnet is the magnetic source of the equivalent magnetic circuit. The magnetic field generated by the magnetic source firmly binds the ferrofluid in the radial sealing gap formed by the pole teeth on the pole piece and the shaft in the sealing gaps to form dense "protective film". The ferrofluid generates a magnetic field force that is resistant to the pressure difference between the two sides. Thereby achieving the effect of pressure seal.

There are two assumptions when designing the magnetic circuit for a ferrofluid seal structure of a construction machinery hydraulic cylinder: Ignore the effects of magnetic flux leakage and edge effects. According to the first law of the magnetic path Kirchhoff.

$$-\Phi_1 + \Phi_2 + \Phi_3 = 0 \tag{2}$$

The total magnetic flux that passes through (or enters) any of the closed faces is always equal to zero. Φ_1, Φ_2, Φ_2 are expressed as the magnetic flux in the magnetic pole, the intermediate magnetic pole, and Magnetic flux in the magnetic pole end of the atmospheric pressure. among them:

$$\Phi_1 = 5B^1_{g1}S^1_{g1} \tag{3}$$

B^1_{g1} represents the magnetic flux density in the radial sealing gap between the first pole tooth and the reciprocating shaft; S^1_{g1} represents the annular area of the radial seal gap formed by the high-pressure measuring first pole tooth and the reciprocating shaft.

$$\Phi_2 = 5(B^6_{g1}S^6_{g1} + B^{11}_{g1}S^{11}_{g1} + B^{16}_{g1}S^{16}_{g1} + B^{21}_{g1}S^{21}_{g1} + B^{26}_{g1}S^{26}_{g1} + B^{31}_{g1}S^{31}_{g1} + B^{36}_{g1}S^{36}_{g1}) \tag{4}$$

B^6_{g1} (B^{11}_{g1}, B^{16}_{g1}, B^{21}_{g1}, B^{26}_{g1}, B^{31}_{g1}, B^{36}_{g1}) shows the magnetic flux density between the 6th(11th,16th,21th,26th,31th,36th) pole teeth on the 2th(3th,4th,5th,6th,7th,8th) pole piece and the radial sealing gap formed by the reciprocating shaft; S^6_{g1} (S^{11}_{g1}, S^{16}_{g1}, S^{21}_{g1}, S^{26}_{g1}, S^{31}_{g1}, S^{36}_{g1}) shows the annular area between the 6th(11th,16th,21th,26th,31th,36th) pole teeth on the 2th(3th,4th,5th,6th,7th,8th) pole piece and the radial sealing gap formed by the reciprocating shaft.

$$\Phi_3 = 5B^{41}_{g1}S^{41}_{g1} \tag{5}$$

B^{41}_{g1} represents the magnetic flux density in the radial sealing gap between the 41st pole tooth on the 9th pole piece and the reciprocating shaft; S^{41}_{g1} represents the annular area in the radial sealing gap between the 41st pole tooth on the 9th pole piece and the reciprocating shaft. It is known by $\Phi_{m1} = \Phi_1$:

$$\Phi_{m1} = \mathrm{B}_{m1} S_{m1} = 5B_{g1}^{1} S_{g1}^{1} \tag{6}$$

Φ_{m1}, B_{m1} and S_{m1} respectively represent the magnetic flux, the magnetic flux density and the area of the first annular permanent magnet. It is known by $\Phi_{m2} = \Phi_3$:

$$\Phi_{m2} = \mathrm{B}_{m2} S_{m2} = 5B_{g1}^{41} S_{g1}^{41} \tag{7}$$

Φ_{m2}, B_{m2} and S_{m2} respectively represents the magnetic flux, the magnetic flux density and the area of the 8th annular permanent magnet. The total magnetic potential acting in any closed magnetic circuit is always equal to the algebraic sum of the magnetic field drop of each segmented magnetic circuit. Similar to Kirchhoff 's second law in circuits, this law is called the Kirchhoff 's second law of magnetic circuits. It is known by $\sum_{k=1}^{n} H_k l_k = \sum N_k k$:

$$F_1 = H_{m1} L_{m1} = 5B_{g1}^{1} S_{g1}^{1} \left[R_{p1} + R_{a1} + R_{m1} + \frac{(R_t^1 + R_{g1}^1)}{5} \right] + 7 \times \left[5B_{g1}^{6} S_{g1}^{6} \left(R_{p2} + \frac{R_{t6} + R_{g1}^{6}}{5} \right) \right] \tag{8}$$

In the above formula, F_1 represents the magnetic potential of the high-voltage side permanent magnet; H_{m1}, L_{m1} represent the magnetic field strength and length of the first permanent magnet; R_{p1}, R_{g1}^1, R_{a1}, R_{m1}, R_t^1 respectively represent the magnetic resistance of the first pole piece, the magnetic resistance of the radial sealing gap of the first pole tooth, Reluctance of the reciprocating shaft, The magnetic resistance of the first permanent magnet and the magnetic resistance of the first pole tooth. R_{p2}, R_{g1}^6, R_{t6} respectively represent the reluctance of the second pole piece, The magnetic resistance of the 6th pole tooth in the radial seal gap, and the reluctance of the 6th pole tooth.

$$F_2 = H_{m2} L_{m2} = 7 \times \left[5B_{g1}^{6} S_{g1}^{6} \left(R_{p2} + \frac{R_t^6 + R_{g1}^{6}}{5} \right) \right] + 5B_{g1}^{41} S_{g1}^{41} \left[R_{p9} + R_{a8} + R_{m8} + \frac{(R_t^{41} + R_{g1}^{41})}{5} \right] \tag{9}$$

In the above formula, F_2 represents the magnetic potential of the low-voltage side permanent magnet; H_{m2}, L_{m2} represent the magnetic field strength and length of the 8th permanent magnet; R_{p9}, R_{g1}^{41}, R_{a8}, R_{m8}, R_t^{41} respectively represent the magnetic resistance of the 9th pole piece, The reluctance of the radial sealing gap of the first pole tooth of the 9th pole piece, Reluctance of the reciprocating shaft, The magnetic resistance of the 8th permanent magnet and the magnetic resistance of the 41th pole tooth. Since the shape, size and performance parameters of the permanent magnets used in the whole structure are the same, it can be known that:

$$F_1 = F_2 = H_{m2} L_{m2} = H_{m1} L_{m1} \tag{10}$$

It can be known that:

$$5B_{g1}^{1} S_{g1}^{1} \left[R_{p1} + R_{a1} + R_{m1} + \frac{(R_t^1 + R_{g1}^1)}{5} \right] = 5B_{g1}^{41} S_{g1}^{41} \left[R_{p9} + R_{a8} + R_{m8} + \frac{(R_t^{41} + R_{g1}^{41})}{5} \right] \tag{11}$$

Multiply equation (7) by equation (9):

$$V_{m2} = \frac{\mathrm{B}_{m2} S_{m2} \times H_{m2} L_{m2}}{\mathrm{B}_{m2} H_{m2}} = \frac{175B_{g1}^{6} S_{g1}^{6} B_{g1}^{41} S_{g1}^{41} \left(R_{p2} + \frac{R_t^6 + R_{g1}^{6}}{5} \right) + 25B_{g1}^{41} S_{g1}^{41} B_{g1}^{41} S_{g1}^{41} \left[R_{p9} + R_{a8} + R_{m8} + \frac{(R_t^{41} + R_{g1}^{41})}{5} \right]}{\mathrm{B}_{m2} H_{m2}} \tag{12}$$

In order to save costs, The thickness and length of the permanent magnets need to be minimized. Thereby obtaining a reduction in volume and weight, Improve the utilization rate of permanent magnets, The permanent magnet should be operated at the maximum magnetic energy product,It can be known that:

$$V_{m2}=\frac{B_{m2}S_{m2}\times H_{m2}L_{m2}}{B_{m2}H_{m2}}=\frac{175B_{g1}^{6}S_{g1}^{6}B_{g1}^{41}S_{g1}^{41}\left(R_{p2}+\frac{R_{t}^{6}+R_{g1}^{6}}{5}\right)+25B_{g1}^{41}S_{g1}^{41}B_{g1}^{41}S_{g1}^{41}\left[R_{p9}+R_{a8}+R_{m8}+\frac{(R_{t}^{41}+R_{g1}^{41})}{5}\right]}{(BH)_{MAX}} \tag{13}$$

Equation (7) is divided by equation (9):

$$\frac{S_{m2}}{L_{m2}}=\frac{B_{g1}^{41}S_{g1}^{41}}{7B_{g1}^{6}S_{g1}^{6}\left(R_{p2}+\frac{R_{t}^{6}+R_{g1}^{6}}{5}\right)+B_{g1}^{41}S_{g1}^{41}\left[R_{p9}+R_{a8}+R_{m8}+\frac{(R_{t}^{41}+R_{g1}^{41})}{5}\right]}\times\frac{H_{m2}}{B_{m2}} \tag{14}$$

Multiply equation (13) by equation (14):

$$S_{m2}=5B_{g1}^{41}S_{g1}^{41}\left(\frac{H_{m2}}{(BH)_{\max}B_{m2}}\right)^{-\frac{1}{2}} \tag{15}$$

Equation (13) is divided by equation (14):

$$L_{m2}=F_{2}\left(\frac{B_{m2}}{(BH)_{m2}H_{m2}}\right)^{-\frac{1}{2}}=\left\{7\times\left[5B_{g1}^{6}S_{g1}^{6}\left(R_{p2}+\frac{R_{t}^{6}+R_{g1}^{6}}{5}\right)\right]+5B_{g1}^{41}S_{g1}^{41}\left[R_{p9}+R_{a8}+R_{m8}+\frac{(R_{t}^{41}+R_{g1}^{41})}{5}\right]\right\}\left(\frac{B_{m2}}{(BH)_{m2}H_{m2}}\right)^{-\frac{1}{2}} \tag{16}$$

4 Finite element analysis and its results

In the magnetic circuit of the ferrofluid seal structure, The permanent magnet material is selected from NdFeB. Its coercive force is $H_c=1.05\times10^{6}A/m$, The pole piece and the sealing shaft material are both 2cr13; ferrofluids are selected as ester based ferrofluid and its saturation magnetization is $41.6KA/m$. The permanent magnet has a width of 6.3 mm and the inner (outer) radius of the pole piece is 25.1 (45) mm. According to the magnetic field distributions in the different sealing gaps of the ferrofluid seal which are shown in Fig.3. a, b, c, d, e denote magnetic distribution with 0.1mm,0.2mm,0.3mm,0.4mm,0.5mm sealing gap height.

It can be seen from Fig.3 that the pressure capabilities of the ferrofluid seal significantly decrease with the increase of the radial sealing gap. it can be found that In the structure of different sealing gaps, the maximum magnetic flux density is located under the second pole teeth on the left and right sides. And much higher than the magnetic flux of the left pole piece and the right first pole piece. This is because the left and right first pole pieces are only powered by a single permanent magnet. The middle pole piece is supplied by a magnetic field from the left and right permanent magnets. The magnetic flux in the middle pole piece is the part of the sum of the magnetic fluxes generated by the permanent magnets on both sides. When the radial sealing gap is changed, the pressure resistance of the ferrofluid seal under different sealing gaps can be calculated which is shown in Fig.4.

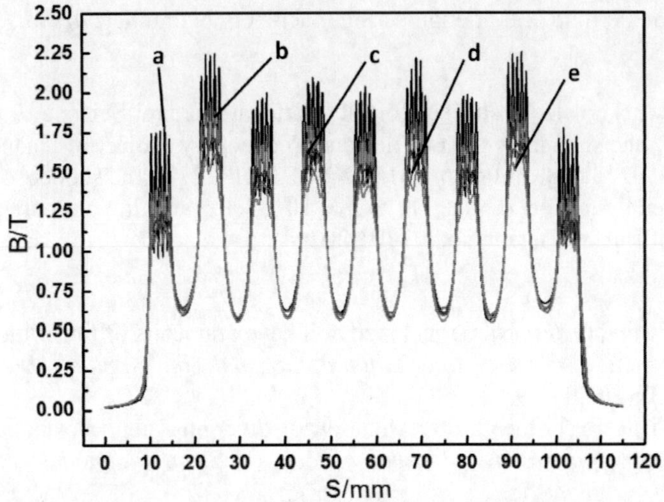

Figure 3. Magnetic field distributions of the ferrofluid seal with different gaps conditions

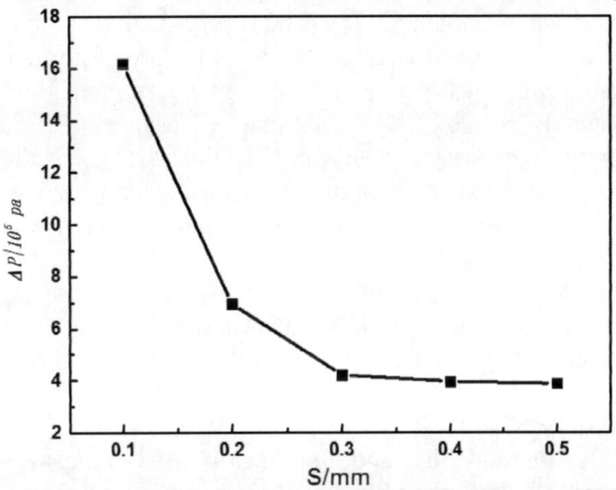

Figure 4. sealing capability (ΔP) under different sealing gap

As can be seen from the above figures, the pressure resistance of the ferrofluid seal of the hydraulic cylinder is continuously decreasing as the clearance increases. This is because the larger the sealing gap is, the larger the magnetic resistance of the sealing gap is. According to the magnetic circuit law. The magnetic flux density in the sealing gap is reduced, and the magnetic induction intensity is correspondingly reduced. This results in a decrease in the pressure resistance of the hydraulic fluid seal of the hydraulic cylinder. At the same time, it can be seen that when the sealing gap is 0.1 and 0.2 mm, the model sealing pressure resistance is higher. Therefore, when the processing conditions permit, try to keep the sealing gap below 0.3mm.

5 Conclusions

Based on the classical magnetic circuit method, the ferrofluid seal structure of engineering machinery hydraulic cylinder is designed. The finite element analysis was carried out and the following conclusions were drawn:

1) The pressure resistance of the ferrofluid seal of the hydraulic cylinder is continuously decreasing as the clearance increases.

2) According to the magnetic circuit method, the detailed parameters of the permanent magnet working at the position of the maximum magnetic energy product are obtained. That is, the thickness of the permanent magnet is 6.3 mm.

3) The test results show that the magnetic circuit design method is feasible. It provides the basis for

the design of the magnetic fluid seal of engineering machinery hydraulic cylinder.

Acknowledgments
The authors gratefully acknowledge the support of National Nature Science Foundation of China (Grant No.51375039), the support of the Science and Technology Project of Guangxi Province (Grant No.2016GXNSFBA380213 and Grant No. 2015GXNSFAA139271), the support of Innovation Project Of Guangxi Graduate Education (Grant No.YCSW2018200), and the support of the Science and Technology Project of Liuzhou(Grant No.2017BC20204).

References
[1] Chen F, Yang X and Gao S. Numerical study on key parameters of ferrofluid seal with large gap and two magnetic sources. *International Journal of Applied Electromagnetics and Mechanics* 55(3) (2017), 1-9.
[2] Yang X and Li D. Experimental investigation of diverging stepped magnetic fluid seals with large sealing gap, *International Journal of Applied Electromagnetics and mechanics* 50(3) (2016),407-415.
[3] Li C, Xu P, He X and Q.Lan, Study on the magnetic fluid sealing for dry Roots pump. *Journal of magnetism and magnetic materials* 289 (2005), 419-422.
[4] K G and Sayles R. Nonlinear Elasticity of Rectangular Elastomeric Seals and Its Effect on Elastohydrodynamic Numerical Analysis. *Tribology International*37(2004), 651-660.
[5] Mizutani Y,Sawano H,Yoshioka H and Shinno H. Magnetic Fluid Seal for Linear Motion System with Gravity Compensator. *Procedia* CIRP088(2015),581-586.
[6] Wang Z and Li D. Theoretical analysis and experimental study on loading process among stages of magnetic fluid seal, *International Journal of Applied Electromagnetics and Mechanics* 48 (2015), 101-110.
[7] Li W and Lam S. Material and Finite Element Analysis of Poly (tetrafluoroethylene)Rotary Seals[J]. *Plast Rubber Compos* 31(2002)(8),359-363.
[8] Radionov A, Podoltsev A and Zahorulko A. Finite-element analysis of magnetic field and the flow of magnetic fluid in the core of magnetic-fluid seal for rotational shaft, *Procedia Engineering* 39 (2012),327-338.
[9] Nikas G and Sayles R. Modelling and Optimization of Composite Rectangular Reciprocating Seals. Proceedings of the Institution of Mechanical Engineers. *Journal of Engineering Tribology*220(2006),395-412.

IOP Conf. Series: Materials Science and Engineering **470** (2019) 012041 doi:10.1088/1757-899X/470/1/012041

Effect of Crack on Dynamic Characteristics of Simple Supported Beam

Xueyan Zhao[1], Weikang Du[1], Chao Cao[1], Yifan Chen[1] and Lijuan Yin[1,a]

[1]College of Engineering, China Agricultural University, Beijing, P. R. China, 100083

[a] Corresponding author: yinlijuanlyc@163.com

Abstract. Simple supported beam is extensively applied in practical engineering structure. Crack is a common damage in simple supported beam and has greatly influences on the dynamic characteristics of the system. In this study, effect of crack on dynamic characteristics of simple supported beam is studied. The model of simple supported beam with crack damage is established by finite element (FE) method firstly. Then, effect of crack location and crack depth on dynamic characteristics of the simple supported beam is studied by comparison with the case when no crack occurs. Results of simulation studies conclude that crack with different location and depth will change natural frequencies differently and the effects of crack location on natural frequencies are related to the vibration model of the simple supported beam. This conclusion provides a good reference for simple supported beam structure performance optimization, damage detection and localization.

1. Introduction

In engineering practice, many mechanical and civil structural systems, such as frame of sugarcane harvester [1], Bridge [2], and pipeline [3], can be represented by simple supported beam. All such structural systems are prone to suffering fatigue crack damage due to long service time, improper use, or hostile working environments, which can cause the whole machine system abnormally. Therefore, it is very important to study the dynamic characteristics and failures of the simple supported beam.

The fatigue crack damage will decrease the local stiffness of the system and changes the system's natural frequency, vibration mode and other system parameters, resulting in deterioration of system performance. Therefore, many studies are concerned with effect of crack on system dynamic characteristics. For example, extensive researchers derived the relationship of crack parameters and modal parameters by analytical method. Guo [4] derived the modal frequency and vibration mode calculation formula of the beam by using the first-order perturbation theory, and analyzed the influence of crack depth and position on the characteristic modal parameters of multiple open cracked beams. The results show that the frequency changes caused by the cracks have obvious patterns. Zhang [5] derived the free vibration equation of the cracked beam structure, analyzed the influence of the change of crack position and depth on the natural frequencies, and then developed characteristic curve intersection method. Ma [6] derived the transfer matrix of the whole beam with the temperature and geometric parameters of cracks, and found that the natural frequencies of the simply supported steel beam were mostly affected by the temperature load and geometric parameters.

Besides, numerical simulation method is also used to study effects of crack on beam system dynamic characteristics. Song [7] established a finite element model of simply supported beam with vertical internal cracks, and proved that the natural frequency of simply supported beams decreases with the increase of crack length, and the difference between simply supported beams and healthy simply supported beams increases gradually. The vertical internal crack will cause the local stiffness of the simply supported beam to change in the section where the crack is located, and the affected area increases with the increase of the crack length. Zou [8] carried out theoretical analysis, numerical simulation and experimental study on the crack damage of simply supported beams. The results show that the modal frequencies of the simply supported beams decrease with the increase of crack depth. Sun [9] studied the bending deformation of simply supported beam under the uniform load and the concentrated force and force couple at the location of crack. The results show that there is a sharp point in the deflection of the beam and a jump in the corner distribution. The response of the beam deflection to the load is generally in the form of a double-fold line, which corresponds to the open and closed state of the crack. Bakhtiari-Nejad [10] predicted the first natural frequency of simply supported fractured beams by the analysis estimates of Rayleigh method. Vigneshwaran [11] calculated the stiffness and natural frequency of multi-cracked beams, and obtained the modal shape and FRF of the cantilever beam with fracture and respiratory cracks, indicating that the shape of the model changes significantly due to the presence of cracks.

Most of above researches are mainly concerned with one crack parameter, location or depth, but both location and depth should be considered at the same time. Therefore, the effects of both location and depth of crack on the dynamic characteristics of the simple supported beam are investigated systematically in this paper. The paper is organized as follows. After this introduction, finite element model of a simple supported beam with a crack is established in Section 2. Effect of crack location and crack depth on dynamic characteristics of simple supported beam is studied by comparison with the case when no crack occurs in Section 3. Finally, the conclusions are presented in Section 4.

2. Finite element model of simple supported beam with a crack

In this paper, a traditional supported beam with a straight crack is investigated, as shown in Figure 1. In the figure, one end of the simple supported beam is full fixed, namely, all freedom of this end is fixed, the other end of the simple supported beam is semi-fixed end, namely, only freedoms in y-direction and z-direction are fixed. L, B and H denote the beam length, width and thickness, respectively; l denotes the distance between the crack tip and the full fixed end in x-direction; h is the crack depth. In order to present the crack location and depth more clearly, relative crack location and depth are exploited. The dimensionless relative crack location Lr is defined as Lr =l/L and dimensionless relative crack depth is defined as Hr=h /H.

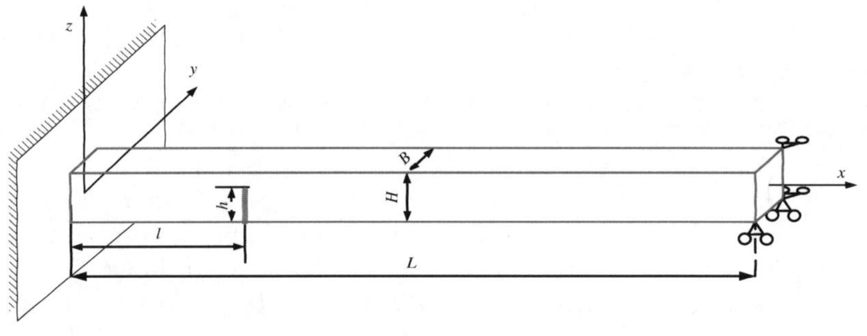

Figure 1. Schematic of a simple supported beam with a straight crack

2.1. Finite element model of crack

The ANSYS/Workbench software is used to establish the Finite Element (FE) model of the cracked simple supported beam. In order to simplify the modeling process and reduce the element number, two assumptions about the crack are made as follows:

(1) Crack is assumed to be straight crack and non-propagating, and the crack surfaces are assumed to be smooth and planar;

(2) Crack is assumed to be breathing crack, which is considered to be a frictionless contact problem where the contact elements are located on the semi-fixed end and the target elements on the fixed end.

V-crack is one of the most common crack types in engineering practice and is employed in this paper. And the crack geometric model is obtained by removing a small portion of the material in the position of the crack and then forming the V-gap, whose main parameters are crack location, crack depth, crack width and crack vertex angle. Crack location indicates the position where crack occurs. The crack depth is the depth of the V-gap, the crack width is equal to the length of the bottom of the V-gap, and the crack tip angle is the angle between the two sides of the V-gap. It must be pointed out that the crack gap should be as small as possible so as to make the crack model closer to the actual situation.

The main body of the crack is meshed by tetrahedral meshes, and the mesh refinement is conducted in a local sphere region near the middle of the crack. As is displayed in Figure 2, the radius of the sphere is 6mm, and the grid size inside the sphere is 1mm.

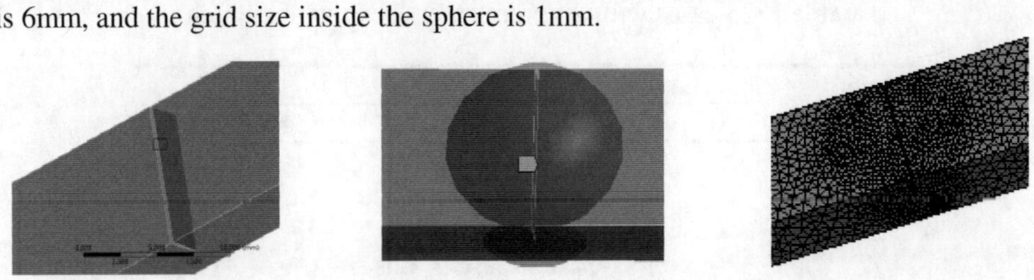

(a) Contact region (b) Lobcal sphere region (c) Mesh refinement result

Figure 2. Finite element model of crack

2.2. Finite element model of the simple supported beam

In order to simplify the modeling process, the material properties of the cracked simple supported beam are assumed to be linearly elastic. The structural material of the simple supported beam is stainless steel, young's modulus 207 GPa, poisson's ratio 0.3 and density 7850 $kg{\cdot}m^{-3}$. The size parameter of the cracked beam is assumed to be: B=10mm, H=5mm, L=100mm.

In finite element model of the simple supported beam, all freedom of full fixed end is constrained and is shown in Figure 3 (a), and the freedom on y-direction and z-direction is constrained and is shown in Figure 3 (b). The finite element model of the cracked simple supported beam is meshed by the Quad shell element with the smallest unit size 0.5 mm. When the relative position of the crack is 0.5 and the relative depth is 0.5, finite element model of the simple supported beam is shown in Figure 3 (c). The total number of elements and nodes are 74405 and 108914, respectively, and the minimum Jacobian of the grid element is 1.

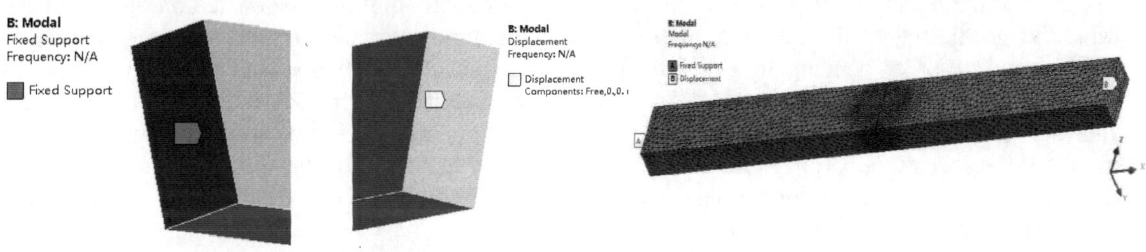

(a) Full fixed end (b) Semi-fixed end (c) FE model of the simple supported beam

Figure 3. FE model of the simple supported beam

3. Simulation analyses

In this section, the effects of crack parameters, including crack location and crack depth, on the beam dynamic characteristics are analyzed. The crack is assumed to locate on Lr = 0, 0.2, 0.4, 0.5, 0.6, 0.8 and 1, respectively, and the crack depth is assumed to Hr= 0, 0.1, 0.2, 0.3, 0.4 and 0.5, respectively. Hr= 0, Lr = 0 and 1 means there is no crack. Therefore, 26 cases are selected to analyze the effects of different parameters on the beam dynamic characteristics, and the corresponding crack parameters are shown in Table 1. In ANSYS/Workbench, there are 5 types of modal analysis methods including Direct Method, Iterative Method, Supernode Method, Superspace Method, and Unsymmetric Method. The Space Method and Iterative Method are suitable for solving large symmetric matrices, and the Subspace Method and the Subnode Method have higher accuracy and are applicable for solving the higher order model. The Asymmetric Method is mainly applied to acoustic or fluid solid coupling analysis, while the Direct Method is faster and it can also meet the requirement of solving precision. Therefore, Direct Method is chosen to conduct modal analysis.

Table 1. 26 cases with different crack location and depth

Relative location	Relative depth Hr 0	0.1	0.2	0.3	0.4	0.5
0 and 1		Case 1 (no crack)				
0.2	Case 1 (no crack)	Case 2	Case 7	Case 12	Case 17	Case 22
0.4		Case 3	Case 8	Case 13	Case 18	Case 23
0.5		Case 4	Case 9	Case 14	Case 19	Case 24
0.6		Case 5	Case 10	Case 15	Case 20	Case 25
0.8		Case 6	Case 11	Case 16	Case 21	Case 26

3.1. Modal analysis of the simple supported beam without crack

In order to understand the dynamic characteristics of the simple supported beam in the normal working status, the modal analysis of the simple supported beam without crack is studied in this section. The first 6 order natural frequencies and the vibration mode are shown in Table 2 and Figure 4. It can be seen from Table 2 that the first 6 order natural frequencies are between 1758.8 Hz and 11511 Hz, and vibration mode covers 1st-order bending, 1st-order torsion, 2nd-order bending, 3rd-order bending.

Figures 4 (a) and (b) indicate that the first and second vibration modes are 1st-order bending in y-direction and z-direction, respectively, and maximum vibration amplitude is located on the 3/5 of the whole length from the full fixed end. Figures 4 (c) and (d) indicate that the third and fourth vibration modes are 2nd-order bending in z-direction and y-direction, respectively, and maximum vibration amplitude is located on the 1/3 and 4/5 of the whole length from the full fixed end. Figure 4 (e) indicate that the fifth vibration mode is 3rd-order bending in z-direction, and maximum vibration amplitude is located on the 1/7, 1/2 and 4/5 of the whole length from the full fixed end. The sixth vibration mode is 1st-order torsion, and maximum torsion amplitude is located in the middle of the whole length as shown in Figure 4(f).

Table 2. The first 6-order modes of beam without crack

Order	Natural frequency (Hz)	Location of maximum amplitude	Vibration mode	Order	Natural frequency (Hz)	Location of maximum amplitude	Vibration mode
1	1758.8	3/5	1st-order bending in y-direction	4	10435	1/3 and 4/5	2nd-order bending in y-direction
2	3414.3	3/5	1st-order bending in z-direction	5	11438	1/7, 1/2 and 4/5	3rd-order bending in z-direction
3	5607.9	1/3 and 4/5	2nd-order bending in z-direction	6	11551	1/2	1st-order torsion

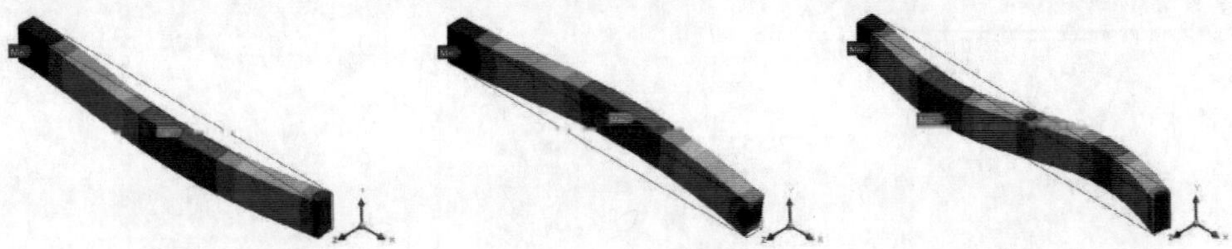

(a) First order (b) Second order (c) Third order

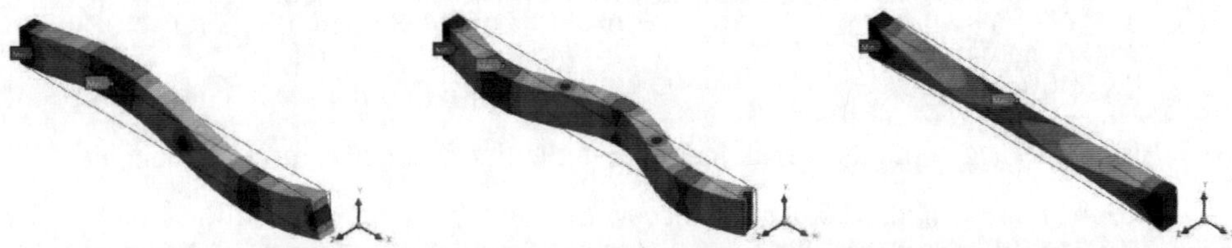

(d) Fourth order (e) Fifth order (f) Sixth order

Figure 4. The first 6-order vibration modes of beam without crack

3.2. Modal analysis of the simple supported beam with crack

In order to study the influence of the crack parameters on the dynamic characteristics of the beam, the modal analyses of the simple supported beam with cracks in different positions and depths are carried out. The first 6-order natural frequencies in 26 different cases are shown in Figure 5 by taking the crack location as variable.

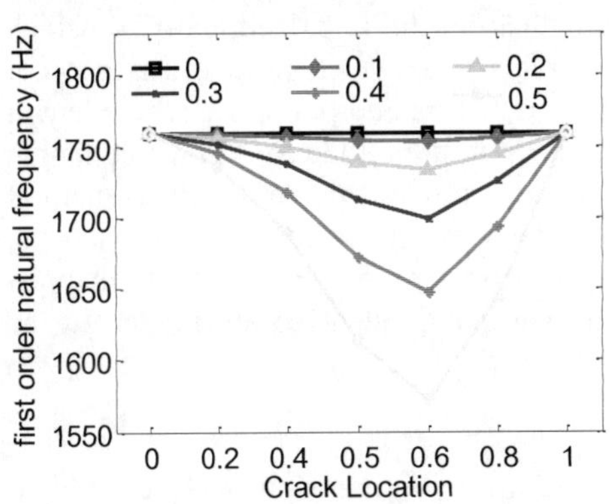

(a) First order

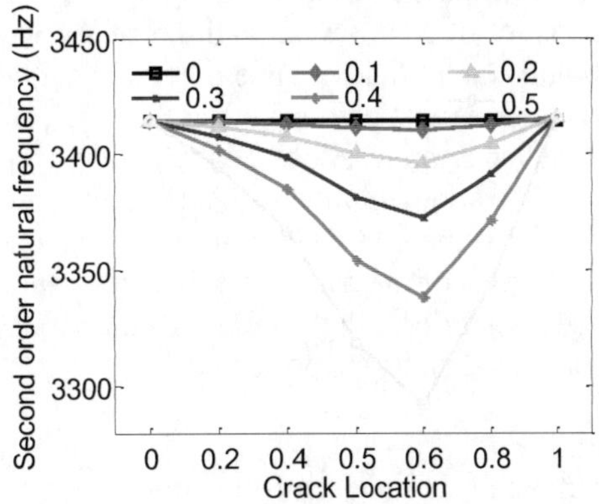

(b) Second order

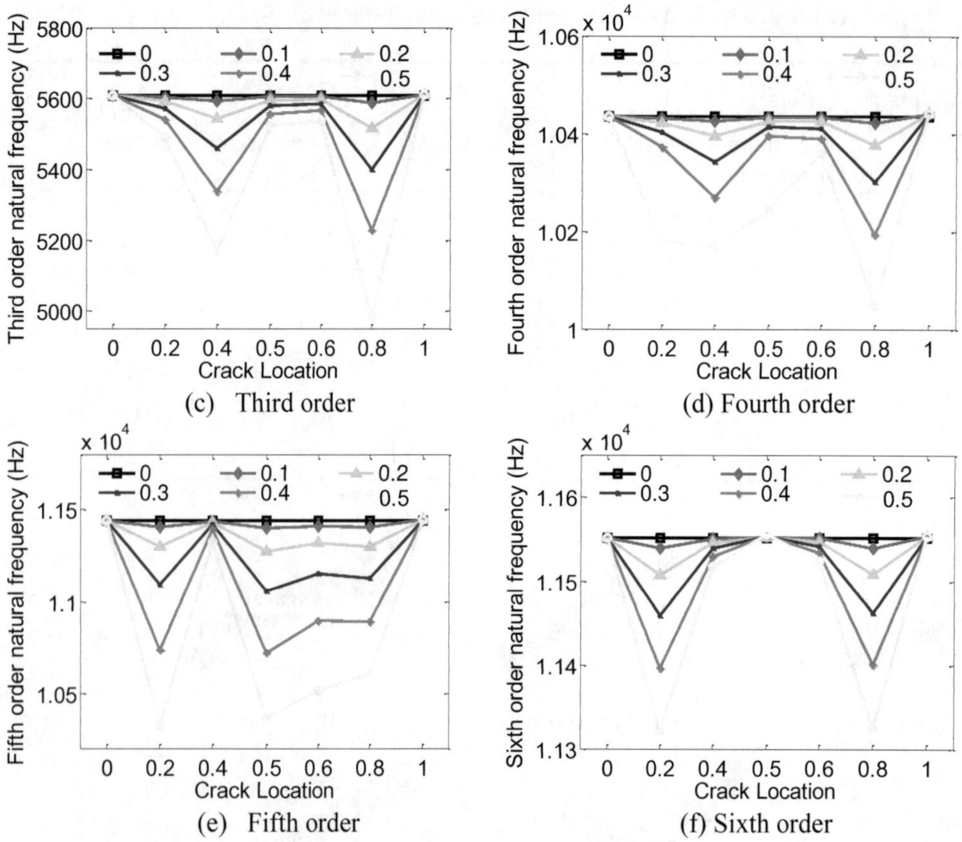

Figure 5. Effect of crack depth and crack location on the first 4-order natural frequencies

It can be seen from Figure 5 that when cracks occur, the natural frequencies of the first 6 orders become smaller, whatever the crack depth or the crack position changes; and at the same crack location, deeper cracks causes smaller natural frequencies. But when cracks occur at different locations, the effect of crack depth on the natural frequencies is quite different.

Figure 5(a) and (b) indicate that the changes of first and second natural frequencies are most obvious when the crack is located on the 0.6 of the whole length from the full fixed end where maximum vibration amplitude happens in both cases. Figures 5(c) and (d) indicate that the changes of third and fourth natural frequencies are most obvious when the crack is located on the 0.4 and 0.8 of the whole length from the full fixed end near which maximum vibration amplitudes happen in both cases. Figure 5(e) demonstrates that the changes of fifth order natural frequencies are most obvious when the crack is located on the 0.2 and 0.5 of the whole length from the full fixed end near which maximum vibration amplitudes happen. Figure 5(f) demonstrates that the changes of sixth order natural frequencies are most obvious when the crack is located on the 0.2 and 0.8 of the whole length from the full fixed end and least obvious when the crack is located in the middle of the whole length where maximum vibration amplitudes happen. To sum up, the most obvious decreases of the natural frequencies always happen where maximum vibration amplitudes happen in bending mode and the least obvious decreases of the natural frequencies happen where maximum torsion amplitudes happen in torsion model. Therefore, the effects of crack location on natural frequencies are related to the vibration model of the simple supported beam.

4. Conclusions

Both crack model and the model of the simple supported beam are investigated, and corresponding finite element model of the simple supported beam with crack damage is established in

ANSYS/Workbench software. Modal analysis of 26 cases with different crack location and crack depth are conducted so as to study effect of crack location and crack depth on dynamic characteristics of the simple supported beam. It is found that when cracks occur, the natural frequencies of the simple supported beam become smaller. The effects of crack location on natural frequencies are related to the vibration model of the simple supported beam.

Acknowledgements
The authors would like to express thanks to the Beijing Natural Science Foundation (No: 3184053) and the National Natural Science Foundation of China (No: 11702318) for their financial support of this research.

References

[1] Zhou J H, Li S P and Mo H N 2018 Test on influence of sugarcane harvester‘s dynamic properties on the broken rate of ratoons Journal of Agricultural Mechanization Research 40(07):150-156
[2] Suparp S and Joyklad P 2013 A study on simple beam bridge responses due to thai truck loads Procedia - Social and Behavioral Sciences 88
[3] Hou X L, Wang J L, Sun H and Wang R T 2015 A differential iterative algorithm for solving fluid-structure interaction problem and application in fluid-conveying pipeline Journal of Shenyang Jianzhu University(Natural Science) 31(01):149-157
[4] Guo Z G and Sun Z 2013 Modal analysis of euler-bernoulli beam with multiple open cracks based on perturbation method Journal of Vibration and Shock 32(10):1-6+30
[5] Zhang J W 2009 A study on the method based on vibration of damage identification of structures Changsha University of Science and Technology
[6] Ma Y J, Chen G P and Yang F 2017 Modal analysis of a simply supported steel beam with cracks under temperature load Shock and Vibration
[7] Song X H and Liu J 2016 Free vibration analysis of a simply supported beam with a vertical embedded crack J. Journal of Mechanical Strength.38(04):892-895
[8] Zou Z Q and Wu B 2014 Vibration-based damage detection for crack simply beam J. Ship Electronic Engineering.34(09):155-158
[9] Sun J L and Yang X 2015 Bending Deformation analysis of the Euler-Bernoulli beam with effect of switching crack gap J. Chinese Quarterly of Mechanics.36(04):703-712
[10] Bakhtiari-Nejad F, Khorram A and Rezaeian M 2014 Analytical estimation of natural frequencies and mode shapes of a beam having two cracks J. International Journal of Mechanical Sciences.78
[11] Vigneshwaran K and Behera R K 2014 Vibration analysis of a simply supported beam with multiple breathing cracks J. Procedia Engineering.86

The Critical Axial Force Model in Drilling of Carbon Fiber Reinforced Composites under Different Stacking Modes

M Li[1], W Tian[1], H L Sun[2], Z Liu[1], L Zhang[1], W H Liao[1] and F F Hua[1]

[1] College of Mechanical and Electronical Engineering, Nanjing University of Aeronautics and Astronautics, Nanjing 210016, P.R. China
[2] AVIC Chengdu Aircraft Industrial (Group) Co., Ltd., Chengdu 610092, P.R. China

E-mail: fengyasonglm@nuaa.edu.cn

Abstract. Delamination is a failure mode in which the initial crack propagates along the interlayer due to the axial force pushing the uncut material beyond the interlaminar shear strength during drilling. Therefore, it is necessary to study the effect of axial force on the delamination of carbon fiber reinforced composites (CFRP) during drilling. The critical axial force model of different stacking modes is established based on linear elastic fracture mechanics, thin plate bending theory, composite mechanics and superposition principle. The experimental results show that the theoretical values are in good agreement with the experimental values and can be used to predict the critical axial force to suppress the formation of delamination defects.

1. Introduction

In order to meet the lightweight and durability goals of modern aircraft structural design, laminated components are widely used in aircraft wing and tail, which are composed of carbon fiber composite materials (CFRP) and metal materials (aluminum alloy and titanium alloy) [1,2]. The quality of connection holes is directly related to connection strength, stiffness and safety [3,4]. The integrated laminated drilling reduces the complicated process of "Split specimen-Deburring-Reassembly", which has ensured the accuracy and accuracy of hole size, and has attracted wide attention.

There are many reasons for the drilling defect in the drilling process of CFRP, delamination is one of the most important processing defects. From a macro perspective, it is mainly caused by the drilling force, especially the axial force and the temperature produced in the process of machining, the former has a greater influence than the latter. The internal stress of the interlayer increases with the increase of drilling axial force, fiber layering due to the fracture of the resin when reaching or exceeding the strength of the resin [5]. At present, many scholars have done a lot of research on axial force during drilling. Qi et.al [6] established critical thrust force predicting modeling for delamination-free drilling of metal-CFRP stacks, the location and expansion rate of the composite material are determined. A theoretical critical axial force model for tool/material contact considering the variation of energy release rate of I type crack propagation with the relative fiber angle between laminates is established by Lachaud et.al [7], experiments show that the theoretical model under uniformly distributed load is the closest to the test results. Pierre et al [8] established mathematical models under different load states considering the influence of tools geometry on stress state of materials, the results of the experiment show that the test data are in good agreement with the data under the composite loading state. Rahme et.al [9] further studied the relationship between drilling axial force and process parameters on the basis of studying critical axial force, and put forward the concept of critical feed rate, the experiment shows that the CFRP does not delaminate when the feed rate is lower than critical feed rate, which provides a good way to control drilling delamination by changing process parameters. Ohzeki et al [10] monitor and feedback the axial force signal in real-time during drilling CFRP, and change the feed parameter according to the change of the axial force signal to control the axial force of drilling, so as to avoid export layering defects.

This study will establish a mechanical model of critical axial force by considering the different stacking sequence of CFRP-Al material, that is, different drilling directions. The delamination mechanism at the exit of CFRP was investigated and provides a basis for predicting delamination defects of laminated components.

2. Critical axial force model

2.1. Stacking sequence:Metal→CFRP

When drilling the laminates from the metal layer, which means that the metal plate is above the CFRP plate, the mechanism of composite material delamination is the same as that of single drilling CFRP, because there is no support for the lower layer of CFRP. Not only should we consider the contact conditions of the connection parts, but also we should reasonably simplify the axial force. So the study assumes that the drilling area is a circular thin plate subjected to axisymmetric load, the circular plate is referred to in Figure 1. According to the theory of thin plates in elasticity, its equilibrium differential equation is:

$$\frac{\partial^2 M_{xx}}{\partial x^2}+\frac{\partial^2 M_{yy}}{\partial y^2}+\frac{\partial^2 M_{xy}}{\partial x \partial y}=-q \tag{1}$$

Because the contact area between the drill and the workpiece is axisymmetric, the stress / strain relationship model can be expressed as:

$$\begin{Bmatrix} M_{xx} \\ -M_{yy} \\ -M_{xy} \end{Bmatrix} = \begin{pmatrix} D_{11} & D_{12} & D_{16} \\ D_{21} & D_{22} & D_{26} \\ D_{16} & D_{26} & D_{66} \end{pmatrix} \begin{Bmatrix} -\dfrac{\partial^2 \omega}{\partial x^2} \\ -\dfrac{\partial^2 \omega}{\partial y^2} \\ -2\dfrac{\partial^2 \omega}{\partial x \partial y} \end{Bmatrix} \tag{2}$$

where,M_{xx}, M_{yy}, M_{xy} is bending moments in different directions. q is uniformly distributed load on uncut sheet. D_{ij} is bending stiffness.

According to the theory of composite laminates, the stiffness matrix of a single plate can be drawn from the following equation:

$$\left|D_{ij}\right| = \sum_{k=1}^{n}\left(\overline{Q_{ij}}\right)_k \left(\frac{Z_k^3 - Z_{k-1}^3}{3}\right) \tag{3}$$

The CFRP used in this paper are orthotropic, then D_{16}=D_{26}=0, and the stiffness matrix is:

$$[Q] = \begin{pmatrix} Q_{11} & Q_{12} & 0 \\ Q_{21} & Q_{22} & 0 \\ 0 & 0 & Q_{11} \end{pmatrix} \tag{4}$$

The equation (2) and (3) are substituted by differential equations (1).

$$D_{11}\frac{\partial^4 \omega}{\partial x^4}+2\left(D_{12}+2D_{66}\right)\frac{\partial^4 \omega}{\partial x^2 \partial y^2}+D_{22}\frac{\partial^4 \omega}{\partial y^4}=-q \tag{5}$$

$$\omega(r)=\frac{q\left(a^2-r^2\right)}{64D} \tag{6}$$

where, $q=\frac{F_z}{\pi a^2}$ and $D=\frac{1}{8}\left(3D_{11}+2D_{12}+4D_{66}+3D_{22}\right)$

According to the principle of virtual displacement and linear elastic fracture mechanics, the total energy done by uniformly distributed load equals to the sum of the elastic strain energy and the crack growth energy of the material [11,12]. The equation is:

$$\delta W=\delta U+\delta Ud \tag{7}$$

In the equation, δW is the total virtual energy produced by uniform load per unit area, δU is the change of elastic strain energy of the material, and δd is the crack propagation energy per unit area, reflecting the ability of the material to resist fracture damage.

The total virtual energy produced by uniform distributed load:

$$W=\int_0^{2\pi}\int_0^{a} q\cdot w(r)r\cdot drd\theta=\frac{F_z^2a^2}{192\pi D} \tag{8}$$

$$\delta W=\frac{\partial W}{\partial a}\partial a=\frac{F_z^2a}{96\pi D}\delta a \tag{9}$$

Variation of elastic strain energy:

$$\delta U=\frac{\partial U}{\partial a}\delta a=\left(\frac{F_z}{16D}\right)^2\cdot\frac{a}{\pi}\left[\left(\frac{D_{11}+D_{22}}{2}\right)+\left(\frac{D_{12}+D_{66}}{3}\right)\right]\delta a \tag{10}$$

The crack propagation energy per unit area:

$$\delta Ud=\frac{\partial Ud}{\partial q}\delta a=2G_{IC}\pi a\delta a \tag{11}$$

Among them, G_{IC} is the I-type crack propagation energy per unit area.

The equation (9) (10) (11) was substituted (7), the critical axial force can be introduced:

$$F_z=8\pi\left(\frac{G_{IC}D}{(1/3)-\left(D^{'}/8D\right)}\right)^{1/2} \tag{12}$$

Where, $D^{'}=\frac{D_{11}+D_{12}}{2}+\frac{D_{12}+D_{66}}{3}$

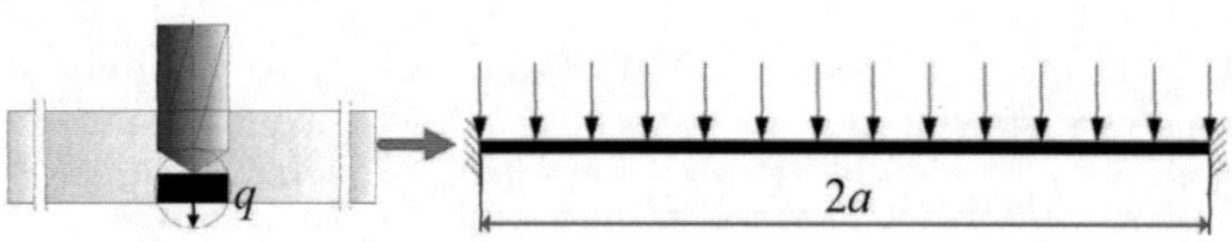

Figure 1. Uniform distributed load model for tool / workpiece contact

2.2. Stacking sequence:CFRP→Metal

The stress characteristics of composite plates are quite different from those of individual drilling when the stacking sequence is CFRP to Metal. The overall stress condition of the composite plate will be affected due to the additional supporting force. In addition, the displacement and deformation of laminated plates also affect the critical axial force of lamination, so the shape of the plate, boundary conditions, fixture conditions and other factors should be considered when establishing the model of critical axial force. The stress state is shown in Figure 2. At this point, the borehole area can still be regarded as a circular thin plate with fixed edges.

The concentrated force generated by a chisel edge:

$$F_c=\xi F \tag{13}$$

Among them, F_c is the total axial force. ξ is the ratio of the axial force produced by the transverse edge part to the total axial force, also called the proportion coefficient.

The offset of each point on the circular plate is:

$$\omega_r(r) = \omega_{F_c}(r) - \omega_{P_{re}}(r) = \frac{F_c - P_{re}}{16\pi D}\left[2r^2 \ln\frac{r}{a} + \left(a^2 - r^2\right)\right] \tag{14}$$

In the equation, r is the extreme radius of circular plate and D is the flexural rigidity of the uncut layer. From the conclusion in document [8,13], P_{re} can be obtained from the following equation:

$$P_{re} = F_c - 2\sqrt{3}\pi\sqrt{G_{IC}D} = \xi F - 2\sqrt{3}\pi\sqrt{G_{IC}D} \tag{15}$$

Because the composite plate is subjected to axial force and supporting force, the displacement of every point on the plate is determined:

$$\omega_c(r) = \frac{F - P_{re}}{16\pi D_c}\left[2r^2 \ln\frac{r}{b} + \left(b^2 - r^2\right)\right] \tag{16}$$

Among them, D_c is equivalent flexural rigidity of composite plates can be obtained by equation (3).b is the diameter of plates.

When the metal plate is thin plate (the ratio of plate thickness to plate diameter: $\delta/2b \le 1/8$), the displacement of metal plates at various points is:

$$\omega_m(r) = \frac{3\left(1 - v^2\right)P_{re}}{4\pi E h^3}\left[2r^2 \ln\frac{r}{b}\left(b^2 - r^2\right)\right] \tag{17}$$

As shown in Figure 2, the displacement of the uncut region is the same as that of the metal plate based on the hole edge of the composite plate, that is:

$$\omega_r(r=0) + \omega_c(r=a) = \omega_m(r=0) \tag{18}$$

The critical axial force is obtained by substituting equation (14) (15) (16) (17) into (18):

$$F_z = \frac{2\sqrt{3}}{\xi - K}\sqrt{G_{IC}D} \tag{19}$$

Where, $K = \dfrac{\dfrac{\xi a^2}{D} + \dfrac{2a^2 \ln\frac{a}{b} + \left(b^2 - a^2\right)}{D_c}}{\dfrac{9\left(1 - v^2\right)b^2}{2Eh^3} + \dfrac{a^2}{D} + \dfrac{2a^2 \ln\frac{a}{b} + \left(b^2 - a^2\right)}{D_c}}$

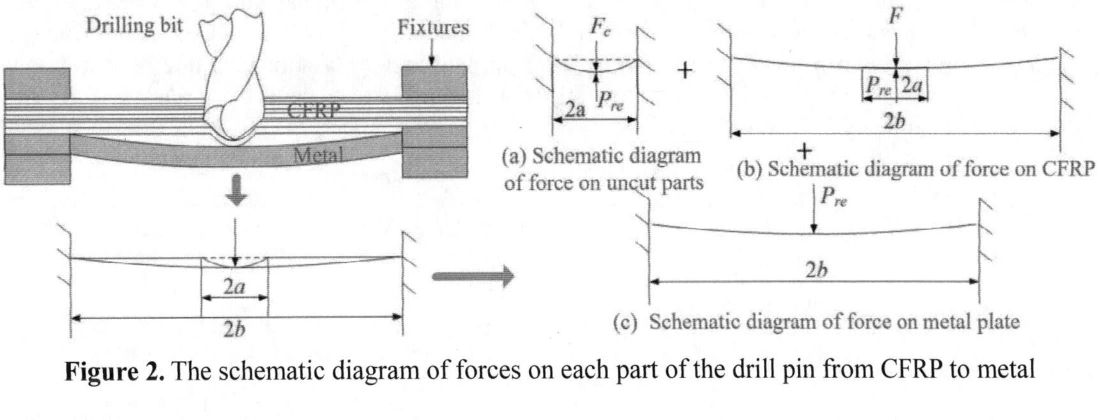

Figure 2. The schematic diagram of forces on each part of the drill pin from CFRP to metal

3. Experimental research

3.1. Experimental devices and materials

The experimental equipment is LGmazak430Al CNC machining center, the thrust force and torque during machining were measured using piezoelectric dynamometer (Interface1216), two ME SGA/A charge amplifiers and PC are used for data acquisition and analysis, as shown in Figure 3. Carbon fiber-reinforced composite materials (CCF300) of 200×200×3.07mm and aluminum alloy (Al7075-T7) of 200×200×3.1mm were used for experiments, its mechanical and physical properties are shown in table 1. The CFRP is orthogonal woven structure, laid up according to an orientation of [0°/90°], which is composed of 24 plies, each layer thickness h=0.125 mm, Properties of each layer of material as shown in Table 2. The tool is diamond coated carbide twist drill with a diameter of 5mm.

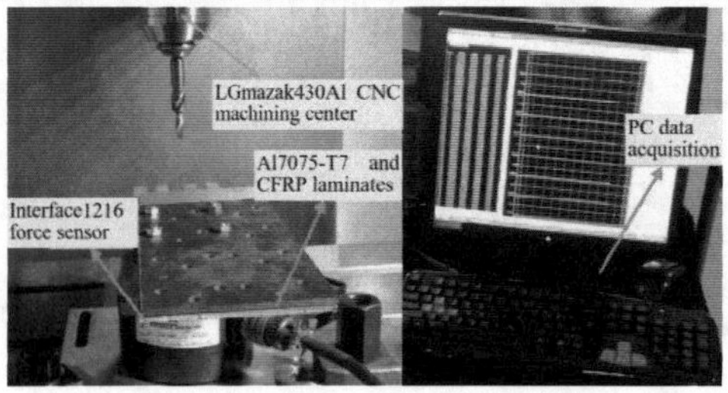

Figure 3. Experimental platform

Table 1. Mechanical and physical properties of CCF300 and Al7075

Material	Mechanical and physical properties			
	Tensile strength(MPa)	Elasticity modulus (GPa)	Density(g/cm^3)	Elongation(%)
CCF300	3950	232	1.76	1.7
Al-7075	572	71.7	2.81	12

Table 2. Mechanical properties each layer of CCF300

Mechanical properties	CCF300
Shear strength /MPa	89
Longitudinal elastic modulus /GPa	137
Transverse elastic modulus /GPa	8.7
Longitudinal and transverse shear modulus /GPa	4.2

3.2. Experimental design

In order to verify the relationship between delamination and axial force the relationship model between axial force and machining parameters should be established, and the fitting equation of axial force should be obtained. Firstly, using the orthogonal experimental design method, each experiment condition was repeated five times to get consistent values and avoid accidental error. The parameter level of spindle speed and feed rate is shown in Table 3. Then, the spindle is not rotated during the experiment, only a vertical downward fixed load was applied and a constant feed rate of 10mm/min was used ,the axial force gradually increases until delamination defects occurs in order to verify the accuracy of the critical axial force model.

Table 3. Level table of cutting parameters

Name	Units	Level 1	Level 2	Level 3	Level 4	Level 5
Spindle speed	n/(r·min^{-1})	2700	3500	4300	5100	5900
Feed rate	f/(mm·r^{-1})	0.03	0.04	0.05	0.06	0.07

4. Results and discussion

4.1. Effect of cutting variables on thrust force and torque

The thrust force and torque time-history curve when drilling CFRP/Al stacks as shown in Figure 4. The axial force and torque increase gradually from the drill point to the composite material, and then stabilize when it enters the composite material completely. When the drill tip reaches the transition zone of the laminated member, the cutting force increases suddenly and reaches the maximum value after drilling into the aluminum alloy completely. The axial force and torque gradually decrease to zero when the drill bit is drilled from the bottom of the aluminum alloy. The axial force of the aluminum alloy at stable stage is about 1.5 times of the axial force of the CFRP.

The effect of the process parameters on the thrust force of CFRP and Al7075-T7 when drilling CFRP/Al stacks as shown in Figure 5 and Figure 6, respectively. It's obvious that the spindle speed and feed rate have a certain effect on cutting force, but the influence of feed rate is greater than that of spindle speed. The variance analysis of cutting force is made in order to judge the degree of influence, as shown in Table 4. The main effect has a significant effect on the cutting force, and the interaction between them has a certain influence on the cutting force, but the main effect is not obvious.

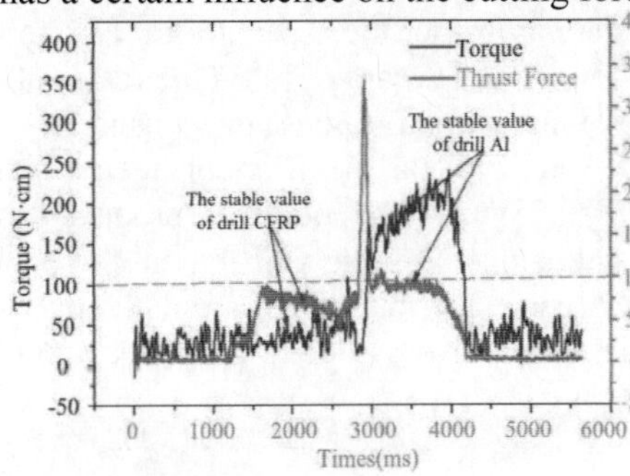

Figure 4. The thrust force and torque t time-history curve when drilling CFRP/Al stacks(Spindle speed n=3500rpm, feed rate f=0.05mm/r)

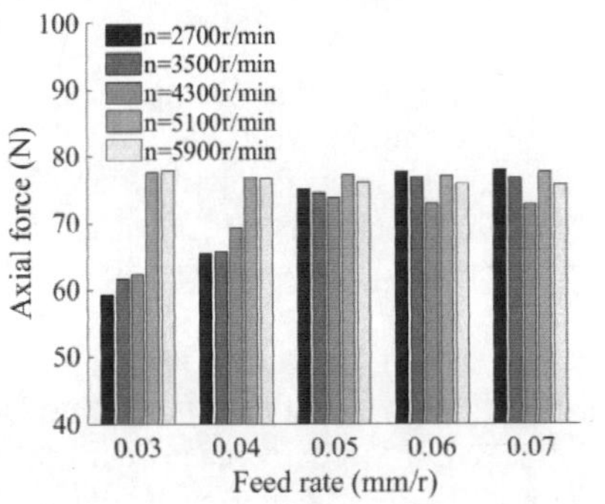

Figure 5. The effect of the process parameters on the thrust force of CFRP.

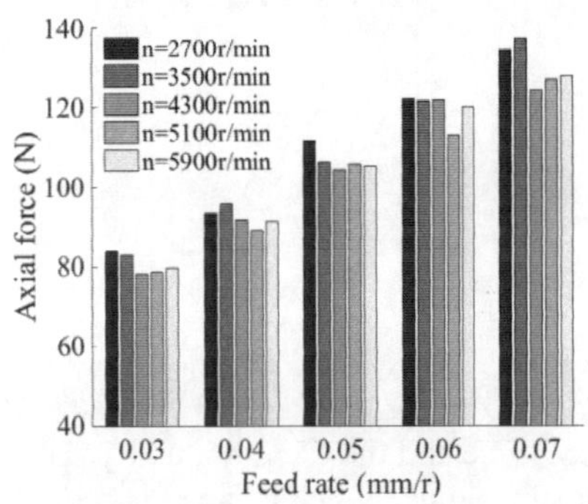

Figure 6. The effect of the process parameters on the thrust force of Al.

Table 4 Variance analysis of cutting force

Difference source	*SS*	*df*	*MS*	*F*	*P*-value
Spindle speed	186.7896	4	46.69741	7.996328334	0.000967541
Feed rate	7977.123	4	1994.281	341.4948581	2.87539E-15
Error	93.4377	16	5.839856		

4.2. Effect of cutting variables on thrust force and torque

According to the law of axial force changing with cutting parameters, there is a nonlinear relationship between axial force and spindle speed, feed rate and their interaction. Therefore, quadratic regression model can be used to fit the empirical equation, the fitting equation as follows:

$$F_{CFRP} = 53.36 - 0.0055n + 2.4598\times10^{-7}n^2 + 885.87f - 5692.9f^2 - 0.0585nf \ (R^2=0.9712) \quad (20)$$

$$F_{Al} = 88.3323 - 0.0125n + 1.1038\times10^{-6}n^2 + 344.38f + 5792.9f^2 - 0.0074nf \ (R^2=0.9363) \quad (21)$$

Among them, F is axial force, n is spindle speed, f is feed rate, C_0-C_5 is an undetermined constant.

The determination coefficients of the fitting equation are all greater than 0.9, indicating that the fitting results are ideal. It can be concluded that the regression equation can be used as an empirical equation to predict the axial force within the selected process parameters. In practical application, higher spindle speed and lower feed rate should be selected to reduce the lamination defect of CFRP. At the same time, reducing the cutting force can weaken the impact on the tool in the drilling process and thus prolong the service life of the tool.

4.3. Critical axial force in different stacking modes

4.3.1. Critical axial force when drilling Al to CFRP stacks. A comparison between the critical axial force model value and the experimental result as shown in Figure 7 when the stacking sequence is Al to CFRP. The critical axial force also increases linearly as the number of uncut layers increases. The experimental value is always smaller than the model value, this is because the axial force is uniformly distributed on the main cutting edge and the chisel edge under the assumption of uniform load, which weakens the pushing effect of the concentrating force on the cross edge on the material and makes the critical axial force in the model is higher. The experimental values are consistent with the model values when the number of uncut layers of CFRP is small (n< 6), it is considered that the model is more precise. When the uncut layers exceeds 6 (H=0.75mm), the model will produce a great error with the experimental value. The main reason is the establishment of the critical axial force model is mainly for thin plates, the pushing force of the cutting force on the material is strengthened when the uncut part exceeds 0.75mm, the occurrence of delamination is not only due to the propagation of cracks, but also to the stress concentration.

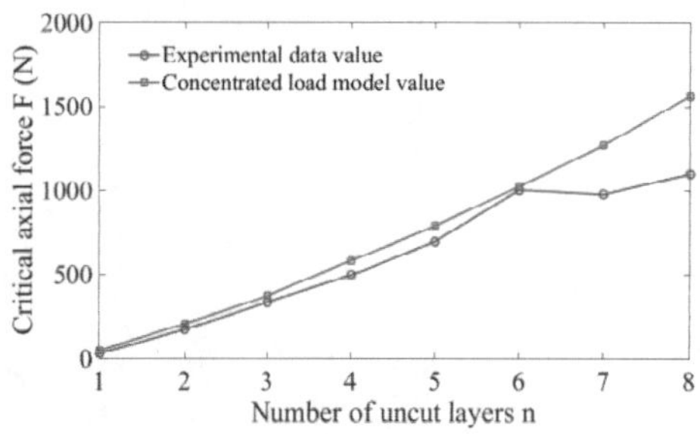

Figure 7. The comparison diagram between the critical axial force model value and the experimental result when the stacking sequence is Al to CFRP

4.3.2. Critical axial force when drilling CFRP to Al stacks. Figure 8 is the relationship between the critical axial force and the fixed boundary radius. The critical axial force decreases with the increase of fixed boundary radius, which means that delamination is more likely to occur at the exit of CFRP with the increase of fixed boundary radius. It can be seen from equation(19)that the critical axial force is directly proportional to the thickness of h, the undrilled thickness of CFRP decreases gradually with the increase of drilling depth, and the critical axial force for delamination decreases. To avoid stratification of the production, cutting edges of the tools should be sharp enough to cut quickly.

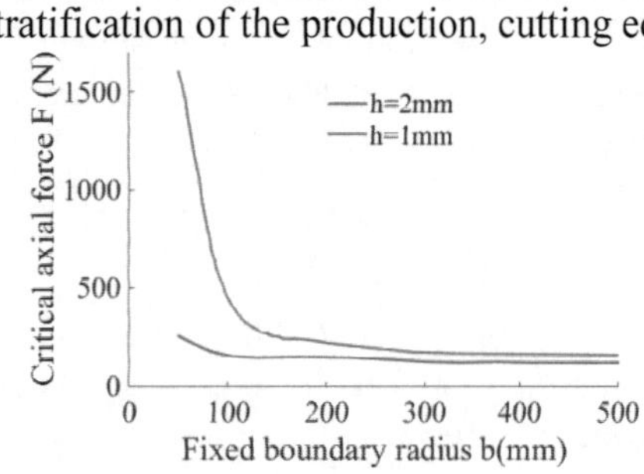

Figure 8. The relationship between the critical axial force and the fixed boundary under different thickness.

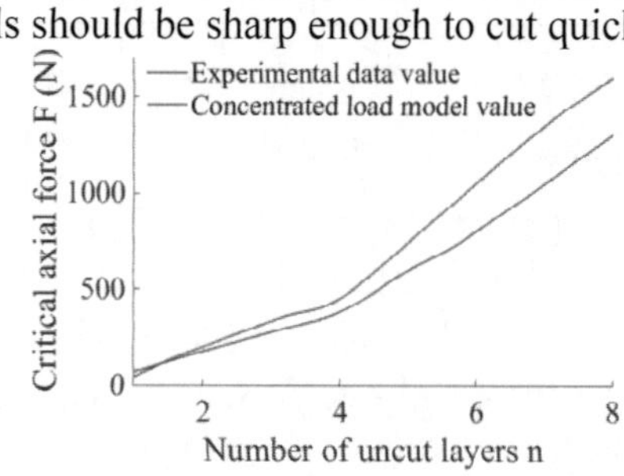

Figure 9. The comparison diagram between the critical axial force model value and the experimental result when the stacking sequence is CFRP to Al.

From Figure 9, axial force decreases as the number of uncut layers decreases when the number of uncut layers is less than 4, and the difference between theoretical value and experimental value is very small. No matter how thick the thickness of the uncut layer is, the experimental data are always slightly higher than the theoretical value. This is mainly due to the fact that the ratio of the concentrated load generated by chisel edge to the total cutting force in the model is less than the ratio in actual cutting. ξ is related to the structural form of the drill bit and the geometric angle of the cutting edge. In general, ξ is 0.4~0.6, the value of this model is 0.5.

5. Conclusions

In this study, the effect of different stacking sequences of Metal/CFRP on the delamination of CFRP was analyzed, and the mechanism of delamination defects in two different stacking sequences was studied. The study on controlling the delamination of composite materials enables the following conclusions. The spindle speed and feed rate have a certain effect on cutting force, but the influence of feed rate is greater than that of spindle speed. The experimental values are consistent with the model values when the number of uncut layers of CFRP is small ($n < 6$), when the stacking sequence is Al to CFRP. Contrary, when the stacking sequence is CFRP to Al, the critical axial force is related not only to the thickness but also to the radius of the fixed boundary. The axial force is the main factor for the delamination of composite materials. The deformation of the metal plate will inhibit the delamination of the composite material, and its critical axial force will be greatly improved when the metal plate is supported below the composite material.

Acknowledgments

The authors would like to thank National Science and technology major project "high-end CNC machine tools and basic manufacturing equipment"(**Grant No. 2014ZX04001071**)for sponsoring this project.

References

[1] Tian W, Hu J, Liao W, Hu J and Zhang L 2016 *Chin. J. Aeronaut.* 29 283
[2] Ramulu M, Branson T and Kim D 2001 *Compos. Struct.* 54 67
[3] Zitoune R, Mansori M E and Krishnaraj V 2013 *Wear* 302 1560
[4] Zitoune R, Krishnaraj V and Collombet F 2010 *Compos. Struct.* 92 1246
[5] Liu D F, Tang Y J and Cong W L 2015 *Compos. Struct.* 91 1268
[6] Qi Z, Zhang K, Li Y, Liu S N and Cheng H 2014 *Compos. Struct.* 107 605
[7] Lachaud F, Piquet R, Collombet F and Laurent S 2001 *Compos. Struct.* 52 513
[8] Rahmé P, Landon Y and Lachaud F 2011 *Int. J. Adv. Manuf. Technol.* 52 611
[9] Rahme P, Landon Y, Lachaud F and Piquet R 2015 *Int. J. Adv. Manuf. Technol.* 76 1546
[10] Ohzeki H, Hoshi H and Arai F 2012 J. *Adv. Mech. Des. Syst. Manuf.* 6 53
[11] Ho-Cheng H and Dharan C K H 1990 Journal of Engineering for Industry. 112 237
[12] Reddy J N 1994 *Mechanics of Composite Materials* Composites (USA:Kluwer Academic Publishers) p 33
[13] Caprino G, Santo L and Nele L 1998 *Compos. Pt. A-Appl. Sci. Manuf.* **29** 888

Design and Simulation Technology of Piezoelectric Ultrasonic Transducer with Sandwich Composite Horn

Yi Xuan[1], Aimin Wang[2] and Niansong Zhang[1, a]

[1]College of mechanical engineering, Nanjing University of Science and Technology, Nanjing,China
[2]College of mechanical engineering, Beijing Institute of Technology, Beijing, China

[a] Corresponding author: zns@ustc.edu.cn,+86-13270737066

Abstract. The structural characteristics of piezoelectric ultrasonic transducer with sandwich composite horn are expounded,the vibration characteristics are theoretically analyzed,the velocity distribution equation and stress distribution equation of transducer under longitudinal vibration are obtained,the frequency equation is given,and the piezoelectric ultrasonic transducer with compound amplitude-changing rod is designed.Then the modal analysis and harmonic response analysis of the transducer are carried out by using the finite element simulation software Workbench,comparison between the results obtained and the theoretical values,the error is very small, which meets the need of engineering practice and verifies the correctness of the derivation design.

1. Introduction

An ultrasonic transducer is an energy conversion device that converts alternating electrical signals into acoustic signals or acoustic signals into electrical signals in the range of ultrasonic frequencies[1].At present, the transducer is mainly divided into two types: magnetostrictive transducer and piezoelectric transducer. Because the electro-acoustic conversion efficiency of piezoelectric transducer is much higher than that of magnetostrictive transducer, piezoelectric transducer is generally used in ultrasonic transducer. In ultrasonic vibration system, the transducer and the amplitude-change rod are usually connected as a whole by bolts, and the length is usually the wavelength of a longitudinal wave.Many scholars at home and abroad have studied this structure[2-4].

In this paper, an analytical method is used to design a transducer with a variable amplitude bar, which can greatly improve the energy transfer efficiency of the transducer and reduce the redundancy of the whole vibration system. Then the design of the transducer is simulated by finite element software to verify the consistency between the design and the simulation results.

2. Piezoelectric ultrasonic transducer with sandwich composite horn

2.1. structural analysis

The structure of the sandwich piezoelectric ultrasonic transducer is shown in Figure 1, including: front cover plate, piezoelectric ceramic, electrode, back cover plate, prestressed bolt and insulating pipe.Piezoelectric ceramic stack consists of a plurality of piezoelectric ceramic wafer mechanical series, with independent positive and negative electrodes.The electrical parallels between ceramic chips are opposite to the polarity of adjacent ceramic plates.A flange is used to connect the transducer to the external mechanism to meet the needs of practical application of power ultrasound.

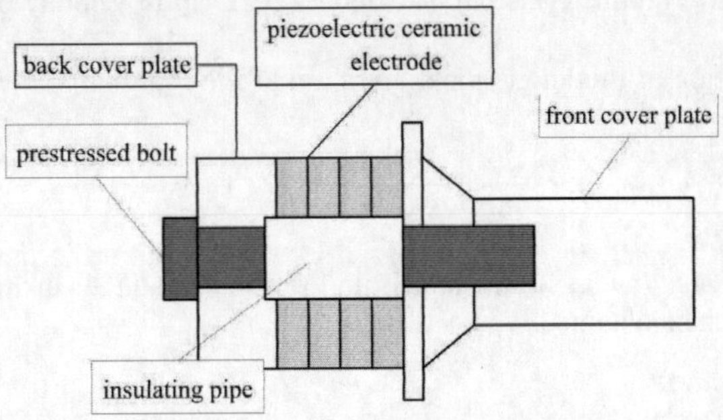

Figure 1. Structure diagram of piezoelectric ultrasonic transducer with sandwich composite horn.

Its main characteristics are[5-6]: (1) Piezoelectric ceramics have great compressive strength, prestressing force is given by prestressed bolt, on the one hand, the stability of transducer can be enhanced when the environmental strength changes; On the other hand, it ensures that the components are in a compression state under the condition of high power drive, thus avoiding the rupture caused by the expansion of the ceramic plate. (2) The number and connection of the piezoelectric ceramic pieces are open to choice. Thus the transducer can be designed in a wide range of impedance and frequency. (3) Changing the material and size of the front and rear metal cover plate can effectively control the performance parameters such as the front and back vibration speed ratio and the effective electromechanical coupling coefficient of the transducer.

2.2. Theoretical analysis

When an object vibrates in an elastic medium, it will cause the vibration of the medium. For all solids, the vibration wave can be regarded as an elastic body. There is an elastic relation between the points of the medium, and the vibration of the object propagates in the elastic medium.It is assumed that there is a uniform thin rod with arbitrary variable cross section and mechanical loss is omitted. When the cross sectional area of the rod is much smaller than the wavelength the plane longitudinal wave propagates along the axis of the rod.As shown in Figure 2,any variable cross-section bar,the axis of symmetry isx,optionally, a small volume element(x, x + dx),the tensile stressacting on it is $\frac{\partial \xi}{\partial x} dx$.

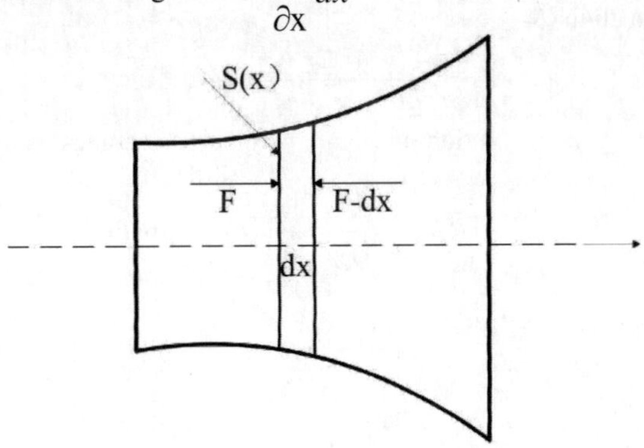

Figure 2. Any variable cross-section bar.

According to Newton's law, the kinetic equation can be obtained:

$$\frac{\partial (S\sigma)}{\partial x} dx = S\rho \frac{\partial^2 \xi}{\partial t^2} dx \tag{1}$$

In the form, S is the cross-sectional area function of the bar, ξ is particle displacement function, σ is stress function, ρ is density of rod material,E is young's modulus.

Under the condition of resonance, the wave equation of longitudinal vibration of the rod with variable cross section can be obtained:

IOP Conf. Series: Materials Science and Engineering **470** (2019) 012043 doi:10.1088/1757-899X/470/1/012043

$$\frac{\partial^2 \xi}{\partial x^2}+\frac{1}{S}\times\frac{\partial S}{\partial x}\times\frac{\partial \xi}{\partial x}+k^2\xi=0 \tag{2}$$

In the form, k is circular wave number, ω is circular frequency, c is the velocity of the longitudinal wave propagating through the rod.

The vibration velocity equation of the longitudinal vibration of the rod with variable cross section can be obtained by taking the equation (2):

$$\frac{\partial^2 v}{\partial x^2}+\frac{1}{S}\times\frac{\partial S}{\partial x}\times\frac{\partial v}{\partial x}+k^2 v=0 \tag{3}$$

2.2.1 uniform equal-circle cross-section bar. If the area is uniform and equal,the equation of vibration velocity of a rod with uniform section can be obtained:

$$\frac{\partial^2 v}{\partial x^2}+k^2 v=0 \tag{4}$$

The general solution of equation (4) is:

$$v(x)=A\sin kx+B\cos kx \tag{5}$$

The elastic force and stress are:

$$F(x)=\frac{ES}{j\omega}\frac{\partial v}{\partial x}=-j\rho c S(A\cos kx-B\sin kx) \tag{6}$$

$$\sigma(x)=\frac{E}{j\omega}\frac{\partial v}{\partial x}=-j\rho c(A\cos kx-B\sin kx) \tag{7}$$

2.2.2 Circular section conical bar. The large end radius of the circular section conical rod is R1, the small end radius is R2, the length is l, and the expression of the radius and area of the section at any position is as follows:

$$R(x)=R_1(1-\gamma x) \tag{8}$$

$$S(x)=S_1(1-\gamma x)^2 \tag{9}$$

$$\gamma=\frac{R_1-R_2}{R_1 l} \tag{10}$$

Replace equation (9) with equation (3):

$$\frac{\partial^2 v}{\partial x^2}+\frac{2\gamma}{\gamma x-1}\frac{\partial v}{\partial x}+k^2 v=0 \tag{11}$$

By solving equation (11), the general solution of vibration velocity of conical rod with circular section is obtained as follows:

$$v(x)=\frac{1}{x-1/\gamma}(A\sin kx+B\cos kx) \tag{12}$$

The elastic force and stress are:

$$F(x)=-j\rho c S\left\{\begin{array}{l}A\left[\frac{1}{x-1/\gamma}\cos kx-\frac{1}{k(x-1/\gamma)^2}\sin kx\right]-\\-B\left[\frac{1}{x-1/\gamma}\sin kx+\frac{1}{k(x-1/\gamma)^2}\cos kx\right]\end{array}\right\} \tag{13}$$

$$\sigma(x)=-j\rho c\left\{\begin{array}{l}A\left[\frac{1}{x-1/\gamma}\cos kx-\frac{1}{k(x-1/\gamma)^2}\sin kx\right]-\\-B\left[\frac{1}{x-1/\gamma}\sin kx+\frac{1}{k(x-1/\gamma)^2}\cos kx\right]\end{array}\right\} \tag{14}$$

2.3. Dimension Design

The following analysis is made on the sandwich type piezoelectric ultrasonic transducer with compound amplitude change rod. The model is simplified as shown in Figure 3. The 1 is the rear cover plate, the 2 is the piezoelectric ceramic plate, the 3,4,5 is the composite horn, and as the front cover plate, the joint surface is located on the joint surface of the piezoelectric ceramic stack and the front cover plate. It is assumed that the front surface vibration velocity of the transducer oscillator is v_f and that the rear surface vibration velocity is v_b.

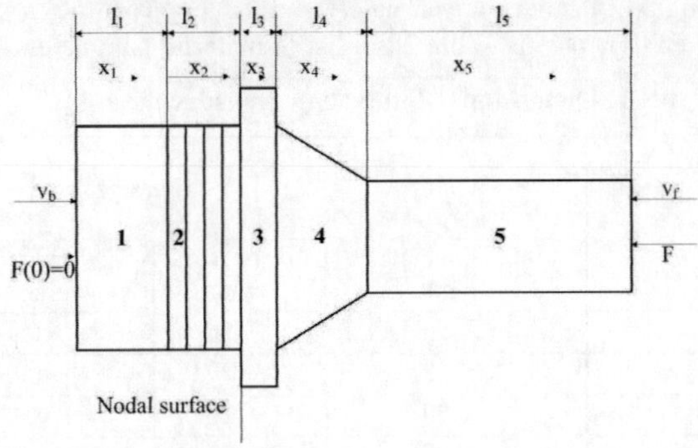

Figure 3. Model schematic diagram of piezoelectric ultrasonic transducer with sandwich composite horn.

The frequency equation on both sides of the nodal plane is derived below. First consider the left part of the nodal plane. The boundary conditions are as follows:

$$\begin{cases} v_2(l_2)=0 \\ v_2(0)=v_1(l_1) \\ F_2(0)=F_1(l_1) \\ v_1(0)=v_b \\ F_1(0)=0 \end{cases} \tag{15}$$

Substituting equations (5)(6)(7) gives the frequency equation on the left side of the nodal plane:

$$\tan k_1 l_1 \tan k_2 l_2 = \frac{Z_2}{Z_1} = \frac{\rho_2 c_2 S_2}{\rho_1 c_1 S_1} \tag{16}$$

The right part of the nodal plane with boundary conditions as follows:

$$\begin{cases} v_3(0)=0 \\ v_3(l_3)=v_4(0) \\ F_3(l_3)=F_4(0) \\ v_4(l_4)=v_5(0) \\ F_4(l_4)=F_5(0) \\ v_5(l_5)=v_f \\ F_5(l_5)=-Z_w v_f=0 \end{cases} \tag{17}$$

Substituting equations(12)(13)(14) gives the frequency equation on the left side of the nodal plane:

$$\begin{aligned} \frac{Z_5}{Z_4}\tan k_5 l_5 = {} & \frac{Z_4 k_4 (l_4 - 1/\gamma)}{Z_3 k_4 \cot k_3 l_3 \sin k_4 l_4 - \gamma Z_4 \sin k_4 l_4 + Z_4 k_4 \cos k_4 l_4} \times \\ & \times \left[\left(\frac{Z_3 k_4 \cot k_3 l_3 - \gamma Z_4}{Z_4 k_4} \right) \left(\frac{k_4 (l_4 - 1/\gamma) \cos k_4 l_4 - \sin k_4 l_4}{k_4 (l_4 - 1/\gamma)^2} \right) - \frac{k_4 (l_4 - 1/\gamma) \sin k_4 l_4 + \cos k_4 l_4}{k_4 (l_4 - 1/\gamma)^2} \right] \end{aligned} \tag{18}$$

The forward and rear vibration speed ratio of the transducer is deduced below, we can see that the boundary condition at the node is the continuity of elastic force from figure 3:

$$F_2(l_2) = F_3(0) \tag{19}$$

Substituting equation (6)(15)(17):

$$\begin{aligned} \frac{v_f}{v_b} = {} & \left(Z_3 k_4 \cot k_3 l_3 \sin k_4 l_4 - \gamma Z_4 \sin k_4 l_4 + Z_4 k_4 \cos k_4 l_4 \right) \times \\ & \times \frac{Z_2 \cos k_1 l_1 \sin k_3 l_3}{\gamma Z_3 Z_4 k_4 \sin k_2 l_2 \left(l_4 - 1/\gamma \right) \cos k_5 l_5} \end{aligned} \tag{20}$$

Figure 3 shows the dimensions of the sandwich composite horn piezoelectric ultrasonic transducer, as shown in Table 1. The diameter of the large end of the conical segment of the front cover plate is 50 mm, and the diameter of the small end is 20 mm. In the table ,the length of the front cover plate is calculated by equation (16) (18) except the length of the back cover plate 1 and the length of the front cover plate cylinder 5. The other dimensions are all given dimensions.Table 2 is a list of material parameters.

Tabel 1. Dimensions of ultrasonic transducer parts

Component		Parameter			
		External diameter /mm	Bore size /mm	Lengh /mm	Material
Back cover plate		50	18	16.5	Steel
Piezoelectric Ceramic chip		50	20	6.0	PZT-8
Electrode sheet		50	20	0.2	Brass
Front cover plate	**Flange plate**	65	18	5.0	TC4
	Conical segment			16.5	TC4
	Cylindrical segment	20		46.5	TC4

Tabel 2. Material parameter table

Material	Parameter			
	Elastic modulus /Gpa	Density /kg*m-3	Poisson ratio	Acoustic velocity /m*s-1
Steel	209	7800	0.28	5150
PZT-8	65	7600	0.31	2950
TC4	110	4430	0.33	5000

3. Numerical simulation analysis of piezoelectric ultrasonic transducer with sandwich composite horn

3.1. The establishment of finite element model

According to the design experience, the overall structure of the transducer is often simplified to a certain extent in the analysis process, and the influence of bolts and electrodes is ignored in this analysis, that is, the front cover plate of the transducer when modeling. Both the back cover plate and the piezoelectric ceramic plate are considered as solid bodies, and the parallel connection of piezoelectric elements in the circuit is not considered. The finite element analysis of the designed transducer is carried out by workbench. After a series of preprocessing, such as simplification of transducer model and definition of performance parameters, the model was meshed and the finite element model was established. The finite element model of the transducer is shown in Figure 4.

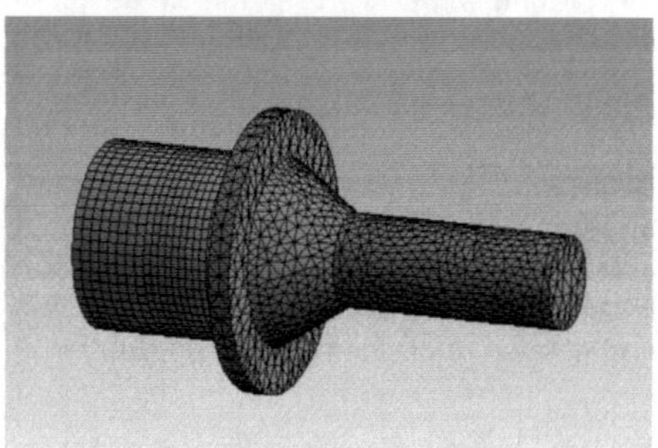

Figure 4. Finite element model of transducer.

3.2. Modal analysis technique

In modal analysis of the transducer, the positive and negative electrodes are briefly connected to obtain the resonant frequency of the transducer, that is, all the nodes on the surface of the piezoelectric plate are coupled with the voltage degree of freedom and the coupling voltage. In addition, the displacement constraint of the degree of freedom is applied on the nodal surface of the transducer. In modal analysis, a direct solver controlled by program is used. The order of solution is 10 and the frequency of solution is 15-25 kHz. The natural frequency and vibration mode are shown in Table 3.

Tabel 3. Natural frequency and vibration mode of transducer

Order	Natural frequency	Mode of vibration
1	18394	winding
2	18399	winding
3	19695	direction
4	20482	winding
5	23384	winding
6	23434	winding

It can be seen from Table 3 that the natural frequency and mode shape of the third mode meet the requirements of the design. The first to sixth order modal analysis clouds are shown in Figure 5.

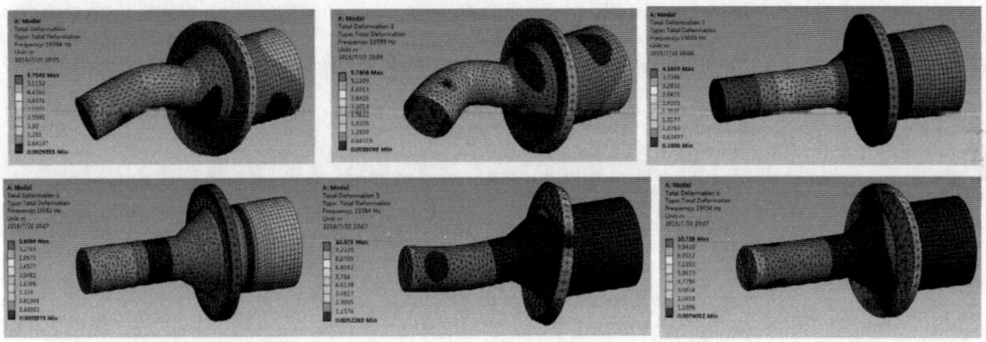

Figure 5. First to sixth order Modal Analysis of Transducers

Table 3 and figure 5 show that the frequency of longitudinal vibration of the transducer is 19695 Hz, which is close to its theoretical design value 20000Hz, and its error is about 1.52%.This shows that the theoretical design of the transducer holds true.

3.3. Harmonic response analysis technique

In the case of external excitation, harmonic response analysis is a method to determine whether resonance can occur in the frequency range of a mechanism. The harmonic response of the designed transducer is analyzed, and the voltage applied on the electrode is the load of harmonic response analysis. A sinusoidal alternating voltage is applied to the piezoelectric ceramic stack, a 0 V voltage is applied to the negative electrode and a 100 V alternating voltage is applied to the positive electrode. According to this excitation method, the front end amplitude of the front cover plate of the transducer varies with the frequency, as shown in Figure 6.

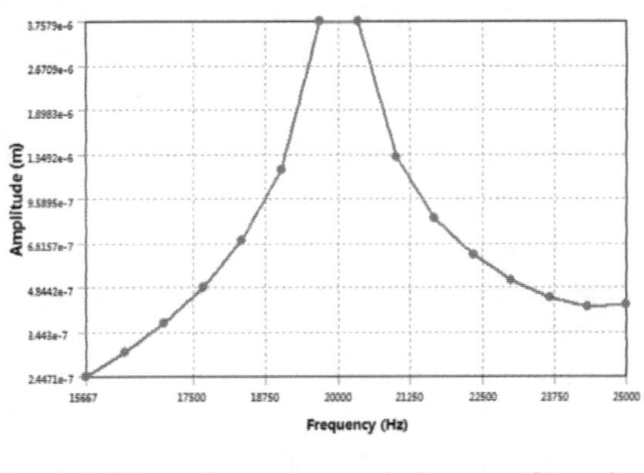

Figure 6. Harmonic response analysis curve of transducer

From Figure 6, it can be seen that when the frequency is 20333Hz, the center node displacement of front end of front cover plate of the transducer is the largest, and the amplitude of longitudinal vibration can reach about 4 micron, which has met the requirement of the required amplitude for ultrasonic vibration application. At the same time, in the safe range of excitation voltage, according to the different process requirements, without replacing the transducer, the value of the transducer excitation voltage can be adjusted by the ultrasonic power supply, and the amplitude of the front face of the front cover plate of the transducer can be finally adjusted.

4. Conclusion

Using the boundary condition analysis method, the vibration characteristics of the sandwich-type composite horn piezoelectric ultrasonic transducer are theoretically analyzed, and the vibration velocity distribution equation, stress distribution equation and resonance frequency equation of the transducer are obtained. The size parameters of the transducer are solved, which solves the problem of the design of the piezoelectric horn transducer of the composite horn.

The finite element simulation software Workbench is used to analyze the harmonic response of the theoretically designed transducer. The results show that the natural frequency of the longitudinal vibration of the transducer is within 2% of the theoretical value, which verifies the theory. The correctness. Subsequent processing will be performed to produce the actual transducer, further verifying the correctness of the theoretical design.

References

[1] Abdullah A, Shahini M, Pak A. An approach to design a high power piezoelectric ultrasonic transducer[J]. Journal of Electroceramics, 2009, 22(4):369-382.
[2] Canning S, Walker A J, Roach P A. A Mathematical Model of a Novel 3D Fractal-Inspired Piezoelectric Ultrasonic Transducer[J]. Sensors, 2016, 16(12):2170.
[3] Liu S, Xu L, Zhang Z, et al. The sandwiched radial composite piezoelectric ultrasonic transducer[J]. Shengxue Xuebao/acta Acustica, 2014, 39(1):104-110.
[4] Cheng C, Zhu F, Bo F U. Modeling and Analysis of Sandwich Piezoelectric Ultrasonic Transducer Based on ANSYS[J]. Mechanical Engineer, 2015.
[5] Wang Y, Wang Z, Song Y, et al. Impedance Characteristic Analysis and Matching Design of Piezoelectric Ultrasonic Transducer[J]. Piezoelectrics & Acoustooptics, 2016.
[6] Peng J, Chao C, Tang H. Piezoelectric micromachined ultrasonic transducer based on dome-shaped piezoelectric single layer[J]. Microsystem Technologies, 2010, 16(10):1771-1775.

Design of Data Acquisition System for Silicon Micro Machined Resonant Accelerometer

XU Yun[a], QU Yaohui and LUO Zhihang

School of Mechanical Engineering & Automation, Zhejiang Sci-Tech University, Hangzhou, Zhejiang Province 310018, China

[a] ntxuyun@126.com

Abstract: The output of the silicon micro machined resonant accelerometer is a frequency signal, in which the measured acceleration has been converted. Thus, it can realize a better performance. The output of the silicon micro machined resonant accelerometer in this paper is the differential frequency signal. The resolution of it is ±1mg, the scale factor is100Hz/g, the measuring range is ±50g, and the resonant center frequency is 20 kHz. Aiming at the above technical indexes, and based on multiple-period synchronous method, the data acquisition system is designed in this paper. The data acquisition system contains the signal shaping module, the counter module, and the microcontroller and the communication module. The test experiment shows that the designed data acquisition system can satisfy the measurement requirement of the silicon micro machined resonant accelerometer.

1. Introduction

Silicon micro machined accelerometers can be divided into the capacitive accelerometer, the piezoresistive accelerometer, the piezoelectric accelerometer, the resonant accelerometer, etc. [1-4]. Based on different detection principles, the inertial force can be converted to different electrical signal, such as voltage signal, frequency signal and so on. Resonant accelerometer can convert the detected acceleration to the frequency signal [5]. It has the advantage of stability and reliability. What's more, the output frequency of the accelerometer will not distort during the signal transmission, and there's no need for A/D converter [6]. Therefore, design of data acquisition system for silicon micro machined resonant accelerometer can be simplified.

In this paper, a data acquisition system is designed to measure the frequency output by the silicon micro machined resonant accelerometer. The organization of this paper is as follows. First, the working principles for the silicon micro machined resonant accelerometer and multiple-period synchronous method are introduced in section 2. Second, each module for the designed data acquisition system is analyzed. And then, experiment tests have been carried out and discussed. Finally, the conclusion is drawn.

2. Working Principles

2.1 Working principle for silicon micro machined resonant accelerometer

Ideally, according to the measuring principle, when there's no acceleration, the force of the two resonant beams in the resonant accelerometer will be the same, and then the nature frequencies for the two resonant beams will also be the same [7]. It can be expressed by:

$$f_1 = f_2 = f_0 \tag{1}$$

Where, f_0 is the resonant center frequency. f_1 and f_2 are the frequency output by the two resonant beams.When there's the acceleration, the frequency output by the two resonant beams will be changed as:

$$f_1 = f_0 + \Delta f \qquad f_2 = f_0 - \Delta f \tag{2}$$

Then, the acceleration can be obtained from equation (2) by:

$$2\Delta f = f_1 - f_2 = Ka \tag{3}$$

In equation (3), the frequency difference of the two resonant beams has a linear relationship with the input acceleration. Thus, it should detect the frequencies of the two resonant beams, and then the acceleration can be obtained [8].

2.2 Working principle for multiple-period synchronous method

Multiple-period synchronous method is one of the most important methods for frequency measurement [9]. The gate time can be auto-adjusted to be an integer times of the measured signal's period, thus it is synchronize with the measured signal [10], and ±1 error from the traditional frequency method can be avoided which improves the measured accuracy obviously. The waveform graph for frequency measurement principle of multiple-period synchronous method is illustrated in figure 1.

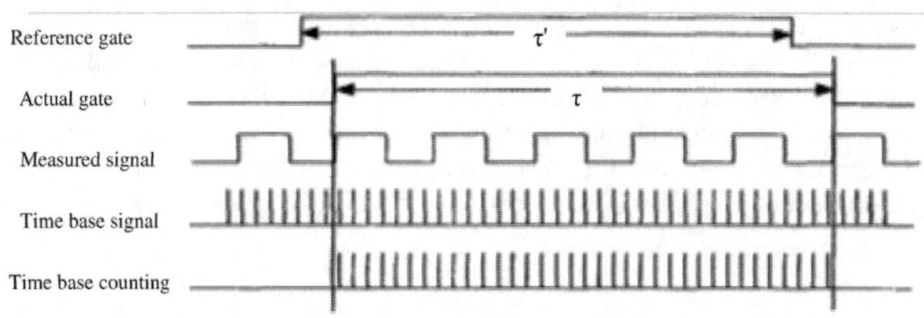

Figure.1 principle of multiple-period synchronous method

First, open the reference gate, and the counters will be triggered by the rising edge of the measured signal. Counter A will count the measured signal and counter B will count the time base signal. Then, close the reference gate, the counters will stop counting when the rising edge of the measured signal comes. Suppose the time for opening the actual gate is τ, and the time for opening the reference gate is τ', then there will be a difference between τ and τ'. According to the working principle of the multiple-period synchronous method, this difference will no more than one period of the measured signal.

Suppose N_x is the count number for the measured signal, and N_0 is the count number for the time base signal, f_0 is the frequency for the time base signal. During the gate time τ, the frequency for the measured signal can be obtained by:

$$f_x = N_x f_0 / N_0 \tag{4}$$

The counter is triggered by the measured signal. Therefore, there are integral times of the measured signal during the actual gate time, and ±1 error from the traditional frequency method will be avoided. Calculate the differentiation for equation (4) by:

$$df_x = -N_x f_0 \bullet dN_0 / N_0^2 \tag{5}$$

Where, $dN_0=\pm 1$, and $\tau = N_x/f_x$, then the measuring resolution of this method is:

$$df_x / f_x = \pm 1/(\tau f_0) \tag{6}$$

Suppose, the crystal oscillator for the time base signal is 40MHz, the gate time τ=10ms, the frequencies of the two measured signals are 22.5 kHz and 17.5 kHz, then, the measurement errors for them are:

$$\begin{cases} df_1 = \pm f_1 / (\tau f_0) = \pm 22.5kHz / (10ms \bullet 40MHz) = \pm 0.05625Hz < 0.1Hz \\ df_2 = \pm f_2 / (\tau f_0) = \pm 17.5kHz / (10ms \bullet 40MHz) = \pm 0.04375Hz < 0.1Hz \end{cases} \quad (7)$$

Measurement error analysis shows that the proposed frequency measurement error will be less than 0.1Hz, which will satisfy the measurement requirement.

Equation (6) illustrates that the measuring resolution is not relate to the measured signal, but relate to the actual gate time and the time base signal. Thus, equal precision measurement can be realized, and the resolution will be improved when prolong the actual gate time or increase the frequency of the time base signal.

2.3 Scheme for the data acquisition system

According to the working principle of multiple-period synchronous method, the scheme of data acquisition system for silicon micro machined resonant accelerometer is shown in figure 2.

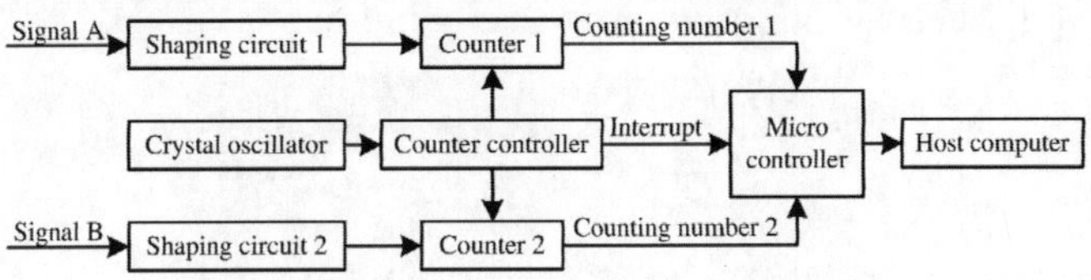

Figure.2 Scheme of data acquisition system for silicon micro machined resonant accelerometer

Two frequency signals output from the silicon micro machined resonant accelerometer will be transmitted to the counters through the shaping circuits. Then, the counter will be triggered and the counter number will be transmitted to micro controller, which will carry out the data processing and obtain the measured acceleration from the observed frequency. The host computer will receive the measured acceleration for the test experiment.

3. Hardware design of the data acquisition system

The hardware of the data acquisition system shown in figure 2 contains the shaping circuit, counting circuit, micro controller. Counting circuit will be realized in CPLD (Complex Programmable Logic Device).

3.1 Shaping circuit

Shaping circuit will transform the output signal output by silicon micro machined resonant accelerometer to be the pulse signal. And it is realized by LM311.

3.2 Counting circuit

Counting circuit will be realized by CPLD. It contains 2 function parts. (1) Function of frequency division. It will divide the frequency of the crystal oscillator and generate a period signal with 10ms, which will be used as the interrupt signal for the micro controller. (2) Function of Counting. It will count the frequency of f_1 and f_2 output by the silicon micro machined resonant accelerometer.

In this paper, EPM550T100 is used to realize the above functions. And it is simulated in QUARTUSII software by VHDL language. Suppose the period of the two frequency signal is T_{ain}=52us (f_{ain}=19.230769 kHz) and T_{bin}=50us (f_{bin}=20 kHz), then the simulation waveform is shown in figure 3.

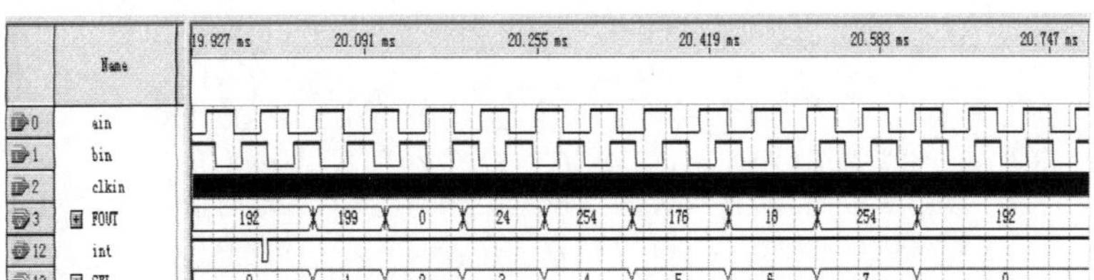

Figure.3 the simulation waveform of CPLD

In Figure 3, signals of ain and bin are the two frequency signals that need to be counted. Signal of clkin is the time base signal with the oscillation frequency of 40MHz. Signal of int is the interrupt signal. When interrupt signal comes, the counting number will be send to micro controller. Signal of SEL is the input of CPLD, and FOUT is the output of CPLD, which will connect to micro controller. When there's the interrupt signal, micro controller will send 0-7 to CPLD, and then it will read the counting number from CPLD. As show in figure 3, when interrupt comes, the counting number for signal ain and bin are 192 and 199 respectively, which means that there are 399360 and 398000 clock cycles of the crystal oscillation. Then, the frequencies of ain and bin are f_{ain}=19.230769 kHz, f_{bin}=20 kHz. It can satisfy the measurement requirement.

3.3 Micro controller

Micro controller will read the counting number from CPLD, and then it will calculate the input acceleration from the measured frequency. Then the acceleration will be read out. ATmega16 is used in this design. The designed data acquisition system is shown in figure 4.

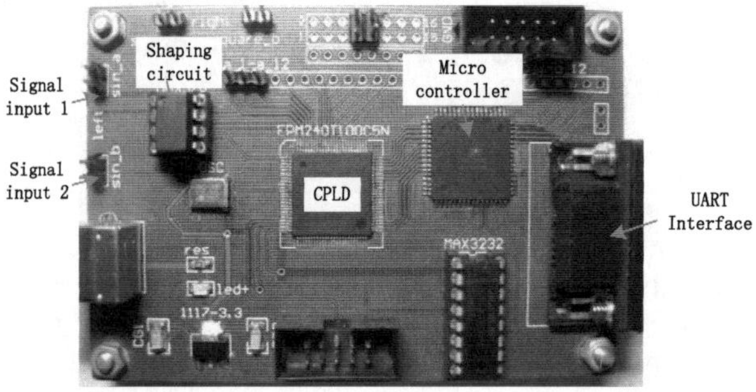

Figure.4 Designed data acquisition system

4. Experiment

4.1 Test by the input from signal source

Square waves will be generated by signal source to simulate the differential outputs from silicon micro machined resonant accelerometer. After the shaping circuit, the signal will be counted and then micro controller will output the counting number to the host computer. Table 1 shows the test results.

Table 1 Test results of data acquisition system by signal source

Input acceleration(g)	f_1 (Hz)	f_2 (Hz)	Measured f_1 (Hz)	Measured f_2 (Hz)	Frequency difference(Hz)	Measured acceleration(g)
50	22500	17500	22500.04	17499.97	5000.07	50.0007
40	22000	18000	22000.02	17999.99	4000.06	40.0006
30	21500	18500	21500.05	18499.96	3000.09	30.0009
20	21000	19000	21000.04	18999.96	2000.08	20.0008
10	20500	19500	20500.01	19499.96	1000.05	10.0005
0.2	20010	19990	20010.02	19989.98	20.04	0.2004
0.02	20001	19999	20001.02	19998.97	2.05	0.0205
0.002	20000.1	19999.9	20000.12	19999.87	0.25	0.0025
0	20000	20000	20000.02	19999.97	0.05	0.0005
-10	19500	20500	19499.96	20500.01	-1000.05	-10.0005
-20	19000	21000	18999.96	21000.04	-2000.08	-20.0008
-30	18500	21500	18499.96	21500.05	-3000.09	-30.0009
-40	18000	22000	17999.96	22000.02	-4000.06	-40.0006
-50	17500	22500	17499.97	22500.04	-5000.07	-50.0007

Table 1 shows that there are measurement errors for the test results of data acquisition system by signal source. These errors mainly come from the actual gate time, which will calculate the time base signal, and there will be ±1 error for the counter of time base signal. According to the measurement errors analyzed in equation (7), the above measurement errors are consistent with the errors analyze.

4.2 Test by silicon micro machined resonant accelerometer

The output signals of silicon micro machined resonant accelerometer will be shaped by the shaping circuit, and then they will be counted by CPLD. In our test experiment, multimeter made by Agilent will be used to compare with the test results from the designed data acquisition system. The measurement period of the designed data acquisition system is 10ms, and the measurement period of the multimeter is about 12ms. By the average processing, the test results of the designed data acquisition system and the multimeter are illustrated in figure 5.

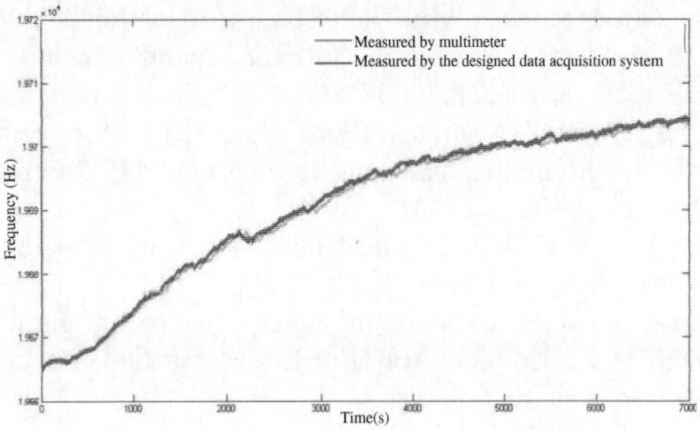

Figure.5 Result comparison for multimeter and the designed data acquisition system

In Figure 5, one frequency output by accelerometer is measured for the comparison. It is obvious that the measurement curve from the designed data acquisition system is coincided with the measurement curve from the multimeter, thus, it can be used for the future application.

5. Conclusion

A data acquisition system is designed in this paper for the silicon micro machined resonant accelerometer. First, the working principles of the resonant accelerometer and the multi-period synchronous method are introduced with the measurement errors analyzed. Next, the hardware design is illustrated and the function of each module has been explained. And then, the experiments have been implemented. And the experiment results have been compared, which obviously indicates that the designed data acquisition system can meet the requirement. The designed data acquisition system can be used for the future application.

Acknowledgments

The author would like to thank all reviewers for their helpful suggestions on this article which will help us in the further research.

Funding

This research was supported by the Young Researchers Foundation of Zhejiang Provincial Top Key Academic Discipline of Mechanical Engineering of Zhejiang Sci-Tech University (Grant Number: ZSTUME02B14), the Initial Scientific Research Foundation of Zhejiang Sci-Tech University (Grant Number:17022148-Y).

Reference

[1] Mo Y, Du L, Qu B, et al. Damping ratio analysis of a silicon capacitive micromechanical accelerometer[J]. Wireless Sensor Network, 2017, 9(05): 178.
[2] Messina M, Njuguna J, Palas C. Mechanical Structural Design of a MEMS-Based Piezoresistive

Accelerometer for Head Injuries Monitoring: A Computational Analysis by Increments of the Sensor Mass Moment of Inertia[J]. Sensors, 2018, 18(1): 289.
[3] Raaja B P, Daniel R J, Sumangala K. A Simple Analytical Model for MEMS Cantilever Beam Piezoelectric Accelerometer and High Sensitivity Design for SHM (structural health monitoring) Applications[J]. Trans. Electr. Electron. Mater.(TEEM), 2017, 18(2): 78-88.
[4] Zou X, Seshia A A. A high-resolution resonant MEMS accelerometer[C]//Solid-State Sensors, Actuators and Microsystems (TRANSDUCERS), 2015 Transducers-2015 18th International Conference on. IEEE, 2015: 1247-1250.
[5] Wang Y, Ding H, Le X, et al. A MEMS piezoelectric in-plane resonant accelerometer based on aluminum nitride with two-stage microleverage mechanism[J]. Sensors and Actuators A: Physical, 2017, 254: 126-133.
[6] Wang Y, Zhang J, Yao Z, et al. A MEMS Resonant Accelerometer With High Performance of Temperature Based on Electrostatic Spring Softening and Continuous Ring-Down Technique[J]. IEEE Sensors Journal, 2018, 18(17): 7023-7031.
[7] Zhang J, Wang Y, Zega V, et al. Nonlinear dynamics under varying temperature conditions of the resonating beams of a differential resonant accelerometer[J]. Journal of Micromechanics and Microengineering, 2018, 28(7): 075004.
[8] Zhang J, Su Y, Shi Q, et al. Microelectromechanical resonant accelerometer designed with a high sensitivity[J]. Sensors, 2015, 15(12): 30293-30310.
[9] Wang X, Li Z W, Wonham W M. Dynamic multiple-period reconfiguration of real-time scheduling based on timed DES supervisory control[J]. IEEE Transactions on Industrial Informatics, 2016, 12(1): 101-111.
[10] Vejdan S, Sanaye-Pasand M, Malik O P. Accurate dynamic phasor estimation based on the signal model under off-nominal frequency and oscillations[J]. IEEE Transactions on Smart Grid, 2017, 8(2): 708-719.

IOP Conf. Series: Materials Science and Engineering **470** (2019) 012045 doi:10.1088/1757-899X/470/1/012045

Increase of Refueling Systems Efficiency of Rocket Launching Sites by Utilization of Cold Energy of Cryogen

Daria Aksenova*, Artem Novikov*, Aleksandra Shimanova*, Artem Shimanov* and Dmitry Uglanov*[a]

*Samara University, Department of Heat Engineering and Heat Engines, 443086, Moskovskoye shosse, 34, Samara, Russia

[a] Corresponding author: dmitry.uglanov@mail.ru, Tel: +79879036026

Abstract. An important point in the use of cryogenic substances is their storage and transportation taking into account their phase state. Cryogenic substances can be used in liquid, gaseous and solid state. In all cases, it is possible to utilize the cold energy potential of the cryogen, either to reject heat from the cooling object during its regasification, or to obtain mechanical work and / or electricity. Therefore, the actual task is to improve the efficiency of rocket launching sites by using cold energy of cryogenic substances.

1. Introduction

The implementation of flight to outer space in addition to creating spacecraft and rocket requires solving many complex scientific and technical problems in preparation for the launch. A complex of such problems include production, transportation to the launching site, storage with minimal losses of cryogenic components of rocket fuel and liquefied process gases, as well as refueling fuel tanks of carrier rockets and booster blocks.

The idea of using the cold energy of cryogenic liquids has recently become more widespread [1-5]. This is primarily due to the increasing volumes of production of cryogenic substances, as well as the general trend of energy saving and the energy potential existing in the cryogen in the form of previously used energy for its liquefaction.

The main advantage of using gases in the cryogenic-liquefied state is the convenience of their storage, as well as transportation and consumption (including motor fuel for vehicles). At one of the steps of its application gasification is performed – transformation from the liquid state into the gaseous one via environmental heat supply. It is recognized, that the use of environmental heat is not an energy consuming process. However it is worth considering, that for LNG liquefaction some substantial energy was spent earlier (about 1kW·h of power per 1 kg of LNG).

The rapid development of the economy, industry and transport in the last 25-60 years, associated with scientific and technological progress, led to a serious increase in energy consumption and, as a result, increase of consumption of hydrocarbon fuels. The most promising fuel is cryogenic fuel - LNG, consisting mainly of methane, which allows to solve the problems of shortage of hydrocarbon fuel. The volumes of production, transportation and consumption of this gas are growing at a rapid pace every year. A study by the International Energy Agency shows that the volume of international trade in liquefied natural gas will reach 40% of total natural gas production in 2020, and the annual growth rate will be 2% -3%. LNG consumption will increase to 10% in China, moreover, LNG imports will grow by more than 3 times [6-7].

The estimated loss of cold energy in the cryogenic systems of rocket launching sites shows that the total value of the power of cold energy losses is about 9 MW. The total available work with the use of this energy in the effective cycle of the heat engine will be 156 MW · h.

Low temperatures of cryogenic liquids cause inevitable heat inflows from the environment, which in combination with small heat values of phase transitions, as well as a small range of existence in the liquid state, and they lead to a continuous change in their parameters, enhance the possibility of phase transitions and cryogen losses.

In this connection, the development of methods for the calculation, design and testing of storage systems and transportation of cryogen at launching sites using its cold energy to generate additional energy and improve operational efficiency is an urgent task.

2. Reservoir with cryogenic filling and calculation of its working processes

As shown above, thermal compressing is the most effective way of cold energy storing of cryogen [5]. Therefore, one of the most important elements in power plants using cold energy are cryogenic recervoirs. Particular interest is the tank of high pressure, in which the cryogen can be filled, regasified and accumulate energy of pressure. Such tank can be a reservoir with cryogenic filling (Figure 1)[5,8].

Figure 1. The structural design of the reservoir with cryogenic filling

A feature of this tank is that it can be charged with both a gases and a cryogen. In the case of refueling reservoir with cryogenic filling(RCF) with a gases it works as a normal high-pressure tank, and in the case of filling with a cryogenic component equal in mass, gasification takes place already inside the reservoir, which allows refueling the RCF at lower pressures. The internal thermos volume for the cryogenic component prevents thermal shocks and softens the conditions for thermocyclic strength of the structure.

3. Physico-mathematical model of the processes in RCF

The behavior of the working fluid in the RCF is considered when the liquid phase of the cryogenic substance is discharged from the thermos volume, taking into account the heat inflow from the environment

After filling of cryogen into the internal (thermos) volume of the RCF (Fig. 2) with heat inflow from outside, the behavior of the liquid and gas in the gas cavity may be different. The physical processes in such a system are difficult to predict, and they will depend not only on the design of the resevoir, but also on the initial state of the tank and the external conditions. In the case of evaporation of a liquid and its inflow into the gas cavity, the heat supplied to the gas is determined by the difference between dQ_c and dQ_l. It is obvious that for a stationary process of heat transfer by heat conduction $dQ_c = dQ_l = dI_c$. In the case of a quasistationary (in the time interval $\Delta\tau$) process during the inflow of cold mass dm_x part of heat dQ_c will be spent on changing the internal energy of the incoming mass of gas dm_x.

$$dQ_c - dQ_l = dU_x. \tag{1}$$

This process is similar to that considered by V.P. Isachenko [5] in the problem of the thermal conductivity of a porous wall with a stationary transit flow of liquid through the pores. In this case it is legitimate to make the same assumptions that the temperature of the main and newly incoming gas at the coordinate r is the same, and the heat transfer can be represented as the thermal conductivity of the main gas and the heat exchange between it and the incoming gas masses dm_{xr}.

The change in the internal energy of the newly supplied mass of gas into each gas layer dm_{xr} is defined as:

$$du_{xr} = dm_{xr} C_{Vr} dt_r, \tag{2}$$

where $m_x = \int_r^{\mathrm{т}r_{б}} m_{xr}(r)dr$.

Here the definition of the distribution $m_{xr} = f(r)$, by analytical way is also very difficult and not accurate. Therefore, calculations were performed using real table values of the thermophysical properties of the working substance (nitrogen), which showed that the mass dm_x entered in the gas cavity is distributed in each gas layer by volume $dV = 2\pi r dr$ almost equally.

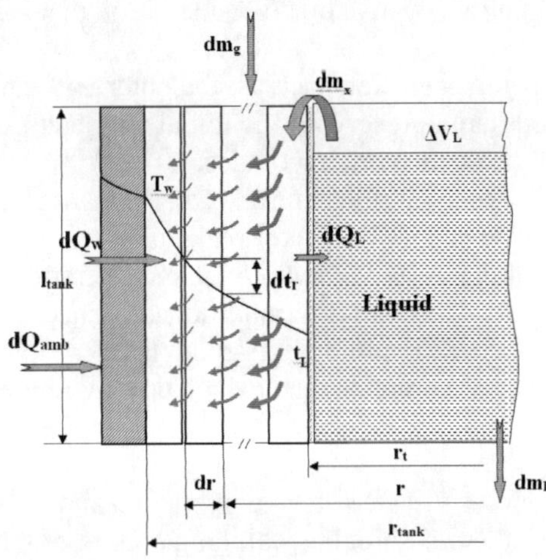

Figure 2. Scheme for the formulation of the problem and the development of processes for calculating algorithm in RCF

So dU_x is determined by the integral sum du_x on all layers.

Taking into account the foregoing, the mass-weighted average value of the gas temperature in the gas cavity will correspond to the arithmetic mean value:

$$\bar{t} = \frac{t_c + t_l}{2}. \tag{3}$$

Up to an average temperature level $\bar{t}$ the incoming mass of cold gas dm_x is heated and the change in its internal energy as a part of the change in the internal energy of the whole mass of gas in the cavity will be:

$$\Delta U_x = dm_x (C_V \bar{t} - C_V t_l), \tag{4}$$

where values C_V are taken at appropriate values t.

Thus, for a considered point in time at a given linear gas temperature distribution along the radius it is possible to determine dQ_c and $dU_x \approx \Delta U_x$.

The equation of energy for a gas cavity with an external heat input and the flow of a mass of gas from the liquid cavity of the tank will be:

$$dQ_{amb} + dQ_c - dQ_l + i_x \cdot dm_x + i_g \cdot dm_g = dU_x, \tag{5}$$

where $dQ_c - dQ_l-$ is heat which is equal to the difference of the input and output heat flow and it goes to dU_x, such

$$dQ_c - dQ_l = dU, \quad (6)$$

dU – change of internal energy of gas in the cavity due to external heat and incoming cold gas mass.

Thus, dU as the total differential can be defined as follows:

$$dU = \overline{C_V} \cdot m \cdot dT + \overline{C_V} \cdot T \cdot dm_x, \quad (7)$$

where m and T respectively, the mass and the gas temperature in the gas cavity.

Substituting (6) into (5) gives the following equation:

$$dQ_{окр} + dU_x + i_x \cdot dm_x + i_г \cdot dm_г = dU \quad (8)$$

or taking into account (7):

$$dQ_{amb} + dU_x + i_x \cdot dm_x + i_g \cdot dm_g = \overline{C_V} \cdot m \cdot dT + \overline{C_V} \cdot T \cdot dm_x. \quad (9)$$

4. Algorithm for calculating the parameters in RCF with the liquid phase rate flow of the cryogen

Based on the above ratios, an algorithm was developed and a calculation program was compiled.

The first stage of the work consisted in setting up a "block of auxiliary procedures," in which variables were placed. Determination of it can be made through analytic relationships and two-parameter functions.

After that, a number of parameters were set, such as: the outer and inner diameters of reservoir, the length of reservoirr, the initial pressure in reservoir, the initial temperature of reservoir wall, etc.

Next - the implementation of the cycle, which includes:

- first initialization of all cycle variables; check conditions for continuation of a cycle or exit from it, depending on the chosen "cyclicity" operator; execution of the sequence of operations in the cycle - the body of the cycle; updating cycle variables after each iteration

At the beginning of each cycle, the cyclicity condition are checked, taking as a basis the data obtained as a result of the previous iteration, the program updates the variables and initializes them as first ones for the subsequent cycle, or completes the calculation by redirecting to the output form and plotting.

Also during the calculation:

- the time $\Delta\tau$ was chosen as the step; the temperature and mass of the gas, as well as its specific volume in the gas cavity, the pressure, the wall temperature of the tank, the temperature of the liquid and the total volume realised from the liquid in the thermos are calculated twice during one iteration at the beginning of the cycle [j] and at [j+1]; the value of the wall temperature is determined on the basis of the loss of enthalpy of the reservoir wall, and is refined from the change in internal energy in the gas cavity by the inflows of low-temperature vapors dm_{gh} from thermos; the gas temperature at the beginning of the first iteration at j = 0 assumes the average value between the wall temperature of the tank and the charged liquid, then at the end of the cycle, it recalculates after mixing with the incoming portion of the low-temperature gas from the thermos. The obtained value of the gas temperature will be taken for T_g at new calculation step; the pressure at the end of the cycle is determined by the two-parameter function of the specific gas volume v_{gg} and the adjusted values of the gas temperature T_{gk}; the fluid temperature is refined according to the reference data in accordance with the pressure values in the new step; the refined value of each of the parameters or their value at the end of the cycle (after the time elapses) is taken as the initial value of the variable in the next cycle.

Based on the above algorithm, a program was written in the integrated software development environment – Embarcadero DelphiXE 5 under Windows 8.

5. Cryogenic pump-gasifier with RCF and power plant for obtaining additional energy

In the future, the use of a cryogen in the energy complex is necessary to carry out its gasification. During this process, the heat exchanger uses heat of the environment or specially supplied heat in the gasifier. To reduce the gasification time, as well as to generate additional energy, this work propose to use a cryogenic pump-gasifier (Fig. 3).

The cryogenic pump-gasifier consists of two circuits: a gasification circuit and an additional energy generating circuit (steam power plant)

The device is operated as follows:

From RCF 1, when the valve 2 is open, the cryogen enters the gasification circuit. Passing through the plunger pump 3, the pressure of the cryogen increases, and it enters the heat exchanger-evaporator 4, takes heat from the working substance of additional energy generating circuit, and returns to RCF 1.

Work of additional energy generating circuit is carried out as follows. The working substance enters the heat exchanger-condenser 4, where it gives heat to the cryogen, and then it enters the pump 5. The pump 5 compresses the working fluid, as a result of which it is heated, and enters the heat exchanger-evaporator 6, in which it is completely gasified. The obtained gas enters the turbine 7, where the heat is converted into mechanical work, and then into electrical energy by means of an electric generator 8. The generated electric energy is stored in the battery 9. Electric energy is used to drive the plunger pump 3 and it goes to the electric consumers.

This system can work on various cryogens: nitrogen, oxygen, hydrogen, methane, helium. All these gases can be used for purging pipelines both for their cooling down, and for warming up, as well as for fuel.

Table 1. Estimation of the efficiency of the use of different working substances in the circuit for obtaining additional energy

	Renkin cycle						Briton cycle					
	q_1, kJ/kg	q_2, kJ/kg	l_c, kJ/kg	G, kg/s	W, kW	ηt, %	q1, kJ/kg	q2, kJ/kg	lц, kJ/kg	G, kg/s	W, kW	ηt, %
Methane	755	490	265	0,28	74,2	0,35	329	280	46	0,5	23	0,15
Argon	156	125	31	0,88	27,3	0,2	77	63	20,4	2,37	48,4	0,18
Ethane	800	535	265	0,29	76,9	0,33	239	215	33,2	0,55	18,3	0,1
Krypton	141,5	98	43,5	1,32	57,4	0,31	-	-	-	-	-	-
Neon	-	-	-	-	-	-	146	118	35,2	1,23	43,3	0,19

In the circuit for obtaining additional energy, it is possible to use power plants operating in the Rankine or Brighton cycle. On the basis of known calculation methods, the feasibility of such a solution was estimated for different working substances suitable for operating in such temperature regimes.

The general expression for the thermal efficiency of the cycle is

$$\eta_t = \frac{l_c}{q_1} = 1 - \frac{q_2}{q_1}, \tag{10}$$

where l_c – specific cycle work q_1 – amount of heat supplied to the working substance, q_2 - amount of heat rejected from the working substance.

The power produced by the turbine can be defined as

$$W = l_c \cdot G, \tag{11}$$

where G – consumption of working substance in the circuit for obtaining additional energy.

An estimate of the amount of additional electric energy that can be obtained is given in Table 1. Analysis of the calculation shows that it is most effective to use steam power plants operating in the Rankine cycle on methane or ethane.

Thus, the proposed technical solution will allow to reduce the gasification time of LNG in a reservoir, to obtain a high-pressure gas, and also to use cold energy of cryogenic substance to generate electric energy necessary for driving a plunger pump, as well as for another electrical consumers.

6. Cryogenic system for fuel cooling at launching sites of a rocket based on the RCF

IOP Conf. Series: Materials Science and Engineering **470** (2019) 012045 doi:10.1088/1757-899X/470/1/012045

Based on the methodology for calculating the processes occurring in the RCF, with the help of the developed program at the level of a numerical experiment, calculations of the main thermophysical parameters of the working fluid in the RCF, which is part of the fuel cooling system on the launching site, were performed.

In the liquid nitrogen filling system of the launching site [9-10], the fuel cooling mode is performed by supplying liquid nitrogen from the reservoir when nitrogen gas is supplied under pressure from the nitrogen gas storage. It is proposed in this system to use RCF as a pressure accumulator for nitrogen (Figure 4).

The change in the parameters in the tanks is shown in Figures 5 - 6.

In this case, the RCF refueling takes place with liquid nitrogen, which allows filling RCF at lower pressures, and its gasification takes place already inside the reservoir, allowing to obtain high-pressure gaseous nitrogen in RCF, which is part of the fuel cooling system at the launching site.

7. Conclusions

As a result of the theoretical studies, methodology and technique for the calculation, design and testing of RCF with a liquid phase flowrate of a cryogen has been developed.

The following results and conclusions were obtained:

1. A structural design of cryogenic pump-gasifier with RCF and power plant for obtaining additional energy is proposed;
2. Stctural design of powerplants for obtaining additional electric energy at spacer rocket launching sites are proposed;
3. A physico-mathematical model of the processes in the RCF with a liquid phase flow rate of the cryogen was developed;
4. The calculation of the energy characteristics of power plants operating on the Rankine cycle and the Brighton cycle, which are part of the cryogenic pump-gasifier, is made and allows to increase its efficiency was made.
5. Analysis of calculations showed that the most efficient is a cryogenic pump-gasifier using a power plant operating on the Rankine cycle with the following working substances: methane (0.35% efficiency, the power produced by the turbine is is 74.2 kW) and ethane (efficiency 0.33 %, the power produced by the turbine is 76.9 kW).
6. The algorithm is developed, as well as the calculation program in the integrated software development environment – Embarcadero Delphi XE 5 for modeling thermophysical processes.

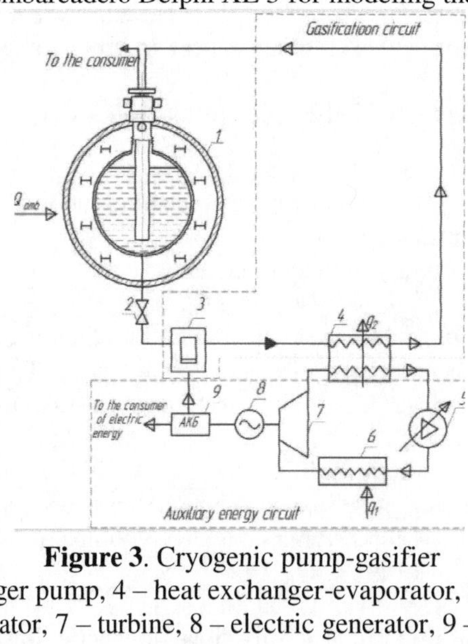

Figure 3. Cryogenic pump-gasifier
1 – RCF, 2 – valve, 3 – plunger pump, 4 – heat exchanger-evaporator, 5 – pump, 6 – heat exchanger-evaporator, 7 – turbine, 8 – electric generator, 9 – battery

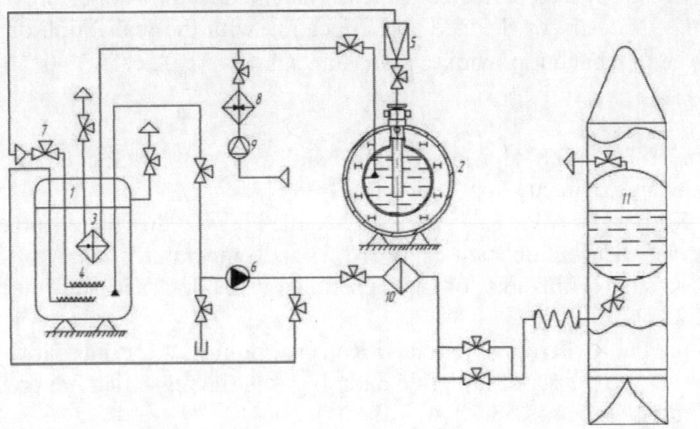

Figure 4. Cryogenic system for fuel cooling at launching sites of a rocket based on the RCF: 1 - fuel reservoir; 2 - RCF, filled with liquid nitrogen; 3 - heat exchanger; 4-barboter of nitrogen gas; 5 - gas reducer; 6 - pump; 7-support armature; 8 - air heater; 9 - compressor; 10 - filter; 11 - fuel tank

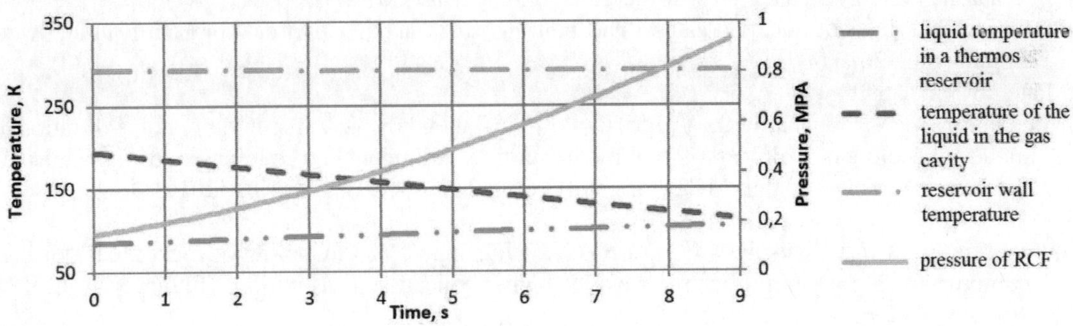

Figure 5. Change in the parameters in the spherical reservoir used to cool the fuel in the launching site (RCF volume is 448 m3, the volume of the thermos container is 120 m3, the outer diameter is 9.5 m, wall thickness is 0.5 m, refueling weight is 88.3 tons of nitrogen, refueling temperature is 84 K, filling pressure is 0.2 MPa, consumption of liquid nitrogen is 2 kg / s)

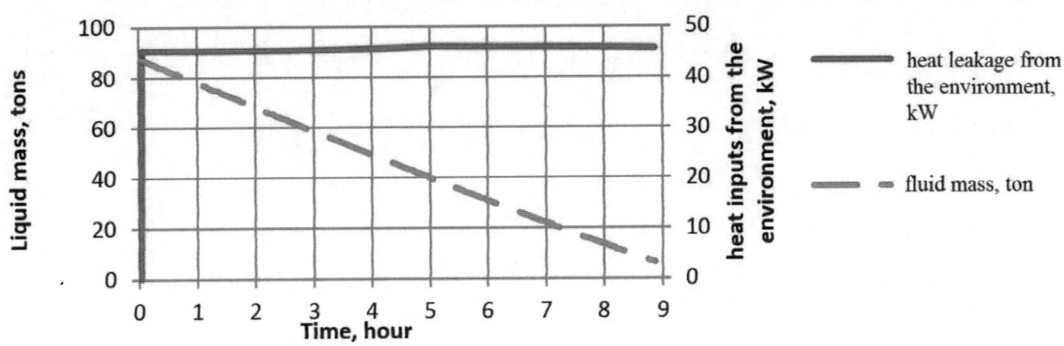

Figure 6. Change in the mass of fluid and heat flows from the environment to the wall of a spherical RCF used to cool the fuel in the launching site (RCF volume is 448 m3, the volume of the thermos is 120 m3, the outer diameter is 9.5 m, the wall thickness is 0.5 m, refueling weight is 88.3 tons of nitrogen, filling temperature is 84 K, filling pressure is 0.2 MPa, consumption of liquid nitrogen is 2 kg / s)

Acknowledgments

This work was carried out by lead performer of research and development effort with financial support of Russian Federation Ministry of Science and Education with the realization of Government decree №218 according to contract about providing and using subsidy № 02.G25.31.0150 from 01.12.2015.

References

[1] Kumar S, Kwon H-T, Choi K-H, Lim W, Cho J H, TakK, etal LNG:an ecofriendly cryogenic fuel for sustainable development. ApplEnergy2011;88: 4264–73.

[2] Burtsev, S.A., Karpenko, A.P., Leontiev, A.I. A method for distributed production of liquefied natural gas at gas-distribution stations. (2016) High Temperature, 54 (4), pp. 573-576.

[3] S Hirakawa , K Kosugi. Utilization of LNG cold energy [J], International Journal of Refrigeration; 1981,4 (1):7-21.

[4] M. Romero Gómez , R. Ferreiro Garcia, J.Romero Gómez, J.Carbia Carril Review of thermal cycles exploiting the exergy of liquefied natural gas in the regasification process Renewable and Sustainable Energy Reviews 38(2014) 781–795.

[5] E.V. Blagin, A.I. Dovgyallo, S.O. Nekrasova, D.V. Sarmin, D.A. Uglanov, Applied Thermal Engineering Estimation of the energy efficiency of cryogenic filled tank using in different system and devices. 2016. Volume 101, 25 May 2016, Pages 537-544

[6] International Energy Agency. World Energy Outlook 2012 ;2012.

[7] Development status of liquefied natural gas industry in China [Jj.Department of Energy and Power Engineering; 2010 (4).

[8] Patent RU2163699.

[9] Alexandrov A. A., Goncharov R. A., Igritsky V. A., Chugunkov V. V. method of choice of rational modes of cooling of hydrocarbon fuel by the starting equipment before refueling of fuel tanks of the launch vehicle. Vestnik MGTU im. N. Uh. Bauman. Ser. Engineering. 2011. No. 1. P. 40-46.

[10] Aleksandrov A. A., Denisov O. E., Zolin A.V., Chugunkov V. V. Cooling of rocket fuel starting equipment using liquid nitrogen // news of higher educational institutions. Engineering. 2013. No. 4. P. 24-29.

Automatic Text System Based on PLC for Walk-Through Metal Detector

Zheng Haidong[a], Jiang Lingling, Wu Jianyi, Wang Wei, Wang Xiaokan and Li Zhiguo

NO.2 huoju road, changping district, Beijing, China

[a]zhenghd@fiscan.cn, +86 18515522061

Abstract. In recent years, the safety situation around the world is very serious. Each country has higher quality requirements for walk-through metal detector, and the newly issued standards are stricter. It has required precise and detailed testing work. Therefore, an automatic test system was designed to meet the requirements. The system was an AC servo system based on PC+PLC, with the control core of PLC and three-dof motion platform composed of servo motor and linear module to perform slide components. The paper focused on the overall architecture, structure design, working process, hardware composition, PLC programming and software functions of the system. The system has no interference to the testing work of metal detector with good testing ability and high positioning accuracy. The system improves the accuracy of the testing, and provides a friendly man-machine interface, which greatly reduces the working intensity of testers.

1 Introduction

In recent years, the safety situation around the world is very serious. Each country has higher quality requirements for walk-through metal detector, and the newly issued standards[1] are stricter. It has required precise and detailed testing work, the technical indicators and test methods have been greatly adjusted. Only the automatic testing system can achieve the accuracy and repeatability which is difficult to be achieved by the traditional testers. The US standard NIJ is also equipped with the automatic testing system.

Therefore, an automatic test system is required to meet the requirements and replaced the traditional testers[2-7]. The tester only needs to operate the software to make the test object passing the metal detector with specified speed and position, and record the result automatically. An automatic test system is designed in this paper, which is using OMRON PLC as the core controller to control the servo motors with pulse mode. The system frees testers from repetitive manual operations and greatly improves positioning and motion accuracy.

2 Literature review

2.1 System design

The automatic testing system is mainly composed of PLC, PC, servo motors, linear module, limit switch, origin switch and industrial camera. Block Diagram of the system is shown in figure 1. The actuator of the system is a three-dof motion system, which is composed of a linear module and a servo motor in the three directions of X/Y/Z. All three linear modules are driven by gears and belts. The

Z-axis is in the vertical direction. The X-axis is horizontal and parallel to the metal detector. The Y-axis is perpendicular to the metal detector and passes through the detection area of the metal detector. Main function of the system are complete the whole testing process stipulated by orderd standard and record the optical alarm signal of the metal detector automatically.

2.2 Structural design

The mechanical structure of the system is shown in figure 2, which is mainly composed of the following modules: base, Z-axis component, X-axis component, Y-axis component, fixture, control cabinet, camera, PC. The base is made of steel to ensure the stability of the system.

The difficulty design of this system is that, according to the national standard, there cannot be any metal materials except test objects within the range of 1.5m from the metal detector. Therefore, Y-axis components are required to be made from non-metallic materials, Including belt, bearing, spring, fastener and all other parts.

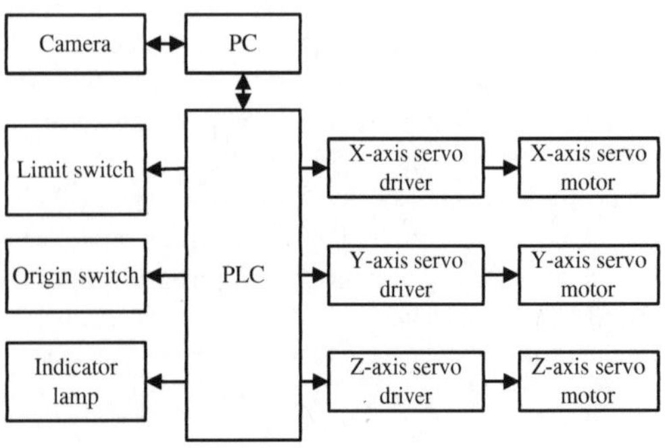

Figure 1. Block Diagram of the system

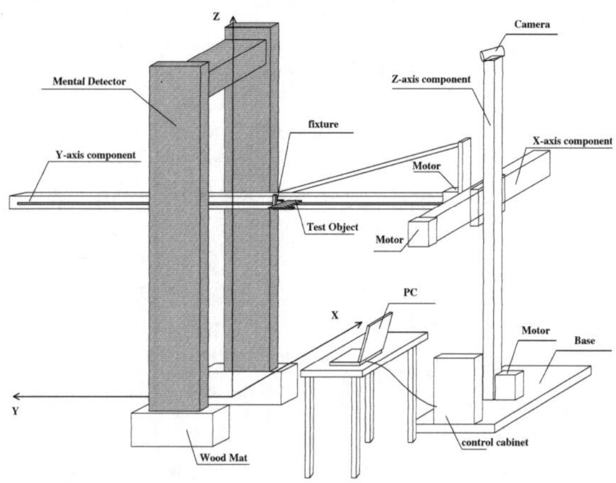

Figure 2. Mechanical structure of the system

2.3 Workflow

The workflow of the test system is shown in figure 3. Details are as follows:

1) After the system is energized, all ports are initialized, including the level state of limit switches and origin switches.

2) Open the software on PC, check the position of the motors automatically, and execute the reference point search program.

3) Set parameters and select standards on PC, enable automatic identification and alarm, test data storage and other functions.

4) Click to start test, the PC sends the parameters to the PLC, and then the PLC sends the pulse to the servo motors to control the triaxial movement accordingly. When the software prompts "whether to change a posture to continue the test", change the posture of the test object and complete the test.

5) When the software prompts "whether to save the test data", click "yes" to save the test data into the database file, and task completes.

2.4 Hardware composing

This system adopts OMRON SYSMAC CP series[8-9] PLC controller as the control core, and Schneider Lexium 23D series servo driver +BCH series servo motor[10]. The servo system selects the position control mode and converts the pulse signal and direction signal into angular displacement and rotation direction, which drives the servo motor to move to the specified position at the specified speed. The I/O distribution of PLC in the system is shown in figure 4.

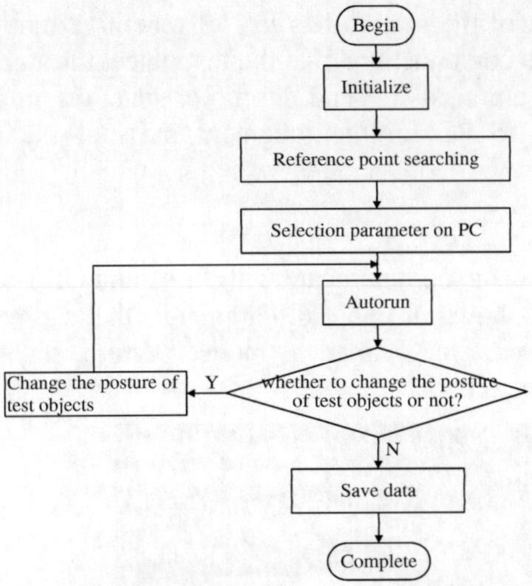

Figure 3. Workflow of the system

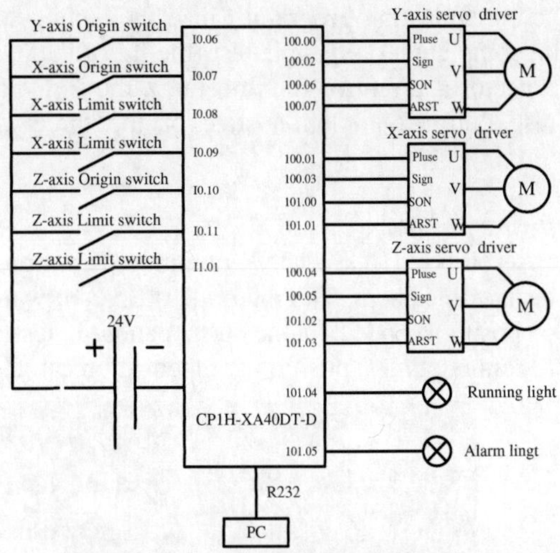

Figure 4. The I/O distribution of PLC

2.5 Programming of PLC

PLC development software "CX-programmer" provides flexible programming style (trapezoid diagram, function block diagram, statement table) and lots of instruction set, which can complete complex function easily and quickly. According to the workflow of the system, the program is divided into two parts: initialization program and motion control program. Each program was written separately, which was simple structure, clear thinking, easy to debug and modify.

2.5.1 Initialization program Set the initial position of each axis, Initial state of limit switch and origin switch in the initialization program. Each time the autorun is started, load the configuration, set motors enable, complete the current speed identification, position identification and direction identification of the motion axis.

2.5.2 motion control program Reference point search program: Reference point search is realized through the limit switch and the servo motor's encoder. After the software initialization is completed, it will run automatically.

Motion control program: In the autorun mode, the three axes start simultaneously, the speed of X/Z axis is fixed value, and the speed of Y axis is determined by the instruction sent by PC. The displacement of the three axes is also determined by the instruction sent by PC. The data is written into the register and sent to the servo driver in the form of pulse frequency + direction signal and drive the servo motor to the specific stroke. At the same time, there is a stop motion program, press the stop button in the software, the three axes stop immediately. The trapezoid diagram is shown in figure 5.

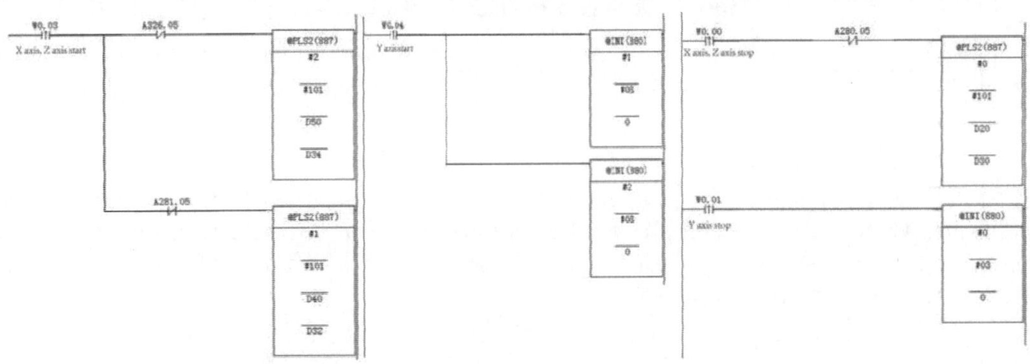

Figure 5. Trapezoid diagram of motion control program

IOP Conf. Series: Materials Science and Engineering **470** (2019) 012046 doi:10.1088/1757-899X/470/1/012046

PC communication program: Since the testing work of the system has lots of repetitive motions, after the completion of a single test (the jig of the Y-axis component carries the test object through the metal detector and returns to the origin), the PLC program needs to send the motor state, origin state and pulse output completion state of the three axes to PC for the judgment basis of the next instruction.

2.6 Software functions

PC uses serial port (HOST LINK communication protocol) to communicate with PLC, and USB cable to connect with camera. The interface of PC software is shown in figure 6. Parameters that can be set include: position, speed, mode, test standard, test object, etc. In manual mode, posture, scanning interval, running times, position compensation can also be set.

Figure 6. The interface of PC software

3 Results

According to the structure and frame of the system, complete the installation of the whole machine and the connection of cable. After several tests, when the system passes through the metal detector without load, the security gate does not have an alarm, which proves that the system has no interference in the testing of metal detector and can meet the basic conditions of the test. The system can make precise point movement in the X/Y/Z directions, can carry the test object through the mental detector according to specific position, speed and posture. The system can automatically record the alarm lamp of the metal detector through the camera, and generate the test record form. The performance index of the system is shown in table 1.

Table 1. The performance index of the system

direction	positioning accuracy(mm)	stroke (mm)	Speed(ms^{-1})
X	0.2	0-800	0.5
Y	0.5	0-2700	0-2
Z	0.2	0-2000	0.5

After the test, the system can generate the test form in excel format by clicking "export" button in the software. The test form records the test object's name, posture, speed and position in the test process. The form contents are the alarm number of the test process, 0 when no alarms are made, and empty when no tests are made, as shown in figure 7.

Position	Posture	GB-T2			GB-T4-1			GB-T4-2			GB-T6-1		
		0.2	1	2	0.2	1	2	0.2	1	2	0.2	1	2
1-190M	Transverse	0	0	0	10	10	10	10	10	10	10	10	10
	Vertical	0	0	0	10	10	10	10	10	10	10	10	10
	Longitudinal	0	0	0	10	10	10	10	10	10	10	10	10
2-170M	Transverse	0	0	0	8	9	8	10	10	10	10	10	10
	Vertical	0	0	0	10	10	10	10	10	10	10	10	10
	Longitudinal	0	0	0	10	10	10	10	10	10	10	10	10
3-145L	Transverse	0	0	0	5	10	9	10	10	10	10	10	10
	Vertical	0	0	0	10	10	10	10	10	10	10	10	10
	Longitudinal	0	0	0	10	10	10	10	10	10	10	10	10
4-145M	Transverse	0	0	0	9	10	8	10	10	10	10	10	10
	Vertical	0	0	0	9	10	9	10	10	10	10	10	10
	Longitudinal	0	0	0	10	10	10	10	10	10	10	10	10
5-125M	Transverse	0	0	0	10	10	10	10	10	10	10	10	10
	Vertical	0	0	0	10	10	10	10	10	10	10	10	10
	Longitudinal	0	0	0	10	10	10	10	10	10	10	10	10
6-100L	Transverse	0	8	0	10	10	10	10	10	10	10	10	10
	Vertical	0	4	0	10	10	10	10	10	10	10	10	10
	Longitudinal	0	6	0	10	10	10	10	10	10	10	10	10
7-100R	Transverse	0	9	0	10	10	10	10	10	10	10	10	10
	Vertical	0	0	0	10	10	10	10	10	10	10	10	10
	Longitudinal	0	2	0	10	10	10	10	10	10	10	10	10
8-75M	Transverse	0	0	0	8	10	10	10	10	10	10	10	10
	Vertical	0	0	0	10	10	10	10	10	10	10	10	10
	Longitudinal	0	0	0	10	10	10	10	10	10	10	10	10
9-40L	Transverse	0	0	0	10	10	10	9	9	8	10	10	10
	Vertical	0	0	0	10	10	10	10	10	10	10	10	10
	Longitudinal	0	0	0	10	10	10	10	10	10	10	10	10
10-40R	Transverse	0	0	0	10	10	10	10	10	10	10	10	10
	Vertical	0	0	0	10	10	10	8	8	9	10	10	10
	Longitudinal	0	0	0	10	10	10	10	10	10	10	10	10
11-10L	Transverse	0	0	0	10	10	10	10	10	10	10	10	10
	Vertical	0	0	0	10	10	10	10	10	10	10	10	10
	Longitudinal	0	0	0	10	10	10	9	9	9	10	10	10
12-10R	Transverse	0	0	0	10	10	10	10	10	10	10	10	10
	Vertical	0	0	0	10	10	10	10	9	10	10	10	10
	Longitudinal	0	0	0	10	10	10	9	8	9	10	10	10

Figure 7. Test form of the system

4 Conclusions

This paper adopts mainstream industrial control products for system integration. Based on PLC and servo control module, an automatic test system for walk-through metal detector is constructed. The actual test proves, the system has good testing ability and positioning accuracy, which improves the accuracy of testing and greatly reduces the working intensity of testers. This paper has carried out a beneficial exploration in the field of metal detector's testing, which has good practical application. The high efficiency, stability and security of the system are guaranteed, and a good human-computer interaction interface is provided, which has been highly praised by the industry personnel.

References

[1] Standardization Administration of China. *GB15210 – 2003 General specification for walk-through* metal *detector* [M].2003.

[2] Jia Desheng. PLC application development utility subroutine [M]. *Beijing: POSTS AND TELECOM PRESS*. 2006.

[3] Wang Donghui, Geng Lu, Cheng Wushan. Intelligent Control Positioning System Based on Machine Vision and PLC [J]. *Instrument Technique and Sensor*, 2012(**11**):83-85.

[4] Huang Jingwen. Design and Optimization of the PLC-based Three-degree of Freedom Numerical Control System [J] . *Process Automation Instrumentation*, 2016(**37-5**):24-26,31.

[5] Wang xiaoling. PLC Control of Industrial Manipulator [J]. *Mechanical & Electrical Engineering Technology*. 2004, 33(**9**):32-36.

[6] Li yong-cheng. The application of PLC-based the stepper motor control in industrial

manipulator [J]. *Science & Technology Information*, 2008(**18**):30-31.
[7] Xu kai. Development & Market of Metal Detector. *CHIINA SECURITY & PROTECTION*, 2004(**1**):52-53.
[8] OMRON Corporation. *CP1H CPU Operation Manual* [M].2008.4.
[9] OMRON Corporation. *SYSMAC CP-CP1H CPU Programming Manual* [M]. 2008.4.
[10] Schneider Electric. *Lexium 23D Ac servo actuator product manual* [M]. 2013.1.